Presented in clear and accessible language with wonderfully supportive graphics, Roberts offers the reader a voyage through the stages of human knowledge. He then examines the outstanding mysteries of modern physics, the phenomena that lie outside the boarders of our current understanding (dark energy, dark matter, the Big Bang, wave-particle duality, quantum tunneling, state vector reduction, etc.) and suggests that the next step in our intellectual journey is to treat the vacuum of space as a superfluid—modeling it as being composed of interactive quanta, which, in a self-similar way, are composed of subquanta, and so on. With this proposition Roberts imbues the vacuum with fractal geometry, and opens the door to explaining the outstanding mysteries of physics geometrically.

    Roberts' model, called quantum space theory, has been praised for how it offers an intuitively accessible picture of eleven dimensions and for powerfully extending the insight of general relativity, eloquently translating the four forces into unique kinds of geometric distortions, while offering us access to the underlying deterministic dynamics that give rise to quantum mechanics. That remarkably simple picture explains the mysteries of modern physics is a way that is fully commensurate with Einstein's Intuition. It is a refreshingly unique perspective that generates several testable predictions.

EINSTEIN'S INTUITION

Visualizing Nature in Eleven Dimensions

THAD ROBERTS

Einstein's Intuition: Visualizing Nature in Eleven Dimensions
First Edition
Softcover with black and white images
Printed by CreateSpace, an Amazon company
First Print Date, Dec 11, 2015
Based on Hardcover with full color images, printed by Lulu
Also available in iBook format via Apple & Audiobook via Audible.

Copyright © 2015 by Thad Roberts

All rights reserved. Published in the United States by Thad Roberts, in association with the Quantum Space Theory Institute (QSTI), a worldwide collaboration dedicated to exploring and further developing quantum space theory (qst), and to get beneath the formalism of quantum mechanics.

ISBN 978-0-9963942-4-6

Illustrations by Jeff Chapple

Cover design by Jeff Chapple, Angela Arvizu & Thad Roberts
Cover art by Jeff Chapple

www.EinsteinsIntuition.com

To the first kiss.

To standing on the precipice of the vast unknown;
Where the sands of time dance in the winds of confusion,
And the comforts of ignorance are left behind.

To transcending the first of all illusions, trembling as you leap.

To falling into the most feared chasm with a racing heart.

To discovering your wings.

To the way everything changes from a new perspective.

Table of Contents

Prologue — i

Preface — iii

Part One—Returning to a Conceptual Approach — 1

    Chapter 1 — Ascending the Footholds of Rationality — 3

        Our Maps, A Brief History — 7
        Two Pieces of the Puzzle — 15
        Fitting the Pieces Together — 18

    Chapter 2 — Rethinking Space and Time Again — 23

        Newton's Bucket and Absolute Space — 26
        Einstein's Absolute Spacetime — 31
        Modern Clues for an Ultimate Reference Frame — 32

    Chapter 3 — Dimensions — 39

        Exploring "Dimensions" — 41
        Additional Dimensions — 43
        Curvature and Other Dimensions — 45

    Chapter 4 — The Quantized Nature of Spacetime — 55

        The Sands of Space and Time — 57
        The Case for Quanta — 60
        The Investigation Begins — 66

Part Two—The Framework of Quantum Space Theory — 71

    Chapter 5 — Absolute Volume — 73

        Passing Through Flatland — 74
        Excavating the Foundation — 78
        Absolute Volume — 80

    Chapter 6 — Space — 87

        The Wilderness of Eternity — 91
        Distance and Measure — 92
        Tensors — 94
        Focusing Out on the Vacuum — 95

    Chapter 7 — Time — 101

        Questioning Time — 104
        Making Time Real — 106
        The Temperament of Time — 108
        Supertime — 110

    Chapter 8 — The Speed of Spacetime — 117

        Cosmic Ripples — 119
        Speed and Resolution — 123

  Lorentz Contraction   125
  Speed Outside of Space   126

 Chapter 9 — Warped Spacetime   129

  Rethinking Gravity   132
  Beyond Newton's Apple and Einstein's Rubber Sheet   139
  Depicting Warped Time   141
  Gravity's Rainbow   143

 Chapter 10 — The Bucket   149

  Newton's Question   151
  Einstein's Question   153
  Transforming the Question   155

 Chapter 11 — Dimensional Analysis   157

  Complex Magic   159
  Imagine a Number   161
  Spinor Visuals Compared   162
  Infinite Dimensional Cascades   165
  Architectural Detail   168

Part Three—Physical Reality in Eleven Dimensions   175

 Chapter 12 — The Questions of Quantum Mechanics   177

  Upsetting Our Worldview   178
  The Dual Nature of Light and Matter   181
  The Photoelectric Effect and the Uncertainty Principle   186
  The State Vector   187
  The Bell Theorem: Blurring Local Realism   190

 Chapter 13 — Beneath Quantum Mechanics   197

  A New Scene of Thought   198
  Paradigm Shift   199
  Beneath Quantum Mechanics   200
  The State of Space   203

 Chapter 14 — Quantum Tunneling & Entanglement   209

  Euclidean Misdirection   210
  Quantum Tunnels   212
  Stepping Out of Spaceland   215
  Schrödinger's Cat   217
  Entanglement   219

 Chapter 15 — Black Holes   223

  New Solutions   223
  Entropy and Radiation   227

 Chapter 16 — The Constants of Nature   233

  Geometric Origins of the Constants of Nature   236
  All Universes Have the Same Constants   242

Chapter 17 — Deterministic vs. Stochastic — 247

- Level of Description — 251
- The Geometry of Determinism — 258
- Interpretations — 260
- Consequences of Determinism — 262

Chapter 18 — Emergent Reality — 271

- Supervenience — 272
- The Emergence of Structure — 275
- Bottom-up Structure — 278
- Eliminating Illogical Infinities — 280

Chapter 19 — The Hierarchy Problem — 283

- Probing the Hierarchy Problem — 283
- Geometric Unification — 287

Chapter 20 — Beyond Forces — 291

- A New Perspective — 292
- Gravity — 295
- The Weak Nuclear Force — 297
- Electromagnetism — 298
- The Strong Nuclear Force — 304

Chapter 21 — Quantized Vortices — 309

- Lord Kelvin's Idea — 310
- The Higgs Mechanism — 312

Chapter 22 — Superfluidity — 317

- Superfluidity — 317
- Deriving Schrödinger's Wave Equation — 319
- Analogue Gravity — 323

Chapter 23 — Illuminating Dark Matter — 329

- In the Dark — 330
- A Natural Explanation — 333

Chapter 24 — Bohmian Mechanics — 337

- Coming out of the Dark — 338
- Finding Bohm — 339
- Deriving Bohm's Formalism — 343
- Going Beyond Bohm's Formalism — 348

Chapter 25 — Symmetry and Symmetry Breaking — 357

- Symmetry — 358
- Symmetry Breaking — 359
- Eras of Symmetry — 360
- Lorentz, Galilean and Other Symmetries — 362

Chapter 26 — Entropy 369
    Mixing Things Up 370
    Holograms 375
    Asymmetric States 377

Chapter 27 — Genesis 381
    The Universe of Imagination 382
    Before the Bang 384
    Eternal Recurrence 388

Chapter 28 — Dark Energy 393
    The History of Dark Energy 394
    Expanding Space 396
    Measuring Redshift 400
    Another Way to Explain Redshift 401

Chapter 29 — Intellectual Astronauts 405
    Breaking the Conceptual Barrier 406
    Gaining our Wings 408

Chapter 30 — The Wilderness of Intuition 411
    Avoiding Moral Shipwreck 412
    Personal Inspiration 416
    Seeking Explanation 419
    Parsing Consciousness 422

Afterword **429**

Acknowledgements **433**

References **435**

Appendix A – Approaches to Quantum Gravity **449**

About the Author **452**

Prologue

In my youth, I was enthusiastic about all of science, but by the time I was twenty, I was very disappointed, especially with physics, because no one could answer "why?" I wanted to do more than memorize a multitude of equations—I wanted to understand how it all fit together. When they said it was unknowable, that Nature's secrets are ultimately beyond human comprehension, I decided there were more exciting things for me to do with my life.

After four decades, this book has rekindled the fire that college physics dampened. Now my passion for inquiry has returned and a new sense of vigor has filled my life. If you are a truth seeker, or just curious, or feel disconnected, or discouraged by science, or are a physics dropout like me, you should read this book. In fact, I encourage everyone—all ages and backgrounds—to read and think about what Thad Roberts has written here.

If you have more questions than answers, if you know some of today's accepted answers are wrong and costly, or if you feel that the world around us should be explainable, then turn the pages to take a journey of thought that will transform you. If you have been discouraged by the study of science, or have been told that you aren't smart enough to understand science, then the new map Thad is proposing will change your world.

If you want to better connect to Nature—to understand time, light, matter, and wave-particle duality as viscerally as you understand waterfalls, you should read this book. If you want to contribute to the quest to make a better map of existence, or if you appreciate technological advances and want them to continue, pushing for cleaner energy, smaller/faster electronics, etc., then read on.

Part One reviews the history of physics, taking us on a journey through the evolution of ideas and perspectives in physics. It leaves us at the threshold of our ignorance, staring down the outstanding mysteries in physics. Part Two describes a new model of reality based on the proposition that we actually live in an eleven-dimensional superfluid. Intrigued? Part Three uses that new model to examine the big mysteries of physics—offering an intuitive grasp of each of those effects.

This book has changed my life. I hope you find it as transformative as I did.

David B. Mckenzie

Preface

"There is no finer sight than that of the intelligence at grips with a reality that transcends it."
Albert Camus

"One cannot help but be in awe when one contemplates the mysteries of eternity, of life, of the marvelous structure of reality."
Albert Einstein[1]

Before the New World was discovered, a legend, first put to pen by Plato, foretold of a magnificent realm whose shores hid in the void that stretched beyond the Pillars of Hercules (the Strait of Gibraltar). This mysterious Atlantic shire stood as a symbol of harmony. It was described as a place full of transcendent treasures where human potential could expand beyond traditional limits. The gateway to that clandestine land was said to be a port so rich in symmetry that all of its elaborate details came together to form an aesthetic marvel with an unmistakable sense of artistic magic.

The mere possibility that such a golden city existed was enchanting, yet there was something more to this legend, something stirring and profound that resonated every time it was told. The very idea that out there, beyond the mists, an entire continent was still waiting to be discovered was bone chilling. If it were true, it would mean that despite all the knowledge humanity had amassed, there was much more to learn. It would mean that every previous rendition of the world that we had so faithfully relied upon was wildly incomplete, that there was far more to the world than anyone had previously imagined. In the end, it would require mankind to completely rewrite its trusted maps.

This legend offered a possibility that intrepid explorers could not ignore—the chance to participate in a transcendental quest. It amplified the dream of connecting to certain and universal truths, by offering humanity a way to touch the underlying mystery and to actively extend their perceptions beyond the horizon. To those that would sail off into the glimmering mirages that capture the evening sun, this whisper of hope became the Sirens' song. Over time, the lure and passionate curiosity characterized by this legend entered the common tongue and became known as the call of Atlantis. Answering this call was to embrace the heart of Plato's legend. Those who did came to believe that by ascending through the stages of rationality the attainment of enlightenment is a real possibility; that ultimately we can escape the cave of ignorance and learn to grasp what lies beyond the shadows.

For the most part, the legend of Atlantis was considered a heretical myth. The very notion that the world contained entire continents yet to be discovered was considered laughable and blasphemous. European maps of the world clearly showed three continents—no more. Faith in the accuracy of these maps had won wars and guided men home from faraway places. Consequently, the rulers of every land kept their personal maps under lock and key and considered them their most prized possessions. These maps gave them perspective, framed their world, and defined their place in it. Any claim that their maps were wrong was taken as an attack on their model of reality. In spite of this, the legend of Atlantis lived on.

The boldest pages of our past are colored by the achievements and discoveries of gallant individuals that have directly participated in the unraveling of our world's mysteries. By challenging convention and following their intuitions toward a richer, more complete map, these figures have brought us new insights. Of the historical figures that have shared in this experience I will mention the two that most appropriately set the stage for the work herein. These particular explorers have made profound impacts on the maps that frame our modern worldview. Their insights have led to many of the discoveries that ultimately motivate the higher-dimensional map we will be introducing and exploring in this book.

The first of these individuals may have secretly believed that the tale of Atlantis was more than just a myth. He may have harbored the intuition that beyond the horizon there was more to be found than his inherited maps resolved; that somewhere in that rhythmic oceanic trance, Plato's city of gold was shimmering in the sunlight. Seven years after Marsilio Ficino translated Plato's legend of Atlantis into Latin, the Queen of Spain agreed to finance his expedition.[2] The officially recorded goal of that expedition was to find a shorter trade route to the rich continent of Cathay (modern China), India, and the fabled gold and Spice Islands of the East. The man who wrote the proposal for this expedition, and then audaciously sailed out into the Atlantic abyss, was Christopher Columbus.[3]

It is conceivable that Columbus used a combination of rumors, legends, and the maps at his disposal to secretly calculate where the land of Atlantis was most likely to be. Despite his preparations, his projection contained a fortuitous calculation error that would guide his fate. The size of the Earth, according to his maps, was far smaller than its actual size, because the calculations they were based on had not been converted from Arabic miles—which are significantly longer. As a consequence, Columbus conceived the lands of the East being much closer than they actually were. Nevertheless, his officially proposed route would take him in the general direction of his hidden goal.

History records that while underway Columbus deviated from his proposed India-bound course toward the northwest for several days. This deviation suggests that he was searching for something not on his official itinerary. He was sailing into the unknown, fueled only by a dream, and taking a chance that would forever change mankind's perception of the world. This maneuver almost ended in mutiny.

Although Columbus never found Atlantis, or a shorter route to India, his voyage did show that our trusted maps can be incomplete. In this, his intuition was wholly vindicated. There was indeed an entire continent beyond the Atlantic waters awaiting discovery. There was much more to the world than the maps of his era portrayed, and the missing portions were discoverable.[4]

Hundreds of years later, another incomplete map was confronted. Instead of charting the various lands divided by the waters, this map charted the very parameters of physical reality. The man who challenged this old map had a dream to discover a framework wherein all the laws of Nature were simple, harmonious, and unified. He believed that such a map must ultimately be within intuitive grasp. He recognized that the old map, Newtonian mechanics, was no longer capable of charting the ever-increasing array of human observations. To him, this meant that there must be new parameters, new islands of thought waiting to be discovered, and he set out to unearth them. The call of this quixotic quest defined his entire life. The man, of course, was Albert Einstein. Einstein was far more than the father of relativity or the grandfather of quantum mechanics; he was the author of a new legend—the legend of a new Atlantis.

The map that Einstein was searching for—the one that would reveal this new 'Atlantis'—is called a unified field theory. Those that could not hear the underlying call of Atlantis often mistook Einstein's goal to be to obtain a unified and simplified mathematical representation of gravity and electromagnetism, no small task in itself, but Einstein's aim was much higher. His was a dream of being able to peer into the fabric of physical reality—to ontologically access and fully comprehend its structure, beauty, and the bounds of its potential. His goal embodied the highest aspiration of experiencing the ultimate connection to Nature and obtaining the most elegant understanding of what it means *to be*.

Figure i Einstein's Quest, Portrait by Yousuf Karsh

Einstein poetically wanted to be able to touch what lay beyond the horizon. His intuition told him that this goal was within human reach and his explorer's spirit endowed him with the passion to continue his quest throughout his entire life. On the day before he died, Einstein called for paper and scribbled some calculations in a last hope to complete the map. "He knew he was dying. He knew he would not be able to complete the calculation. He did it anyway. The problem still mattered, and he still cared."[5] This was his legend.

Over the years, Einstein's legend infused the world. Newspapers heralded each of his publications with enthusiastic anticipation; building the rumors that Dr. Einstein had discovered a key insight that allowed him to unveil some of Nature's deepest secrets. When his papers were released, people flocked to see the new equations, even though most of them considered the menagerie of symbols to be completely incomprehensible. The Prussian

Academy printed a thousand copies of one such paper and released them on January 30, 1929. They promptly sold out. The Academy had to print three thousand more. When one set of those pages was pasted in the windows of a London department store, crowds of people who were drawn to the call of this new Atlantis gathered in the cold, pushing forward for their chance to glimpse the complex mathematical treatise. It didn't matter that the thirty-three arcane equations were unintelligible to most of them. What mattered was that a great mind was attempting to bring humanity a transcendent treasure—a map of the mind of God.

As it turned out, those equations didn't complete the map. Nevertheless, as a symbol of the growing recognition of the importance of Einstein's efforts, Wesleyan University in Connecticut paid a large sum to purchase the handwritten manuscript. The papers were deposited in the University library as a treasure.[6]

Like Columbus, Einstein never reached his end goal. Nevertheless, his discoveries did show that the Newtonian map was incomplete, that entire dimensional *continents* were still waiting to be discovered. Einstein resurrected the call of Atlantis and left us with the intrepid task of sailing the tumultuous waters of perception as we attempt to discover those continents.

To help guide us toward our destination, Einstein constructed "rubber sheet diagrams" portraying curved spacetime. Those diagrams are quite useful, but they are also incomplete. Because they offer no pictorial explanation for warped time, and only map the curvature of two space dimensions, these diagrams only allow us to incompletely access what is embodied by the math of general relativity. Nonetheless, that partial picture, combined with the full formalism of general relativity, dramatically improves our understanding of Nature. It reveals space and time as relative entities, gravity fields as geometric distortions in the vacuum and, most importantly, it releases us from some of our most long held assumptions about reality.

Einstein's collective insights (the discovery of the photoelectric effect, which contributed to the foundation of quantum mechanics, his explanation of Brownian motion, which verified the existence of atoms, and his masterpieces of special and general relativity) sired a technological revolution that gave rise to our modern world. As a result we now have transistors, atomic bombs, lasers, bar-code scanners, digital alarm clocks, charged-coupled devices in digital cameras, broadband Internet access, iPhones, solar panels, GPS, fiber optics, remote controls, televisions, DVDs, cancer radiation treatments, smoke detectors, the chemistry of colloids, which is the progenitor of our modern roads, and many of our modern pharmaceuticals from statins to Viagra and more. His insights also set into motion the science of cosmology, the study of ultimate origins, which produced the Inflationary Big Bang theory and enabled us to comprehend far more about the evolution of our universe and our place in it than ever before.

Despite the dramatic effects these things have had on our every day lives, it is important to recognize that all of these advancements were simple pit stops toward Einstein's true destination. They were the fabled spices, not the city of gold.

Because of the clarity he attained by peering into the fabric of reality further than anyone before him, Einstein never wavered in his belief that a deeper truth was attainable. He had glimpsed the edge of that truth by "lifting a corner of the great veil."[7] Because of this, he spent his life in opposition to those whose goal was to define reality as incomprehensible. These opponents aimed to reduce Einstein's legend into mere myth, by claiming that the human mind has inbuilt limitations that can never be overcome. Many of them have declared that even if a complete map of Nature exists in principle, it would be forever beyond our ability to comprehend in practice. They have been feeding into the self-destructive notion that "good science" cannot be mixed with emotion or spirituality—that emotion and spirituality can only be grounded in the supernatural. Somehow they never heard the music. They never felt the call of Atlantis.

Above all else, Einstein's life stands to refute this restrictive view. He once said, "...the serious scientific workers are the only profoundly religious people," because "science can be created only by those who are thoroughly imbued with the aspiration toward truth and understanding."[8] He profoundly understood that the path of science is ultimately paved by the desire to attain a deeper connection to Nature, to discover a clearer, more comprehensive description of reality, to unveil the causal story. Without this passionate desire, scientific progress comes to a screeching halt. "When this feeling is missing, science denigrates into mindless empiricism."[9]

Realizing that this spiritual fuel, this emotional fire, is integral to science, Einstein embraced and fanned his internal flame. Because of this he grew to see reality in a way that surpassed the vision of all those who had come before him. Consequently, he was the first to obtain a deeper clarity—an experience that connected him with the divine.

> *"...the cosmic religious experience is the strongest and the noblest driving force behind scientific research."*
>
> Albert Einstein[10]

To differing degrees, many people have felt echoes of the deeper connection that Einstein talked about. We experience them in moments scattered throughout our lives. The first time a child beholds the fossil of a ferocious dinosaur, or gazes into the Trapezium of Orion's Great Nebula through a telescope, a powerful connection with the vastness of time and space is experienced as an overwhelming feeling of awe and exhilaration. The first time we witness the mesmerizing rhythm of a comb jellyfish, and when we first hear the melodic trill of a Meadowlark, our intellectual horizon expands and our intuition becomes charged with a potential to grow.

Whenever we lose track of our physical boundaries, whether we grasp a piece of the Moon in our hands, or experience the ambrosial touch of love, we catch a glimpse of that deeper connection—and a flicker of our 'magnificent insignificance'. Einstein's 'cosmic religious feeling' is not inherently limited to punctuated intervals. As a direct link to the divine it has the potential to asymptotically increase until it reaches a constant level of clarity. This is the goal: our quest is to discover the map of reality and to meld our intuition with Nature's true form. From this union we will discover that physical law itself is divine, and God will be unmasked as the ultimate manifestation of Nature's order. The trade winds of Einstein's efforts blow us in this direction; they motivate us to continue the quest. The journey is more than a dream to complete the map of Nature; it is more than an aesthetic desire for symmetry and mathematical beauty. The scientific quest is about obtaining a direct link to the divine structure of the cosmos—to touch God, and to understand truth.

Figure ii Einstein in Santa Barbara, 1933: Einstein archives

> *"The supreme task of the physicist is to arrive at those universal elementary laws from which the cosmos can be built up by pure deduction. There is no logical path to these laws; only intuition, resting on sympathetic understanding of experience, can reach them."*
>
> Albert Einstein[11]

Because this quest necessarily guides us into uncharted waters, our journal entries tend to seek clarity by making use of poetic reference. As a consequence, our discoveries are often confused (by those not on the quest) with something supernatural. But they are not supernatural. Such a misconception would be selling the experience of the quest, and Einstein's 'religious' foundation, short. Einstein was not a theist, nor was he a deist. His "cosmic religion" reflected his devotion to the task of discovering Nature's hidden structure and his sense of adoration and awe for the infinite potential of that process. He is considered a pantheist.[12] He said, "I believe in Spinoza's God who reveals himself in the orderly harmony of what exists, not in a God who concerns himself with fates and actions of human beings."[13]

Einstein's repeated use of the word God, despite his knowing that many would be unable to comprehend his intended meaning, was unavoidable. He was as incapable of speaking of the Cosmos in a technical connotation as a young boy is of recounting his first kiss in monotone. His connection to this deeper reality—to God—was the entire point. Those who miss this message, that it is about obtaining a deeper connection to reality, but still try to follow the path of discovery are, as Lee Smolin writes, "reaching for a beautiful flower but missing the beauty of how it is that flowers come to be."[14]

This quest embodies the preeminent mystery. By definition, it aims to surpass the restrictions that stem from fundamentalism. It is the search for a vivid unfurlment of Nature's secrets unfettered by dogma. It is a quest to attain a new form of common sense, an elevated intuition, by which an intimate understanding of the foundations of the mysterious naturally bestows upon us a fellowship with the infinite.

Einstein invited us to make his dream our universal quest. He blazed a new path and made it possible for all of us to take part in the great intellectual adventure, but like all adventures it has its perils. If we partake in it we will be required to face our ignorance. We will have to brave the thick fog of chaos as we drift across a sea of confusion. Eventually we will even need to challenge our most fundamental beliefs about the realm we mean to understand. But by doing this we will be part of a timeless voyage, pursuing an echo of the most ancient intuition, and actively searching for the underlying mystery. This, in and of itself, is reason enough to join the quest, for it is exquisitely human to search for philosophical truth. As Nietzsche said, "It is time for man to set himself a goal. It is time for man to plant the germ of his highest hope. His soil is still rich enough for it."[15]

Those considering joining this journey should first heed one warning. Threads of an underlying mystery have already been discovered. They have revealed wonders that challenge the imagination and opened chasms that defy empiricism. But the map that connects all of these discoveries is still missing. We are currently lost and confused. Because of this many people have begun to abandon the journey. They have outright given up. They say that if what we are searching for exists, it cannot be grasped by human imagination. To them, additional dimensions, quantum mechanical uncertainty, wave-particle duality, and nonlocality are destined to be forever beyond our grasp.

This attitude continues to grow because, for the most part, the sails of our ship have remained slack. Since the death of Einstein our efforts to unveil a complete framework—a model of spacetime that brings the mysteries of our world within the reach of human intuition—have diminished. What we are after is intellectual transcendence. The fact that we have not yet achieved it is no reason to give up. There is always a chance that an imaginative insight will restore the wind to our sails. A new picture of reality, the map we have been after, might simply be waiting for someone to challenge an assumption that has never been seriously challenged before. If this is the case, then the conceptual portal we have been searching for might be just one fathom away.

In memory of the dying wish of a great dreamer, and in honor of his intuition, adventurous spirit, and passionate belief in a deeper truth, now is the time to hoist our sails. Now is the time to join the voyage. It is up to us to continue the quest, to brave the dimensional cascades, and to question the fabric of our old map. It is up to us to set out to discover the holy grail of modern physics—the new Atlantis.

In the spirit of Einstein's intuition, let us launch our intellectual quest from daring shores—starting from the assumption that the structure of spacetime is far richer than we have presumed it to be. Let our voyage of discovery be based on careful examinations of the mysteries of modern physics, creative attempts to get beneath assumptions that might be holding us back from understanding those effects, and the process of sifting through a flurry of ideas by checking our axioms against the mysteries we mean to explain. Although this book will carry us through that process by exploring the depths of a specific geometric proposal (that spacetime is a superfluid with a fractal structure), the end goal of our investigation will be to fill the sails of our ship with many new and imaginative ideas, and to teach us how to realize the full power of those ideas by learning how to properly trim the sails that capture them.

As we explore new isles of thought we are looking for ones that give us a way to visualize Einstein's curved spacetime in a richer way. We are searching for a perspective that will allow us to marry the brilliance of Einstein's intuition with the paradoxical, and sometimes nonsensical, visions of quantum mechanics; creating an intelligible, visualizable system that facilitates a deep understanding, both ontologically and epistemologically, of what is actually going on behind the veil. The treasure we seek is a geometric description of spacetime that can be shown to be deductively responsible for the mysterious effects of both general relativity and quantum mechanics.

In this book we will be exploring genuinely new ideas, a new set of assumptions about the geometric structure of space and time (introducing axioms that give rise to a new candidate for the theory of quantum gravity called quantum space theory), and we will be investigating whether or not those particular assumptions carry us toward our goal of obtaining a greater understanding. My hope is not to convince you beyond any reasonable doubt that Nature adheres to the structure proposed herein. Rather my hope is that this investigation encourages you to personally join the quest, to challenge assumptions that you have always taken for granted, to immerse yourself in the unknown, to actively participate in the great mysteries, and to devote yourself to making sense of them. This book chronicles how I have begun that quest for myself. Should it be a useful guide in your intellectual journey, I shall consider it a success.

[1] Einstein to William Miller (1955, May 2). Quoted in *Life magazine*. Calaprice, 261; Walter Isaacson. Einstein, p. 548.

[2] With fastidious resolve Columbus made proposals, and switched his allegiance, several times. First to the duke of Anjou in France, then to the king of Portugal, the duke of Medina-Sedonia, then the count of Medina-Celi, and finally to the king and queen of Spain. All of these proposals were denied, but after appealing their first rejection, the king and queen of Spain eventually granted him three ships. Jared Diamond. (2005). *Guns, Germs and Steel—The Fates of Human Societies*. New York: W. W. Norton & Company, p. 412.

[3] In 1485 Marsilio Ficino translated Plato's works into Latin. Columbus made his first formal proposal to John II, King of Portugal that same year. Seven years later Columbus sailed to the Americas. Whether or not Columbus was actually motivated by a desire to discover Atlantis, or even whether or not he read Plato's story has been debated. There is no reliable record. Nevertheless, von Humboldt, "whose intellectual portrait of Columbus remains unrivalled" notes the absence of any mention of Atlantis from Columbus's writings but nevertheless maintains that Columbus "took pleasure in Solon's reference to Atlantis" (von Humboldt. *Historie de la géographie du nouveau continent*, 1:167). Pierre Vidal-Naquet & Janet Lloyd (1992, Winter) *Critical Inquiry*, Vol. 18, No. 2. Atlantis and the Nations, p. 309. Columbus was not very secretive about how he felt about gold. "Gold is the most exquisite of things," said Christopher Columbus. "Whoever possesses gold can acquire all that he desires in this world. Truly, for gold he can gain entrance for his soul into paradise." Christine King. (1978, November 30). The gold of El Dorado. *New Scientist*, p. 705. This was not an uncommon opinion. The Spanish Conquistadors were prepared to commit genocide to find the city of gold known as El Dorado (Ibid.).

[4] Vikings such as Lief Eriksson had visited North America five centuries before Columbus' voyages, and the Polynesians had been trading their chickens for sweet potatoes with Native Americans for at least a hundred years before Columbus. This has been verified by DNA analysis of chicken bones found in the Americas, which have been dated between 1300 and 1424 CE and are clearly of Polynesian, not Spanish, origin. These encounters did not significantly affect European maps of the world. The enrichment of the European maps is Columbus' great achievement. See: Elizabeth Matisoo-Smith. (2007). University of Auckland and Proceeding of the National Academy of Sciences, DOI: 10.1073/pnas. 0703993104.

[5] Thomas Levenson. (2004, September). Einstein's Gift for Simplicity. *Discover*, p. 45.

[6] Walter Isaacson. (2007). *Einstein: His Life and Universe*, p. *343*.

[7] Einstein praised de Broglie's work effusively, saying that it had "lifted a corner of the great veil." Walter Isaacson. (2007). *Einstein: His Life and Universe*, p. 327.

[8] Walter Isaacson. (2007). *Einstein: His Life and Universe*, p. *390*.

[9] Einstein to Marice Solovina. (1951, January 1). In Solovina, 119; Walter Isaacson, *Einstein*, pp. 462–463. Einstein also said, "…I maintain that the cosmic religious feeling is the strongest and noblest motive for scientific research". (1954). *Ideas and Opinions*.

[10] *New York Times Magazine*, November 9, 1930, 1–4. Reprinted in *Ideas and Opinions*, 36–40; (2005). The New Quotable Einstein, Collected and edited by Alice Calaprice, p. 199.

[11] Principles of Research, address by Albert Einstein (1918). Physical Society, Berlin, for Max Planck's sixtieth birthday.

[12] "Pantheists don't believe in a supernatural God at all, but use the word God as a non-supernatural synonym for Nature, the Universe, or for the lawfulness that governs its workings." Richard Dawkins. *The God Delusion*. (2006). New York: Houghton Mifflin Company, p. 18.

[13] Richard Dawkins. (2006). *The God Delusion*. New York: Houghton Mifflin Company, p. 18. He also said, "I have never imputed to Nature a purpose or a goal, or anything that could be understood as anthropomorphic. What I see in Nature is a magnificent structure that we can comprehend only very imperfectly, and that must fill a thinking person with a feeling of humility. This is a genuinely religious feeling that has nothing to do with mysticism." Ibid., p. 15.

[14] Lee Smolin. (2004, September). Einstein's Lonely Path. *Discover*. p. 40.

[15] Friedrich Nietzsche. (2005). *Thus Spoke Zarathustra*. Translated by Clancy Martin. New York: Barnes & Noble Classics. p. 13.

Part One—Returning to a Conceptual Approach

"The true lover of knowledge naturally strives for truth, and is not content with common opinion, but soars with undimmed and unwearied passion till he grasps the essential nature of things."

Plato

Chapter 1 — **Ascending the Footholds of Rationality**

"Look into nature, and then you will understand it better."
Albert Einstein

"Perhaps what we mainly need is some subtle change in perspective—something that we all have missed…"
Roger Penrose

Vermilion cliffs. First light on Thanksgiving Day.

Morning dew dripped from the scarce blades of grass, gently filling the air with the nostalgic aroma of wet earth. In slow motion the atmosphere danced about us, wafting the pungent fragrance of nearby sagebrush to our nostrils. The sky reluctantly gave up the last of its stars, but it would take another three hours for the Sun to complete its climb over the towering rocks that surrounded us. Our tents were sprawled within a small field, one that could claim no more than an inch of topsoil. In this thin blanket small ants busied themselves waging war. Above, two white butterflies erratically drifted through invisible eddies. Crouching down on the ground I saw tiny yellow dots resolve into the four-petaled flowers of the Violet Sagebrush, no more than one centimeter in diameter. The coolness of the night was beginning to fade. There was no time to waste.

The towering rocks ripped the skyline into a jagged curiosity rendering us unnoticed in their formidable shadows. Our excitement built, as we collapsed our tents and carefully balanced the sixty pounds of supplies that filled each of our backpacks. After strapping on our gear we followed a small sandy footpath. Within minutes it led us to a crevasse—a gateway that would begin our journey. Its proportions betrayed the grandeur it protected, but our hearts quickened with the knowledge that this four-foot wide threshold guarded a forty-mile maze of twisting rock. Inside, a magical glimpse of Nature awaited us. We paused for a moment and listened to the faint whispers coming from the mouth of our trail. Then, with wide eyes, the six of us entered the world's longest slot canyon.

Our bulky packs had transformed us into a single file line of clumsy giants, barely able to squeeze through the rock walls. Petrified swirls of orange and red jutted inward and then outward, occasionally wedging our packs so tightly that we could give our weight to the canyon walls and dangle our feet below. The trail beneath us was sandy and cool to the touch. The echoes of our footsteps became malleable, changing their tone and cadence with every twist and turn. Each section molded the timbre and attenuation of our movements in its own way. Desert varnish dripped down the sandstone canvas, covering it with oozing streaks of black, a gift from the bacteria high above who spent their lives basking in the sunlight of the canyon's rim waiting for the next regenerating rainstorm. Ancient stories of great hunts and perilous dangers were highlighted on the walls in the form of petroglyphs. Foretelling. Warning. This place was a forgotten rite of passage, a portal into eons long past, a gateway to another set of rules.

Here everything was serene. Every step was riddled with an unfamiliar blend of sensation. It felt like we were inside Nature's hourglass. A steady flow of sand trickled down from the sliver of sky above. Every sound twisted and turned before fading into the background choir of echoes. And at any moment everything could be turned upside down.

As the path descended, the walls climbed higher and higher and the world we knew disappeared. There was no wind, but we could feel the air resisting our intrusion. There was no direct sunlight, yet we were surrounded by brilliant patterns of orange and red. Step after step the walls continued to climb. Overhead we spotted large decaying trees that were forcibly wedged sideways between the rock walls. They were inescapable omens, not-so-subtle reminders of the flash floods that routinely carved this beauty. They testified of the violent and unpredictable power that etched this place and the towering wall of water that could be upon us at any moment.

This was a landscape in eternal flux. Each footprint was a first, every vista pristine. The rocks smelled of childhood memories mixed with dreams of exploring Mars. The promise of piercing the veil of Nature's deepest secrets hung pregnant in the air, waiting for us to round the next bend.

Shadows danced throughout the day, resisting the sun's attempt to glimpse the path below us. The deepest scars kept the complexity of this realm hidden from the prying orb above. The more we descended, the more time betrayed us. Before we knew it the cloudless filament of blue above faded and stars began to reclaim the strip of sky. We lighted our path with headlamps and pressed forward. When we came upon a small sand bar, we finally stopped and made camp. Then, as a little surprise for the two of us that were Americans, our self-appointed leader, who was also the field guide for our dinosaur expeditions, began to cook prepackaged turkey and instant potatoes for a celebratory Thanksgiving dinner.

The one-pound stove performed perfectly, but it was defenseless against the constant percolating sand from the world above. Our cook was convinced that trying to avoid its inevitable taunting was an unnecessary inconvenience. Although his pot had a lid, he didn't even bother using it. He said that a half pound of dirt would help fill us up and that we wouldn't even notice its presence if we chewed without letting our teeth touch—a trick he learned in Madagascar. Apparently the technique required some practice to perfect.

As we woke, the morning air had such a bite to it that we might as well have been on Mars. The only immediate sign that we were still on Earth was a single patch of sagebrush, which was reluctantly doubling as a makeshift clothesline. We had draped our socks over the bush late the previous night hoping to air them out. It didn't work quite as we had hoped. All of our socks were now frozen and shaped like Dr. Seuss pretzels. Mia, the youngest in our group and an outdoor adventure writer, grabbed her socks and struck them against a rock to flex some of the ice out. The collision sounded like the tapping of a metal axe. It was funny until we realized that Mr. Sandy Potatoes wasn't likely to let some frozen socks get us behind schedule. At this thought we scrambled in vain to thaw them out.

After we devoured some prepackaged food we began familiarizing ourselves with the unique screams that people make when they try to wedge their feet into socks reinforced by small, sharp threads of ice. That was all the encouragement we needed to get moving.

The canyon had widened to about fifty feet from wall to wall. A small stream braided its way through the trail, filling the air with soothing echoes of gurgling water. Overhead, raven puppet masters squawked with laughter at the earthlings trapped in their maze below.

The turns were more rounded now, the straight-aways longer. The open spaces made us feel even smaller. We were like tiny ants making our way between two unabridged dictionaries spaced just a couple of fingers apart. The braids of water grew more and more tightly woven, concentrating in the middle of our trail. The soft dry sand slowly became hard packed and damp. Everything started to wake. All around us we could feel a deep vibration. The air was filling with life, moving just enough to rustle the hair on the back of our necks. As we walked, the vibration became audible as a faint rumbling sound. With each step it grew louder and the rustling air developed into a breeze. It quickly became clear that we were approaching the source of all this commotion.

After rounding one more bend, we found ourselves standing before a long corridor of towering rock that was convincingly auditioning for the next Indiana Jones movie. In the far off distance our trail was truncated by another wall of rock. Pressing on, our legs began to thaw and the details of the constriction slowly began to resolve. The canyon suddenly merged into a major artery (only twenty feet wide at this point). Here the trail disappeared under a foot and a half of icy water whose cresting echoes resonated throughout the rock corridors for miles. Stepping into the frigid currents, I was overwhelmed with the sense that I had just entered a realm that was completely unaware of any standards or imposed conformities.

Turning right at the junction we followed the flowing water. I felt completely out of place in this strange underworld lair. Water was swirling around my numbed legs, reverberating as it buffeted against the rocks ahead. Echoes were growing louder and louder, filling in the melodic treble of Nature's most recondite song. This masterpiece was far more vibrant than anything I had imagined. The ground was water, the sky was rock, and everything came together like a bizarre surrealistic painting in progress. It was unfamiliar and mysterious.

By lunchtime we reached a semi-dry sand bar featuring a rock-sculpted bench. A jet of cold clean water, as thick as the stream of a garden hose shot out from a canyon wall and over two weathered seats. I removed my backpack and tried to take it all in.

> *"The most beautiful experience we can have is the mysterious. It is the fundamental emotion that stands at the cradle of all true art and science. He to whom this emotion is a stranger, who can no longer wonder and stand rapt in awe, is as good as dead, a snuffed out candle."*
>
> *Albert Einstein*[1]

This was my first experience hiking through a slot canyon. I had never before seen Nature in this way. It was so different from what I had expected that I had difficulty imagining how I would fully explain this alien world when I got home. I wondered how I could accurately depict the rich beauty of this secret realm to someone who has no context by which to ground that description. This question led to more questions.

Is it possible to reveal the beauty of Nature without translating that beauty into the terms of human senses? Is it possible to convey what Nature looks like without constructing a picture? After I pondered these questions, I realized that in order for us to wrap our intuition around the natural realm we must find a way to relate that realm to our senses. Literally, if we want to know what Nature looks like then we have to construct a picture. As Steven Strogatz eloquently puts it, "without direct visualization we are dynamically blind."[2]

To explore this point, suppose that I took a digital picture of what we dubbed 'The Fountain of Buckskin Gulch,' and then presented the digital information of that picture, the raw sequence of ones and zeros, to someone. Would that untranslated information help them see the fountain? This is more than just a question of lexicon, semantics, or syntax—it is a matter of connection. In other words, if I tried to present a facet of Nature's beauty to someone without translating that information into a display that can be directly experienced by at least one of the senses, then how could I ever expect the recipient of that information to fully comprehend that beauty?

Einstein addressed this issue more poetically when he said, "Knowledge exists in two forms—lifeless, stored in books, and alive in the consciousness of men. The second form... is the essential one."[3] We can only obtain this second form when we extend the reach of our intuition into the depths of Nature's secrets. But in order to do this we need a conceptual portal that is capable of unveiling a richer map.

This realization highlights a fundamental problem in the approach taken by modern physics. For the past several decades, theorists and mathematicians have been working on constructing a framework of Nature that is capable of mathematically combining the descriptions of general relativity and quantum mechanics under the same rubric. (We will discuss these theories in detail later on.) These efforts have focused on the task of organizing Nature's data into a self-consistent assembly—like the ones and zeros of a digital picture. The problem is that this inductive approach does not encourage, let alone require, the discovery of a conceptual portal.

Even if physicists were to one day conclude that their assembly was mathematically correct, it would not actually increase our ability to truly comprehend Nature unless it was translated into some sort of picture. Therefore, because it is really the picture that we are after, maybe it is time for us to consider whether or not our efforts will bear more fruit under a different approach. Specifically, to maximize our chances of completing our goal of intuitively grasping Nature's complete form, maybe we should follow the lead of young Einstein and return to a conceptual approach. Perhaps it is time for us to place our focus on constructing a richer map of physical reality. If we don't, then all of Nature's elaborate arrangements may very well remain forever hidden in obscure mathematics and impenetrable sequences of data.[4]

As I sat at the fountain surrounded by melodic purls and dancing caustics, these thoughts echoed through my mind. It suddenly became clear to me that what we need is a new picture of Nature—one capable of depicting its deepest symmetries and beauty. We need a map that can introduce our senses to what lies beyond their experiences. We need an insight that transforms our intuition and opens our eyes to the breathtaking simplicity that underlies the world we know *and* the world of bewildering mysteries. It must unify everything around us and make sense of it all. But how do we attain such a map? How do we lift that veil of ignorance?

Let's begin our search for the answer to that question by examining the history of the map we have inherited.

Our Maps, A Brief History

Our ideas, concepts, explanations, and questions are encased within a mental framework—a picture of reality that structures our interpretation of the world around us. This framework is the foundation by which we understand physical reality and the filter by which we process our sense of self. The model, however, is not static—it evolves over time.[5] As an example of this, consider how the Greeks conceptualized the heavens until the classical period (a conceptualization shared by the peoples of the Near East until the Hellenistic period, India until the Gupta period, and China until the 17th century) (Figure 1-1).

Figure 1-1 An early map of the heavens

In this model the universe was conceptualized as closed, which meant that it had a natural finite limit, and the world was believed to be flat. In this picture the outer boundary of the knowable universe was depicted as a great sphere that encircled the Earth. This sphere was understood to be the edge of the universe. All the stars in the heavens were seen as embedded in this sphere.[6] As it rotated, this sphere would continuously drag the stars within it from the east to the west.

A later version of this geocentric model (in which the Earth is depicted as spherical) is called the Ptolemaic model because it was popularized when Claudius Ptolemy published "Almagest" circa 150 CE. In its first incarnation, this model explained the observed daily and yearly motions of the stars, and portrayed the universe as finite and permanent, but it was not yet complete. In order to explain the observed motions of the sun and the moon, two more celestial spheres had to be included in the model (Figure 1-2).

Figure 1-2 Three levels of heaven

Although this change complicated the model, it was generally well accepted because the existence of three celestial levels intuitively jived with humanity's experience of the heavens. The sun, the moon, and the stars each belong to their own order of brilliance; therefore, many people found it easy to accept that they also belong to their own celestial sphere. These "three levels of heaven" portrayed the universe in keeping with the Ptolemaic belief, which held that circles and spheres were the geometries of the divine. This sentiment has been preserved in the texts of the great Western religions.

This tri-layered depiction of the cosmos[7] was believed to be a complete model of the cosmos until some curious minds pointed out that five special 'points of light' moved about in the heavens independent of the three celestial spheres. Although they appeared to be stars, they did not remain fixed within the sphere of stars. Because of this, these lights were given the name *planet* (the Greek word *planete* means wanderer and *planan* means to lead astray). The motions of these "planets" were complex and difficult to describe. They drifted about the heavens, moving from constellation to constellation, and sometimes even reversed their directions. To the fastidious observer they "seemed to move of their own volition, which is perhaps why they were identified with gods".[8]

The presence of these 'gods' in the heavens meant that the celestial model needed some reconstructing. The motions of these planets (Mercury, Venus, Mars, Jupiter, and Saturn)[9] severely complicated the celestial model (Figure 1-3). In order to describe all heavenly motions, the model now required seven levels (which became the seven days of the week) in addition to the original outermost sphere of stars.[10]

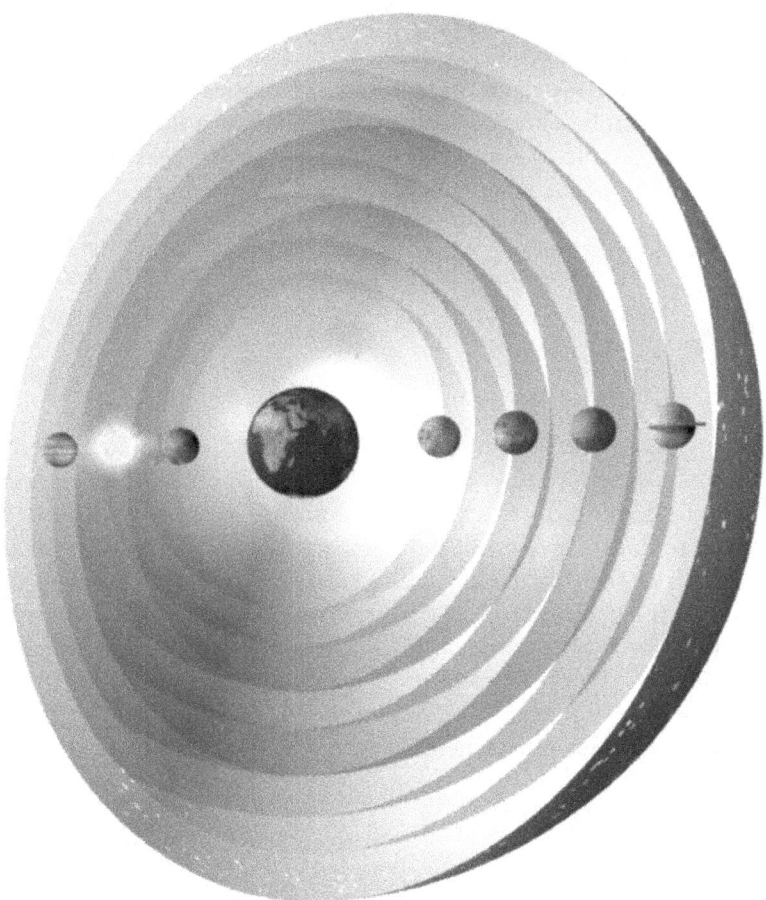

Figure 1-3 Seven levels of heaven with a spherical Earth

This reconstruction presented a problem. There was no longer intuitive support for that number of levels, or for the epicycles that were subsequently added to account for observations of things like retrograde motion. The heavens were no longer simple, intuitive, or what mathematicians might call *beautiful*. Nevertheless, these additional levels did produce a map that accounted for all observed heavenly motions, which allowed humankind to continue seeing itself as the center of the universe. The only problem was that this multi-leveled map of the heavens was getting fairly ugly. Nature was getting increasingly and unexplainably complicated.[11]

Parsimony, a foundational principle of science, encourages us to look for elegance and simplicity in our descriptions of Nature. Indeed, the well-known precept of Occam's razor advances the notion that we should prefer simplicity over complexity (all other things being equal).[12] Such considerations led a Polish man by the name of Nicolaus Copernicus (1473–1543) to challenge the multi-leveled Ptolemaic model in favor of a simpler description—one that depicts that the Earth revolving around the Sun (a heliocentric model) instead of the Sun revolving around the Earth.[13] This beautiful model did not need epicycles to explain the motions of the planets.

Fearing that his work would lead to professional ridicule and "controversy" among the masses, Copernicus refrained from bolstering his idea. Almost one hundred years later an Italian astronomer named Galileo Galilei (1564–1642) recognized the mathematical simplicity and beauty of Copernicus' heliocentric model and publicly supported it. In response to this, the Catholic Church sentenced Galileo to a life of confinement. Despite its efforts, the church was unable to completely suppress this simple idea. The seeds of curiosity had been planted and the heliocentric model began to gain ground.

"Copernicus, that fool, who would reverse the entire art of astronomy…
Joshua bode the sun and not the earth to stand still."

Martin Luther

Long before these models of heaven were debated, the shape of the Earth was a controversial topic. In the 6th century BCE the Greeks noticed that when a ship approached, they first saw its sails coming over the horizon and only later saw its hull. They recognized that if the Earth were flat this wouldn't happen. Navigators were also quite familiar with the fact that the position of the North Star changed in the night sky as they traveled north or south. Near the equator the North Star (Polaris) lies on the horizon, but as one approaches the North Pole, the North Star climbs higher and higher until it is directly overhead. This occurrence is what defines latitudinal position. If you are at 37° north then the North Star will appear 37° above the horizon from your position. If the Earth were a flat disk it would be very difficult to explain this phenomenon.

In 340 BCE Aristotle famously argued for a spherical Earth, rather than a disk shaped Earth. In his book *On the Heavens* he noted that during lunar eclipses (when the Moon passes into Earth's shadow) the shape of Earth's shadow is always circular. But if the Earth were a flat disk its shadow would not always be a circle—unless the eclipse always marked identical alignments between the Sun, Earth, and Moon—they don't. This, he argued, meant that the Earth must be a sphere.

In the heliocentric model it is natural to conclude that the Earth is spinning about its own axis, but in the Earth-centered model the heavens are seen as spinning around the Earth. Therefore, in order to determine which model is correct, all one has to do is come up with an experiment capable of proving that the Earth is either stationary or spinning. Can you think of an experiment that is capable of this differentiation?

Today, we can observe such an experiment first hand in planetariums and science museums around the world. It consists of a large pendulum suspended from the ceiling that is swinging back and forth slowly knocking over the Dominos that encircle it (Figure 1-4). Scientists call this a Foucault pendulum because in 1851, the French physicist, Jean-Bernard-Léon Foucault was the first to construct such a pendulum at the Pantheon in Paris. To see how this simple set-up can distinguish a spinning Earth from a stationary one, let's imagine taking a Foucault pendulum to the North Pole during the season of perpetual night (the northern hemisphere's winter). When we get there we observe that the stars are constantly and slowly circling about us in a counterclockwise manner. In order to determine whether this apparent motion is due to a spinning Earth or to the spinning heavens, all we have to do is align the plane of our pendulum's swinging motion with any star we fancy (Figure 1-5). As the hours pass, we discover that the motion of the pendulum remains aligned with the star we've chosen, but the orientation of the swinging pendulum, with respect to the ground, changes over time. It makes a complete cycle in one day (Figure 1-6).

Figure 1-4 Foucault's Pendulum

Because there are no identifiable forces pushing or pulling on the pendulum's orientation, we must conclude that the stars are fixed, that the pendulum's orientation is not changing, and that its apparent motion is due to the spinning of the Earth beneath it.[14] From this conclusion we are compelled to accept the validity of the heliocentric model. A new perspective must be embraced.

The heliocentric revolution introduced a major shift away from a culturally entrenched medieval worldview—a worldview dominated by hierarchical submission, unquestioning duty, and faith; where the *levels of heaven* were seen as depicting the importance of hierarchical social order. Under the old levels of heaven model it was natural to succumb to the idea that birthright determined your fate because a similar fixed hierarchy was seen reflected in the heavens. Each celestial body remained forever entrenched in its own domain.

With the new model, humanity was no longer at the center of the universe, and the reasoning individual was no longer restricted by caste. Objective, methodical reason, and rational inquiry replaced unquestioning faith as the path to truth. Fear in God was replaced by the courageous goal of discovering the laws of Nature. This transition gave birth to a new era, fueled by the power of human reason, with the goal of lifting the great veil to improve the human condition.

Figure 1-5 If the heavens are spinning, then the pendulum will always trace out the same path above the ground.

Figure 1-6 If the Earth is spinning, then the pendulum's motion will change with respect to the ground.

The great human quest now had a vessel—a method for discovering truth—that had been forged by the accumulation of history's highest insights. Like the ship of Neurath's simile, which is continuously being built and repaired by the sailors it is carrying across the sea, the vessel of science is capable of improving its methodology as it assists our exploration of the realms that lie beyond the reach of our senses.

In many ways, the first official captain of that modern vessel was Sir Isaac Newton. Newton was born one year after Galileo died. He grew up to become the world's first theoretical physicist (although, in his day, the profession was called "natural philosopher"). During Newton's remarkable life, he added calculus to René Descartes' arithmetized geometry, which enabled him to simplify his expression of gravitational attraction and richly capture the heliocentric framework within his formulation. Because of its precise predictive power, Newtonian mechanics was accepted as the new, universal and complete model (map) of physical reality, officially replacing the Ptolemaic model, which was overthrown by Copernicus and Galileo.

Newton's model simplified the natural realm into a deterministic picture governed by a few interlinked rules. It gave us more than the ability to explain and predict observations (orbiting bodies, colliding bodies, etc.) in terms of forces, accelerations, velocities, mass, position, and so on. It provided an elegant unification of natural phenomena.

For example, before Newtonian mechanics the phenomenon of motion and the phenomenon of heat were considered unique and separate. The Newtonian lens reveals a relation between the two because it explains heat in terms of the average motion of atoms in a medium, and sound in terms of how collective molecular motions contribute to propagating distortions in a medium.

Phenomena that had long been thought to have completely separate origins were tied together by a deeper understanding of Newton's laws of motion.[15] This bolstered the possibility that Nature is logically composed and that it is ultimately comprehensible. Of course, this wasn't the end of the story. As we looked closer and closer into Nature we began to find phenomena that could not be explained by Newton's map.

Meticulous observations, like the discovery of Mercury's perihelion advance (first observed in the 1840s) really called the validity of Newton's map into question. The perihelion is the location on a planet's elliptical orbit that is closest to the Sun. If there were nothing else in the solar system besides the Sun and Mercury then the orientation of Mercury's elliptical orbit would remain fixed in space. But in our solar system, which has many other massive bodies, the combined gravitational fields interact in such a way as to cause all the elliptical orbits to precess (Figure 1-7). For Mercury, this perihelion precession turns out to move ahead of where Newtonian mechanics says it should be by about 43 seconds of an arc per century (3,600 arc seconds equals one degree).

Newtonian mechanics predicts that Mercury's orbit should precess at a rate of 532 arc seconds per century, yet it precesses at a rate of 575 arc seconds per century. When astronomers first discovered this discrepancy they assumed that it was due to an undiscovered planet gravitationally tugging on Mercury—similar assumptions had led to the discovery of Neptune. Urbain Le Verrier first described this effect in 1859, calculating where the mystery planet should be and named it Vulcan after the Roman god of fire.

For the next 20 years astronomers from around the world searched for a planet inside the orbit of Mercury. Observing such a planet is rife with difficulties. It requires pointing your telescope at or near the Sun, which leaves you susceptible to the production of false reflections inside your optics. It is also very difficult to distinguish a round sunspot from a planet in transit. Over the years, as we might expect, many false sightings were reported.

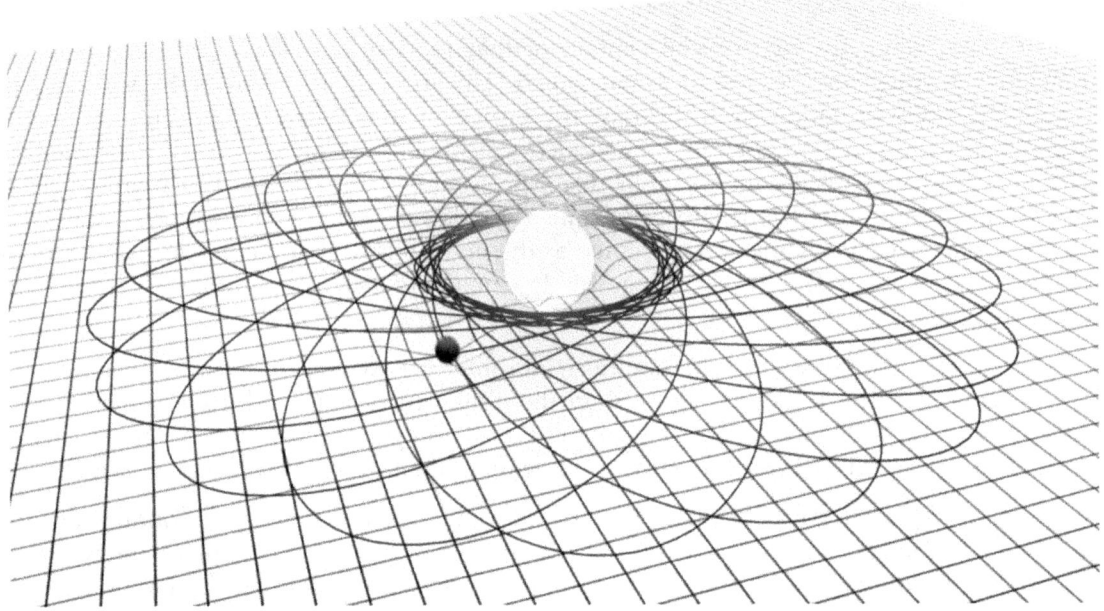

Figure 1-7 Precession of Mercury's perihelion

Le Verrier died in 1877 believing that he had discovered another planet. After his death the search for Vulcan dwindled and scientists eventually concluded that Vulcan did not exist. Without another planet to tug on Mercury's orbit, the completeness of Newton's map of Nature had to be called into question. Faith in that map could no longer be absolute.

As the years passed, more and more observations amassed that challenged the completeness of Newton's map. Then, in 1915, Einstein formulated a new map (general relativity) that predicted Mercury's perihelion advance without any reference to another planet. With this stroke of genius Newton's old map was officially dethroned. But Einstein's new map is only partially intuitively accessible, and it does not accurately map the microscopic realm. Despite its successes, this new model leaves us hungry for a more complete answer.

Since the fall of the Newtonian classical perspective we have been walking deeper and deeper into a jungle of confusion—discovering more and more aspects of Nature that defy the intuition of our old map. We no longer have a complete map of Nature's parameters. What we have instead are two maps (models) that are, at best, approximations, or fragments, of a larger, complete picture of Nature. The ultimate dream of a physicist is to discover the complete picture—a map that naturally fits together the predictions of quantum mechanics with the explanation of general relativity.

The theory of general relativity (a theory can also be referred to as a map or a model) builds a picture from which we can partially explain phenomena found in the macroscopic (large scale) world, including Einstein's discoveries of warped spacetime, black holes, time dilation, and so on. This is a rich and beautiful theory, but its explanatory power remains fragmented. There are two reasons for this. First, as it is presented Einstein's map can be only partially used to intuitively penetrate the phenomena it describes. Second, the theory cannot be extended to accurately describe the microscopic realm.

On the other hand, phenomena found in Nature's microscopic realm are exclusively translated to us via the mathematics of quantum mechanics—a formalism that makes fantastically accurate statistical predictions, but fails to provide any ontological clarity or carry any explanatory import. The canonical rendition of that formalism is, as Franck Laloë puts it, non-intuitive and conceptually relatively fragile.[16] It is so plagued to the core with conceptual difficulties, that in 1927 Niels Bohr said, "Anyone who is not shocked by quantum theory does not understand it." And forty years later Richard Feynman said, "Nobody understands quantum theory."

To further complicate matters, different foundational assumptions can be used to derive the same formalism, each with their own take on the nature of reality. As a consequence, we have no consensus concerning the interpretation of the theory or its foundations. As Claude Cohen-Tannoudji states "a really satisfactory and convincing formulation of the theory is still lacking."[17]

The conceptual difficulties beneath quantum mechanics originate from the object it uses to describe physical systems—the state vector $|\psi\rangle$. "While classical mechanics describes a system by directly specifying the positions and velocities of its components, quantum mechanics replaces those attributes with a complex mathematical object $|\psi\rangle$, providing a relatively indirect description."[18] What exactly does it mean to say that a system is better represented by a state vector, than by a specification of its component's positions and velocities? What does a state vector represent in reality?

The most difficult part of ontologically penetrating quantum mechanics is figuring out the exact status of $|\psi\rangle$. Does it describe physical reality itself, or does it convey only some (partial) knowledge that we might have of reality? Is it fundamentally a statistical description, describing ensembles of systems only? Or does it describe single systems, or single events? If we assume that $|\psi\rangle$ is a reflection of an imperfect knowledge of the system, then shouldn't we expect that a better description exists, at least in principle? If so, what would this deeper and more precise description of reality be?[19]

To ask this question, to remain open to the possibility that on some deeper level there is a more complete description, is to be at odds with the standard interpretation of quantum mechanics. This is the case because the standard interpretation doesn't just fail to touch base with an intuitive representation—it attempts to forbid one.[20] It brutely asserts that the "transition from the possible to the actual—is inherently unknowable."[21] But there is no reason to logically commit to that claim. It remains possible that a more complete description exists, and that the peculiar effects of quantum mechanics can be tied to a conceptual picture.

Two Pieces of the Puzzle

General relativity and quantum mechanics make different claims about how the world works. Our inability to augment and piece together these two theories into one coherent map is the reason that we haven't been able to formulate a picture of reality that encases all known phenomena. We are conceptually adrift, incapable of uniting all of Nature's dissonant details, and without a trustworthy map for our journey. Our goal is to come into possession of a complete map.

The incompatibility between the two fragmentary descriptions of Nature we use today is reflected in the fact that the Western intellectual world has acquired its heritage from two distinct intellectual modes—ways of thinking that can be traced back to independent historical and geographical origins. These modes set the foundations for how we formulate concepts and relate to the external world. Because their constructions are different, the two modes often misinterpret each other. One of those modes germinated modern science while the other gave birth to the mythological texts of the great Western religions. Because of this, many people consider these modes of thought diametrically opposed. But this conclusion fails to recognize the symbiotic contributions that these modes make to our evolving worldview.

The first mode, which is passed down from the Greek tradition of thought, assumes that ultimate reality possesses permanent physical characteristics. This approach is concerned with unmasking the ultimate *nouns* of reality and mapping Nature based on the permanent and unchanging parameters of those nouns. It is classical by design and assumption, and it emphasizes a spatial dependence for reality. It assumes that if you remove space from the framework, reality dissolves—literally nothing remains. But with space, objects can possess definite properties: position, mass, extension and so on.

The second tradition, the Hebraic mode of thought, assumes that ultimate reality, that which is to be considered real, is an action—a *verb*.[22] Reality, according to this way of thinking, must be explained and mapped through interactions, evolution, and change. It follows that the forms of Nature are born of interplay—which means that change is the only immutable, fundamental construct of Nature. Consequently, this view cannot describe Nature without action—without time.

Following the Greek tradition of thought, science has achieved many breakthroughs. Atomic theory has followed the cascading structure of matter down to smaller and more fundamental parts. It has found that all the objects we experience around us: the air, water, rocks, trees, dogs, and humans, are all composed of atoms. Furthermore, we have determined that every atom is made up of more fundamental parts, namely protons, neutrons, and electrons, and that protons and neutrons are in turn constructed of even finer and more basic parts called quarks. More recently, string theory has suggested that all primary particles are made of tiny strings whose different vibrational modes cause all the varying masses, charges, and spins found in matter. The search for these hypothetical strings represents the ultimate goal of the reductive Greek tradition—to discover the ultimate *noun* from which the universe is constructed.

The Hebraic tradition is built upon an entirely separate foundation. Its time dependent framework underlies the ideologies of the great Western religions, which often label the fundamental *verb* of Nature as *God*. This view ontologically organizes reality into a construct of fundamental interactions. Therefore, in order to accurately depict Nature, this mode concentrates on discovering and understanding the most elemental *verb*.

History has seen many conflicts that stem from the foundational differences between these two modes of thought. Supporters of each camp have long argued that either the Greek tradition or the Hebraic tradition must be right. But modern discoveries suggest that Nature is surprisingly a mixture of the two. In fact, when Einstein married space and time into a single entity called spacetime, the spatially focused framework of the Greek tradition became melded with the temporally focused framework of the Hebraic tradition. It now appears that with no space, we have no time, and with no time, we lose the potential for spatial forms—without spacetime our physical reality possesses no description.

Somehow the Greek and Hebraic traditions were both right.

As it turns out, the two traditions were also both right in another way. The Greek tradition claims that ultimate reality is classical—that the base constituents of Nature possess definite properties like position and velocity at all times, and that every action in Nature is strictly deterministic. The Hebraic tradition, represented most formidably in modern science by several interpretations of quantum mechanics, claims that ultimate reality is probabilistic, that before an event occurs, a range of possibilities exists for the outcome of that event—meaning that definite properties are replaced by probabilities of properties.

The framework we are about to construct will show that although these two claims appear to be strict opposites, from another perspective they can, in fact, both be seen as correct. To discover that vantage point we are going to challenge one of humanity's oldest assumptions about space and time. We are going to assume that instead of being a continuous medium, space is a superfluid medium composed of quantized units.

The level of understanding that we are looking for requires us to properly follow questions that are capable of guiding us to a deeper understanding of Nature. But to do this we need to be able to identify which questions are relevant. This sort of proper identification was Einstein's shining strength. He was able to focus on the three most important issues of his day: Is the fundamental structure of matter continuous or atomic? Can Nature's laws of motion and Maxwell's equations of electromagnetism be reconciled? And, is light quantized? He then used these questions to author an entirely new construction of physical reality. If we are to complete that process then we too must properly identify questions capable of revealing Nature's deeper structure. For this purpose let us continue reviewing the route of our scientific journey thus far and then discuss how to make the next conceptual leap.

The most satisfying aspect of Newtonian physics was that it gave us intuitive access to the world around us. Now that we have discovered phenomena in Nature that Newton's model cannot account for, we are obligated to search for a new physics.

Einstein dethroned the framework of Newtonian physics with a monumental breakthrough that developed from a visually intuitive picture in his mind. Mathematics and observations later solidified his conceptual solutions, but those equations and measurements hold little value compared to the deeper understanding and pictorial insights that remain their foundation.[23]

Thanks to Einstein's insight we can now at least partially visualize the world of general relativity. As a result, some of the mysterious nature of the universe can now be mapped and intuitively accessed. That understanding remains partial because Einstein failed to replace Newton's model with another "complete" and intuitive model. The math he used was remarkably elegant, but the picture he associated with that math was only capable of mapping one plane of reality at a time (Figure 1-8). The model he introduced also failed to account (even non-intuitively) for quantum mechanical phenomena (new discoveries in his day).

If we are to complete this map and extend the reach of our intuition, we must, once again, challenge our assumptions about what space and time are.

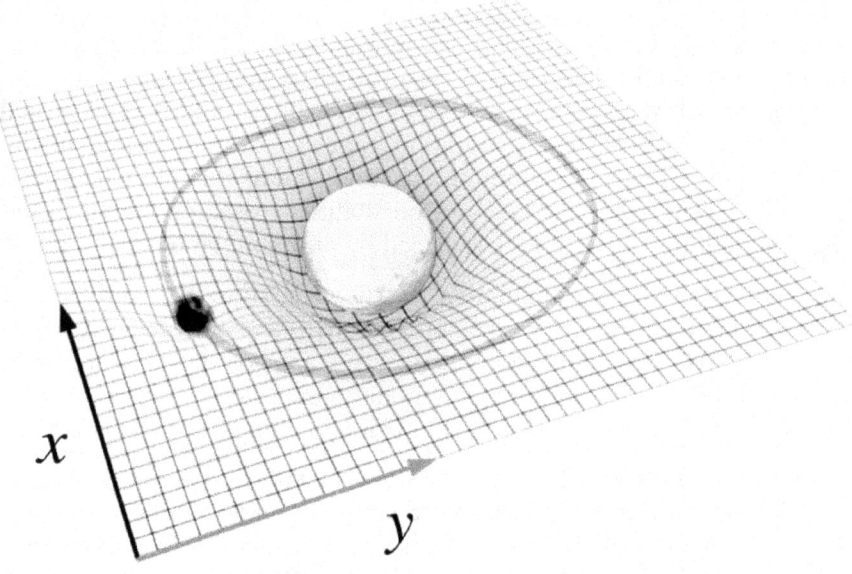

Figure 1-8 Traditional depiction of a warped two-dimensional slice of space.

Fitting the Pieces Together

> *"In the last three decades in his life, Einstein failed to create the unified field theory largely because he abandoned his conceptual approach, resorting to the safety of obscure mathematics without any clear visual picture."*
>
> *Michio Kaku*

Today's most advanced theories (superstring theory, M-theory, loop quantum gravity, supersymmetry, etc.) have all been unsuccessful at unifying quantum mechanics and general relativity. Mathematical attempts to meld these two descriptions—to reduce the four forces of Nature (gravity, electromagnetism, and the strong and weak nuclear forces) into an all-encompassing framework—produce equations so long and complex that no one entirely understands them. As Brian Greene puts it,

"the mathematics of string theory is so complicated that, to date, no one even knows the exact equations of the theory. Instead, physicists know only approximations to these equations, and even the approximate equations are so complicated that they as yet have been only partially solved."[24]

It's like having a gargantuan partial set of digital codes with no information on how to translate that information into a picture. What's even more disturbing (or intriguing), is that the supercomputers we have tasked with analyzing the structure of these codes have given us reason to believe that patterns are available for translation in higher dimensions. This suggests that to accurately map Nature we may have to use a framework that possesses more than the three dimensions of space and the one dimension of time that we are familiar with. There may be more to space and time than we have assumed. What could this possibly mean? How can we ever expect to understand something that is described in more than the four dimensions of space and time that we are familiar with?

As we ponder that question, it is useful to recognize that Einstein's view of gravity suggests, albeit subtly, that additional dimensions may be necessary to understand gravity. It describes gravity as a geometric effect—a consequence of the way massive objects distort the shape of spacetime. Distorted spacetime alludes to the existence of additional dimensions because those distortions require translation in reference to other dimensions, which Einstein's rubber sheet diagrams play off of. Spatial curvature implies a quality that extends into something other than the familiar three spatial dimensions. But are these additional dimensions physically real? We might say that it is okay to allow them to exist in the abstract mathematical sense, but it's impossible to visualize more than three dimensions. Isn't it? They can't physically exist. Can they? Even if they do exist, how could we ever understand them? Given that Einstein had to suppress a familiar dimension of space in order to visualize the dimension that allows depiction of a plane's curvature, how can we ever hope to visualize many dimensions at once?

This kind of thinking is exactly what is holding us back. It is our belief that we are limited to simultaneously visualizing only three spatial dimensions at once (length, width, and depth) that is keeping an intuitive picture of reality hidden from us. It is possible that once we cross this intellectual chasm, *Atlantis* will no longer be able to hide and the mysteries of Nature will be revealed.

In our four-dimensional (three space dimensions and one time dimension) models of physical reality, space and time are still laced with confusion. We can't even explicitly define them. But, as we will soon discover, the assumption that the vacuum is a superfluid leads to simple and powerful definitions for both space and time. As a result, this assumption has the potential to give us creative access to the mysteries of physics.

"All the great empires of the future will be empires of the mind."

Winston Churchill, 1953

Just as I cannot meaningfully express the beauty of the fountain of Buckskin Gulch, without appealing to human senses and portraying some sort of image, we cannot comprehend Nature's deepest beauty without discovering its deeper structure. In this book I will probe for that deeper structure by exploring a set of assumptions about the structure of reality that make room for the currently unexplained mysteries in physics (wave-particle duality, the uncertainty principle, nonlocality, the state vector, dark energy, dark matter, etc.). The correctness of any set of assumptions is constrained by their ability to give rise to an elegant picture of Nature and offer us a richer understanding of the phenomena that those constructions were designed to capture.

In order to attain this picture, map, or all-encompassing framework, we must return to a conceptual approach and, once again, rethink the very parameters of space and time. To accomplish this task let's examine how great thinkers have done this in the past and then see where our inquisitive freedom can take us. Even if the specific picture we explore falls short, the procedure we are about to follow gives us all the tools we need to eventually unveil a better map of Nature.

[1] This quote originally comes from an article written by Einstein that was published in Forum and Century Magazine in 1931.

[2] Steven Strogatz. *The Next Fifty Years*. p. 123.

[3] Albert Einstein. (1973). *Ideas and Opinions.* London: Souvenir Press; first published 1954, p 80.

[4] Einstein explained the important distinctions between these two approaches in a 1919 essay he wrote called "Induction and Deduction in Physics."

"The simplest picture one can form about the creation of an empirical science is along the lines of an inductive method. Individual facts are selected and grouped together so that the laws that connect them become apparent… However, the big advances in scientific knowledge originated in this way only to a small degree… The truly great advances in our understanding of Nature originated in a way almost diametrically opposed to induction. The intuitive grasp of the essentials of a large complex of facts leads the scientist to the postulation of a hypothetical basic law or laws. From these laws, he derives his conclusions." Einstein. (1919, December 25). Induction and Deduction in Physics. Berliner Tageblatt, CPAE 7:28; Walter Isaacson. *Einstein*, p. 118.

[5] As an early example of this, let's consider the Pythagoreans—a secret group who followed a mysterious figure from Greek mathematics named Pythagoras (c. 475 BCE). The Pythagoreans thought that the whole of the cosmos could be described in terms of the whole numbers: 1, 2, 3, etc. This underpinned their understanding of physical reality and guided their inquiries about the natural realm. Eventually, however, this led to problems. Around 500 BCE arguments ensued within the Pythagorean circle about a number that is now known as the square root of two ($\sqrt{2}$). The Pythagoreans were concerned with this number because of its geometric significance. They had initially assumed that the value of $\sqrt{2}$ could be described as a ratio of two whole numbers, but a clever argument was made that disallowed this possibility. According to legend, this argument was constructed by Hippasus of Metapontum, who had been ordained to the inner circle of the cult. Hippasus' argument meant that the Pythagoreans had to accept the fact that the square root of two could not be expressed as a fraction of integers.

Tragically with the birth of irrational numbers came the death of their discoverer. To the Pythagoreans, irrational numbers represented an idea so dangerous that it created a crisis that reached to the very roots of their cosmology. In an attempt to somehow make this crisis go away and to insure that Hippasus wouldn't be able to divulge the secret to someone outside their circle, the Pythagoreans abducted Hippasus and drowned him on the high seas.

Today irrational numbers and many other profound ideas are so completely integrated into our formalization of mathematics that it is easy to overlook how valuable the information we have inherited is. Men and women have dedicated their lives, and some have lost their lives, attempting to give us the ideas that outline our modern world—ideas like the square root of two. The concept of 'zero' is another one of those ideas. In its early history the Catholic Church banned Hindu-Arabic numerals—the 0 through 9 we use today—throughout much of Italy until the fourteenth century because it regarded the concept of zero as dangerous to its theology. Richard Elwes. (2007, July). From e to Eternity. *New Scientist*, p. 38; Stephen Hawking. *God Created the Integers*; Jared Diamond. *Guns, Germs, and Steel*, p. 235.

[6] The naked eye can see up to 7000 stars in the darkest and clearest skies.

[7] The word *cosmos* comes from the Greek word *kosmos*, which means "the ordered whole," or "world".

[8] Ian Stewart. (2006, March). Ride the Celestial Subway. *New Scientist*.

[9] Uranus is barely visible to the naked eye. Usually only the trained observer can find it. That is why it is not included in this list. Neptune cannot be distinguished without magnification.

[10] As a reflection of how significant these curiosities have been to mankind, note that the objects occupying their own celestial level in the Ptolemaic model are still reflected in our modern days of the week.

Days of the Week		Celestial Body	
English	Spanish	English	Spanish
Sunday	Domingo	**Sun**	Sol
Monday	Lunes	**Mo**on	Luna
Tuesday	**Mar**tes	**Mar**s	**Mar**te
Wednesday	**Mierc**oles	**Merc**ury	**Merc**urio
Thursday	**Ju**eves	**Ju**piter	**Jú**piter
Friday	**Vier**nes	**Ven**us	**Ven**us
Saturday	**Sa**bado	**Satur**n	**Satur**no

The days of the week were originally named after these heavenly bodies by the Babylonians. The Romans adopted this representation. We quite literally have a seven-day week because of this model. The week was set up so that each day was spent in worship of a corresponding level of heaven—as each level was occupied by a different god. Saturn occupied the highest of those levels, which is why Saturn-day (Saturday) was the most holy day—the day everyone went to church. In 321 CE, Emperor Constantine switched the Sabbath to Sunday instead of Saturday. Before all of this the Romans were also using an eight-day week (the seven levels of heaven plus the eighth level of background stars).

English retained its reflection of the seven levels of heaven, but some of the gods were switched for their local counterparts—the Anglo-Saxon words for the gods of Teutonic mythology. Mars was switched with Tiu or Tiw, the Anglo-Saxon name for Tyr, the Norse god of war. Odin or Woden replaced Mercury and Woden's day eventually became Wednesday. Jupiter, also called Zeus, was switched out with his lightning bolt throwing counterpart Thor. So Jove's day became Thor's day (Thursday). Friday was derived from Frigg's day after the goddess Frigg, who like Venus represented love and beauty in Norse mythology.

[11] Religious texts that stem from this era, such as the Koran, reflect these changes by their mention of the "seven levels of heaven". Older texts still hold onto the "three levels of heaven".

[12] This maxim asserts that assumptions introduced into an explanation must not be multiplied beyond necessity. This is often stated as: all else being equal the simplest explanation is usually the correct one. It is attributed to the 14th century logician and Franciscan friar William of Occam (Ockham was the village in the English county of Surrey where he was born). Parsimony has become a reliable tool of modern science. In fact, modern hypothesis can, in general, be effectively evaluated for merit based on their elegance, simplicity and beauty. Our experiences have taught us that a theory that successfully mimics Nature is, at least in some mathematical, symmetrical way, simple, elegant and beautiful. In this, it has proven to be an extremely useful guide, but it is still no substitute for insight, logic and the scientific method. "As arbiters of correctness only logical consistency and empirical evidence are absolute." Einstein artfully stated his version of Occam's Razor as: "Everything should be made as simple as possible, but not simpler."

[13] Aristarchus of Samos first proposed the idea of a heliocentric universe in the third century BCE.

[14] For simplicity we are ignoring air resistance and currents. The experiment will be even more precise if we perform it within a vacuum.

[15] Richard Feynman. (1988). *QED, The Strange Theory of Light and Matter.* Princeton University Press, p. 4.

[16] Franck Laloë. *Do We Really Understand Quantum Mechanics?* p. xi.

[17] Ibid. Quoted in Foreword by Claude Cohen-Tannoudji. p. ix.

[18] Ibid., p. xii.

[19] Ibid.

[20] The formalism of quantum mechanics that goes under the name of the Copenhagen interpretation "should probably more correctly be called the Copenhagen non-interpretation, since its whole point is that any attempt to interpret the formalism in intuitive terms is doomed to failure…" A. J. Leggett. (2002). Testing the limits of quantum mechanics: motivation, state of play, prospects. *J. Phys. Condens. Matter* **14**, R415-R451.

[21] N. D. Mermin. (1993). Hidden variables and the two theorems of John Bell. *Rev. Mod. Phys.* **65**, 803–815; in particular see §III. This is logically unfounded because it denies the possibility of other valid interpretations—of which there are many. Most notably it denies the possibility of a deterministic interpretation, like Bohm's interpretation.

[22] For an in-depth analysis of this division of thought see: Thorlief Roman. (1970). *Hebrew Thought Compared with Greek.* New York: W. W. Norton & Company.

[23] In the fall of 1919, Einstein received an urgent telegram informing him that astronomers had observed evidence of the bending of light by the Sun's gravity, validating a key prediction of his general theory of relativity. He handed the cable to a student, who began congratulating him. "But I knew that the theory is correct," he interrupted, and she asked, what if the observations had disagreed with his calculations? "Then I would have been sorry for the dear Lord," Einstein answered. "The theory *is* correct." Richard Panek. (2008, March). The E Factor. *Discover,* pp. 20–21.

[24] Brian Greene. (2003). *The Elegant Universe: Superstrings, Hidden Dimensions, and the Quest for the Ultimate Theory.* New York: Vintage Books, p. 19.

Chapter 2 — **Rethinking Space and Time Again**

> *"We are so locked into the world of our own senses that, although we readily understand and fear a loss of vision, we cannot conjure a picture of a visual world beyond our own. It is humbling to realize that evolutionary perfection is a will-o'-the-wisp and that the world is not quite what we imagine it to be when we measure it through a lens of human self-importance."*
>
> Timothy H. Goldsmith

> *"What is essential is invisible to the eye."*
>
> Antoine de Saint-Exupéry

Neutral Buoyancy Laboratory, NASA's Johnson Space Center, Houston, Texas.

I am drifting, floating just inches above the dazzling white surface of the International Space Station (ISS). This brilliant playground calls to me, connecting me to a sensation discovered during childhood after building my first fort—a place where I could keep my top-secret projects safe, a place where I could observe the rest of the world as an outsider. Simply being here excites my nerve endings and heightens my senses. In the background there are invisible eddies of resistance and my skin tingles with a constant stir.

As I look at the structure before me it is impossible to tell whether I am moving or it is moving. All that exists of the motion between us is relational; no other meaning survives. Another yellow handle comes within reach. I extend my hand and gently tug on it to redirect my course. I can feel the billowing cylinder rotate beneath me right on cue. I have to remind myself to breathe.

I continue gliding slowly from one handle to the next as if I were playing out the stanza of an eloquent symphony. Arm over arm I move over this surface as the music in my head builds toward its crescendo. Although I am watching this space ship move beneath me, I suspect that an onlooker would describe me as a small bug encircling the branch of a tree. That is, if they allowed themselves to compare one of history's most impressive construction projects to the branch of a tree.

In the middle of this stanza I hear the crackly voices of Mission Control over my bone phone. They are detailing the Orbital Replacement Unit (ORU) procedure as the Mission Specialists make their way from the Pressurized Mating Adapter (PMA). One of those astronauts is my dive partner's father.

After completing our NASA nitrox certification, my dive partner, Brad, and I are undertaking our first mission. My beating heart is constantly expressing how big of a deal this is to me. Even access to the deck above is tightly restricted, but now, as official flight leads, we are floating with astronauts around the ISS with a project of our own. The sensation is exhilarating.

The Neutral Buoyancy Laboratory (NBL) contains the world's largest indoor pool (202 feet long, 101 feet wide, and 40 feet deep). It is a satellite to NASA's Johnson Space Center (JSC) in Houston, Texas. This pool harbors exact scale mock-ups of the ISS, the Hubble

Space Telescope (HST), and the Space Shuttle Cargo Bay, which are used to simulate mission EVA's (Extra Vehicular Activities or "space walks").

When the NBL was originally built, NASA had some difficulties procuring the appropriate water allocations for it. Consequently, it took over a month to fill the pool using a garden hose. Now the entire volume, along with its carefully balanced chemicals, is filtered every twenty-four hours.

As the astronauts continue their simulation Brad and I begin our task. Our "cowboying around," which is what it is called when an EVA is performed without a tether, is not just an attempt to fulfill a childhood dream; we are photographing several of the external ISS components and its general profile for a catalogue we are composing. The ISS is reconfigured daily to replicate the stage of construction that each simulation crew will encounter in space. Photos of the intermediate stages will be a useful reference. Proposing this project, and volunteering to carry it out, gave us a good excuse to get into the pool every day.

I am carrying a bulky underwater digital camera and snapping pictures of this inspirational behemoth as it floats beneath/above me. It takes surprisingly little imagination to pretend that I am actually in space. Everything is neutrally buoyant—just floating about. The Cargo Bay of the Space Shuttle is visible off in the distance, and when the conversations with Mission Control cease, an eerie silence surrounds me. The colors are different too—not quite like they would be in space, but different enough to spark a feeling of the unfamiliar. It is a feeling that pours over my body and passes right through me.

As I recall my dreams of being in space I am overcome with the desire to discover what it feels like to be drifting off into the heavens without the possibility of retrieval. Knowing that I am not wearing a tether (and allowing myself to believe that I am in space instead of a pool), I grip another protruding yellow handle and accelerate toward the edge of the cylindrical laboratory. I see the massive structure move beneath me. Handle-to-handle, I pull and push. Then, as I launch off the edge of the structure, I turn and watch home base drift farther and farther away.

That's when it hit me. That's when I really figured out what it means to say that velocity is entirely relational. I had expected to experience what it was like to be drifting away from the ISS to my inevitable end, but instead I witnessed the ISS drifting away from me. This was somewhat surprising. For some reason, every time I had imagined what this experience would be like I had visualized it from the reference frame of the ISS. Now I was seeing it through my own eyes—from my own reference frame. The experience deeply rooted my intuition in the foundational principle in physics which tells us that in Nature all inertial frames are on equal footing, that one constant velocity reference frame is just as valid as any other.

Galileo Galilei connected to this principle inside the cabin of a ship.[1] Einstein used the train station in Bern, Switzerland to relate his connection to it. I had learned from their insights and had completely accepted the principle of inertial frames as a fundamental truth. But until I actually saw the ISS drift beyond my reach, my intuition had not absorbed it. I had not grasped the enigmas that come with this truth. I had not wrestled with the mysteries that surround this simple property of spacetime. I had never asked *why* it is that all inertial frames are equal. This simple question turns out to be very profound.

The greatest mysteries of the physical realm are but echoes of our ignorance of the true nature of space and time. Although they underlie all of our experiences and form the very metric of Nature, space and time have remained so clandestine that we haven't definitively defined them. It is time for us to get beneath this nebulous understanding. It is time for us to crown our search for ontological clarity, to open the door of a wondrous world accessible to us by the power of scientific imagination, and to learn to see what is invisible to the eye. In order to do this we must focus in on the core of our ignorance. We must recognize the root of our confusion and wrestle with questions that reflect that root.

This is not an easy thing to do; in fact, it is extremely difficult. The brilliant physicist Kip Thorne used a superb example that highlights why it is so difficult for us. He notes that Hendrik Lorentz and Henri Poincare both produced valuable insights that could have easily led them to discover Einstein's new vision, but neither of them took that final step. Why? The answer, according to Thorne, is that both men "were groping toward the same revision of our notions of space and time as Einstein, but they were groping through a fog of misconceptions foisted on them by Newtonian physics"[2].

Einstein, by contrast, was able to cast off Newtonian misconceptions. His willingness to start his investigation from scratch, whether or not it meant abandoning the foundations of Newtonian physics, "led him, with a clarity of thought that others could not match, to his new description of space and time."[3]

The lesson here is that if we are serious about questioning things, we need to question even the structural foundations that lie underneath our assumptions. We need to be willing to rebuild the entire metric of physical reality—should our investigation require it. Only then can we reach into the depths of our ignorance. Only from this state of mind can we truly press on with our journey.

In this spirit, let's ask the most foundational questions we can—questions about the metric of spacetime. What is space? What is time? These questions appear to be entirely embryonic, and it seems that the answers should be readily evident, but they aren't. Many hypothetical solutions to these questions suggest that there is a realm beyond our experience and imagination. To unveil that realm, we need to enter the debate over the essence of space and time.

Newton spearheaded this intellectual journey under the direction that space and time are real—that they are physical entities. When Mach took command, he reversed our course by insisting that space and time aren't real physical entities at all. After that, Einstein redirected us to an entirely new heading by redefining what we mean by space and time. His course carried us to waters that had never before been charted. For a while, the dream of discovering a richer map filled our sails. But this optimism didn't last for long. After a few short years Einstein reluctantly relinquished his command to the tyrannical whims of quantum mechanics. From that point on we have been randomly changing course, nauseously flickering from one heading to the next with each new moment.

The wind still blows but our sails rarely capture it. It has become increasingly obvious that we are lost in the middle of a disorienting ocean, spinning about a heavy anchor.

To reach our desired goal we need to lift that anchor, re-establish a heading for our intellectual quest, and to use our full sail to propel us toward a new destination. In order to do this, we need to figure out where we are and how we got here. We need to trace out the ideas that have guided us to this point, and then we need to find out what assumptions those ideas are based upon. After we have done this, we will concern ourselves with scrutinizing the map that falls out of those assumptions. It is by this process that we will learn how to pick a new direction, trim our intellectual sails, and recapture the wind. Here we go.

Newton's Bucket and Absolute Space

"There is no coherent sense in which nothingness could be the subject for the qualities of length, breadth and depth."

al-Farabi [4]

In 1689 Sir Isaac Newton described an experiment in which he used a simple bucket of water to acutely focus the riddles of space and time.[5] This experiment demonstrates a profound set of truths about spacetime, and it exposes us to questions that will ultimately lead us to a deeper understanding of reality, but the argument herein can be a bit subtle. To properly grasp the material, to avoid missing out on how deeply Newton's bucket challenges our innate concepts of reality, the reader is advised to carefully absorb each step in the following discussion.

To repeat Newton's experiment we take a bucket partially filled with water and suspend it from a rope. Next we twist the bucket until the rope is tightly twisted, hold the bucket still until the water inside is stationary, and then we release the bucket (Figure 2-1).

Sounds simple right? Well, the implications of what happens next aren't so easy to understand. At first, the bucket begins to spin, but the water remains mostly stationary. Consequently, the surface of the water remains flat and smooth. However, as the bucket continues to pick up speed, its motion is communicated to the water via friction, and the water starts to spin. As it does, the water's surface takes on a concave shape.

Figure 2-1 Stationary water has a flat surface

The exact reason that this thought experiment is puzzling can be a bit illusive. But this experiment is capable of bringing to focus the central enigmas of space and time. You might be tempted to say that there is no mystery at all, that the shape of the water's surface changes because the bucket is spinning; that is, the points on the bucket's rim are accelerating (always changing direction). Therefore, for the same reason that we feel pressed into the seat of our car when we accelerate, the water is pressed toward the walls of the bucket. Because the only place the water can go is up, the water climbs the walls making the surface of the water assume a concave shape (Figure 2-2).

This description is fine, but our goal is not to describe how the water takes on the shape it does—our goal is to explain what it means to say that something is accelerating (spinning is an example of acceleration). Accelerating according to what, or whom?

Figure 2-2 Spinning water has a concave surface

Why is it that observers from any imaginable reference frame will agree that the water's surface is not flat? Shouldn't acceleration be a relational concept? Shouldn't we have to mention a comparative reference frame in order to claim that the bucket is spinning? After all, we always have to mention a reference frame when we are describing something's position, or its velocity. The very meaning of *position* or *velocity* comes into focus only when we have a reference frame by which to define them. Shouldn't acceleration share this property?

If acceleration did depend upon a reference frame, then it would be possible to pick a reference frame from which the acceleration of the water in the bucket would be zero. The surface of the water should look flat from that reference frame. But is there any kind of reference frame from which the surface of the water would appear flat?

It appears the answer to that question is no—and this is the mystery. All positions can be defined as the zero position (the origin) within each reference frame. And there is always a reference frame that will allow us to define any particular velocity as the zero velocity (to define any object with constant velocity as not moving). But there appear to be no reference frames that allow us to set a nonzero acceleration to zero. In other words, there is no reference frame (inertial or accelerated) from which an observer could claim that the water's surface is flat. How do we make sense of this?

Some people have argued that this condition reveals a need for a reference frame that is special—a single reference frame that allows us to define acceleration in comparison to it. To secure such a reference frame we must have some rational explanation for how that reference frame has come to be the frame from which all accelerated forms of motion are compared, and we have to be able to explain why quantities like position and velocity are not fixed by this special reference frame.

To entertain the possibility that this special reference frame exists, let's imagine what would happen if we removed all external references. Could an object spin if it was the only material object in the universe? If that object was a bucket of water, could the surface of that water maintain a concave shape? Or, would the very notion of spinning (acceleration) disappear without external references—without something to compare it to?

If acceleration is a relative quantity, then we should be able to use the relative spinning motion between the bucket and any other object in the universe to explain the concave shape of the water. (In Mach's treatment acceleration is uniquely defined in reference to all of the objects in the universe.) But without external references the water cannot maintain a concave shape. In other words, if acceleration is a relative measure, like velocity, then without some exterior object by which to compare it to, the bucket could not meaningfully possess acceleration—just as it cannot meaningfully possess a velocity without comparison.

If this is the case, then the shape of the water inside of the bucket should depend upon the existence and motions of exterior references—even if those references are light-years away. This would be very difficult to explain. A problem closer to home is that if acceleration were truly a relative measure, then the apparent shape of the water in Newton's bucket should depend upon the observer's state of acceleration. This is the quintessential property of relative measures—their magnitudes are defined by comparison.

It is impossible to distinguish between two relational frames—to pick one over the other as being the "actual" frame. To explore this point, imagine an object moving through the atmosphere with great speed, and being deformed due to air resistance (if the object is a person, imagine her hair and clothes being pulled back). From one perspective this deformation is a result of the object's great velocity. There are other valid reference frames by which to describe this effect, but the fact that velocity is relative means that there is no reference frame in which the object is not deformed such that the clothes and hair look as if nothing is happening. As we change reference frames the effect does not change, rather our explanation of the effect changes. For example, in the reference frame in which the velocity of the object is zero, the explanation for the deformation is that the atmosphere is rushing past the object.

With the bucket, the situation is similar. If acceleration is relative, there must be a reference frame in which the acceleration is zero. In this reference frame, centrifugal force can no longer be an explanation for the concave shape of the water. So one must look for another explanation. According to Mach, the water comes up because in this reference frame, all the masses in the universe are rotating around the bucket like crazy, and their gravitational attraction causes the water to come up. This commits him to the idea that in an otherwise empty universe, the water in a rotating bucket does not come up at the sides.

Note that if acceleration is relational, then there should be no distinguishable physical difference between the claim that the bucket is spinning in a stationary universe, or the bucket is stationary in a spinning universe. The effect (the shape of the water's surface) is present in both situations, but the explanation differs in each. However, acceleration is relational only if distinguishing between these two claims is arbitrary.

It turns out that in the real world this relational condition does not hold for acceleration. As in our earlier discussion of the Foucault pendulum, there *is* a physical difference between a system in which the stars are stationary and Earth is spinning, and a system in which the stars are spinning and the Earth is stationary. We can distinguish between these descriptions because they lead to different results. Therefore, acceleration cannot be truly relational.

If acceleration were a relative measure, then it should be the case that if we were to spin ourselves around the rotating bucket—changing the relative amount of acceleration between the bucket and us—the water should appear to us to take a different shape (climb higher or lower up the walls). When we do this, however, we find that the water's shape doesn't depend upon the observer's state of acceleration.

From this we conclude that acceleration is not a relative measure. Relative measures like position, time, and velocity have a quality that sets them apart from non-relative measures—we can't feel them. We feel acceleration so it must be non-relational. This, plus the fact that we cannot affect the shape of the water's surface in our spinning bucket by changing the magnitude of our acceleration around the bucket, rules out the possibility that acceleration is a relational property. Recognizing acceleration as a non-relational property leaves us at the door of a rather big task—identifying Nature's unique reference frame. How can we do this?

This is a tricky question. Newton tried to answer this question by saying that space itself comprises the ultimate reference frame, which he called "absolute space". This claim elevates space into a thing, a noun, instead of an abstract idea, but it doesn't reveal the properties of that noun. Nevertheless, if we briefly put aside our discomfort about not knowing what this absolute space is, we will notice that the existence of an ultimate reference frame would provide us with an intuitive solution to the questions Newton's bucket raised. An object's absolute motion is defined in reference to the ultimate reference frame. If absolute space is that reference frame then an accelerating object is accelerating with respect to absolute space. This makes it trivial to understand the shape of the water's surface in Newton's bucket.

As the bucket begins to spin, it does so with respect to absolute space, but the water, at first, remains stationary with respect to absolute space, which is why its surface remains flat. As the bucket's motion increases and is communicated to the water via friction, the water takes on a concave shape because it is spinning with respect to absolute space. Then, when the bucket stops spinning as the rope twists up tightly, the water is still spinning with respect to absolute space—explaining why it retains its concave shape.

So, according to Newton, absolute space itself is the ultimate reference by which motion can be defined. However, when it came to explaining what this absolute space is, Newton didn't have an answer.[6]

Newton's proposed version of an ultimate reference frame has an interesting limitation. It allows us to explain why acceleration is nonrelational—why we can feel it—but at the same time it requires that, contrary to our experience, position, time, and velocity should also be nonrelational. This mismatch needs to be reconciled. Within a framework of continuity, an ultimate reference frame is expected to provide an origin for position, time, velocity, acceleration, and any other description of motion through the framework's dimensions. So how do we explain the fact that acceleration is defined in relation to this ultimate reference frame, while position and velocity are not?

If there is an ultimate reference frame, shouldn't we be able to use it to define all types of motion? Shouldn't this reference frame define an absolute position, velocity, acceleration, and so on, for all objects? At first blush, we might expect the answers to these questions to be yes, but affirmatively answering these questions contradicts the fact that position, velocity, and time *are* relational. Given this, and the fact that acceleration is nonrelational, what we really need is a structure of spacetime that maps out a unique reference frame for accelerated motion while simultaneously forbidding us to apply that reference frame to position and velocity. To do this we are going to need to have a deeper understanding of what space and time are, and what physical characteristics they possess.

Accelerated frames are somehow different from frames of constant velocity. We can feel acceleration, but we can't feel velocity. This difference needs to be explained. A model of spacetime that evokes an ultimate reference frame should naturally include a satisfactory explanation for why position, time, and velocity only manifest as relative measures in that model. As we search for our complete map, we need to keep this requirement in mind.

Almost two hundred years after Newton introduced his bucket, and in response to Newton's inability to reveal how the notion of absolute space can be made compatible with the observation that position, velocity, etc., are relational, Ernst Mach echoed the opposing sentiment—that absolute space doesn't exist after all. When it came to Newton's bucket, Mach felt that a spinning bucket must be spinning compared to some "stationary" external object, otherwise there is no meaning to the word *spinning*. Because Mach believed it self-evident that space is not a physical entity—a belief that was strongly held by his predecessor Leibniz—space could not be that reference.[7]

In this view, space and time become the means by which we make measurements or comparisons between objects, or events, and nothing more. They become bookkeeping devices and not real entities. This implies that in a universe with no fixed references the meaning of a spinning bucket would be as nonsensical as a hungry buffalo nickel. Therefore, if Mach is right, the water in a bucket of an otherwise empty universe will always possess a flat surface.[8]

Whereas Newton held that a bucket in an otherwise empty universe could spin with respect to the ultimate reference, Mach firmly disagreed. In Mach's framework there was no such thing as an ultimate reference frame. The consequence of this is that all measures of space and time (position, time, velocity, acceleration, jerk, snap, crackle, pop, and so on) lose all meaning in the absence of other objects by which to make comparisons. This follows from the assumption that space and time are strictly relative measures that have no independent physical essence.

With this idea Mach completely redirected our intellectual heading, but this course merely traded one problem for another. Mach's view fails to address the fact that we can clearly distinguish accelerated frames from purely relational frames because we can feel acceleration. More importantly, the fact that the shape of the water inside a spinning bucket does not change when we alter our accelerated reference frame means that acceleration cannot be a relational measure. This condition demands the existence of some sort of ultimate reference frame. Newton's concept of absolute space (which was based on the rules of Euclidean geometry) may not possess the exact character that such a reference frame requires, but in the end there still must be an ultimate reference frame.

Two hundred and sixteen years after Newton introduced his spinning bucket, Albert Einstein published his theory of special relativity (eleven years later, he published general relativity) and had a profound impact on the debate on what space and time actually are. He argued that space and time are somehow united into a single entity called spacetime, which can possess distortions, folds, and warps. He demonstrated that an object's movement through space affects how it moves through time and visa versa. Movement through time is directly tied to movement through space—and the shape of space.

With these revolutionary insights Einstein took the helm and changed our heading once again. He transformed the question of whether space and time are real physical entities to the question of whether spacetime is a real physical entity. Following his intuition, he directed us toward the answer to that question by showing that spacetime can possess different amounts of curvature—a measurable real property. Suddenly, we could no longer doubt its physical existence. Spacetime would forevermore be known as a real physical entity, an undeniable noun.

Einstein's Absolute Spacetime

Like Newton's model, Einstein's model of physical reality evokes an absolute benchmark, an ultimate reference frame in Nature.[9] According to general relativity, a bucket in a universe otherwise empty of material things can be accelerating or spinning. Spacetime provides the reference by which we can define this acceleration because of the intimate correlation it depicts between motion through space and motion through time.

If an object travels through spacetime in a consistent unchanging manner, then it is not accelerating. However, if an object changes its motion through spacetime—by changing its direction or its speed—then the object has accelerated. Spacetime is the benchmark for acceleration, because any change in an object's experience of time demands a reciprocal change in its experience of space, and vise versa. This trade-off between space and time is what allows spacetime to be an ultimate reference frame. Einstein labeled this absolute benchmark "absolute spacetime."

To make this a little clearer, consider the following: objects can move through time and space, but their combined movement through time and space is always equal to the speed of light (c). At the two ends of the spectrum an object can be moving only through *space*, wherein it doesn't progress through time at all, or only through *time*, wherein it doesn't progress through space at all. Between these limits, as an object takes on different velocities, it trades travel through time for travel through space, but it does not change its total magnitude of spacetime travel because traveling one second through time is equal to traveling 299,792,458 meters through space.

Einstein's concept of absolute spacetime is an improvement over Newton's concept of absolute space, but it cannot be the complete answer because it does not reveal why other measures in Nature are strictly relational. It gives us an ultimate reference frame (a spacetime field of zero curvature), which explains why acceleration is nonrelational, but it does not give us an explanation for why position, velocity, etc., are relational quantities.

This is as far as we have come in our quest to expose Nature's ultimate reference frame. Our current geometric description of spacetime enables us to explain acceleration's nonrelational character, but it leaves us incapable of explaining why time, position, velocity, etc., remain relational measures—why they are not also uniquely fixed by that reference frame.

To go further we need to have a richer understanding of spacetime than we currently do. We have established that spacetime is something, but what is it? Space is part of it, time is part of it, warps and ripples are some of its properties, and it is the reference by which acceleration gets its meaning. But what is this thing we call spacetime? How are we to fully map or understand it? What is it about the structure of spacetime that does not allow position and velocity to be strictly defined? Why does spacetime, which plays the role of ultimate reference for acceleration, fail to play the role of ultimate reference for position, time, and velocity?

While we ponder what spacetime is, let's discuss some of the clues about space and time that have been discovered more recently. (Answers to the questions posed in this chapter require an introduction to our new model of spacetime. They can be found in Chapter 10.)

Modern Clues for an Ultimate Reference Frame

Quantum physics has found that the ultramicroscopic realm is suffused with quantum jitters. What does this mean? The usual answer tends to include talk of fields and/or vacuum fluctuations, both of which seem to avoid a graphic explanation by answering with terms just as confusing. This isn't done with any intent to mislead. The truth is that we are still missing a picture of spacetime that is rich enough to capture these concepts, so any talk about quantum jitters (or any of the other quantum mechanical occurrences) tends to be technical or mathematical. Nevertheless, these observations can serve as glimpses into the structure of spacetime. They can give us clues about how the structure of spacetime must be—clues that will assist us in our goal of constructing a complete map.

Hendrik Casimir envisaged one of those clues. He predicted that two uncharged metal plates (or mirrors) will move toward each other when they are placed in a vacuum and are arranged parallel to each other. Because the gravitational force between these two plates is far too weak to explain this movement, and nothing other than space is included in the system, this effect is very intriguing.

To explain this motion, Casimir suggested that the quantum fluctuations of space itself are analogous to a pressure caused by the combined motions of many molecules. Based on this assumption, he calculated that when the two plates are placed extremely close to each other, the "molecular pressure" of space should slightly decrease between the plates because of the respective differences in "molecular motion" inside and outside the plates (Figure 2-3). In other words, if we treat the vacuum as a fluid that is made up of interactive parts, then the collective motion of those parts can be thought of as phonons (metric distortions in the vacuum substrate). Between the plates, phonons are restricted to wavelengths that are less than or equal to the spacing of the plates. This means that there are more phonons outside the plates than inside, which sets up an effective pressure differential that "pushes" the plates together.[10] The plates clash together like a pair of tiny cymbals and the system ends up with less space between the plates. Casimir claimed that the interactive geometry of space itself would cause this motion. We now refer to it as the Casimir effect.

Figure 2-3 The Casimir Effect

Although Casimir made this prediction in 1948, equipment sensitive enough to measure this effect wasn't technologically available until 1996. During this time span, Casimir's prediction was widely assumed to be just a quirk of mathematics. Then, in 1997, Steve Lamoreaux produced a convincing demonstration of the effect.[11] Today, "dealing with the Casimir effect has become a matter of urgency for nanotechnologists".[12] The Casimir effect strongly argues that quantum field jitters are the result of the interactions of some theoretical molecules or atoms that somehow compose the medium of space.[13]

This is important because physicists have discovered that as we approach the microscopic realm spacetime loses its function as the ultimate reference frame. In the subatomic realms, space doesn't even approximate Newton's backdrop. This is a significant problem, because if we no longer have an ultimate reference frame, then all of the questions introduced by Newton's bucket are back up in the air. So long as our understanding of spacetime dissolves on the microscopic scales, we will remain in this cloud of confusion. This is why it is important for us to study the clues that the microscopic realm can offer. If we can use them to depict a new picture of Nature, that picture might reveal the ultimate reference frame. The clarity that would come from such a coherent theory is what we are after.

Einstein's vision of human transcendence requires that we accept nothing less than a theory that gives a completely coherent account of individual phenomena. Working toward such a theory requires that we become aware of all of the unique phenomena in Nature that require explanation and that we actively investigate those phenomena. Every unexplained occurrence tells us something about the shortcomings of our existing fragmentary maps (or descriptions) of physical reality. Most of those clues point toward the need for stricter scrutiny of the microscopic realm. This is where our unexplained mysteries originate, and this is where we will find our most valuable clues by which to rewrite a richer, complete map of physical reality. Let's investigate some more of those clues.

In 2005, Theodore A. Jacobson and Renaud Parentani showed that "the propagation of sound in an uneven fluid flow is closely analogous to the propagation of light in a curved spacetime." This suggests that, "spacetime may, like a material fluid, be granular and possess a preferred frame of reference that manifests itself on fine scales".[14]

Further support of this inference comes from Stephen Hawking's famous argument that black holes are not truly black. Back in the 1970s, Hawking predicted that black holes emit thermal radiation, but relativity demands that any radiation emitted from the surface of a black hole will be infinitely stretched as it propagates away—making it impossible to measure. This infinite stretching assumes that spacetime is infinitely divisible. But if we treat spacetime as granular, then we can depict it as a fluid system. When we do this, "The fluid's molecular structure cuts off the infinite stretching and replaces the microscopic mysteries of spacetime by known physics".[15]

This approach would support Hawking's claim, but so far no one has come up with a framework for physical reality that depicts a granular structure for spacetime. One reason for this may be that such a framework must be what physicists call a background independent formulation. This means that the framework cannot presuppose the fluctuations of quantum fields, or the vibrations of string theory, to be stuck within spacetime. Instead, this formulation is required to explain quantum effects as the result of interactions within a spaceless and timeless framework. By definition this requirement can only be met in a higher-dimensional model, but to date, higher-dimensional models have escaped intuitive depiction.

Another clue we have about the microscopic realm is that theoretical minimum discrete values for space and time exist.[16] According to this theoretical claim we cannot infinitely divide a region of space, or an interval of time, into smaller and smaller amounts because we will eventually arrive at a scale where further division of those parameters no longer makes reference to space or time. Space cannot be divided into units smaller than the Planck length (l_p), and time cannot be divided into units smaller than the Planck time (t_p).

To make sense of this consider a more familiar analogy. We cannot infinitely divide a chunk of pure gold into smaller and smaller chunks because eventually we reach the limit of one gold atom. Any further division forces us to transcend the very definition of gold. Therefore, we cannot meaningfully talk about less than one gold atom in terms of gold.

Today there is a plethora of evidence supporting the physical existence of these minimum limits. The Planck constants are universally accepted values within the formulation of quantum mechanics. The Swedish mathematician Oskar Klein originally picked the Planck length in 1926 as a unique value because it is the only length that could naturally appear in a quantum theory of gravity. Because gravity is directly connected to the shape of space, this value seemed a necessary requirement. The Planck time is a unique value because it is the only value that can be combined with the Planck length to yield c, the speed of spacetime—otherwise known as the speed of light.

The existence of these Planck values restricts all measures of distance and time to whole number multiples. In space two objects can be a distance of 77 Planck lengths apart, but they cannot be 77.5 Planck lengths apart. Two events can be separated by 33 Planck chronons of time, but they cannot be separated by 33.5 Planck chronons.

These clues suggest that the vacuum may be best thought of as a fluid—a medium that has granular structure. (Later, we will narrow this claim to the suggestion that spacetime is a particular kind of fluid—a superfluid.) This point deserves some rumination, because if we are going to take this condition seriously, then we are going to have to allow the literal physical existence of additional dimensions (regions that define what is between the atoms of space). A quantized structure for spacetime means that the full map of Nature must be dimensionally richer than we have assumed. If we figure out how to comprehend and explore those dimensions, then a whole new realm might open up to us. But before we can even start to comprehend or explore unfamiliar dimensions, it is pertinent that we understand exactly what a dimension is. Therefore, we turn now to define and explore what physicists mean by *dimensions*.

[1] "Shut yourself up with some friend in the main cabin below decks on some large ship, and have with you these same flies, butterflies, and other small flying animals. Have a large bowl of water with some fish in it; hang up a bottle that empties drop by drop into a wide vessel beneath it. With the ship standing still, observe carefully how the little animals fly with equal speeds to all sides of the cabin; and, in throwing something to your friend, you need throw it no more strongly in some direction than another, the distances being equal; jumping with your feet together, you pass equal spaces in every direction. When you have obtained all these things carefully, have the ship proceed with any speed you like, so long as the motion is uniform and not fluctuating this way and that. You will discover not the least change in all the effects named, nor could you tell from any of them whether the ship was moving or standing still." Galileo Galilei. (1632). Dialogue Concerning the Two Chief World Systems. Translated by Stillman Drake, p. 186; Walter Isaacson, *Einstein*, pp. 108–9.

[2] Kip Thorne. (1995). *Black Holes and Time Warps: Einstein's Outrageous Legacy*. New York: Norton, p. 79.

[3] Ibid. Quote from *Einstein* by Walter Isaacson, p. 133.

[4] al-Farabi. (1951). Farabi's Article on Vacuum, N. Lugal and A. Sayili (ed. and trans.), Ankara: Turk Tarih Kurumu Basimevi.

[5] Isaac Newton. (1934). *Principia, Scholium on Absolute Space and Time* Florian Cajori, trans., Berkeley: University of California Press; reprinted in *The Scientific Background to Modern Philosophy*, Edited by Michael R. Matthews, Hackett Publishing Company Indianapolis/Cambridge, 1989, pp. 139–146: Cohen, I. Bernard. *The Newtonian Revolution*. Cambridge: Cambridge University Press, 1980; Manuel, Frank E. *A Portrait of Isaac Newton*. Cambridge, Massachusetts: Harvard University Press, 1968; Westfall, Richard S. *Never at Rest: A Biography of Isaac Newton*. Cambridge: Cambridge University Press, 1980.

[6] Technically, Newton originally tried to explain gravity in terms of an intervening medium known as the ether or aether. The force of gravity, he argued, was due to the "circulation" of ether and the different ether densities and in *Principia* he attempted to explain the elasticity and flow of this ether. Later he changed his theory of gravitation to the one we recognize today, involving a *force* and laws of motion, but remained unsatisfied with how this *force* lacked a mediator. L. Rosenfeld. (1969). Newton's views on Aether and Gravitation. Archive for History of Exact Sciences. 6.1: 29-37. Web. (2013, June 4). Isaac Newton. (1679, February 28). *Isaac Newton to Robert Boyle*.

[7] Leibniz said, "I hold space to be something merely relative, as time is… I hold it to be an order of coexistences, as time is an order of successions." H. G. Alexander. (1956). The Leibniz-Clarke Correspondence. Manchester University Press (1956), 3rd paper, §4; Olaf Dryer. (2004, April 13). *Relational Physics and Quantum Space*. arXiv:gr–qc/0404054v1.

[8] Of course, a universe containing only a bucket of water would not possess enough gravity by which to keep the water from floating out of the bucket. So in this case, because we mean to discuss acceleration in general, imagine instead that you were inside a large bucket touching its wall. If the bucket were spinning, you would feel pressed into that wall—thrown outward from the center of the bucket. Mach's claim is that without another reference by which to define the spinning of the bucket it cannot be spinning. Therefore, in this view, it is impossible in an otherwise empty universe, to feel a pull toward the walls of the bucket.

[9] Ironically, Einstein began his intellectual endeavor by trying to prove that Mach was correct in his relational approach.

[10] In quantum mechanics everything has a wave-particle duality. Everything, therefore, has an associated wavelength.

[11] The publication on this demonstration can be found at: Physical Review Letters, DOI:10.1103/PhysRevLett.78.5

[12] Saswato Das. (2008). New York City.

[13] Even without the Casimir effect as an explanation, vacuum energy would still hold as a valid and secure claim through the well-established phenomenon known as Lamb shift. The inference goes like this: because predictions for the wavelengths of light absorbed and emitted by molecules (which only match observation if physicists assume that vibrating molecules contain zero point energy) can be extended to explain how "vacuum fluctuations alter the frequencies of light that hydrogen atoms absorb and emit," zero-point energy must be inherent in vacuum fluctuations. The "same basic theory that works for molecules says that the vacuum contains zero-point energy too, there is no reason to believe otherwise." David Shiga. 2005, October). Something for Nothing. *New Scientist*, pp. 34–37.

[14] Theodore A. Jacobson & Renaud Parentani. (2005, December). Black Holes. *Scientific American*. p. 70.

[15] Ibid.

[16] These values are called the Planck length (l_P), and the Planck time (t_P). There are also maximum discrete values for mass, charge, and temperature called the Planck mass (m_P), Planck charge (q_P), and Planck temperature (T_P). These maximums are in reference to the minimum limits of space and time.

$$l_P = 1.616252(81) \times 10^{-35} m$$
$$t_P = 5.39124(27) \times 10^{-44} s$$
$$m_P = 2.17644(11) \times 10^{-8} kg$$
$$q_P = 1.875545870(47) \times 10^{-18} C$$
$$T_P = 1.416785(71) \times 10^{32} K$$

If we interpret the medium of spacetime as a molecular or atomic composite, then these parameters can be easily understood as the physical values that relate to the individual "molecules" or "atoms" of that medium. Support for this interpretation comes from the fact that the constants of general relativity and quantum mechanics are natural derivatives of these fundamental constants.

The primary physical constants of general relativity and quantum mechanics are:

$$c = 2.99792458 \times 10^8 \frac{m}{s}$$

$$G = 6.67428(67) \times 10^{-11} \frac{m^3}{kg\, s^2}$$

$$\hbar = 1.054571628(53) \times 10^{-34} \frac{kg\, m^2}{s}$$

(c is the characteristic speed of spacetime (the vacuum phonon speed), colloquially referred to as the speed of light, $\hbar$ is Planck's constant, and G is the gravitational constant.)

These constants can be derived from the fundamental constants of the space quanta in the following manner:

$$\frac{l_P}{t_P} = c, \quad \frac{l_P^3}{m_P\, t_P^2} = G, \quad \frac{m_P\, l_P^2}{t_P} = \hbar$$

Working backwards we can solve for l_P, m_P, and t_P in terms of the general relativistic and quantum mechanical constants (measured values) in this manner:

$$l_P = \sqrt{\frac{\hbar G}{c^3}}, \quad t_P = \sqrt{\frac{\hbar G}{c^5}}, \quad m_P = \sqrt{\frac{\hbar c}{G}}$$

There are many other constants of Nature that appear in our equations for physics, chemistry, electronics etc., that also turn out to be natural composites of the Planck parameters. For example: the magnetic constant (μ_0), the electric constant (ε_0), the Boltzmann constant (k), and the characteristic impedance of the vacuum (Z_0). We will discuss these relationships, and several others, in greater detail in Chapter 16.

Chapter 3 — Dimensions

"It's hubris to think that the way we see things is everything there is."
Lisa Randall

"Nature loves to hide."
Heraclitus

Uinta Mountains, Utah.

Running down the mountainside, concentrating, dodging boulders and trees, our feet displayed the magic of youth by never missing their mark. Maneuvering over jagged gray rocks and stirring up the smell of fallen leaves we each forged our own path. This was our hour to be free: to wonder, to yell out for no reason, to be boys.

As we descended we started to sense a strange aura to the place. Something was different about this place. A blanket of mist danced about the base of its trees, but there was something more. There were secrets here, an essence drawing us in like a mesmerizing murmur.

Soon we were racing to touch every hidden corner of this new landscape. I darted off in my own direction where the mist had thickened into a fog that swirled around my legs as I sliced through it. Every time I disturbed it, the fog became a little more transparent. Noticing this, I hunched down, held still, and watched the fog fill back in. The sky above was textured with a curtain of virga—rain streamers that had escaped the clouds but ceased their fall short of the ground.

Between the virga and the fog I saw something strange. A single leafless tree was vigilantly moving back and forth. I had to investigate. When I reached the base of that tree I saw one of my fellow scouts trying to topple it. Like several others in the forest, the tree was dead. Also, like several other trees in the forest, it was about to be transformed into a twenty-foot javelin. Soon, all of us had one of our own.

With our new weapons balanced in our arms we continued down the mountain pretending to be on medieval horses. Breathing heavy we came upon a large open field of wild grass with a patch of brilliant green highlighted by the sun. It was a green oasis in the middle of a grey forest. The secret we had been searching for was here. We all sensed it. Silently we walked out into the clearing. Then simultaneously, we froze in our tracks. There was something very strange going on, something that we could not yet identify. We slowly looked around. The birds were singing their same songs, the mist was hugging the shadows of the trees around the skirt of the clearing, but something else was going on. We had all felt it. With our curiosity piqued, we silently continued toward the center of the opening. Then we discovered what it was. The ground was moving.

It wasn't an earthquake; that much we knew. Each time we took a step, the thick grass beneath us rippled outward. The closer we got to the center, the larger the waves became. It felt like a stiff waterbed. When we stood close together, the ground beneath us depressed and slowly filled with water. When we walked alone, the ground depressed only slightly, remaining completely dry. We had discovered a hot spring, camouflaged by a thick mat of grass with tightly interwoven roots.

Wanting to know how deep the water was, we scattered ourselves about the middle, then one of the boys pierced the ground with the pointed end of his javelin. We watched as the long pole disappeared into the ground. The boy pulled it back out and, as tradition dictated, instantly came up with a dare for Brian.

Brian was my best friend in Junior High School. One of thirteen children, he was lanky, scrawny, and had a really deep voice. He was always hungry and in need of food money, so he invited dares. Brian also enjoyed the attention.

"I'll pay two dollars to see Brian do a cannonball right here," the boy said. "Me too," said another, "but it has to be a double leg cannonball." We quickly agreed on the terms and shelled out two dollars each into one big pile.

Brian prepared himself with a display of showmanship. We backed away from the selected spot and watched intensely. Fully dressed, he found a good starting point and began to run. Then, when he reached the predetermined location, he jumped high into the air and grabbed both knees.

We all clenched our teeth. It looked like this was really going to hurt. None of us expected what came next. When Brian hit the ground he just disappeared. The grass must have parted beneath him, but there was no splash, no left over hole. He was just gone. If I hadn't already discovered that there was a deep pool of water beneath the grass, I would have been completely convinced that I had just witnessed a person going through a wormhole or a stargate. One moment he was here and the next he wasn't. We were stunned.

A few seconds went by, maybe fifteen, and none of us had moved or made a sound. None of us knew what to do or what to think. Then, one of the boys who was usually quiet unnervingly said, "We killed him." Another didn't seem as worried. "No, we didn't," he said. "He just went into another dimension." "Stick a tree in there," someone suggested. "No," I said. "You'll poke him. He can swim. He's a strong swimmer." I knew this was true and I knew he could hold his breath for over two minutes, but I didn't know if either of those things counted in this situation.

Just as we started to move toward the mysterious spot, an arm jutted out of the ground. Muddy fingers were reaching around pulling handfuls of grass. Looking back, I can't help but wonder what someone would have thought if they had walked up at this moment—especially if it was during Halloween.

Brian pulled himself back out into our realm with little trouble and had a good laugh when he saw our expressions. When we asked why he was down there so long, he said it was much warmer than he expected and he just had to explore. Apparently he didn't think we'd become so concerned. It would be a while before any of us would dare him again.

After the danger and novelty of this experience subsided, I started thinking: what if Brian really had gone to another dimension; what would that even mean? I considered it for a while and realized that I honestly didn't know what a dimension was. I had some idea, but the whole concept became rather confusing when I stared it directly in the face. That's when I figured out that I needed to focus on the riddle of dimensions.

Exploring "Dimensions"

What are dimensions? Colloquially, the word "dimension" is loosely used to reference a thought, a quality, an experience, or the scope or importance of a given topic. But in physics physical dimensions relate to the possible structures of space. In order for a structure to qualify as a possible structure of space it must have a determinate number of dimensions.[1] That number allows us to specify a particular structure and define things like length, or whether or not a minimum length path exists between two points, within that space.

In short, dimensions allow us to describe the structure of space (or spacetime) and, ideally, explain all the phenomena that occur within it as a consequence of that structure. Examining them one at a time, dimensions are independent parameters that locate events within a defined space. Spatial dimensions provide independent information about position, and temporal dimensions give independent information about time.

We are most familiar with spatial dimensions, namely x, y, and z, or length, width and height, but why do we call these parameters dimensions? To expose the answer, let's imagine the x-y plane as a flat sheet that extends to infinity in both directions, pick an origin, and label it $(0,0)$—zero in the arbitrarily chosen x-direction and zero in the perpendicular y-direction (Figure 3-1). We can talk about the x-coordinate of any point on that plane, which is the distance the point is from the origin (the $x = 0$ line) in the x-direction. Similarly, we can label the y-coordinate of any point. However, by using only the variables x and y, we are completely incapable of illuminating anything about any distance above, or below the plane. That is why z is another dimension—because it cannot be expressed through terms of x or y. These dimensions are called spatial because they orthogonally (independently) map space. Each dimension provides unique and independent information about the map. A perpendicular geometric configuration is one way to express the orthogonal relationship.[2]

Our common experience suggests that four dimensions are sufficient to completely express where and when any event occurs. We think a specific event is perfectly nailed down once we have described where it occurs in terms of x, y, z and when it occurs in terms of t. But, and this is an important point, physical reality might be made up of more than four dimensions. Events may require more than x, y, z, t information to be uniquely determined.

Physicists have been discovering hints that suggest Nature's most simple and complete description may actually exist in a realm of higher dimensionality. If this turns out to be true it would mean that more than four independent parameters (dimensions) are required to completely identify precisely where and when events occur in Nature.

In mathematics we combine real numbers and what we call *imaginary* numbers to create complex numbers. Real numbers define possible positions in the familiar dimensions and *imaginary* numbers can be taken to define possible positions in unseen dimensions. Therefore, complex numbers can be seen as encoding a higher dimensional setting, which makes the fact that complex numbers play an indispensable role in the laws of quantum physics quite intriguing. In the complex number system each familiar spatial dimension (x, y, z) is morphed into a complex plane through the addition of an *imaginary* dimension (i, j, k). The resultant planes allow us to mathematically describe a six-dimensional manifold, and each plane can be separately graphed.

When attempting to map the microscopic realm, these so called "imaginary dimensions" are absolutely necessary in order to construct a map that can account for observation. But what does this mean? Do these *imaginary* dimensions have physical existence? Or are they just some kind of mathematical trick?

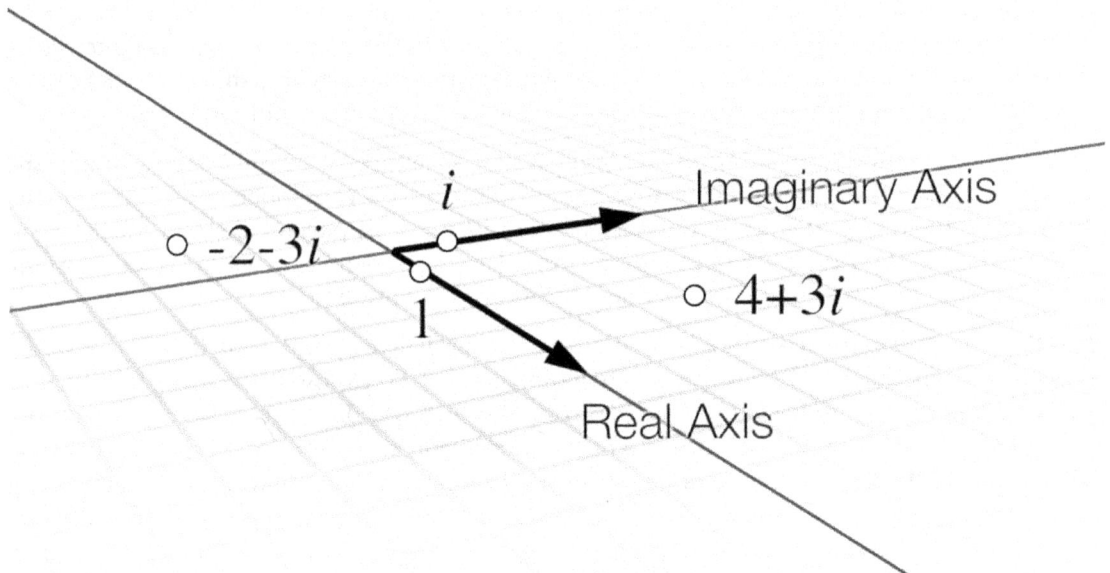

Figure 3-1 A complex plane: Nature's microscopic realm follows rules that can be taken to suggest that it is composed of a union between three complex planes.

While we let that question hang in the air, let's note that we have no physical theory that dictates that there should be only three dimensions of space and one dimension of time. What we do have is an ever-growing collection of observations (most significantly from the fields of cosmology and quantum mechanics) that stubbornly resist explanation within a four-dimensional framework.

Some of these observations can be found close to home. For example, AlCuFe alloys, which are ideal candidates for the nonstick coatings that most of us find on the pans in our kitchens, possess a lattice structure that is not allowed within three-dimensional space. These alloys belong to a class of structures called quasicrystals—structures that can only form their specific geometric lattice patterns in higher dimensional spaces.[3]

All crystals are symmetric lattices of repeating basic elements of atoms or molecules. The number of dimensions those elements have access to determines the type of patterns available for their arrangements. Three-dimensional frameworks allow more geometric arrangements than two-dimensional frameworks do, but these arrangements are still limited.

What is interesting about quasicrystals is that they do not conform to any of the lattice structures allowed in three-dimensional frameworks—yet they exist. In other words, it's only in higher dimensional frameworks that quasicrystals can be described as ordered, repetitive structures (the definition of a crystal), and therefore, it's only with the presence of additional dimensions that their formation can be explained. The very presence of quasicrystals suggests that our current view is laced with dimensional tenuity.

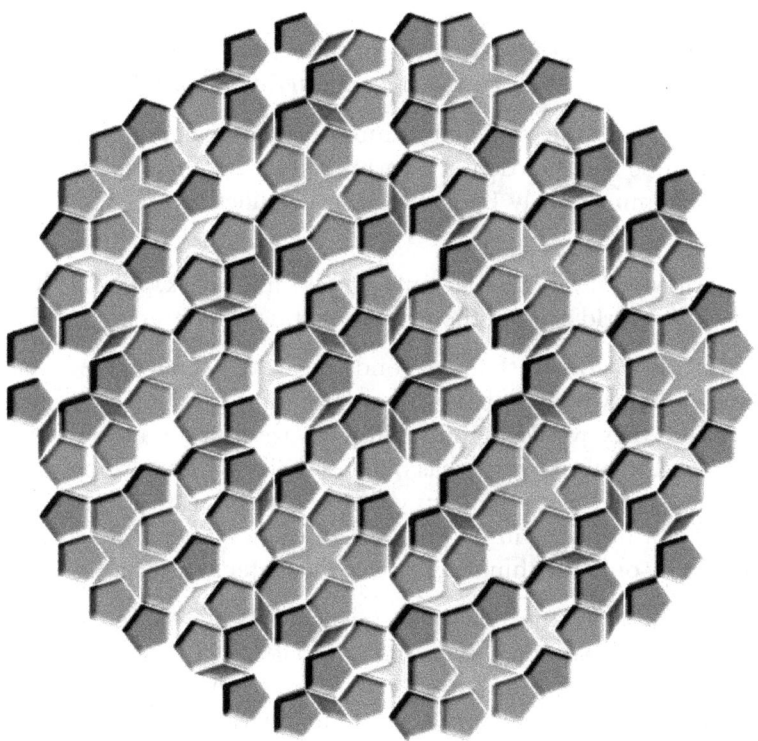

Figure 3-2 This Penrose Tiling is an example of a higher-dimensional repetitive crystalline lattice suppressed to fit our familiar dimensional mold. The suppression hides the fact that the tiles are all identical in shape.

Additional Dimensions

If the fabric of physical reality contains additional dimensions, then gaining intuitive access to those dimensions, and learning how to see them, can enhance our perspective. Such an imaginative leap has the potential to unmask the outstanding mysteries of Nature, and reveal those mysteries as mere artifacts of an outdated, incomplete map.

The Polish mathematician, Theodor Kaluza, affirmed the unifying potential that comes with additional dimensions. In 1919, while working at the University of Konigsberg in Germany, Kaluza took it upon himself to rework Einstein's formula for general relativity after adding an extra dimension to its structure. When he did this he came up with extra equations, which turned out to be the equations Maxwell had written down for describing light.[4] Simply by assuming that the universe contained an additional space dimension, Kaluza discovered a mathematical framework that combined Einstein's equations of general relativity with those of Maxwell's equations of electromagnetism.

Since Kaluza, scientists have become more and more convinced that additional dimensions have the ability to simplify and unify the laws of Nature mathematically. But a complete higher dimensional framework has not yet been constructed. This is largely due to the long held belief that humans are mentally incapable of comprehending a higher dimensional framework—that extra dimensions are impossible to visualize.[5] After all, our entire array of experiences, from our very first moments in life to our most recent experiences, reinforces a conceptual model of three spatial dimensions. So if space possesses more than three dimensions, how can we be expected to comprehend them?

This question reminds me of the inquisitions Michio Kaku made as a young boy. Next to a pond in the Japanese Tea Garden in San Francisco, mesmerized by the brilliantly colored carp that were swimming slowly beneath the water lilies, he recalls letting his imagination wander:

I would ask myself silly questions that only a child might ask, such as how the carp in that pond would view the world around them. I thought, what a strange world theirs must be!

Living their entire lives in the shallow pond, the carp would believe that their 'universe' consisted of the murky water and the lilies. Spending most of their time foraging on the bottom of the pond, they would be only dimly aware that an alien world could exist above the surface. The nature of my world was beyond their comprehension. I was intrigued that I could sit only a few inches from the carp yet be separated from them by an immense chasm. The carp and I spent our lives in two distinct universes, never entering each other's world, yet were separated by only the thinnest barrier, the water's surface.

I once imagined that there may be carp 'scientists' living among the fish. They would, I thought, scoff at any fish who proposed that a parallel world could exist just above the lilies. To a carp 'scientist,' the only things that were real were what the fish could see or touch. The pond was everything. An unseen world beyond the pond made no scientific sense.

Once I was caught in a rainstorm. I noticed that the pond's surface was bombarded by thousands of tiny raindrops. The pond's surface became turbulent, and the water lilies were being pushed in all directions by water waves. Taking shelter from the wind and the rain, I wondered how all this appeared to the carp. To them, the water lilies would appear to be moving around by themselves, without anything pushing them. Since the water they lived in would appear invisible, much like the air and space around us, they would be baffled that the water lilies could move around by themselves.

Their 'scientists,' I imagined, would concoct a clever invention called a 'force' in order to hide their ignorance. Unable to comprehend that there could be waves on the unseen surface, they would conclude that lilies could move without being touched because a mysterious, invisible entity called a force acted between them. They might give this illusion impressive, lofty names (such as action-at-a-distance, or the ability of the lilies to move without touching them).

Once I imagined what would happen if I reached down and lifted one of the carp 'scientists' out of the pond. Before I threw him back into the water, he might wiggle furiously as I examined him. I wondered how this would appear to the rest of the carp. To them, it would be a truly unsettling event. They would first notice that one of their 'scientists' had disappeared from their universe. Simply vanished, without leaving a trace. Wherever they would look, there would be no evidence of the missing carp in their universe. Then, seconds later, when I threw him back into the pond, the 'scientist' would abruptly reappear out of nowhere. To the other carp, it would appear that a miracle had happened. After collecting his wits, the 'scientist' would tell a truly amazing story. "Without warning," he would say, "I was somehow lifted out of the universe (the pond) and hurled into a mysterious nether world, with blinding lights and strangely shaped objects that I had never seen before. The strangest of all was the creature who held me prisoner, who did not resemble a fish in the slightest. I was shocked to see that it had no fins whatsoever, but nevertheless could move without them. It struck me that the familiar laws of nature no longer applied in this nether world. Then just as suddenly, I found myself thrown back into our universe." (This story, of course, of a journey beyond the universe would be so fantastic that most of the carp would dismiss it as utter poppycock.)

I often think that we are like the carp swimming contentedly in that pond. We live out our lives in our own 'pond,' confident that our universe consists only of those things we can see or touch. Like the carp, our universe consists of only the familiar and the visible. We smugly refuse to admit that the parallel universes or dimensions can exist next to ours, just beyond our grasp. If our scientists invent concepts like forces, it is only because they cannot visualize the invisible vibrations that fill the empty space around us. Some scientists sneer at the mention of higher dimensions because they cannot be conveniently measured in the laboratory."[6]

Like the carp of Kaku's pond, we struggle to understand the medium of our universe, inventing forces to make up for the shortcomings of our picture. But with the right picture we might find that the mysterious movements of our 'lilies' have simple explanations.

If extra spatial dimensions exist, where are they? What directions are orthogonal to the directions we have already described? How could there be spatial information that is completely independent of, or orthogonal to x, y, and z? How can it be possible to move in a spatial direction without moving in x, y, or z? What if one of the additional dimensions is another time dimension? When would it be? How could we visualize or comprehend more dimensions in addition to the ones we are familiar with?

As we contemplate these questions let's keep in mind that a dimension provides independent, orthogonal information about the spatial or temporal structure of physical reality. Each dimension maps the natural realm in a completely independent fashion. Additional spatial dimensions must provide information that is entirely separate from length, width, and height (x, y, z). They express entirely new directions.

If we propose a new spatial dimension, it must be possible to move about in that dimension without moving in x, y, or z. Recall that z is an independent dimension because it is possible to move in z without moving in x or y. We must meet this requirement if we are to definitively claim that we have unveiled a new spatial dimension. If we end up with a map that allows geometric movement that doesn't involve movement through x, y, or z, then we can confidently say that this motion takes place within an independent spatial dimension.

Many of us, including myself, were taught that attempts to visualize more than three dimensions are futile because our brains are "incapable of comprehending them." In this book I will outright defy this claim. We will discover that we are capable of simultaneously visualizing more than three spatial dimensions. And when we do, we will gain greater intuitive access to the secret workings of Nature. To reach that end goal, let's begin a process of logical investigation—a process that will non-arbitrarily introduce these extra dimensions and begin to reveal their form.

Curvature and Other Dimensions

Clues of higher dimensions come from our observations of curved spacetime. In order to account for the curvature of spacetime while mapping the universe, it might be necessary to use at least seven independent variables. For example, $x, y, z, \sigma, \mu, \delta,$ and t, where $x, y,$ and z represent orthogonal spatial distances from an origin, t represents time, and the Greek letters σ (sigma), μ (mu), and δ (delta) represent dimensions that enable depiction of the curvature of those three directions.[7]

To elaborate on this point, note that Einstein pictorially made use of additional dimensions by graphically suppressing a familiar spatial dimension and drawing a dimension that allowed him to represent curvature in its stead. He used a visual representation of a rubber sheet being stretched by a bowling ball (Figure 3-3). The bowling ball represents a massive object, like a black hole or the sun, and the stretched membrane of the rubber sheet represents a slice of spacetime's reaction to the bowling ball's presence.

The assumption that you cannot visualize more than three dimensions at once, combined with the assumption that space is a continuum, makes the use of this two dimensional rubber sheet necessary in order to graph curvature. For each familiar plane, one extra dimension is necessary to describe its curvature. Therefore, for three planes (xy, yz, zx which can be thought of as two perpendicular walls and the floor), three extra dimensions are required to explain the complete curvature of space. In order to account for the complete curvature of the (x, y, z) metric, three additional dimensions are necessary. For Figure 3-3 only one dimension is necessary to represent curvature because the spatial distortion it is representing is that of only one plane. It should be clear that this model is not well equipped to help us simultaneously visualize the curvature of three space dimensions (not to mention the fourth dimension of spacetime—time).

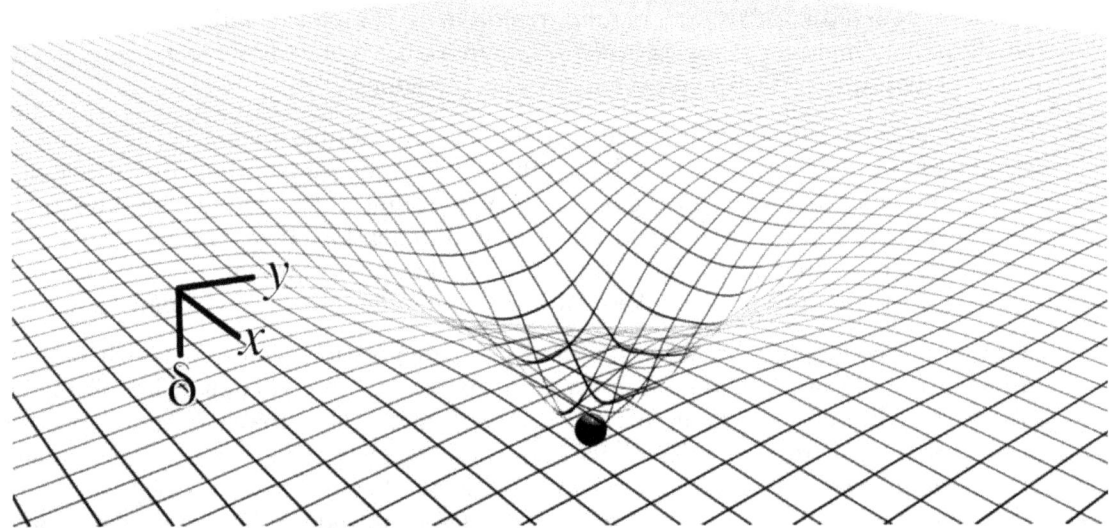

Figure 3-3 A slice of spacetime warping into another dimension

This model possesses other shortcomings that confuse our ability to explain the nature of spacetime. You might look at this diagram and ask: is the weight of the bowling ball causing the rubber sheet to stretch? If this curvature is used to explain gravity, then is it not a circular explanation to visually use the weight of the bowling ball, which is a function of gravity, to describe the cause of curvature and therefore gravity?

Is gravity the cause of gravity? This representation is unsatisfactory because it leaves us with no sense of what is actually responsible for this warping of spacetime. Furthermore, if this diagram is supposed to help us understand the warping of spacetime, how does it give us any representation of warped time? Not only does the rubber membrane offer an analogy that allows us to visualize just a thin slice of space, it fails to offer us an intuitive connection to why time warps in regions of curvature.

If you are familiar with these kinds of representations for curved spacetime, then you might have already noticed that Figure 3-3 differs slightly from the standard representation. In this figure the dimension that is being used to depict curvature is labeled δ, which is meant to highlight the fact that this depiction is conceptually importing a dimension that is outside of familiar space—beyond x, y, and z.

Traditional representations of spacetime curvature make use of an unfamiliar dimension, but they fail to label, or even mention, this additional dimension. While we might say that rubber sheets are meant only as analogies, it is worth noting that the unfamiliar dimension is doing all of the work of those analogies. Because it is counterproductive to offer an analogy as an explanatory tool, and then to completely ignore the element of the analogy that the explanation depends on, we might press for at least some narration of this additional dimension.

We might also note that the extra dimension in these rubber sheet diagrams enables us to depict the curvature of a plane of spacetime, allowing us to partially penetrate the notion of curvature, but it doesn't carry us the full way. To go beyond the limits of this analogy we need to develop a model that is capable of graphically demonstrating the complete curvature of spacetime, one that doesn't suppress any of the familiar space dimensions, or ignore time. Working towards that goal, let's take a closer look at what curvature is meant to portray.

Imagine that we have an observation station on Earth and that we have placed three observation stations in space (in the configuration shown in Figure 3-4a). The four unique observers continuously measure the location of the newborn star Dilabee while also monitoring the positions of the other three observers. Day after day, they see no changes. All four stations agree that there is no measurable velocity between any of the observers or Dilabee. Therefore, the relative positions between these five objects all remain constant and the geometric configuration of the group is static.

One day, however, something confounds this whole set-up. A black hole comes into the picture, traveling on a path that will bring it between Earth and Dilabee. Strangely, as the black hole moves closer and closer toward a position between Earth and Dilabee, Earth's observers see Dilabee's position changing—its distance increases and its angle changes (Figure 3-4b).

When the observers on Earth examine the three stations in space they detect no change in their positions, so they ask the three space station observers to verify that Dilabee has changed position. The three space stations all agree that the Earth-based observations are wrong. They see no change in the star's angle or distance. From their perspectives Dilabee has not moved.

Although this example has been fabricated, this effect is real. It is something that has been detected and measured many times over by scientists around the world. Einstein came up with a geometric way to describe this effect. Note that the example I have used thus far has been two-dimensional. That is, the four observers and the star all lay in the plane of the paper. In Nature this effect is not limited to two dimensions.

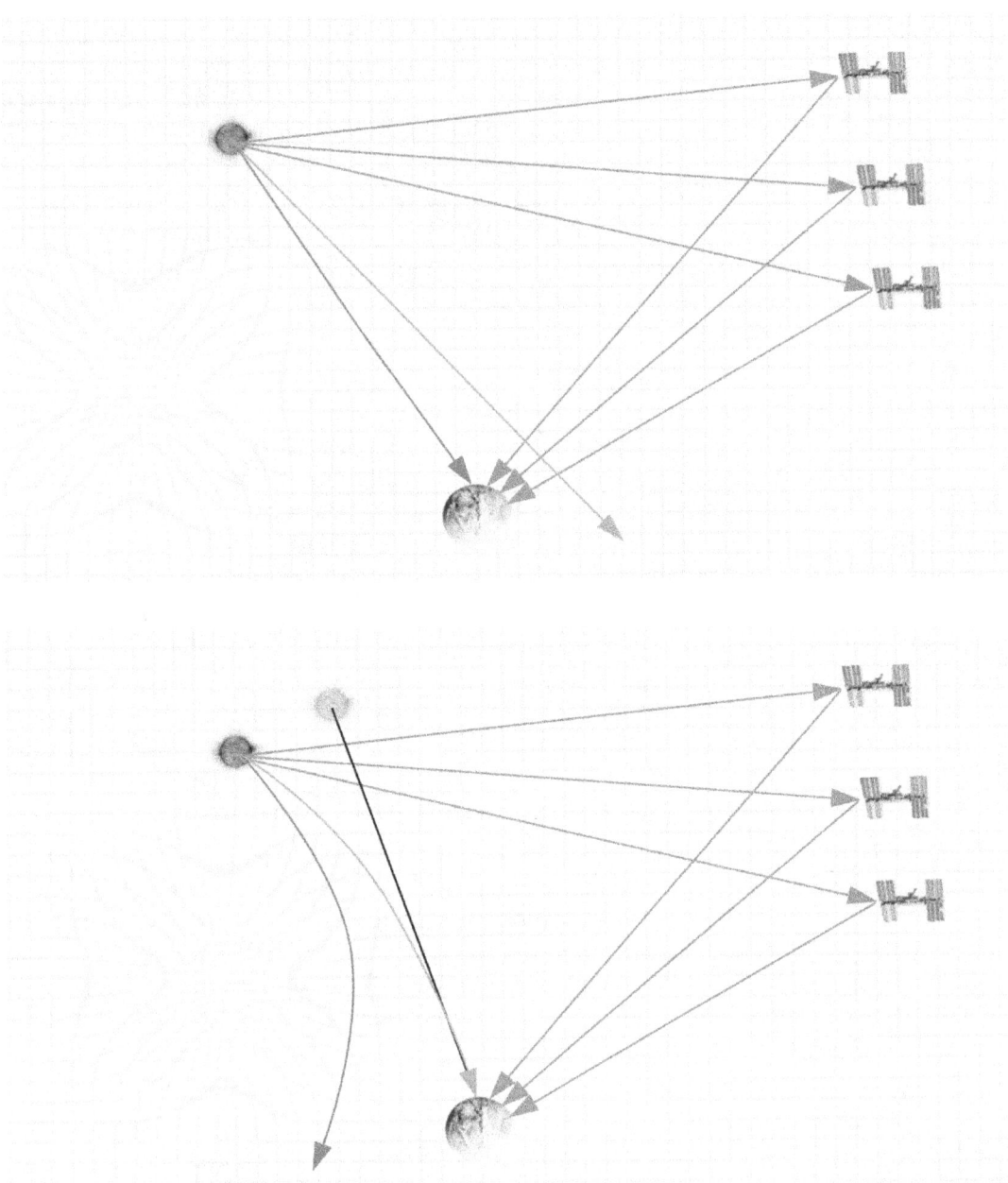

Figure 3-4 a & b   Effects of a black hole
The star's distance and position appears to change from Earth's point of view when the black hole comes into the picture, but the three observation stations detect no change.

Using Einstein's rubber sheet model gives us a sense of what is going on (Figure 3-5). The fact that space is curved around the black hole accounts for the perceived changes in distance and direction. The three space stations do not yet detect any change because they are still in relatively *flat* space.

This is the sense in which rubber sheet diagrams offer us a partial explanation (visual description) of curved spacetime. The goal is to complete that picture. Curvature represents a physically real characteristic of the vacuum. Its magnitude varies smoothly from one region to another. Depiction of curvature is a schematic representation of a change in the *amount of space*, or a change in metric density from one region to the next. The slope of this curvature portrays the magnitude of change at each point, which depends upon the proximity of that point to massive objects and the magnitude of their masses. With this understanding let's consider a volume of space and inquire about this effect we call curvature.

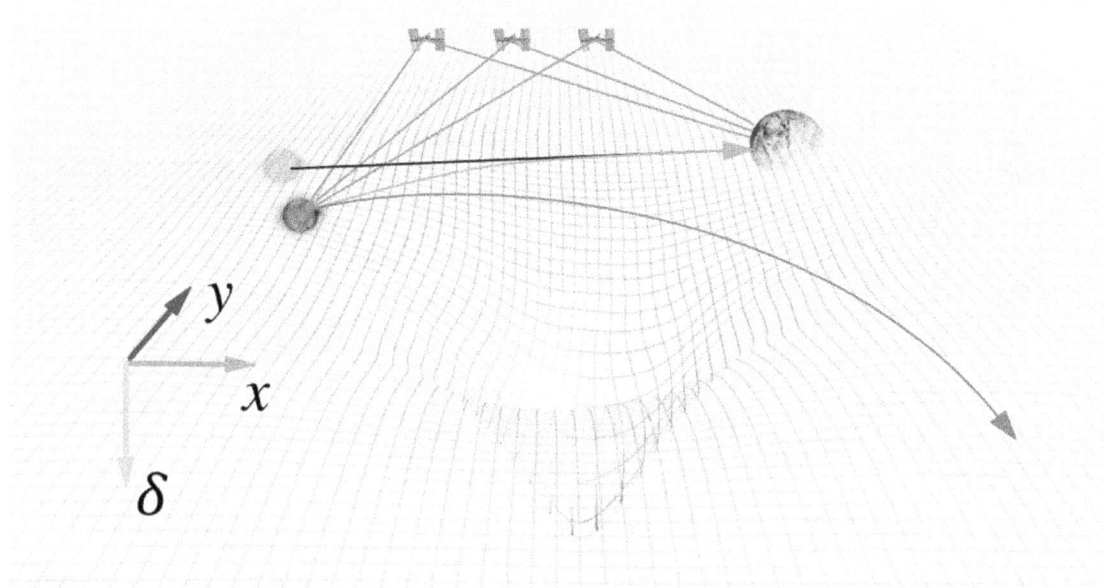

Figure 3-5 Curvature explains gravitational effects
A black hole curves, or warps, spacetime into another dimension.

Imagine that our volume of space is defined as a cube and the sides of that cube are ten astronomical units long. (An AU is the average distance between the Sun and Earth.) At each of the eight corners of the cube there is an observer (Figure 3-6a). These observers keep an open line of communication about their distances. They are all in complete agreement about their fixed positions and zero relative velocities. For example, A measures the distance to B and C and finds that they are 90° from each other and of equal distance. Through simple geometry observer A can determine the distance that B and C will measure between each other: $\sqrt{2}$ times the distance between A and B. Each observer measures the distances to the other seven observers, and then calculates the distances that each of the other observation stations will record for their measurements. All of these calculations and measurements exactly agree.

Now, if a black hole drifts into the center of this cube (Figure 3-6b), what changes? Well, with the black hole at the center the observers measure the same distances along the edges or the faces of the cube.[8] Observers may expect that the distances connecting the furthest corners of the cube will also be identical to what they measured before. But, when they measure distances of paths that pass through the center of the cube—as from C to D—they find that there is a significant increase. There is more space between opposing corners.[9] Although the boundary of the cube has remained the same, there is *more space* inside the cube. How can two cubes with the same boundaries contain different volumes of space?

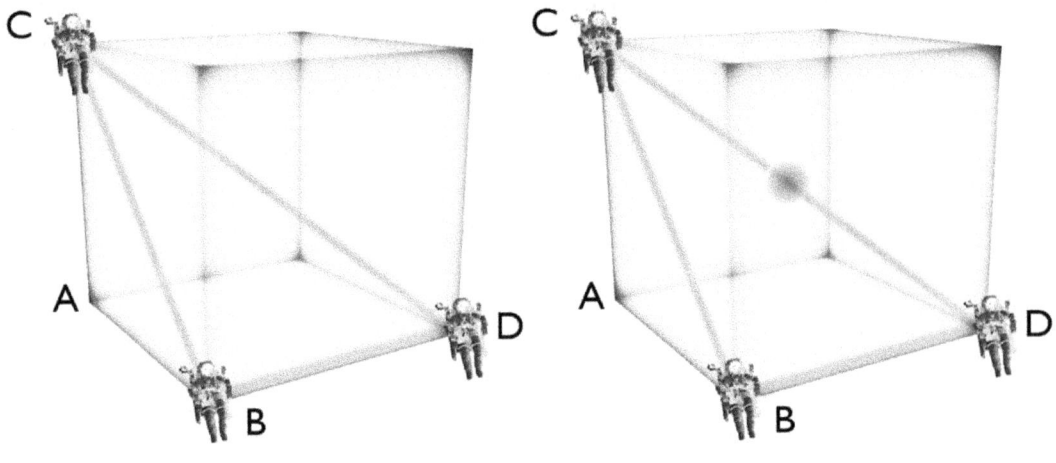

Figure 3-6 Surveying volume
(a) Calculations and measurements of the distance between the eight corners agree. (b) Measurements that utilize paths close to the black hole no longer agree with calculations. They always survey an increased amount of space.

The answer turns out to be quite profound and yet surprisingly simple. Just as the sum of angles in a triangle departs from 180° in curved space, the volumes of cubes can vary when curvature is introduced. To visualize the curved triangles (two dimensional objects), we place them on a curved surface (Figure 3-7). But how can we visualize three dimensional curved objects or regions? The answer makes the postulation of extra hidden spatial dimensions no longer arbitrary or extravagant. It is this question that initiates a chain of deductive reasoning that leads to the inescapable physical existence of extra dimensions.

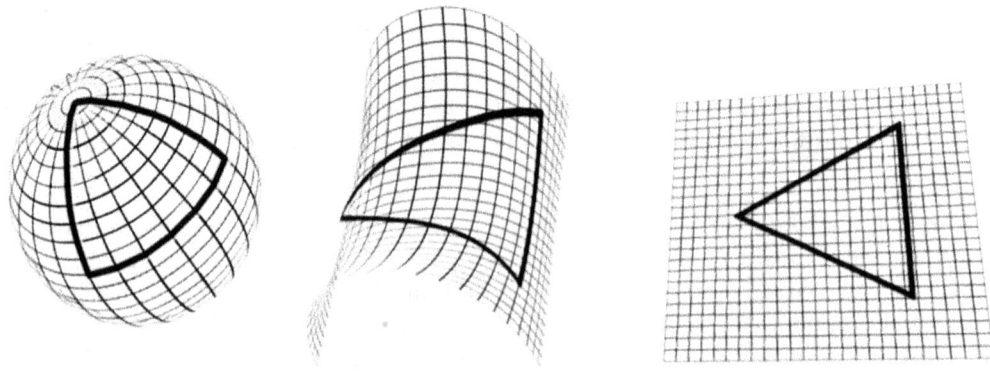

Figure 3-7 Triangle on a sphere, saddle, and plane

To answer this question let's turn to a more familiar example. Let's take two cubes of equal size. One is made of diamond and the other is made of graphite. Both, therefore, contain only carbon. If these cubes are painted black, we may guess that they are in every way identical because we are told that they are composed of the same kind of atoms. But upon picking both cubes up we will quickly surmise that one is heavier than the other. How can one explain this? They are of equal volume and are both made of only carbon, so how could their weights vary?

Naturally we turn to a description of densities. We explain that the cubes are made of *atoms*, or small mass particles, in this case carbon atoms. Inside the diamond cube these particles are packed more closely together than they are in the graphite cube (Figure 3-8). In other words, the diamond cube is denser than the graphite cube. This is why they can be composed of the same kind of atoms, have the same volumes, but possess different total amounts of carbon.

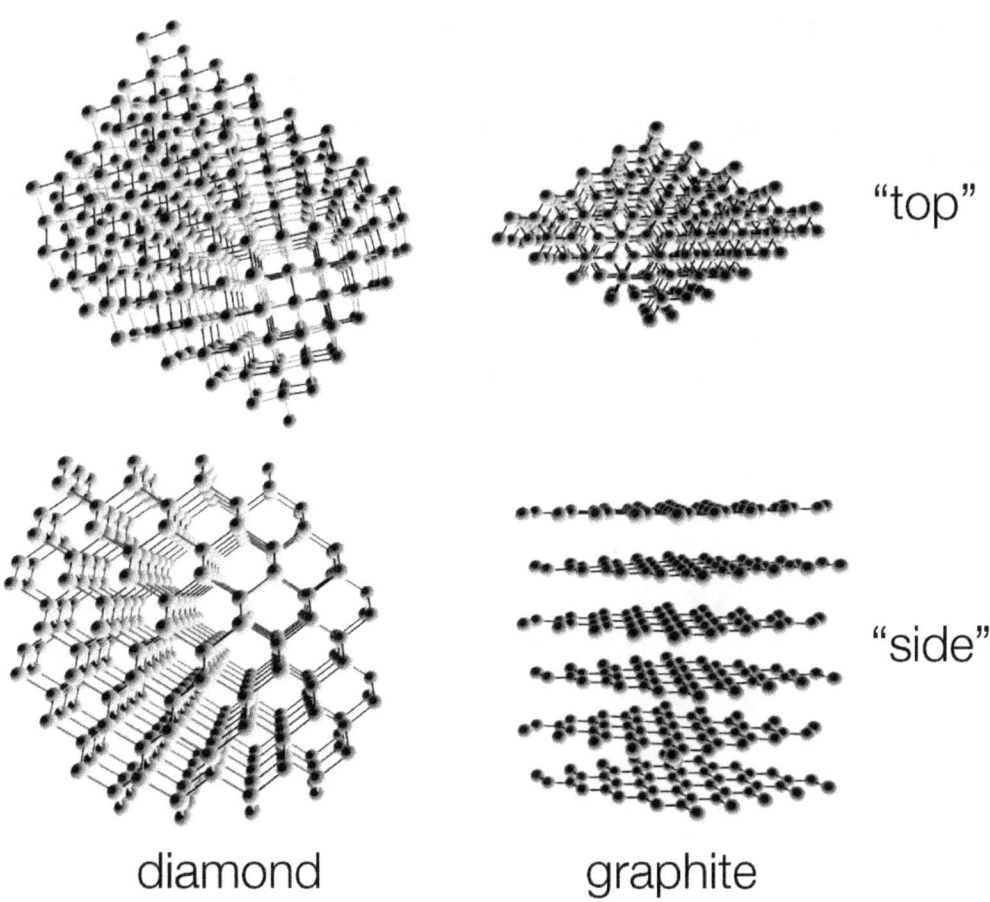

Figure 3-8 Lattice structures of diamond and graphite
When carbon is subjected to a pressure under 20,000 atmospheres it takes on the crystalline lattice structure of graphite. Over 20,000 atmospheres it takes on the crystalline lattice structure of diamond.

It is important to recognize that in order to explain the concept of density we must describe, or comprehend, two things: the particles (atoms) and the medium (space) in which the particles are distributed. If we did not assume or visualize the medium (space) in which the atoms reside, then we could not explain variable densities.

In our example of the spatial cubes, we have a similar situation. Two cubes of equal size that contain a different amount of *stuff*. Only this time the *stuff* we are referring to is space itself. Therefore, when we think through the implications of curved space, we find that curvature is a description of how regions of space are not uniform. It expresses that equal volumes (bounded by identical boundaries) can contain different amounts of space. This suggests that space can have variable densities, which in turn strongly suggests that space itself is particulate and that its pieces are distributed within an even deeper background medium. Because this medium enables the dispersion of the pieces of space, it must possess spatial dimensions that are entirely separate from the dimensions held by those pieces. This is where we will find the additional dimensions.

If this line of reasoning holds up, then spacetime curvature might be more richly explained if we assume that the fabric of spacetime (the vacuum) is a discrete medium. Quantizing the vacuum means that extra dimensions are logically necessary instead of just being arbitrary assumptions, extravagant postulations, or inspired guesses. In Part II of this book, we will explore the full dimensional structure for spacetime that is suggested by vacuum quantization. To prepare us for that, we will now examine some of the clues that have popped up from the microscopic realm—clues that support the hypothesis of vacuum quantization.

[1] Carolyn Brighouse. (2014). Geometric Possibility: An Argument from Dimension. Occidental College, 01/2014; 4(1). DOI: 10.1007/s13194-013-0074-1 Available from: www.researchgate.net/publication/263104059

[2] Independent parameters, or bits of information, uniquely relate to "where" or "when" an event occurs. Color, as it turns out, is something that doesn't fit that bill because it is encoded by the metric when all dimensions are included. Color also doesn't tell us anything about where or when something is.

[3] Lisa Randall. (2006). *Warped Passages: Unraveling the Mysteries of the Universe's Hidden Dimensions*, Harper Perennial, p. 4-5.

[4] The added spatial dimension was posited as circular. This is an important point that will come into play later. Kaluza had produced five extra quantities. Four of these could be used to produce Maxwell's electromagnetic equations. Walter Isaacson, *Einstein*, 2007.

[5] Even today's prominent physicists have a difficult time wrapping their minds around higher dimensions. Many have prematurely jumped onto the 'impossible' bandwagon—claiming that because they haven't visualized higher-dimensional realms, it is impossible. For example, in his recent book *Hyperspace*, Michio Kaku reflected the current tendency to accept this impossibility when he said: "How do we see the fourth spatial dimension? The Problem is, we can't. Higher dimensional spaces are impossible to visualize—so it is futile even to try." Michio Kaku, *A Scientific Odyssey Through Parallel Universes, Time Warps, and the 10th Dimension* (New York: Anchor Books, 1995). Stephen Hawking concurs with this consensus saying that, "It is impossible to imagine a four-dimensional space. I personally find it hard enough to visualize three-dimensional space!" (Hawking, 'A Brief History Of Time,' p. 24.) In Lisa Randall's opinion: "It's not thinking about extra dimensions but trying to picture them that threatens to be unsettling. Trying to draw a higher-dimensional world inevitably leads to complications." Lisa Randall. *Warped Passages*.

Randall believes in the physical existence of additional dimensions; but she doesn't think it is possible to visualize them alongside the familiar dimensions. This attitude has very strong roots in historical philosophy. The modern metaphysical tragedy that clings to this almost unanimously accepted claim can be summed up in Immanuel Kant's (1724–1824) conclusion that "since we lack direct access to 'reality in itself,' we are limited to what we perceive." Diane Barsoum Raymond, *Existentialism and the Philosophical Tradition*.

All of this echoes the sentiments of Werner Heisenberg who set the tone for modern physics with the claim that we should "abandon all attempts to construct perceptual models of atomic processes." Werner Heisenberg. (1971). *Physics and Beyond*. New York, Harper & Row, p. 76.

Nevertheless, despite the historical failure to do so, visualizing higher dimensional realms is not impossible. With the right insight it turns out to be rather straightforward. It is accomplished through realizing levels of dimensional hierarchy by allowing the fabric of spacetime to be composed of stippled constituents residing within a volume of superspatial dimensions. By the end of this book you will be able to visualize more than three spatial dimensions simultaneously.

[6] Michio Kaku. (1995). *Hyperspace: A Scientific Odyssey Through Parallel Universes, Time Warps, and the 10th Dimension*. New York: Anchor Books, pp. 3-5.

[7] Notice that I did not choose (x, y, z, i, j, k, t) despite quantum mechanical laws demanding the existence of imaginary dimensions for each spatial dimension. This is because the

dimensions that allow depiction of curvature are in fact separate from the dimensions tied to imaginary characteristics, which describe quantum mechanical systems. Compactified versions of these imaginary dimensions will come into play later.

The three Greek letters σ, μ, δ were taken as phonetic components of the Sanskrit word Samadhi.

[8] For the purposes of this example I am assuming that the black hole in question is not particularly massive, such that its effects are negligible five AU away.

[9] Light signals also take longer to travel through regions with high curvature because they have more space to traverse. This is known as the Shapiro effect.

Chapter 4 — The Quantized Nature of Spacetime

"God created the integers, all else is the work of man."
Leopold Kronecker

"The unification of general relativity and quantum mechanics may lead us to abandon the idealization of continuous space and time and to discover the 'atoms' of space-time."
Theodore A. Jacobson

"Every phenomenon in quantum mechanics has a quantum aspect which makes it discontinuous."
Gary Zukav[1]

Kaiparowits Plateau, Grand Staircase Escalante National Monument.

Every rock tells a story. Some speak of violent eruptions, cataclysmic impacts, ancient rivers, orogenic earthquakes, or slowly cooling crystalline batholiths, while others whisper about ancient ecosystems. The powerful force of erosion can protect these ancient secrets by burying them under a mountain of rock, but it can also uncover past burials and pulverize them. The problem is that when erosion uncovers a rich tableau of the past it does not pause out of respect for the resurfacing secrets. Unless the record is rescued, it will quickly be destroyed and lost forever.

Hoping to be time-traveling detectives, we followed our geologic maps to the most promising deposit of treasures. Our prospective territory stretched from Lake Powell to Escalante—a scenery full of canyons and gullies and jagged rocks that stretched to the horizon in all directions, and no roads. It was a paleontologist's Mecca, a land riddled with treasures waiting to be discovered.

Reaching our destination required a bit of skilled navigation, over rocky steppes, through washes and dry riverbeds. Our transportation was an old pintle hitch army truck with bad shocks and reinforced tires that we called "the beast." It was rough, squeaky and unforgiving, but it was capable of getting us to this remote location.

The treasures we were after were arguably more valuable than little nuggets of gold, because these rock-encased secrets were capable of offering little glimpses into a time when ferocious battles ensued between history's largest dinosaurs. They enabled us to peer through the hourglass of change that separated the distant past from the present and come to understand the world that the *terrible lizards* lived in. Participating in the search for these treasures was mysteriously all-consuming and thrilling.

We were in a race against time. Scorpions and temperature-obsessed crickets[2] had been the only creatures to witness the secrets that had already been lost to the wind, water and sun. We wanted to change that. Our team of twenty was made up of dinosaur paleontology graduate students, undergraduate students, cooks, drivers, geologists, professional prospectors and volunteers. We were all here at the request of lead paleontologist, Scott

Sampson. For the next few weeks, we would be sleeping in tents with no showers or bathrooms.

We set up camp just over the hill from a promising site, making sure to face the doors of our tents, including the one that served as our kitchen, downwind to protect our things from sandblasting gusts. Under the blinding sun and whipping wind, which had a knack for quickly changing its temperature, we began wandering the desert with pockets full of snacks, water bottles, rock hammers, Talkabouts, and GPS units. Entranced by the hope of unearthing a previously undiscovered species, rough conditions became nuances of the adventure.

As we prospected, we discovered a plethora of dinosaur bones, turtle and croc fossils, and fossilized forest remains. We collected most of these fossils in Ziploc bags, tagging and cataloging them, and recording their GPS locations. Then we discovered a mass deposit site, a locale with a wide assortment of fossils, whose sandstone matrix contained granules of a size and shape of that were compatible with the banked remnants of an ancient meandering river.

Settling in on that site we laboriously began using gas-powered jack hammers, rock picks, five pound sledge hammers, rock hammers, chisels, large screw drivers, wide paint brushes and assorted dental picks to remove the encasement of these tangible secrets. Our excavation revealed several femurs, vertebrae, ribs, ungles (raptor claws), skin impressions and hollow bones that belonged to the non-avian dinosaurs of the theropod clade. There were also a few paleontology gems—rarely preserved fossilized fragments of bones that once made up the skulls of dinosaurs.

The most interesting fossils we uncovered were painted with Vinac, which acts as an adhesive and stabilizer strengthening the fossil from the inside out. When the Vinac dried we covered its smooth surfaces with crumpled paper towels or tissue paper. Then, after mixing plaster and water in five gallon buckets, we soaked paper towels in the mix and then delicately covered the fossils with them. Twenty minutes later, we wrapped the entire block in plaster soaked burlap protecting the fossil in a big cast, which paleontologists call a "jacket." When the jackets were dry, they had to be carried back to our vehicle for eventual transport to our museum. This was often a difficult and awkward task.

At the end of the day we all returned to camp exhausted. One night, as we sat around the fire eating hamburgers and nacho flavored Doritos, Martha, a long-time volunteer with an eight-year-old boy in tow, pulled out her oversized saltshaker and dashed some *"Crazy Uncle Billy's Magic Fire Dust"* over the fire.[3] Immediately, the flames turned brilliant green. Then they slowly and hypnotically returned to their familiar color, prompting Martha to pour some more dust onto the crackling logs. Simultaneously, we watched yet another magical transition in the sky above.

This process carried us through the phases of civil, nautical, and astronomical twilight until the night reached full bloom. Shaula, the stinger of the constellation Scorpius, had climbed to its highest point in the southern sky. Trailing behind it was Sagittarius, who to modern eyes reveals itself as a teapot. Protruding from the spout of this teapot, the Milky Way stretched like a brilliant ribbon all the way through Cygnus the Swan, the Northern Cross, and Cassiopeia, the Queen.[4]

There was a primeval aire all about us. It felt like we were lost in time—as if the green flames drawing our gaze had the power to flicker us between the present and the deep past. The scenery around us reinforced this sensation. There were no signs of modern civilization, no distant glow polluting the night sky, no structures, buildings, street lamps or even roads within the observable horizon. The stars responded to this by becoming so numerous and brilliant that they were almost tactile. The Milky Way was so bright that it cast shadows beneath us as we walked away from the fire.

This serene setting had the potential to connect me to something far more divine than a detailed glimpse of the Cretaceous. It had the power to bring to focus the faint whispers that usually remain completely overwhelmed by a nexus of amplified runaway thoughts. As the heat of the day finished fading and the crickets settled into just the right rhythm, my trance began to focus—guiding me to one thought. Maybe the key to unlocking Nature's most clandestine secrets was hidden in the process that guided us through our past discoveries. Having the feeling that I was on to something, I began to recount some of those questions.

How did all of those dinosaur bones become fossils? How does Crazy Uncle Billy's Magic Fire Dust turn the fire green? How do we explain the sky transforming from daylight blue to sunset orange and then midnight black? How does the sun produce its energy?

I had learned the answers to all of these questions, but no one had ever pointed out to me that all of those answers were linked together by one simple assumption about the world. Perhaps it was too basic to point out, but I had never really paid attention to the fact that all of our modern answers rely on the quantization of matter.

As I waited to witness yet another meteor, I thought about the progress modern science has made since it came up with the conjecture that the world is made of atoms. (The word "atom" comes from the Greek word *atomos*, meaning indivisible or uncuttable.) Then I wondered about the outstanding mysteries—the questions our scientific endeavor has not yet been able to answer. Perhaps, I thought, the next great step in understanding is not all that different from the previous ones. Maybe all we have to do to understand Nature's remaining mysteries is assume that matter isn't the only thing that comes in *atoms*. Maybe the vacuum is also composed of atoms.

The Sands of Space and Time

> *"No great discovery was ever made without a bold guess."*
>
> *Isaac Newton*

By conjecturing that the vacuum is quantized I mean to assume that the medium of spacetime is a fluid medium composed of discrete, separate pieces at its fundamental scale. In a very practical way, this suggests that spacetime is somewhat like the medium of air, in that it is made up of an enormous number of identical interactive particulates that move about within a deeper setting. The pieces of spacetime, which we will call quanta, are "the primordial substratum, underlying our macroscopic spacetime picture."[5]

An immediate consequence of the claim that the vacuum is quantized is that it can have different densities. Under this condition the vacuum loses its ability to function as a background (those densities must be distributed throughout something). In other words, if spacetime is quantized, then it cannot be the ultimate background of physical reality. By assuming that the vacuum is a medium composed of constituents that are situated in, and move through, dimensions that are not part of the x, y, z metric, we necessitate the conclusion that additional variables (additional dimensions) are physically real. At each moment, the state of any particular region of space is defined by the positions and velocities of its quanta. These positions and velocities are defined in additional dimensions—the superspatial dimensions.

Some people find the possibility that there are real spatial dimensions in addition to x, y, z to be so radical that they dismiss it outright—in spite of the vast collection of experimental and empirical evidence that points us in that direction. Such a response is understandable, and expected, but it may also be premature and inappropriate. If we never seriously explored new perspectives, if we always allowed common experience to frame our worldview, then we would still be modeling air and water as we experience them—as continuous media instead of as collections of molecules. A model that attempted to explain all media (air, water, wood, etc.) as continuous would be far more complicated than a molecular one, and its explanatory power would severely suffer. By allowing for the existence of atoms, our model of physical reality becomes far more coherent, simple, and intuitive.

It is important that we understand the significance of this point. Although we never directly experience atoms with our senses, we believe they physically exist because their existence enables us to explain a vast array of mysterious natural phenomena. The explanatory power that comes from the assumption that the world is made of atoms is absolutely staggering. Those explanations allow us to understand: crystalline structures, nuclear fusion in the sun, Brownian motion, friction, heat, color, chemical reactions of all sorts—like the ones that control our digestion—why the sky is blue, why Uncle Billy's Magic Fire Dust turns flames green, how fast a receding galaxy is moving away from us, and on and on.

"If in some cataclysm, all of scientific knowledge were to be destroyed, and only one sentence passed on to the next generation of creatures, what statement would contain the most information in the fewest words? I believe it is...that all things are made of atoms."

Richard Feynman[6]

A beautiful, elegant model that is accessible to our intuition connects all of these phenomena, but the model itself relies on the assumption that the world is made of atoms. Once we make that assumption a simple model emerges that explains a truly enormous number of natural phenomena. This is why several luminaries have claimed that the most important scientific idea to date is that the world is made of atoms.

Because we allow for the existence of a deeper background (spacetime) behind other material media, we have no difficulty comprehending atoms. This intuitive framework (with space as the background) allows us to rationalize how it is that we can pass from one medium (like air) to the next (for example water) without having to reformulate our picture of reality. We simply allow the deeper medium to remain the background and we break up the respective foreground media into quantized collections that move about within that background. We imagine all the foreground media as different collections of small, quantized parts, that is, atoms, molecules, etc. that move about in spacetime.

The switch to a quantized perspective has two significant effects. First, it allows us to simplify our picture of Nature as it applies to our experiences. Second, it enables us to explain a vast range of phenomena indicative of the quantized medium, which are not necessarily linked to our common experiences. This consequently strengthens our confidence in that model.

Applying this to the medium of air, we find that high-pressure and low-pressure variances simply translate into changing densities of air molecules within the background of spacetime. (Note that if we modeled air as continuous, then we would have to introduce some sort of magical curvature tensor just to model those changes in pressure.) Temperature becomes an expression of how energetically the molecules are moving about on average, and wind becomes the net result of the collective motions of those molecules as they undergo the process of equalization from high-pressure regions toward low-pressure regions.

On the deeper level, a quantized depiction of air provides us with an explanation of its less intuitive properties, like its optical absorption and dispersion patterns (explaining, among other things, why the sky is blue), its chemical reactivity, the dynamical mechanics behind its phase transitions, and so on.

Quantized models give us access to two levels of description. One level describes the macroscopic properties of the medium—pressure, density, temperature, and so on, and the other describes the microscopic, or the properties of the individual atoms or molecules that make up the medium and how they interact with each other. Both of these parts stem from the same intuitive quantized picture.

Because we are going to extend the domain in which we apply the *atomic* idea, it is important for us to understand that the simplification that emerges when we express different media of matter (air, water, wood, metal, etc.) as constructs of atoms, is a strong argument that those atoms physically exist in Nature. In general, this is how we test the validity of an assumption we make about the world. If a model grows out of a particular assumption and inherently explains physical reality in a more complete yet simple way, then the idea is to be taken seriously. As we have seen, one of the most valuable ideas ever postulated was that the world is made of atoms. In this book we are going to take that idea and extend it one step further. We are going to postulate that literally *all* media are composed of "atoms"—discrete, quantized, interactive parts. Specifically, we are going to add spacetime (the vacuum) to our list of quantized media.

To determine whether or not this extension pushes our intellectual quest in the right direction, we will have to study the model of physical reality that stems from it and check to see if that model connects an array of seemingly separate phenomena in Nature, simplifies our description of those phenomena, and gives us a way to extend our intuition deeper into the heart of physical reality's ultimate form. The goal is to gain ontological access to the mysteries of physics (the uncertainty principle, wormholes, dark energy, the cause of the Big Bang, what black holes are like on the inside, how the constants of Nature were determined, etc.) by obtaining an intuitive picture that unites general relativity and quantum mechanics. If the framework that grows out of our new assumption does not grant us this access, then it will be appropriate to dismiss the quantization claim altogether. But if a simplistic, encapsulating framework does begin to emerge, and if it gives us an intuitive understanding of those phenomena, then we will gain confidence in the idea that spacetime is quantized.

When Islamic philosophers debated the possibility of an underlying atomic structure for space and time they envisioned atomic space to be something like a chessboard with no "interstitial voids" between the atoms of space. The problem with this atomic model was that it was incapable of reproducing the claims of the Pythagorean theorem, which was esteemed as "the best attested theorem in mathematics."

According to *chessboard atomism* the length of the inscribed hypotenuse must be equal to the number of atoms (cells) that sum down the diagonal of the triangle. But according to the Pythagorean theorem the hypotenuse's length should be the square root of the sum of the two shorter sides squared $a^2 + b^2 = c^2$ (Figure 4-1). The incompatibility between the Pythagorean theorem and the *chessboard atomic model* ultimately led to a rejection of atomism.

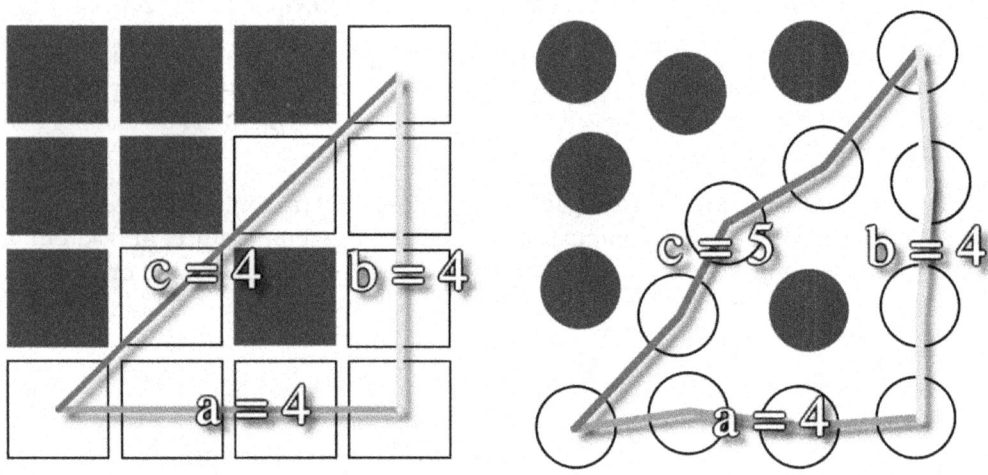

Figure 4-1 Islamic chessboard atomism vs. a proposal with interstitial voids

> *"If at first, the idea is not absurd, there is no hope for it."*
>
> Albert Einstein

This incompatibility is no longer sustained when quantization is understood to mean that the vacuum is a medium composed of quanta that are dispersed throughout a superspatial background. This dispersion allows the Pythagorean theorem to be regained as a macroscopic identity, in the absence of density gradients (curvature).[7]

The Case for Quanta

> *"The holy grail… is the prediction of observable consequences derived from the microscopic quantum structure."*
>
> Jan Ambjørn [8]

As we warm up to the idea of vacuum quantization, let's recall that a quantized medium appears continuous and smooth from large scales, and that it best reveals its atomic properties on scales that approach the size of its constituents. A macroscopic projection of a continuous medium is created when the combined interactions of those constituents are interpreted through an averaging process.

Averages are useful for describing many effects, but by design, they dissolve the underlying details from which they rise. When averages are interpreted as macroscopic emergent properties, they can be very convenient tools. But when they are mistaken as representative of the system's fundamental character, conceptual difficulties are imported. A model that mistakes emergent macroscopic properties for fundamental ones abandons a deterministic evolution for one that can at best be understood as statistical. If the mistake isn't noticed, then it will lead to the claim that the stochastic elements of that evolution represent something real in Nature.

If spacetime is quantized, then we should expect our familiar image of it, an image that has been distorted by an averaging process, to be incapable of accounting for characteristics that directly depend upon the system's underlying quantized structure (like the arrangements of the quanta—its specific state). Beneath that averaged-over picture is a lower level description in which each constituent of the quantized vacuum has an associated velocity vector that denotes how it is moving about in the superspatial void at a given instant. We can take an ensemble of those constituents, average over their vectors and represent the combined tendency of that region with a state vector $|\psi\rangle$. But if we make the mistake of using that average distribution as a fundamental descriptor, instead of seeing it as an emergent property, we end up with a vague theory that is subject to a wide variety of equally valid and nebulous interpretations. The possibility that modern quantum physicists have made this very mistake strongly motivates our investigation of vacuum quantization.

Science best serves humanity in its commitment to the idea that by ascending through the footholds of rationality it is possible for us to improve our understanding of Nature. Quantum mechanics has made it abundantly clear that the universe cannot be made to conform to *local realism* within the confines of four dimensions. For a while now supporters of the standard interpretation of quantum mechanics have twisted this discovery into a doomsday proclamation, using it to reduce science to a data collecting empire, a statistical prediction generator, that no longer concerns itself with understanding.

Despite the fact that local realism is a more fundamental scientific presupposition than the idea that the universe is made of four dimensions, significant figureheads (under the shadow of Niels Bohr) have actively promoted the idea of abandoning our primary scientific commitment. They point out that if local realism is ruled out, then there is no rational or causal structure to discover, and attempting to make sense of a system that does not conform to local realism is by definition a fool's errand. This may be an accurate claim, but it is also an irrelevant claim. Local realism can be restored in models with additional variables. As scientists it is our duty to explore such models.

Local realism combines the expectations of locality and realism. Locality is a very general and basic concept in physics that predates both Galilean and Einstein's relativity. It encodes the idea that the mutual influence of events decreases when their distance increases; that the mutual influence of events directly depends upon their spatial separation. This purely spatial notion (no time is needed) is one of the foundations of all experimental sciences.[9]

Realism is the view that the world described by science represents the real world in a mind-independent way. A scientist might believe that the formalism of quantum mechanics directly represents reality, that it tells us something about what is out there and gives us real insight into reality, or a scientist might believe that the current quantum formalism provides a blurred representation of Nature and that a deeper level, more accurate description exists beneath it. Either way, a realist holds to the idea that the model describes reality on some level, that what is being described actually exists independent of anyone's beliefs, conceptual schemes, linguistic practices, and so on.

Famously, Bell's theorem indisputably shows that we cannot coherently assume local realism, and at the same time, that all predictions of quantum mechanics are correct. This is because a self-contradiction arises when we try to simultaneously hold to those assumptions. This conclusion stands independent of any experimental result, which means that it is a purely logical conclusion.[10] (We will discuss Bell's theorem at length in Chapter 12.)

Because the predictions of quantum mechanics have stood up against the most intricate examinations experimentalists have been able to invent, the conclusion most popularly drawn from Bell's theorem is that Nature does not cohere to the rules of local realism. It obeys laws that are non-local, non-realist, or both. If Nature is non-realist, if our physics theories fail to map it at all, then the question remains—why do we get our predictions right? On the other hand, if Nature is truly non-local then a true map is unavailable.

This line of thinking retains the assumption that our model of Nature is complete, that no *elements of reality* are missing from the canvas of quantum mechanics. I refer here to the EPR (Einstein-Podolsky-Rosen) *elements of reality*, also known as *additional variables* and sometimes misleadingly called *hidden variables*. To be more thorough we should note that an experiment that is taken to suggest a departure from local-realism can also be taken to suggest that additional variables are missing from our model.

With this in mind, we note that if the vacuum is quantized, then the presence of nonlocality references a mismatch between the averaged-over description of the vacuum (used in quantum mechanics)[11] and its actual structure. Vacuum quantization introduces additional variables (describing the positions and velocities of the vacuum quanta), and the lower-level physics that describes this microscopic structure deterministically controls the evolution of the vacuum's state in a local-realist way. In other words, vacuum quantization restores local-realism, without contradicting the fact that quantum mechanics is incompatible with it, because it gets us beneath the canonically smooth, averaged-over, vacuum of quantum mechanics.

Because non-local phenomena reference this geometric mismatch, they manifest with a dependence on resolution, showing up less and less on larger scales. On macroscopic scales the mismatch between the smooth averaged-over description of the vacuum and the vacuum's actual geometry vanishes. As we approach quantum scales nonlocality emerges because the averaged-over picture begins to strongly misrepresent the vacuum's geometry.

If the vacuum is made up of quanta that are geometrically mixing about, mutual influence of events will not directly depend upon their spatial separation. Any model that ignores this mixing (one that assumes a smoothly connected vacuum) will observe the effects of mixing as a violation of local-realism. But a model that resolves the lower-level physics (how all the vacuum quanta are moving about in additional dimensions) retains local-realism.

Going further we note that manifestations of the *uncertainty principle* are also compatible with a quantized vacuum. According to this principle, there is a fundamental limit to the accuracy with which we can know positions and velocities in space. Within a quantized vacuum this is exactly what we would expect. On quantum scales, the individual pixels of Nature have dramatic effects but, like the image of a TV screen, as we zoom out from the pixelated image, individual contributions lose their potency to the average. If we naively assume that the average is a fundamental representation, that it directly represents reality at the lowest level, then effects that originate from the internal quantized structure (such as quantum jitters, quantum tunneling, and quantum entanglement) will manifest as astounding and confusing. However, if we are aware of the vacuum's quantized structure, then those effects transform into expectations.

In addition to the fact that studies of radioactivity have shown that the empty vacuum of space has spectroscopic structure similar to that of ordinary quantum fluids, another hint that the vacuum is quantized comes from cosmological descriptions of the early universe— most famously the inflationary Big Bang model, which invokes a description of spacetime that includes phase transitions and their associated increase in symmetry and entropy. Whenever a medium is capable of undergoing a phase transition it suggests that it is atomic, molecular or, in general, quantized.

For instance, consider the phase transitions of H_2O. A collection of H_2O molecules can undergo phase transitions transforming into ice, water, or steam (Figure 4-2). All three phases share the same molecular composition—H_2O.[12] Out of these phases ice possesses the least entropy (the least disorder) and the least symmetry. The molecules of H_2O inside the ice crystals are arranged in an ordered hexagonal lattice. This fixed arrangement means that the overall pattern of molecules retains its appearance only by rotations of multiples of 60 degrees. This limit on rotational symmetry means that the ice lattice has low symmetry and low entropy.

As the ice melts the molecules of water rearrange into a jumble of uniform clumps. In this state, rotating the system in any direction no longer changes its properties in reference to a lattice, which means that the system has become more symmetrical. Therefore, by melting the ice into water the system has gained symmetry and entropy. As water transitions into steam, the clumps of H_2O, which tend to be arranged with the oxygen side of one molecule facing the hydrogen side of another, break up into completely random orientations. Again, this phase transition is accompanied with an increase in entropy and symmetry.

It follows that if the vacuum is composed of quantum units, then the phase transitions it underwent early on can be explained as changes in the arrangements and associations of those quanta. Therefore, the data that suggests the universe, as a whole, has undergone phase transitions inadvertently supports the idea that the vacuum is quantized. Phase transitions have a natural association with molecular or atomic arrangements. Additional support of this comes from the discovery that the quantum mechanical description for regions of space, called fields, respond to temperature changes just as ordinary matter does. If we increase the temperature of a region of space, we find that the amplitude of the field undulations within that region of empty space increases in the same manner that the atomic motions of a gas increases when heated.

"The universe as a whole acts somewhat like a gas."

Neil deGrasse Tyson[13]

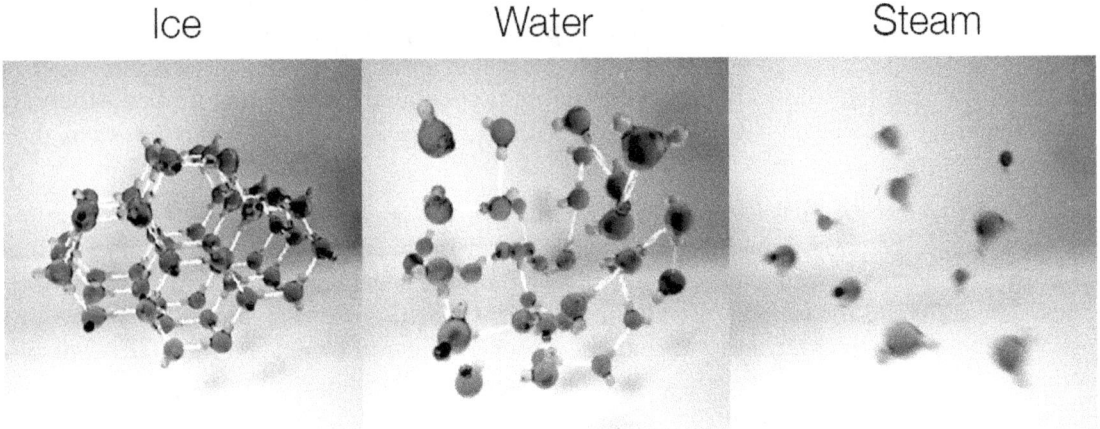

Figure 4-2 The phases of H_2O: Ice is a lattice of molecules—a fixed geometric arrangement, water preferentially aligns the positive and negative ends of the polar molecules, and steam has too much kinetic energy to maintain any alignment.

Another quirk of nature that points towards the possibility of a quantized vacuum is known as the blackbody ultraviolet catastrophe. A blackbody is an idealized object that absorbs all incoming light without reflecting it, heats up, and then begins to emit light. The character of the light it emits is entirely dependent upon its temperature. The "catastrophe" comes from a conflict with observation that arises when one calculates the amplitude of expected emission for the wavelength spectrum (assuming that spacetime is smoothly connected on all scales and therefore allows a continuous spectrum for energy). Such calculations predict a far greater contribution to the blackbody radiation in the shorter wavelengths (higher energies like ultraviolet) than is actually observed (Figure 4-3).

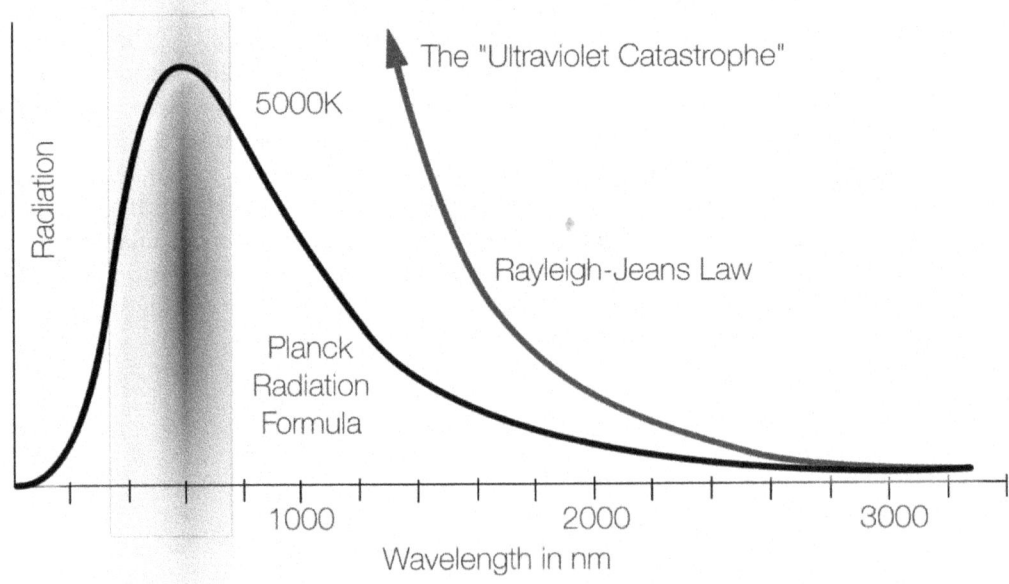

Figure 4-3 Black Body Radiation and the Black Body Catastrophe

Very short wavelengths contribute less than expected, that is red contributes more than blue, which is why fires are commonly more red than blue. The most important thing to note about all of this is that if we recalculate blackbody radiation allowing for quantization (of light's substrate—spacetime), then the discrepancy vanishes. When we do this the ultraviolet catastrophe is automatically resolved because only certain wavelengths (colors) are allowed. This restriction explains why hot objects radiate as they do. When a blackbody is heated, the first visible color it radiates is red because the energy packets of red light are the smallest energy packets in the visible light spectrum. With more heat, higher-energy colors (shorter wavelengths) can be emitted as the discrete (quantized) value of energy for each successive color is reached.[14]

Max Planck took a step quite similar to this when he suggested that light could only be delivered in quantized units. This fundamental unit, now called Planck's constant h, restricts the possible values for the frequency of light to whole number multiples $(1hf, 2hf, 3hf ...)$. Intermediate values of that energy, according to Planck, cannot occur. This is exactly what we would expect if we started from the assumption that space is quantized.

> *"...the hypothesis of quanta has led to the idea that there are changes in Nature which do not occur continuously but in explosive manner."*
>
> Max Planck[15]

Unfortunately, Planck believed that the quantization inherent in his math was some sort of mathematical trick necessary to produce results in agreement with observation. He didn't see it as representing a real property of light or the substrate it propagates through. It wasn't until Einstein's remarkable year that light quanta became thought of as real physical entities instead of mathematical abstractions.[16] Vacuum quantization gives us a way to explain this quantization of light.

There are other reasons to doubt the continuity of space. For example, black holes represent a severe conflict with the notion of continuous space. The existence of just a single singularity demands discontinuity in the fabric of spacetime. In other words, if there are rips in the fabric of spacetime at any level, then that fabric can no longer be accurately described as smoothly connected and continuous everywhere. However, the existence of singularities (black holes) is compatible with the idea that the vacuum is a medium that behaves like a fluid on macroscopic scales. Taking this claim seriously explains why Theodore A. Jacobson, Renaud Parentani, and their colleagues found that "the propagation of sound in an uneven fluid flow is closely analogous to the propagation of light in curved spacetime... [which] suggests that spacetime may, like a material fluid, be granular and possess a preferred frame of reference that manifests itself on fine scales."[17]

We might also note the observation that the Earth receives an overabundance of ultrahigh-energy cosmic rays. Calculations based on special relativity predict that these extremely energetic cosmic rays should only rarely reach Earth because they lose energy as they travel through space. But a Japanese observatory has seen more of these rays than the calculations (based on a continuous metric of spacetime) allow for. Theorists, such as Amelino-Camelia, think that this excess is evidence that spacetime is granular because "graininess" would ease the passage of high-energy particles.[18]

If spacetime truly is granular, if the vacuum is a quantized medium with fluid properties, does it have viscosity? If so, how viscous is the vacuum? To explore this question we note that if the vacuum were to possess a non-zero viscosity, then particles of different energies should disperse and be teased apart as they travel through the vacuum just as waves of different wavelengths eventually disperse and travel at different speeds in water. High-energy photons, which have the shortest wavelengths, should lose energy as they travel across the cosmos, and the larger the vacuum viscosity is the more noticeable this effect should be.

High-energy X-ray and gamma-ray light coming from the Crab Nebula, a supernova remnant about 6,500 light-years away, have given us a way to observationally determine how viscous the vacuum is. Those observations tightly restrict vacuum viscosity to being zero, or very close to zero, because they measured no dissipation whatsoever. Therefore, if the vacuum is a fluid, it must be a fluid without internal dampening. It must have very low viscosity—like a superfluid.

Grøn and Hervik have pointed out that, although it uses different terminology, general relativity carries the assumption that the vacuum is a superfluid.[19] And Alexandre Martins notes that the assumption that the vacuum is a superfluid is supported by experimental confirmations of Newton's first law (that constant velocity particles in space do not slow down as they travel), by experimental verifications that relativistic mass increase depends on velocity and not acceleration, and by the fact that matter possesses the property of resisting acceleration.[20]

The Investigation Begins

"If you really want to grasp the truth with both hands you have to be willing to completely let go of everything you know."

David Cantu

Any of these arguments should be compelling enough to warrant a thorough investigation of the axioms we use to define spacetime's structure, but when we consider all these arguments together (and by no means have we considered them all) the possibility that the vacuum is quantized is worth our pause. With this footing, we shall now begin our construction of a model of physical reality based on the idea that the vacuum is a superfluid—that it has a quantized structure.[21]

To the best of my knowledge, past attempts to incorporate quantization into an axiomatic framework have been treated as either metaphorical or merely mathematical. As a result, none of those investigations have achieved the epistemological or ontological clarity that a deductive scientific theory aims for.

That opportunity motivates the introduction of quantum space theory (qst), which stands as a specific interpretation of the more widely known, but sparsely assembled superfluid vacuum theory (svt). Starting from the assumption that the vacuum is a superfluid composed of quanta we shall now investigate whether or not this assumption sheds new light on the conundrums of modern physics, and offers a deeper understanding of Nature.

In the last years of his life, Einstein proposed giving up the idea that space and time are continuous, but the imagination of his youth had faded and he was unable to visualize such a structure. In reference to this he said, "I cannot imagine how the axiomatic framework of such a physics would appear... But I hold it entirely possible that the development will lead there." He also said, "I consider it quite possible that physics cannot be based on the field concept, that is, on continuous structures."[22]

Einstein quantized the world of matter, now it is up to us to see what happens when we extend that quantization to the vacuum itself. It is time for us to further Einstein's work, and to visualize how Nature appears in higher dimensions.

If you were taught that visualizing more than three dimensions of space simultaneously is impossible, then note that you are about to do the impossible. We are about to break free from the Euclidean limitations that have, until now, kept our intuition at bay. It is time to explore a dimensionally richer map that is capable of translating the great beyond, or as Karl Jaspers might call it, "authentic reality,"[23] to our sensory experience. It is time to set our heading towards a new isle of thought.

"There lies the high adventure for later generations, often mourned as no longer available. There lies great opportunity."
E. O. Wilson [24]

[1] Gary Zukav. *Dancing Wu Li Masters—An Overview of the New Physics*, p. 207.

[2] If you count the chirps from a single cricket (specifically from the snowy tree cricket, which is common in the United States) during the span of 14 seconds and add 40 to the number, you will end up with a number that corresponds to the temperature in degrees Fahrenheit. For example, 33 chirps in 14 seconds means 73 degrees Fahrenheit because 33 plus 40 equals 73.

[3] Evidently this concoction originated from Jimmy Kirkman, the state's paleontologist, but I'm not sure if "uncle Billy" had any relation to Jimmy. Martha worked with Jimmy. We all knew him because he would participate in our digs from time to time.

[4] The sky over Grand Staircase Escalante is nearly the darkest in the country. In fact, it is hardly distinguishable from the sky that stretches over the nearby Natural Bridges National Monument, which was the first park to receive the designation of "International Dark Sky Park" from the IDA (International Dark-Sky Association). The only other park to receive this designation in the U.S. is Cherry Springs State Park in Pennsylvania. On the Bortle scale, which correlates pristinely dark skies to the number one and inner-city light polluted skies to the number nine, Natural Bridges is rated a class 2.

[5] Manfred Requardt. (2003, March 25). A Geometric Renormalisation Group in Discrete Quantum Space-Time. arXiv : gr-qc/0110077v3, p. 4.

[6] Richard Feynman. (2007, June). *Lectures on Physics*, Introduction; Alex Stone, The secret Life of Atoms—Until Recently We Couldn't Even See Them. *Discover*, p. 52.

[7] Avicenna. (1983). Shifa', Kitab as-Sama' at-tabi'i, S. Zayed (ed.) Cairo: The General Book Organization.

[8] Jan Ambjørn, Jerzy Jurkiewicz & Renate Loll. (2008, July). The Self-Organizing Quantum Universe. *Scientific American*. pp. 42–49.

[9] Franck Laloë. *Do We Really Understand Quantum Mechanics?*, p. 52.

[10] Ibid., p. 64.

[11] Quantum mechanics overlays itself on top of Euclidean projections for the vacuum, but captures the notion of quantization in terms of energy. This quantization arises from boundary conditions that allow us to normalize the wavefunction and recover a probabilistic interpretation. For example, the energies of a simple harmonic oscillator are expressed as:

$$-\frac{\hbar^2}{2m}\frac{d^2\psi}{dx^2} + \frac{1}{2}m\omega^2 x^2\psi = E\psi(x)$$

Instead of capturing a discrete relationship for allowed energies this equation shows a continuum of solutions. In order to end up with discrete energies, $E_n = \hbar\omega\left(n + \frac{1}{2}\right)$, the Schrödinger equation needs to be supplemented with another assertion—appropriate boundary conditions that keep it from blowing up at infinity. To normalize the wave function and allow for a probabilistic interpretation we must set the boundary condition that $\psi(0) \to 0$, $as\ x \to \pm\infty$. The solutions that conform to these boundary conditions have a discrete set of energies.

[12] Ice has at least 20 different forms. The dominant crystalline structure of ice found on Earth is called 1h (pronounced "one H"). It is a hexagonal structure in which the molecules have regular spaces between them creating a low density of 0.53 ounces per cubic inch. (A cubic inch of water weighs 0.58 ounces.) The empty space in the lattice structure of ordinary ice (1h) makes it possible to rearrange the lattice in 16 different ways corresponding to 16 different crystalline structures (1h–16h). At temperatures colder than -36.4 °F, water can take on a cubic structure 1c. There are also three principal forms of amorphous ice, which are usually found in interstellar space.

[13] Neil DeGrasse Tyson. *Death By Black Hole*, p. 180.

[14] Gary Zukav. (1980). *Dancing Wu Li Masters—An Overview of the New Physics*. Harper Collins, pp. 50-51.

[15] Neue Bahnen de physikalischen Erkenntnis. (1913). trans. F. d'Albe, Phil. Mag. Vol. 28, 1914; Gary Zukav, *Dancing Wu Li Masters—An Overview of the New Physics*, pp. 50–51.

[16] In 1905, the year often referred to as his annus mirabilis, Einstein used what little spare time his job as a Swiss patent clerk afford him to rewrite the way humanity would see the world. He submitted his ideas to the *Annalen der Physik* in hopes of gaining enough recognition to earn him a teaching position. Evidently he really wanted the job.

On March 17, 1905 Einstein submitted his first paper of the year titled, "On a Heuristic Point of View Concerning the Production and Transformation of Light." Heuristic means a hypothesis that serves as a guide and gives direction in solving a problem but is not considered proven. Today this paper is commonly referred to as his photoelectric effect paper. His second paper was completed on April 30, 1905, submitted to the University of Zurich on July 20, 1905, revised and then submitted to the *Annalen der Physik* on August 19, 1905. It wasn't published until January of 1906. The paper was titled "A New Determination of Molecular Dimensions." In it, Einstein assumed molecules were real physical entities and he calculated their size. On May 11, 1905 Einstein completed his third paper but waited until August to submit it. In this paper Einstein used Brownian motion to verify that the world is made of atoms—something that was highly debated until then. Einstein's fourth paper was titled "On the Electrodynamics of Moving Bodies." The *Annalen der Physik* received this paper on June 30, 1905. This landmark paper gave birth to special relativity and it forever shattered the notion of universal time.

Almost as an afterthought, Einstein wrote another paper as an addendum to the fourth. In this paper titled "Does the Inertia of a Body Depend on Its Energy Content?" Einstein penned the most famous physics equation of all time: $E = mc^2$. (The full equation is $E = \pm \lambda mc^2$ where $\lambda = \frac{1}{\sqrt{1-\frac{v^2}{c^2}}}$.) This paper was received by the *Annalen der Physik* on September 27, 1905 (Walter Isaacson, *Einstein*, p. 94, 101–105, 127, 138, 577). (Friedrich Hasenöhrl, an Austrian physicist published the equation $E = mc^2$ a year before Einstein, but he failed to relate it to a principle of relativity.)

All of these ideas were groundbreaking, but the one Einstein eventually received the Nobel Prize for was his paper on the photoelectric effect—not his theory of relativity. "Bitter nationalist sentiments of the post-World War I era played a role, but basically relativity proved to be too radical a concept for the Nobel committee. In eleven different years, Einstein was nominated over and over only to be rejected. One Nobel committee member wrote, 'Einstein must never receive a Nobel Prize even if the entire world demands it.' The world did demand it, and Einstein was awarded the 1921 Nobel Prize for his contributions to physics and for his 1905 paper on the photoelectric effect. He showed that light behaves not only as a wave but also as a stream of particles, or quanta. The committee directed Einstein not to mention relativity in his acceptance lecture. He did so anyway." Heidi Schultz, "Nobel Efforts," *National Geographic*, May 2005.

[17] Theodore A. Jacobson & Renaud Parentani. (2005, December). Black Holes. *Scientific American*, p. 70.

[18] Robert Kunzig. (2004, September). Testing the Limits of Einstein's Theories. *Discover*, p. 60.

[19] Grøn, Ø. & Hervik, S. (2007). *Einstein's general theory of relativity: With modern applications in cosmology*, Springer.

[20] A. A. Martins. (2012). Fluidic Electrodynamics: On parallels between electromagnetic and fluidic inertia. arXiv:1202.4611 [physics.flu-dyn].

[21] Some examples of theories that mathematically address quantization in one way or another can be found in Appendix A.

[22] Abraham Pais. (1982). Subtle is the Lord, Oxford University Press, New York.

After denying the concept of an immobilized ether, Einstein did return to an ether-like expectation—with different and specific properties that accounted for general relativity, where he associated the metric of space with the physical properties of the vacuum. (Einstein, A., "*Sidelights on Relativity—Ether and the Theory of Relativty + Geometry and Experience*," 1922, Elegant Ebooks, 2004; Kostro, L., *Einstein and the Ether*, Aperion 2000; Granek, G., "Einstein's Ether:Why did Einstein Come Back to the Ether?," *Aperion* **8** (3), 2001; Martins, A. A., "Fluidic Electrodynamics: On parallels between electromagnetic and fluidic inertia," arXiv:1202.4611 [physics.flu-dyn], 2012.)

[23] See *The Way to Wisdom*, by Karl Jaspers, translated by Ralph Manheim (New Haven, Conn.: Yale University Press, 1951) Chapter IV, The Idea of God, pp. 39–51.

[24] E. O. Wilson. (1998). Consilience: The Unity of Knowledge, p. 295.

Part Two—The Framework of Quantum Space Theory

> *"…creating a new theory is not like destroying an old barn and erecting a skyscraper in its place. It is rather like climbing a mountain, gaining new and wider views, discovering unexpected connections between our starting point and its rich environment. But the point from which we started out still exists and can be seen, although it appears smaller and forms a tiny part of our broad view gained by the mastery of the obstacles on our adventurous way up."*
>
> *Albert Einstein*

> *"Every great advance in science has issued from a new audacity of imagination."*
>
> *John Dewey*

Chapter 5 — **Absolute Volume**

"We are therefore quite at liberty to suppose that the metric relations of space in the infinitely small do not conform to the hypothesis of geometry; and we ought in fact to suppose it, if we can thereby obtain a simpler explanation of phenomena."

Georg Friedrich Bernhard Riemann

"Now, these properties were all along naturally inherent in the figures referred to… but remained unknown to those who were before my time engaged in the study of geometry."

Archimedes

"It's as if you have seen beyond the way things are, seen a higher possibility.*"*

Douglas J. Soccio

Humboldt Sink, Great Basin Desert, Nevada.

Our bicycle tires sang in a rhythmic trance, belting out the same low note every time we rode over the etched asphalt. These grooves were designed to alert otherwise distracted motorists that they had veered onto the emergency lane. In Germany, the same principle etches the national anthem into a stretch of the autobahn. As a consequence, every set of tires on that route secondarily becomes a musical instrument. Here on I-80, passing through the deserts of Utah, Nevada, and California, the tires of our mountain bikes would only change their pitch if we pedaled slower or faster.

Occasionally, a string of Harleys would scream past us—loud pistons, leather jackets, bandanas, and the casual thumbs up. But for the most part, our own tires set the rhythm.

My high school sweetheart and I were pedaling from Salt Lake City to San Francisco to complete a fund-raiser we had organized for the research department of the Cystic Fibrosis Foundation. With 80 to 110 miles per day to go, there was plenty of time to fall into a deep trance. The rhythmic pedaling, the continuous chasing of the next mirage, and of course, the endless humming of the road encouraged our reverie.

As we rode through Nevada's Humboldt Sink, I pondered some of Nature's greatest unsolved mysteries. I thought about what Einstein searched so diligently for but never found—a complete map of physical reality. As my legs continued their rhythm, I wondered if it would ever be found. Then it occurred to me that the answer to that question directly depends upon whether or not it is really impossible for the human mind to grasp, or conceptually comprehend, higher-dimensional metrics.

Every physics book I have ever read, and every reputable article or website I have ever gleaned, seemed absolutely convinced that it's impossible to imagine a world consisting of more than three spatial dimensions. This claim is justified by the belief that "our brains are simply incapable of imagining additional dimensions".[1] It is not at all obvious to me that this is true. On the contrary, I think we have shown that we are capable of far more than our utilitarian mold. Humanity has transcended many shortsighted 'limitations,' and has managed to break free from many of its self-imposed conceptual bonds. The fact that we are staring down the next great conceptual barrier, whose imposing presence is related to us through the rather arrogant announcement that "this is humanity's intellectual end," is no reason to believe that we will not overcome this barrier.

Many of the greatest human stories have been about individuals and groups that overturned the dogmatic and deep-rooted beliefs of their day. Achievements often require bold defiance, courage, and an imagination that extends beyond the world-view of those who preach intellectual gloom. Progress comes from those who have the ability to see beyond the limiting rules and concepts that frame the current understanding. Science is supposed to provide a general recipe for this process. As Gary Zukav puts it, "the history of scientific thought, if it teaches us anything at all, teaches us the folly of clutching ideas too closely".[2]

In times past, people would have never believed it possible for humans to comprehend, let alone manipulate, the genetic fabric of life. It was considered impossible for humans to set foot on the moon, take to the sky like birds, or to understand the motions of the heavens. Victorian physicist Lord Kelvin flatly claimed that it was impossible to construct viable "heavier-than-air" craft. And Ernest Rutherford, who discovered the nucleus of the atom in 1911, quickly dismissed the idea that the energy stored in that nucleus could ever be released.[3] The lesson here is that our belief in these 'limitations' reinforces their ability to hold us back.

"There ain't no rules around here! We're trying to accomplish something!"

Thomas Alva Edison

As my cyclic trance continued, I began to realize that the first appropriate step toward overcoming the dimensional conceptual barrier is to recognize that there is no inherent reason that we cannot imagine more than three spatial dimensions at once. Acquiring a way to intuitively grasp additional dimensions is simply a matter of exploring new conceptual portals.

Passing Through Flatland

In 1884, Edwin A. Abbot made a noble effort to open our minds to the possibility of higher-dimensions when he published *Flatland: A Romance of Many Dimensions*. In this book, he depicted a three-dimensional sphere that intersects Flatland and attempts to explain the notions of "above" and "below" to a two-dimensional square. As somewhat of a Flatlander intellectual, the square tried to comprehend, but confused these notions with the directions that made sense in his familiar two-dimensional worldview—*left* and *right*, *forward* and *backward*. Determined to help the square understand, the sphere decided to demonstrate how he can move about in the third dimension by passing through the square's two-dimensional world. As he did this, the square saw the circle he'd been observing get smaller and smaller until it miraculously disappeared from Flatland (Figure 5-1). No matter how hard the sphere tried to enlighten the square, it seemed that the square was just incapable of comprehending.

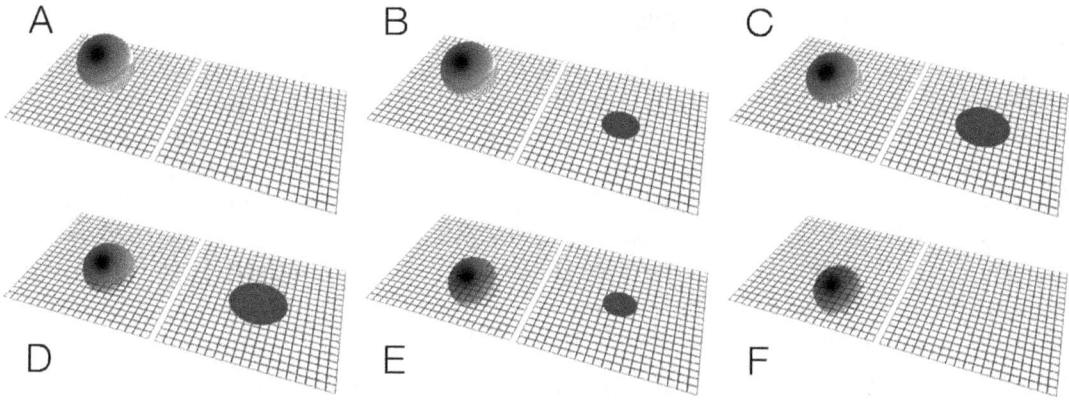

Figure 5-1 A sphere passing through Flatland

Finally, the sphere pulled the square out of his two-dimensional world (against his will) into the three-dimensional world of Spaceland. Suddenly, the third dimension that the square had so desperately tried to comprehend was now visually accessible. This was a completely new perspective. Everything the square had previously believed to make up all of physical reality was only part of an already present, richer whole. The square gazed upon his fellow Flatlanders from this more complete vantage and saw more of them than he had ever before imagined; he saw their insides.

> *"...now my mind has been opened to higher views of things."*
>
> *Edwin Abbot*[4]

From this perspective, occurrences that were fundamental mysteries in Flatland became intuitively accessible and predictable. This fact impressed the square so much that when he heard that there were unexplained mysteries in Spaceland, he began to argue for the possibility of a fourth dimension (and eventually an infinite number of dimensions) to explain them to the sphere. The square used the exact same arguments the sphere had used to convince him that a third dimension existed, but the sphere remained intransigent.

The square was convinced that further wonders, like the unexplained mysteries of Spaceland, would find their solutions in higher dimensions. But no matter how he tried, he failed to come up with an insight that would give him visual or intuitive access to these higher dimensions. The sphere didn't even try to search for this conceptual portal, because as far as he was concerned such an insight was impossible, or as he put it "utterly inconceivable".[5] Repeating what the sphere had told him earlier, the square tried to reason with the sphere by saying, "Doubtless we cannot *see* that other higher Spaceland now, because we have no eye in our stomachs."[6]

Abbott's point was that we are like the intransigent sphere. We cling to the belief that our brains cannot simultaneously visualize more than three space dimensions, and because of this we don't even try. The martyr of his book tried his best to save us from such complacency. He said, "I—poor Flatland Prometheus—lie here in prison... [and] exist in the hope that these memoirs, in some manner, I know not how, may find their way to the minds of humanity in Some Dimension, and may stir up a race of rebels who shall refuse to be confined to limited Dimensionality."[7]

If we are to be that "race of rebels"—if we are going to keep our minds open to the possibility of higher dimensions—then we also need to know how to protect ourselves from being tricked by false claims of success. If someone discovered a conceptual portal to higher dimensions—if they figured out a way to comprehend more than three dimensions of space simultaneously—how would we know that their discovery was genuine? In other words, if someone showed us how to conceptualize a realm of higher dimensions, how could we know that we weren't being tricked? If we thought ourselves to be intuitively grasping more than three spatial dimensions at once, how would we verify that the spatial dimensions we were referring to were really additional spatial dimensions?

To answer that question, we simply have to stare directly at the definition of spatial dimensions. All spatial dimensions facilitate completely unique (orthogonal) motion, by providing completely unique descriptors for position (Figure 5-2). New spatial dimensions must provide ways to move that don't involve motion in the old dimensions: ways to change position without changing position in the familiar dimensions. Therefore, if we discover a way to move about while holding still in x, y, and z (the extended familiar spatial dimensions), then that motion must be taking place in other dimension(s).

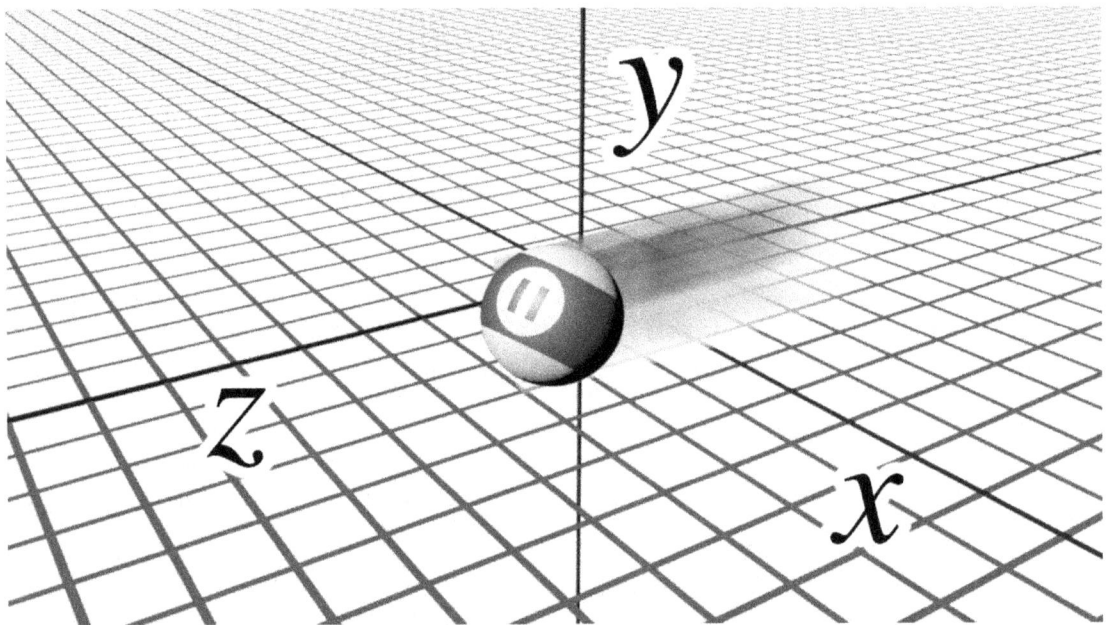

Figure 5-2 Example of motion in z

As an example of this, imagine an object whose x, y, z spatial location happens to be at the center of a stationary hollow sphere. By definition, the distance between this central point and every point on the sphere's surface is equal. Therefore, all we have to do to conclusively locate other dimensions of space is discover a direction in which it is possible for that object to move (change position) while staying equidistant (in the x, y, z sense) from all points on the stationary sphere's surface.

People have long claimed that it is utterly impossible for us to visualize such a thing. Rick Groleau succinctly sums up the modern opinion on this in his article "Imagining Other Dimensions" with the simple words "you can't do it".[8] But the truth is, all we have to do to gain access to additional dimensions is give up the Euclidean assumption of vacuum continuity. We simply have to assume that the vacuum is quantized (Figure 5-3).

When we do this, we end up with a metric that contains three different kinds of volumes. One of those volumes defines the regions inside the quanta (three intraspatial dimensions), another defines the voluminous region that the quanta are dispersed through (three superspatial dimensions), and the familiar x, y, z volume is constructed as a collection of quanta. These quanta represent fundamental locations, or points. The x, y, z tapestry is meaningless unless it references several points. It gains coherent substance as a collection. In a quantized vacuum an object does not move through x, y, z space, it does not change its position in the vacuum, until it goes from occupying one quantum of space to another. To move in a medium you must change locations in that medium.

Figure 5-3 The vacuum up close

We are going to be going over this in great detail in the next few chapters, so if the insight hasn't leapt out at you yet, just hang in there. Einstein once said that "knowledge of the existence of something we cannot penetrate," something that escapes our perceptions but is grasped by reason, is what "constitutes true religiosity." It is this knowledge and this emotion that opens up the sincere explorer to "the profoundest reason and the most radiant beauty." Expanding the reach of our intuition into a richer geometry of physical reality gives us the potential to change everything we believe about the world and ourselves. But undertaking a quest like this—one so heavily guarded with warnings of potentially volatile consequences both intellectually and emotionally—is not for the light-hearted. It is an undertaking reserved only for those that have been smitten with divine curiosity, a calling reserved only for passionate explorers and true searchers.

> *"Of all the communities available to us there is not one I would want to devote myself to, except for the society of searchers."*
>
> *Albert Einstein*

Changing your dimensional perspective can take a bit of practice. But gaining a more encompassing vision of Nature's character is more than worth the effort. Before we discuss how the act of quantizing spacetime provides us with new directions in which to move, without moving in the familiar dimensions of space (or in our sphere example, how it allows us to remain in the exact x, y, z location defined as the sphere's center while moving about in other directions), let's take a closer look at Newton's and Einstein's formulations, and then compare them to this new perspective.

Excavating the Foundation

Beneath Newton's framework of physical reality lies the notion of *absolute space*, an ultimate background, a picture of space that holds it to be homogeneous on all locations and on all scales. From these Euclidean assumptions, Newton built a precise and consistent framework from which he formulated the physical laws he famously published in *Principia* (for example: the law of inertia, $F = ma$, "for every action there is an equal and opposite reaction," etc.).[9] This framework beautifully describes the realm of our common experiences, but it fails to color in the details of Nature that lie outside of those experiences (relativistic and quantum phenomena). For this reason, the assumptions of Newton's map must be called into question.

Realizing this, Einstein restructured the geometric foundations of our map of Nature by replacing *absolute space* with a new notion called *absolute spacetime*.[10] This foundation rejects Newton's claim that the parameters of space are identical at all locations. Instead, it asserts that the geometry of space can vary from place to place (with a dependency on its proximity to mass and the amount of that mass). In this picture, consistency is derived from how objects move through space and time. The more an object moves through space, the less it moves through time, but its total evolution through spacetime remains identical. This condition enables absolute spacetime to play the role of an ultimate background.

In Nature, one second of time turns out to be equal in magnitude to 299,792,458 meters of space. At opposing extremes, one object can move 299,792,458 meters while not moving through time at all, while another object can age one second without moving through space at all. The first object is moving at the speed of light and the second is moving at zero velocity. Because the experience of 299,792,458 meters of space is considered equal in magnitude to the experience of one second of time, it remains coherent to say that the spacetime evolution of both objects is identical. This ability to swap time and space falls out of the assumptions of absolute spacetime and underlies the entire construct of relativity.[11]

Einstein's assumptions about Nature shatter our notion of universal time, which is encoded by the Galilean principle of relativity. This upset is exactly what enables Einstein's framework to explain the relativistic effects that we observe in Nature—the large-scale effects that Newton's map cannot explain.

Einstein's adjustment is a remarkable improvement, but it fails to capture all of Nature's characteristics. For several decades, physicists have been collecting a vast array of observations that cannot be explained by Einstein's map of Nature. Although the majority of these observations are related to the microscopic realm, the fact that they cannot be explained by a foundation of absolute spacetime ultimately dooms the idea that Einstein's model is perfectly precise. Nature's fundamental structure simply cannot be fully represented by his assumptions about spacetime. There must be more to the picture. Nature's true foundation must go even deeper!

The success of any framework of physical reality depends upon the accuracy with which it reproduces the phenomena found in Nature. If the model perfectly matches Nature, then theoretical predictions and observation will always agree. On the other hand, if there are fine details within the model that differ from Nature's actual structure, then observations will eventually be made that cannot be explained by that model. When such observations are made, we must be willing to restructure the foundation of our map in order to progress towards a more accurate representation of Nature.

When Einstein recognized that our understanding of the mortar of physical reality was still incomplete, he accepted the challenge of reshaping that mortar. He was attempting to map the true structure of Nature, without presuming that the structure should be as his predecessors had thought. This quest required him to pry loose that which had successfully remained entombed beneath the deepest chambers of the great pyramid of conviction. To accomplish this task, he began playing around with some new dimensional tinker toys.

Our goal is to follow Einstein's path and to continue the quest by attempting to fully expose a chronometric, higher-dimensional geometry that can account for the effects of both general relativity and quantum mechanics. To do this we must reexamine the mortar of spacetime and, once again, reshape it.

Einstein's notion of absolute spacetime challenges Newton's assumption that the parameters of space are identical in all locations, but it fails to investigate the other pillar of Newton's notion of absolute space—namely, whether or not the parameters of space are identical on all scales. Overcoming this restriction means restructuring the foundation of our map with the notion that the parameters of space are neither identical in all places, nor on all scales. One way to do this is to build a framework of physical reality based upon the idea that space is quantized.

This assumption challenges the validity of both pillars beneath Newton's notion of absolute space. It also introduces us to a new notion of absolute or dimensional consistency. We will call this new foundational principle "absolute volume."

The term *absolute volume* makes reference to the fact that, in a quantized picture of physical reality, the *minimum* number of spatial dimensions belonging to any immediate region will always be equal to three, but the specific dimensions that each locale immediately belongs to may vary. For example, when specifying a precise location we might be talking about a location within the volume of a single quantum (an intraspatial location), or a location within the volume that exists between quanta (a superspatial location), or we might just be referring to a specific quantum of space itself (a spatial location, which is defined in reference to a large collection of other space quanta that fill out a three dimensional volume). This adjustment to our axiomatic structure significantly impacts our attempt to explain the fantastic effects of both general relativity and quantum mechanics. (Note that on macroscopic scales, Einstein's absolute spacetime is a natural derivative of this deeper notion of absolute volume.)

The most important thing about this axiomatic adjustment is that it enables us to regain a picture of physical reality that is entirely intuitive. That picture carries us to some rather interesting philosophical implications, which are fully developed in Part Three of this book.

Now that we have discussed the procedure for restructuring our map of physical reality, let's jump right in and begin exploring the quantized realm that is framed by the notion of absolute volume.

Absolute Volume

"When it comes to the search for a unified theory, the lack of great progress over nearly a century implies a lack of sufficient inspection of underlying assumptions."

Iarens Imanyuel

To initiate our visualization of a quantized picture for spacetime, let's imagine the tapestry of x, y, z space as an enormous collection of identical spheres that, at least to first approximation, elastically collide and interact. The distribution of these spheres is not necessarily uniform (Figure 5-4). Each sphere (at least when time-averaged) is identical in volume and surface area; therefore, each sphere makes a discrete, equal and nonzero contribution to the entire collection of spheres—the medium. Now imagine that each one of these spheres is a discrete, elemental atom of space—or what we will call a quantum of space.

This replaces the concept of a continuum of space with what we shall call an *incontinuum*, a collection of quanta whose connectivity defines the geometric fabric of x, y, z. Within it, the smallest piece of x, y, z space—a 'point'—actually has volume! This adjustment is significant because it allows us to overcome some inconsistencies that plague continuous models (geometries that assume zero-dimensional points at their base).

Before we discuss those inconsistencies, let's examine this model a little closer. Starting with the idea that individual quanta represent discrete pieces of space, the dimensions of space that we are familiar with (x, y, z) come into focus as we consider a collection of these quanta. What about the dimensions within a single quantum, or the dimensions of the void around that quantum? How are we to interpret them (Figure 5-5)? To answer this question, we note that the familiar spatial dimensions (x, y, z) can only be macroscopically focused. They dissolve as we approach the quantum scale. On the quantum scale, every region in this map is either immediately taken up by the intraspatial volume within a quantum of space, or the superspatial volume that defines the void surrounding the space quanta (Figure 5-6). Volumes are always three-dimensional.

If space is quantized, then the precision of location in x, y, z space is limited by the size of the quanta. Despite this, it is easy to imagine locations in a quantized map that exist outside of the space quanta (superspatial locations), and positions within a single quantum of space (intraspatial locations). Neither of these kinds of positions can be addressed by x, y, z information. All three kinds of volume are orthogonally related. As a result we end up with three unique kinds of volume, each making a unique reference to entirely different descriptors of position and geometric arrangement in our map.

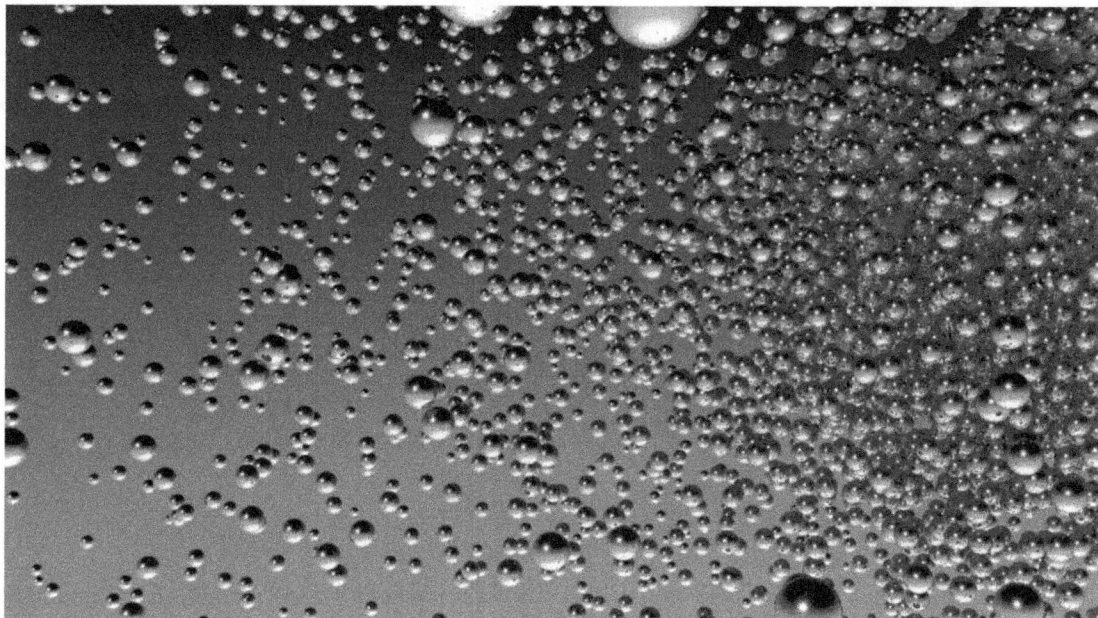

Figure 5-4 Quantization allows for varying vacuum densities

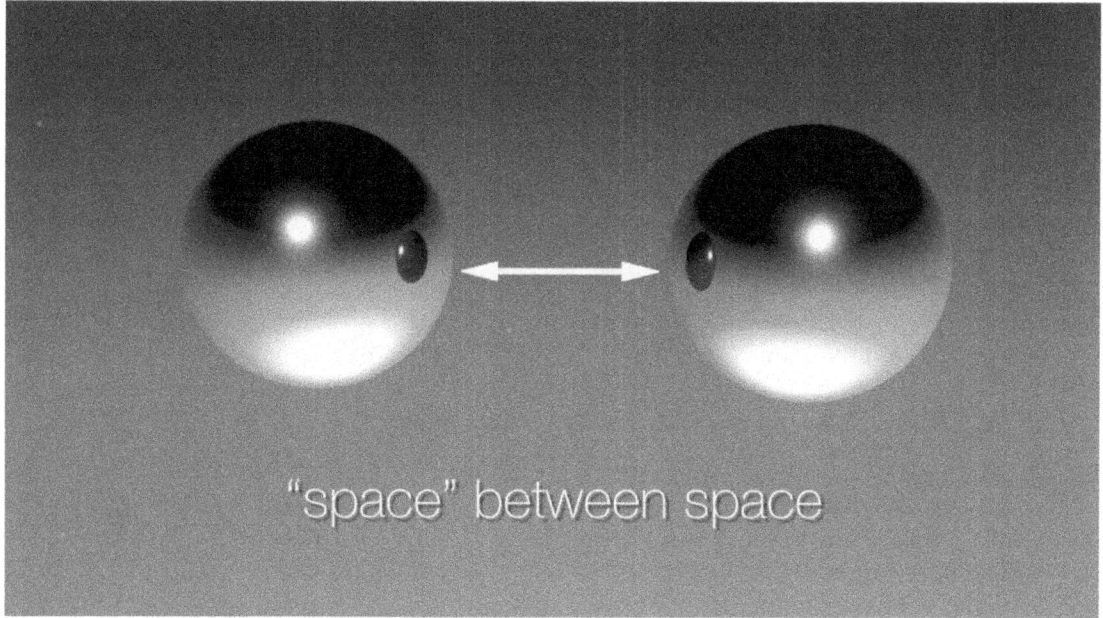

Figure 5-5 'Space' between quanta of space

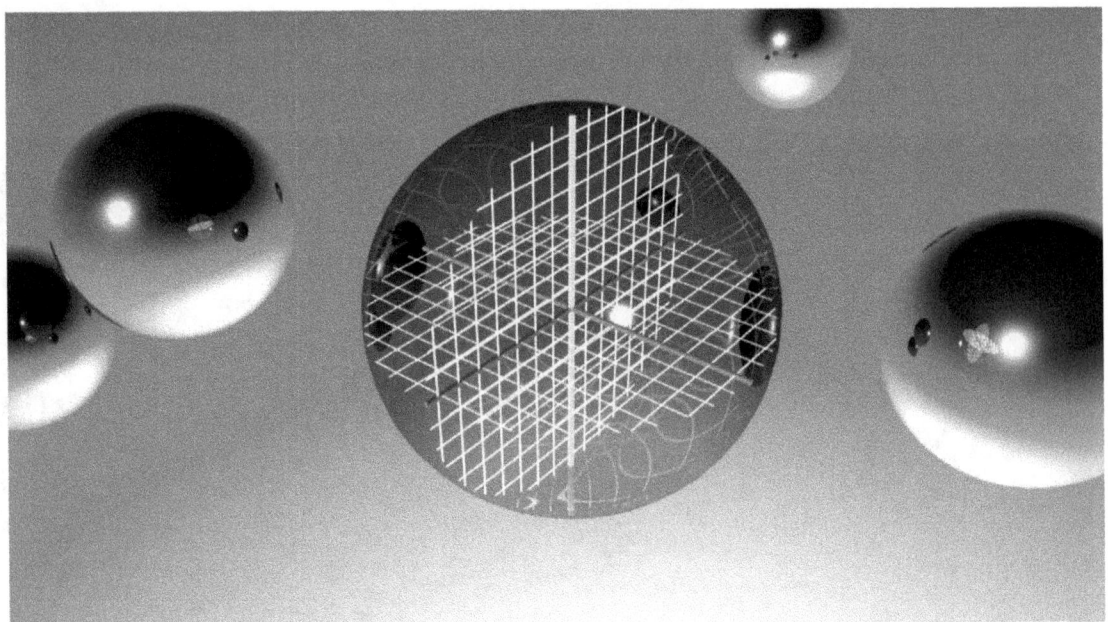

Figure 5-6 'Space' within quanta of space

Positions that lie within the volume of the individual quanta of space belong to *intraspatial* dimensions because they are internal to the familiar x, y, z dimensions of space. Positions that fall within the volume that is outside the quanta belong to *superspatial* dimensions. And when a change in position references a change in occupancy from one quantum in the vacuum fluid to another quantum, we are in the realm of *spatial* dimensions.

This geometric condition projects facets of Nature's character that are nowhere to be found in our previous maps. The question is, do these facets map some of Nature's mysterious qualities?

Continuous geometries, such as the Euclidean or Riemann varieties, have deeply embedded inconsistencies. They assume, as Euclid did, that a point in space has no dimension, a line in space possesses one, a plane of space two, and a volume of x, y, z space contains only three dimensions. Euclid believed that this three-dimensional geometric framework was complete and he proclaimed that there was nothing that enabled or allowed mention of more dimensions. The astronomer Ptolemy more vehemently broadcast Euclid's sentiments, claiming that a fourth perpendicular line is "entirely without measure and without definition. Thus the fourth dimension is impossible".[12]

If we restrict ourselves to Euclidean expectations, Ptolemy's claim seems natural. But Nature is not Euclidean. We might take this as good news, given that the internal structure of Euclidean geometry is incoherent (a weakness shared by all continuous geometries). This incoherence is captured by Euclid's notion of a line, which holds a line to be a one-dimensional construct made up of zero-dimensional points. A zero-dimensional point has zero spatial measure, which, by definition, means that it is not spatially related. Nevertheless, Euclid comes to the conclusion that a summation of these zero-dimensional points constructs a one-dimensional line in space. This does not follow. Any collection of these zero-dimensional points would continue to yield a zero sum because no matter how many times you add zero to zero you always have a total of zero.

This geometric problem extends beyond the construction of a line. For example, in Euclidean geometry a plane is defined as having zero thickness. If we take a large supply of these planes and begin stacking them on top of each other with zero space between them, can we construct a three dimensional volume? The answer is that without assuming some nonzero thickness for these planes, or leaving spaces between them—methods that both presuppose the third spatial dimension—we cannot build a three-dimensional construct. Put simply, a plane of zero thickness stacked with any number of identical planes still yields zero thickness. To claim otherwise is to expect that 0 + 0 + 0... eventually yields a nonzero value.

The lesson here is that we cannot explain the construction of the familiar three spatial dimensions if we assume that the fundamental geometric entity is a zero-dimensional point. We can't build a line in space from these points; therefore, we cannot accurately define what a line in space is. Likewise, we cannot build spatial planes from lines with length but zero width, nor can we build volumes from planes with zero thickness. A consistent and well-defined geometry necessitates fundamental constituents that possess a nonzero volume.

Knowing that all non-zero volumes are three-dimensional, we conclude that the fundamental 'points' or constituents of space (the quanta) are three-dimensional.[13] Therefore, each quantum of space must contribute a non-zero unit value to the metric of x, y, z space. Each quantum represents a fundamental location in x, y, z space, which means that the positions that fill out the volume of each quantum do not relate to positions in the $x, y,$ or z dimensions. Instead, these volumes are characterized by intraspatial dimensions. This is a consequence of vacuum quantization.

With this axiomatic set up, it is refreshingly simple to coherently construct a line segment in space. The quanta represent a non-zero contribution to the vacuum because they contain a three-dimensional volume (the three intraspatial dimensions), which move about within the three superspatial background dimensions. This makes a line in space seven-dimensional.[14] A line in the familiar dimensions of x, y, z space is a directional summation of space quanta that are dispersed throughout a superspatial volume. Thus, a spatial line extends itself through seven spatial dimensions (three intraspatial, three superspatial, and one familiar spatial dimension). A plane in space makes use of eight spatial dimensions (three intraspatial, three superspatial, and two familiar spatial dimensions), and a spatial volume makes use of nine spatial dimensions (three intraspatial, three superspatial, and three familiar spatial dimensions).

In other words, in order to describe a line in space—say in the x direction—or to fully describe the geometry needed to encode a spatial distance between two objects, we must make reference to a collection of space quanta. This collection is seven-dimensional because it involves the directional summation of three-dimensional 'points' of space, which are dispersed throughout the volume of superspace. Extending this, we find that a plane of x, y space is made up of eight dimensions, and a volume of x, y, z space requires nine dimensions.

In flat spatial geometries (geometries that don't contain any spacetime curvature) we may be tempted to overlook the information provided by these additional dimensions, or to represent the quantized vacuum in an averaged-over sense. But doing this forces us to accept a geometric foundation that is either internally inconsistent or completely vague. It also requires us to give up on the idea of ontologically accessing Nature's fundamental structure.[15]

Note that this construction allows us to simultaneously depict space with positive, zero, and negative curvature. *Curvature* references a change in the density of space quanta. A region whose macroscopic quantum density does not change possesses zero curvature. A region containing positive or negative density gradients has positive or negative curvature. This flexibility, the ability of this map to simultaneously accommodate effects held by Euclidean and non-Euclidean geometries, bodes well for its potential to simply portray complex vacuum states.

Vacuum quantization also exquisitely exposes additional geometric degrees of freedom; revealing how it is possible to change position while remaining completely fixed within x, y, z. In other words, the act of quantizing the vacuum resolves new dimensions, imparting new kinds of motion. It opens up space for tiny objects to move around within a single quantum of space (moving in intraspatial dimensions) (Figure 5-7), and for quanta to move around and change their arrangements (moving in superspatial dimensions).[16]

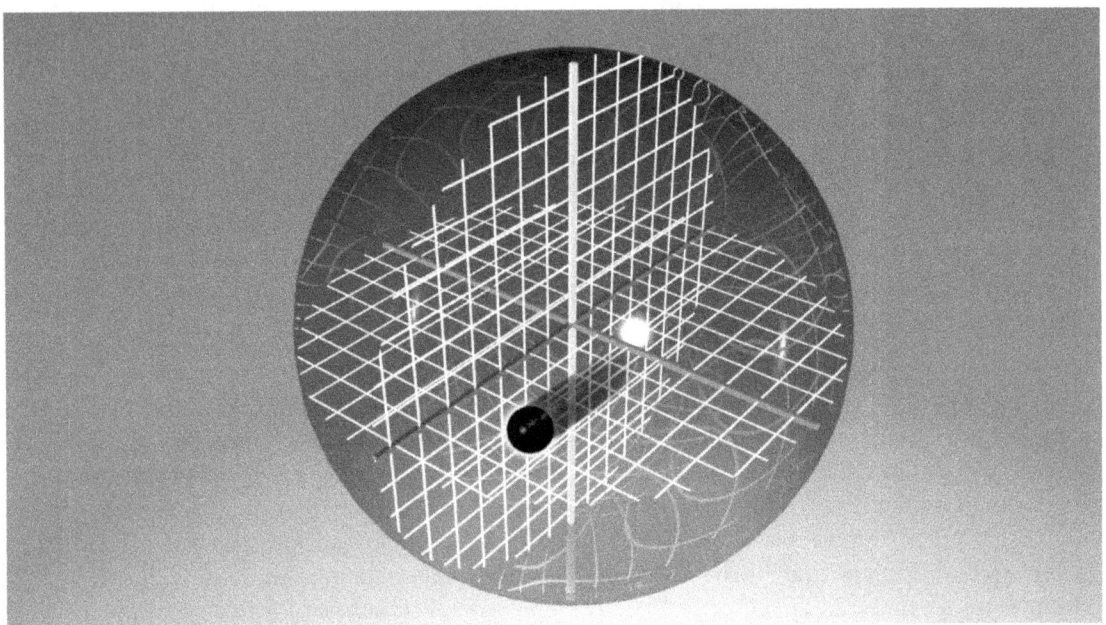

Figure 5-7 Moving about while remaining at the same quantum of space

Furthermore, vacuum quantization introduces microscopic details to our picture of Nature that provide us with fundamental definitions for space and time. If these new definitions enhance our understanding of Nature, then peering beneath the veil of Nature's mysteries may be a matter of intuitively absorbing the implications of vacuum quantization. For, as the great mathematician Georg Friedrich Bernhard Riemann said, "it is upon the exactness with which we follow phenomena into the infinitely small that our knowledge of their causal relations essentially depends".[17] With that in mind, let's explore those new definitions for space and time.

[1] Rick Groleau. (2003, July). *Imagining Other Dimensions.* WGBH. http://www.pbs.org/wgbh/nova/elegant/dimensions.html.

[2] Gary Zukav. (1980). *Dancing Wu Li Masters—An Overview of the New Physics.* Harper Collins, p. 251.

[3] Michio Kaku. (2008, April 5). Never Say Never. *New Scientist*, pp. 36-39.

[4] Edwin Abbot. (1884). *Flatland: A Romance of Many Dimensions*, p. 3.

[5] Ibid., p. 86.

[6] Ibid., p. 87.

[7] Ibid., p. 100.

[8] Rick Groleau. (2003, July). *Imagining Other Dimensions.* WGBH.

[9] In 1684, Newton wrote a manuscript fragment precursing Principia titled 'De moto cororum in nedus regulariter cedentibus' in which he proposed five laws. In Principia he simplified those laws. From Roger Penrose, The Road to Reality, p. 388, 410.

[10] The notion of absolute spacetime was initiated by Einstein's former calculus professor, Hermann Minkowski (who had once remarked that Einstein was a "lazy dog" based on his performance in class). Einstein was reluctant to accept the validity of Minkowski's absolute spacetime, but eventually recognized that it was essential for completely describing his theory of general relativity and mapping gravitational fields. In Minkowski's spacetime, the dimensions of space and time offset each other in such a way that any configuration still yields unity, $dl^2 = dx^2 + dy^2 + dz^2 - c^2 dt^2$. Written this way, this metric expresses a spatial solution because we have chosen a spatial signature where x, y, and z are positive and t is negative $(+, +, +, -)$. A temporal signature $(-, -, -, +)$ would yield a temporal solution, $ds^2 = -\frac{dx^2}{c^2} - \frac{dy^2}{c^2} - \frac{dz^2}{c^2} + dt^2$. Considering all scales of physical reality, additional dimensional considerations must be included to produce a metric of unity. Such higher dimensional unity, or what is called coordinate invariance, is a specific claim of qst and absolute volume.

[11] For non-accelerated objects, the square of an object's experience of space between two events (Δx^2) added to the square of the object's experience of time magnified by the speed of light squared ($c^2 \Delta t^2$) between those two events, will produce a consistent total no matter how fast it is moving. $\Delta x_1^2 + c^2 \Delta t_1^2 = \Delta x_2^2 + c^2 \Delta t_2^2$ In our example, $\Delta x_1 = 299{,}792{,}458\ meters$ and $\Delta t_1 = 0\ seconds$, while $\Delta x_2 = 0\ meters$ and $\Delta t_2 = 1\ second$. Therefore, $(299{,}792{,}458\ meters)^2 + c^2(0^2) = 0^2 + \left(299{,}792{,}458\ \frac{meters}{second}\right)^2 \cdot (1\ second)^2$. The more general equation for an object's expression in spacetime at two different speeds is: $\Delta x_1^2 + \Delta y_1^2 + \Delta z_1^2 + c^2 \Delta t_1^2 = \Delta x_2^2 + \Delta y_2^2 + \Delta z_2^2 + c^2 \Delta t_2^2$.

[12] Michio Kaku. (1995). *Hyperspace: A Scientific Odyssey Through Parallel Universes, Time Warps, and the 10th Dimension.* New York: Anchor Books, p. 34.

[13] When we consider a system made up of a large number of quanta, we can understand what physicists refer to as the vacuum state, which is dependent upon the density or the Lagrangian ($\mathcal{L}$) of the system and defines its dynamics.

[14] By this definition it becomes possible to define a 'point' in space as either three-dimensional or six-dimensional depending upon whether or not our description of the

point includes the three superspatial dimensions around it. If we are talking about more than one point, or quanta of space, then the superspatial dimensions are necessary.

[15] Theoretical physicist Lisa Randall suggests, "In life and in physics, we only register those details that actually matter to us. If you cannot observe detailed structure, you might as well pretend it isn't there." This has been a long held view, and it is in part the reason that quantum effects have remained unexplained. As we will see, if we assume that the small geometric details that are required by a quantized picture of space actually exist, then our inability to map all of Nature's effects simply disappears. For this reason alone we should not be so willing to ignore the fine details of Nature's map. In fact, according to Georg Friedrich Bernhard Riemann this is reason enough to suppose them. Lisa Randall. (2005), p. 39.

[16] Although people had thought it impossible to comprehend what simple objects would look like in higher dimensions, they had deduced that they would likely appear as a series of three-dimensional objects and cast a three-dimensional shadow. This description is quite insightful. According to our model, a hypersphere (a sphere, which is embedded in higher dimension) would indeed appear to be a series of three-dimensional spheres—a collection of quanta.

[17] Stephen Hawking. (2005). *God Created the Integers,* pp. 874-5.

Chapter 6 — Space

"It may seem alarming that our very notion of physical space seems to be of something that evaporates completely as one moment passes, and reappears as a completely different space as the next moment arrives!"

Roger Penrose

"Nature shows us only the tail of the lion. But I have no doubt that the lion belongs with it even if he cannot reveal himself all at once. We see him only the way a louse that sits upon him would!"

Albert Einstein

"... you see you do not even know what Space is."

The stranger in Flatland

Space Shuttle Simulator, Building 5, Johnson Space Center, NASA.

Space has been called "the final frontier"—the medium by which humanity will eventually connect to the most sublime mysteries. This can be interpreted to mean that humanity's ultimate destiny is to stretch across the cosmos, scouring every node of the seemingly endless labyrinth, exploring countless islands of wonder for new insights into Nature's character. This view sees the keys to our intellectual destiny scattered across the heavens, waiting for the explorers of distant generations. It imagines that the expansion of human imagination, the completion of our map, and any chance that we might have to secure a connection to the divinity of Nature, depends upon explorations that we will have to leave primarily to the unborn.

Personally, I find the idea of physically exploring the cosmos to be quite romantic. Undoubtedly, floating about the microgravity caverns of a forty kilometer asteroid, ice fishing Europa's twin around the eighth planet of Rigel 7, gently drifting above the methane lakes of Titan in a plutonium-powered hot-air balloon, watching a triple sunrise from the hot springs of a ringed terrestrial planet, and being chased by a gorilla-roo through the dense jungles of Gliese 581c with a red-dwarf sun high overhead, will be enlightening experiences. Nevertheless, I don't believe that going to the stars is the only way to explore space. In fact, despite my personal love of space travel, I think that the most insightful form of exploration is available to us right now. Right here, at this moment, we can take a journey that requires us to challenge everything that we believe. We can initiate an exploration that consists of more than just new and exciting ways to combine our stimuli. We can fan the spark of a new perspective, and through this we just might give way to an intellectual transformation.

To begin such a journey it is often useful to submerge yourself in unfamiliar domains. This process can stimulate our intellectual capacities and help us rethink yet another detail of the map we mean to color in. We must keep in mind, however, that new experiences can initiate the sort of revolutionary connection we are striving for, but they cannot carry us to the end game. To accomplish that task we must be willing to tear down our previous assumptions and try to see the world from a completely new starting point. What we are after requires more of an imagination boost than we will get from riding one more geyser into the catacombs of space, or discovering a new pattern in the shape-shifting ribbons of light that dance in our plasma streams. Such experiences will retain their value as long as they extend our exposure of the realm we are exploring, but they alone will not carry us to a full understanding of the final frontier. To accomplish that goal, we must spark an in-depth investigation of the essence of space itself. We must logically explore space.

We have taken the first step in this axiomatic transformation by rewriting the geometric assumptions that frame our thoughts. If we are on the right path, then the perspective that develops from this new starting point should eventually give us access to what is going on behind the scenes. And as we experience what it is like to have our outward reach amplified, we should find ourselves filled with a heightened sense of clarity.

The desire to send my feelers of intuition into the conceptual depths of space, like roots penetrating the secrets of a hidden Egyptian tomb, was kick-started by an event that took place in building 5 of NASA's Johnson Space Center. To my surprise, a string of fortunate events culminated with me commanding the Space Shuttle during a launch in the one and only Motion Based Space Shuttle Simulator used to train the astronauts. (Instead of pilot and co-pilot the left and right seats on the Space Shuttle are renamed *commander* and *pilot*.) I was strapped in the left seat with a five-point harness secondarily functioning to restrain my excitement. My pilot, who was a Shuttle Simulator operator, needed to run a full diagnostics check including a complete launch sequence and several landings. When I asked to come along for the ride, he suggested that I command the launch and landings. Needless to say, this was a dream come true.

Inside the cockpit I was visually transported to the launch pad at Cape Canaveral. The nose of the craft was pointed straight up, and the backs of our seats supported all of my weight. Operators at Mission Control announced over our headsets that they were initiating the countdown at T-minus ten minutes. The flight checklist, which was over 200 pages long, was hanging by a wire from the console. Its protrusion was a visual reminder that *down* was straight back. After I found the correct page, my pilot guided me through the prelaunch sequence. "Set **XMIT ICOM** to **VOX/VOX**. Set both **A/G, A/A,** and **ICOM** to **T/R**," and so on. Two thousand switches surrounded us. Some simply had on and off positions while others had mid-switch positions. Directly in front of me, a screen displayed a complete readout with predictor arrows. The hydraulic control stick, replacing the yoke in smaller craft, was impressively smooth and sensitive to the touch.

Using my hand to block the powerful glare of the sun through the windows, I could see the scaffolding and the launch pad superstructure around us. These windows were equipped with projector screens simulating the actual images from previous launches. At T-minus two-minutes, I set **APU AUTO SHUTDOWN 1/2/3** to **INHIBIT**, and my pilot instructed me to punch 505# into the center console. Mission Control confirmed the sequence and at T-minus 31 seconds the on-board computers took over: 28… 27… 26… My mind was racing. I couldn't believe it—I was about to launch into space! 9… 8… 7… At T-minus 6 seconds there was a loud roar as the three main engines fired up far below us. In response to their thrust, the whole vehicle rocked forward then back to vertical as the countdown completed: 3… 2… 1… There was no turning back.

The booster ignited and a sudden thump on our backs accelerated us upward. Everything began to violently shake as if we were riding a hard-tailed bike down an uneven, twisted mountain of stairs. It was impossible to hold back my smile. We automatically began our roll maneuver as soon as we cleared the tower. Forty-five seconds later, we reached max

Q—the maximum aerodynamic pressure on the vehicle that roughly coincides with reaching the sound barrier. (Max Q is the engineering term for maximum dynamic pressure, which signifies the maximum stress loading on a structure. Max Q is also an all-astronaut rock band.)

At this point our engines automatically throttled down, but the vibrations continued to grow stronger due to a pummeling of sonic shock waves on the craft's leading edges. This went on for about 15 seconds until we safely punched through the sound barrier. On the other side of max Q, the vibrations significantly reduced and the engines returned to full power. This increased the surge of acceleration we felt in our backs. Soon we were flying more horizontally than vertically and we were upside down, but because of our acceleration it still felt like *down* was straight back. The blueness of the sky faded rapidly. Seeing stars during the middle of the day punctuated exactly what was happening.

Gradually we could feel our acceleration tapering off as the supply of solid fuel became exhausted. Almost two minutes after lift-off, we heard and felt a *boom* as the explosive bolts fired and released the Solid Rocket Boosters (SRB's). Suddenly the vibrations ceased, but the rockets responsible for moving the boosters away from the shuttle briefly engulfed the entire cockpit in flames.

When the flames disappeared the magnificence of Orion caught my eye. Using my left hand to block the glare from the sun I saw hundreds of stars immediately jump out. "Electric drive," my pilot said, referring to the smooth ride we were now enjoying. "Look," I said, signaling for him to block the sun as I was. "It's the middle of summer and we can see Orion." The constellations were decorated with so many more stars than I was used to. (There were roughly 80 stars in the region of Orion's belt alone.) At first it was difficult to recognize them. When I looked closely I noticed that there was something to those stars that went beyond being little dots of white light. They had an additional brilliance—a gloss of color not seen from Earth.

"What's that?" he asked, pointing to the cluster to the right of Orion. "That's Pleiades," I said, "the cluster known as the seven virgin sisters."[1] From Earth I could see six stars in that cluster on a clear night. Now I could see more than thirty. "The native Americans used to use Pleiades to determine who had good enough eyes to be a scout," I added, "and the Japanese call the same constellation *Subaru*, which means *united*." "Like the car?" he asked. "Exactly, that's what the emblem is supposed to be."

We were still accelerating—being pushed by the three main engines, which were drawing their fuel from the external tank, partially simulating a force of three G's. Over the next six-and-a-half minutes, our speed increased to Mach 25 (17,500 miles per hour). After that, the external tank was depleted of fuel and it separated from the orbiter. Suddenly we were elegantly gliding through the heavens, smooth and steady. We drifted for a few moments watching the entire heavens rotate as I lightly nudged the stick one way then another, causing various thrusters to spew out yellowish-white bursts of vapor. I had dreamed of this experience for years.

Looking back at Earth from space, I found myself overwhelmed with a mixture of longing. I yearned to have this moment last forever. I longed to go farther—yet, strangely enough, I simultaneously longed to have my feet planted safely back on my blue home. Everything felt different. The seemingly endless planet, that I had only begun to explore, had been reduced to a tiny island floating about the heavens. Up here, I couldn't help but think that for the most part, the whole of humanity busily went about their lives without ever looking up—like ants in an ant farm that never questioned what's beyond the glass. Maybe, I thought, we all had just been sticking to our programming like characters in a video game, without focusing enough on the deep questions or attempting to extend our reach.

I realized that on this topic I was guilty of dragging my feet, and this realization threw me into the great human dilemma. Will it be safety and blinders, or danger and second sight? Could I reach beyond my world even if it meant watching it be reduced to a tiny outpost in the great expanse? Drifting in space the answer became clear. The inspiration that comes from realizing that the great beyond is knowable far outweighs the fear of personal insignificance. For me, it became clear that true insignificance would mean not immersing myself in the grandeur of looking up—it would be a result of not questioning what was beyond the glass.

> *"There are moments when one feels free from one's own identification with human limitations and inadequacies. At such moments one imagines that one stands on some spot of a small planet, gazing in amazement at the cold yet profoundly moving beauty of the eternal, the unfathomable: life and death flow into one, and there is neither evolution nor destiny; only being."*
>
> Albert Einstein

> *"That so few now dare to be eccentric marks the chief danger of our time."*
>
> John Stuart Mill

Just as these thoughts finished organizing in my head, the bittersweet moment came. "It's time to go back and land this thing," my pilot said. He walked me through the return to launch site (RTLS) abort sequence and handled the communication with Mission Control. After the shuttle completed its automated S-turns, bleeding off energy to slow down, we began our long descent. As we dropped toward the sound barrier, the entire vehicle began to shudder once again. This is where automation ended. I was instructed to take the stick and start guiding this winged brick down its glide slope. Predictor arrows indicated where my best path should be, so I followed the little arrows and placed the diamond inside the box on the screen in front of me. As we spiraled down, the shape of Florida slowly stretched beyond the horizon and the shores of the Cape gradually came into focus. We were coming down fast—a 20-degree slope—much faster than I had ever experienced in a private craft. It felt like I needed to pull up more, but I stayed on the projected glide slope.

As the runway came into view, I felt my heart quicken. "On altitude and on airspeed," my pilot said into his mic. I was so lost in the experience that it had become completely real to me. Every time I moved the stick I could feel the shuttle's response jerking me in the opposite direction. Mission Control was monitoring the descent, confirming that everything was on track: 300 knots... 290... The runway rushed closer and closer. I adjusted my feet on the rudder and brake pedals. 260 knots... 250... The computer still said everything was on track, but it looked like we were coming down way too fast—210 knots... The ground looked like it was only one story away as we crossed the threshold. My pilot lowered the landing gear as he watched my intense concentration. Finally, the predictor arrows moved upward to initiate the flare. I followed, but was a fraction of a second behind when—*bam*! The main landing gear hit the ground—before I expected—with a descent rate of 10.25 feet per second. It was pretty hard but it didn't seem to faze my pilot.

The cockpit was really high off the ground and the nose wheel was still looking at the sky. I had to jump on the toe brakes and give them a strong push. The nose wheel began to drop, and when it hit the ground it bounced a couple of times as I rode the brakes, trying to keep the vehicle going straight down the runway. My pilot pointed to the red toggle cover for the drogue chute. I flipped the cover open and pushed the protruding square button. The sudden deceleration threw us forward into our harnesses… 25 knots… 20… I stayed on the brakes and kept my hands off the stick… 5 knots… 3 knots… and we came to a complete halt.

What a rush! "Want to try it again?" my pilot asked. My facial expression answered for me. On my third landing I figured out the flare height and achieved a 0.0 descent rate at touchdown.

This experience planted the seeds of a new perspective in my mind. As I watered those seeds, dozens of philosophical questions began to make my thoughts dance. Eventually it became clear to me that humanity was in need of something—something that was capable of releasing them from their habitual worldview. What we needed was a way to escape our entrenched perspective, so that we could expand our own little worlds and connect to the great mysteries of the universe.

What we need is a conceptual model that will allow us to reach beyond the limits of our senses and enable our exploration of reality's hidden structure.

The Wilderness of Eternity

> *"What does a fish know about the water in which it swims?"*
>
> *Albert Einstein*

The epiphany I had about needing a new map furnished as much value in clarity as the shuttle launches provided in adrenaline. Because the details of our picture of Nature define the depth of our perspective of it, and because those depths amplify our connection to all of our experiences, I became increasingly interested in discovering the full texture of space. The details of our map must align with Nature's actual texture, but what is that texture like?

A partial answer to that question is that Euclidean geometry is only an approximation of the actual geometry of physical space. Because regions of spacetime possess irregularities (macroscopically manifested by warped spacetime in the presence of matter, and microscopically ever-present in the form of spacetime ripples and quantum jitters), we know that there must be more to the fabric of Nature than Euclidean frameworks reveal.

Today's physicists routinely evoke the mathematics of many different geometric structures —mapping Nature with more sophisticated approximations of its actual geometry. For example, gigantic configurations of complex particle interactions are described in what's called "phase space," Fourier components find their expression in "momentum space," wave-vectors whirl about in "reciprocal space," and the undulations of quantum mechanical wave functions are mapped within "Hilbert space." All of these geometries are higher-dimensional mathematical constructs that, unfortunately, fail to come with intuitive corollary views.

To do better than this, to fully reveal Nature's propensity for tensors (higher-dimensional vectors and gradients), we need to invest in a geometry that is capable of explaining the non-uniformity of spacetime. Otherwise, we face the possibility that the arcana in physics will remain forever shrouded in mystery and, at best, be known only through complex mathematics. Once we invest ourselves in this new geometry we should check to see if it offers us an intuitive description for how spatial measurements (like distance) derive their meaning from deeper foundations. Only a clear answer to this question will enable us to explore the hidden order of physical reality. With this goal in mind, let's consider how distance, or any measure of space, is defined from the eleven-dimensional perspective that we are proposing.

Distance and Measure

In 1791 the meter was crudely defined as one ten-millionth of the distance from the North Pole to the equator along the line of longitude that passes through Paris—not the most practical definition. In 1889 the meter was redefined as the length of a special platinum-iridium alloy bar held at 0° C, which was stored at the International Bureau of Weights and Measures in Sevres, France. In some ways this definition was an improvement, but when distance really matters—when you care about exactness or the definitive meaning of distance—even this prototype bar is far too crude to be useful. To improve its accuracy and usefulness, physicists redefined the standard for the meter again in 1960, when it was set equal to 1,650,763.73 wavelengths, in a vacuum, of light emitted by the unperturbed atomic energy-level transition 2p10 to 5d5 of the krypton-86 isotope. As Neil deGrasse Tyson jokes, this definition is "obvious, when you think about it".[2]

It eventually became clear that the only way to end up with an exact definition for the meter (or any other distance measurement) was to define it in relation to the parameters that are ingrained in the fabric of physical reality. The most thoroughly examined parameter capable of relating those dimensions to us is c—the speed of light. So in 1983, the General Conference on Weights and Measurements defined the meter as exactly 1/299,792,458 of the distance light travels in a vacuum during one second. But what does this *distance* reference? What does it mean to have a meter of space? How much space is that? And how is it different from having two meters of space? To answer these questions we have to know more about space.

Distance is traditionally defined in vague reference to the "amount of space" between two things, locations, or events. In other words, objects, locations, and events are considered distant from one another if they are separated by some amount of space. The more space between them the more distant they are. But what actually licenses us to say that the amount of space between Salt Lake City and Paris is greater than the amount of space between New York City and Paris? Paradoxically, in smooth metrics it turns out that we are *not* legitimately licensed to say this. Nothing in the metric itself gives us the ability to distinguish the two distances. To define distance we need to have an explicit definition for what space is. If we are going to define distance in relation to amounts of space, then we need some way to distinguish different amounts of space. This is something that smooth metrics, like the Euclidean metric, are incapable of doing.

To explore this point, recall that smooth metrics define space as being made up of zero-sized points. Consequently, they assume that an infinite number of points exist between all distant things, locations, or events. So how do such maps determine how distant two objects are? How do they distinguish greater distances from lesser ones? Exactly how much space does an infinite number of points of space refer to? If it refers to 8148 kilometers in one case and 5834 kilometers in another case, then what distinguishes those cases? Smooth metrics are not equipped to answer these questions, because they lack a clear definition for space, so we ignore their presumptions when communicating about spatial relationships in our day-to-day lives.

Quantizing the vacuum puts our practical application of distance in sync with our theoretical understanding of it, because quantization enables us to acutely define distance and to coherently distinguish different amounts of space. Once we assume the vacuum is quantized, the x, y, z distance between any two x, y, z locations becomes defined as the number of quanta of space arranged between them, multiplied by the spatial contribution of one quantum.

For example, in Figure 6-1 location B is four spatial locations away from location A. Because x, y, z distance is purely a measure of the amount of space quanta between two locations, the x, y, z distance from A to B is four times the value of one space quantum.[3] The amount of superspatial medium between space quanta does not affect x, y, z distance.

In addition to being ontologically clear, this definition for distance does not alter in the presence of spatial density gradients (curvature). Let's see why this is important.

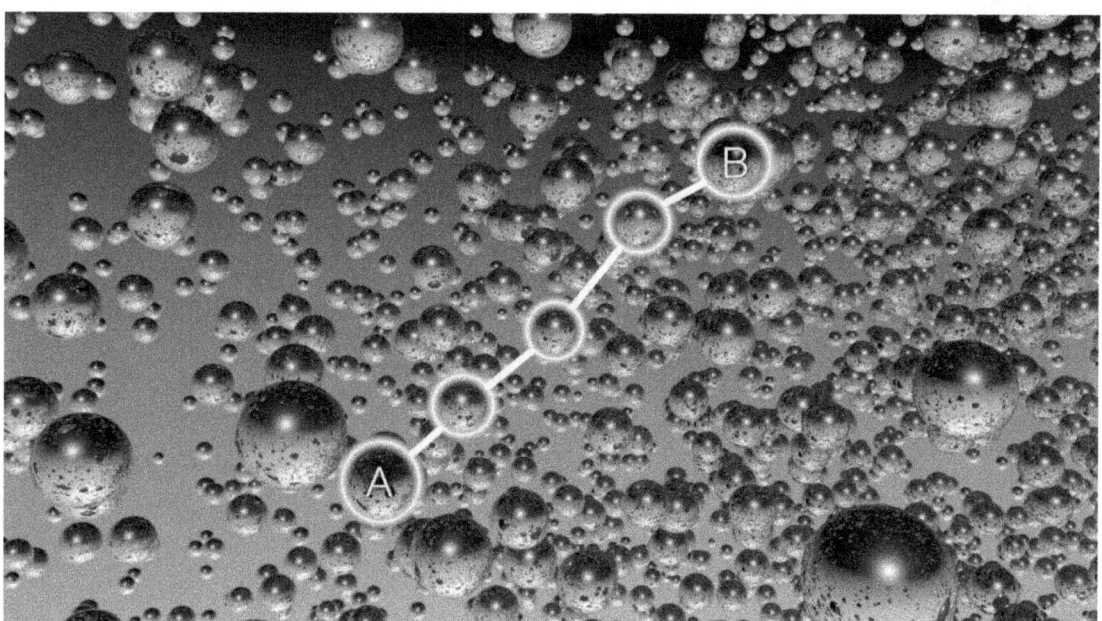

Figure 6-1 Resolving distance on the quantum scale

Tensors

In addition to giving us a natural way to specify and comprehend the meaning of distance—how much space lies between each of the 'points' of space in the map—quantization also removes the need to add tensors to our map. In a quantized depiction, density changes (curvature) can be fully mapped. No tensor is required. To grasp the import here, note that when we want to calculate the distance between two points on a flat map of the world, which actually represents the metric of a curved globe, a metric tensor is needed in order to adjust for how measurements change from region to region—to translate its real geometry to us. The meaning of distance on this flat map varies based on latitude, because the territory space is spread out or diluted as you get farther from the equator. The metric tensor tells us how to interpret that geometric dilution—how to define a relative notion of distance anywhere on the map. This is what tensors do; they translate territories from one geometry to another.

On a flat map of Earth we represent this geometric translation by separating the latitude lines at different intervals. The surface area inside of all of the sectioned off partitions of this grid are meant to represent identical portions, even though the partitions near the poles have significantly larger areas (Figure 6-2). The way that these grid sections stretch or warp on this map is defined by the tensor, which in turn is defined by the difference between the true geometry of the landscape (3D, or spherical) and the geometry on which it is being mapped (2D, or flat).

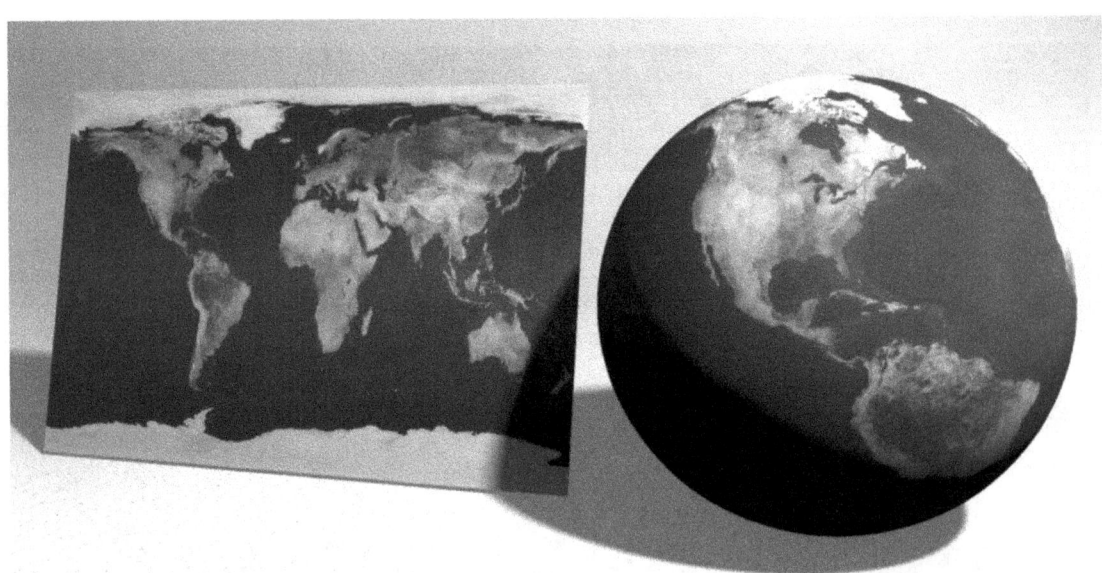

Figure 6-2 A metric tensor translates between the territory of a spherical map and a flat map.

Whenever a map uses fewer dimensions than the landscape it depicts, a metric tensor is required to translate between the geometry of the map and the true geometry of the landscape. If the geometry of the map is identical to the natural geometry, then tensors are unnecessary. With this in mind it is interesting to note that vacuum quantization eliminates, or at least trivializes, the need for a metric tensor when it comes to encoding state evolution and curved space.

If space is a collection of interactive *points*, then it will have a specific state, or configuration, that changes from moment to moment. The arrangement of those constituents and their superspatial velocities will determine how that state evolves. Instead of deterministically capturing this evolution, the state vector $|\psi\rangle$ used in quantum mechanics statistically projects it. This suggests that this state vector is a composite, averaged-over, representation of those vectors. If this is the case, then the state vector should be considered a blurring tensor. The function of this tensor is to (magically) charge the vacuum (a four-dimensional representation of the vacuum) with a propensity for state evolution, but it is incapable of revealing the causal structure behind that evolution.

Another tensor used to bridge a four-dimensional map of spacetime with the geometry of a quantized vacuum is the tensor field responsible for relating spacetime curvature. Both of these tensors are necessitated when we examine the vacuum through the lens of a four-dimensional map. Conversely, when we represent the vacuum as quantized we don't need tensors to endow the vacuum with the propensity for state evolution or curvature. This should stir anyone with an affinity for Occam's razor.

Focusing Out on the Vacuum

According to this model, on the quantum scale there is no vacuum. The vacuum only comes into focus as we zoom out. As we do this, we can take a collection of quanta in an arbitrarily chosen superspatial direction allowing that collection to fill out distance in the *x* direction. We map *y* in a similar way, as a collection of space quanta in orthogonal relation to the chosen *x* direction. Likewise, *z* maps in orthogonal relation to *x* and *y* (Figure 6-3). As more and more available positions are collected, the vacuum comes into focus.

Figure 6-3 Mapping x, y, and z on the quantum scale

As we zoom out, the geometry of the vacuum begins to take on an averaged-over, well defined, macroscopic identity, and the internal structure of the vacuum guides its state evolution. This is analogous to how a macroscopic volume of gas has a well-defined pressure even though its microscopic parameters determine its internal fluctuations. If we believe that the gas is made up of atoms or molecules, then those fluctuations do not contradict the presence of a well-defined pressure. Quantizing the vacuum comes with the same advantage.

A map that does not resolve the internal structure of the vacuum (e.g. a four-dimensional map) will have no means of determining how the geometric state of the vacuum will specifically evolve. Information about x, y, z, t tells us nothing about the superspatial velocities of the space quanta. If we had access to the full picture, if we knew how the quanta within the system of interest were arranged (spread out through superspace) and what their superspatial velocities were, we would be able to precisely predict the evolution of that system's state from moment to moment. Without that information the best we can do is attempt to represent the vacuum's general tendency to evolve by tacking a generalized state vector $|\psi\rangle$ to our description of it.

This state vector represents a blurred average of the vectors that encode the superspatial velocities of the quanta of space that make up the region of interest. This average explicitly assumes zero curvature. Within a four-dimensional projection, the state vector is a necessary descriptor of the vacuum state because, in an averaged-over sense, it encodes the fact that points in space do not remain fixed (connected to any other point in an unchanging geometric fashion). It encodes a general propensity for state evolution. Without switching to a higher-dimensional picture, without quantizing the vacuum, this is the best we can do.

"One cannot step twice into the same river, for the water into which you first step has flowed on."

Heraclitus

"Space can be empty of bodies without being empty of being."

Democritus of Abdera paraphrased by Douglas J. Soccio[4]

In addition to giving us a way to resolve a causal structure behind state evolution, quantization also reveals why there is no physical way to distinguish a state of rest from that of uniform motion within the vacuum. In a quantized vacuum we cannot convincingly claim that a specific point in space (like the electron at the end of your longest eyelash) is or is not located at the same point in space as it was one moment ago.

The absence of a unique constant velocity reference frame within space is reflected by the fact that we could describe the trajectory of that special point (at the end of our eyelash) as not moving. Or we might describe its trajectory by taking into account the motion of the spinning earth—moving nearly 355 meters to the east in every second if it were near Denver. That's faster than the speed of sound, which is ~346 meters per second.[5] If we chose to account for the motion of earth around the sun, then our selected point is moving almost thirty thousand meters (~18.5 miles) in one second.[6]

We could continue to alter our description of that trajectory by taking into account our solar system's motion toward the constellation Hercules, which is about 1.9 kilometers per second (1.2 miles per second), or by including our solar system's motion about the center of the Milky Way, which finds us trailing behind the orbit of Deneb at nearly 1/1000 c (231 kilometers per second, or 144 miles per second).⁷ But why stop there? We could go for a more complicated description of motion and include how the Milky Way moves within the local group,⁸ or the motion of the local group around the center of the Virgo cluster in relation to the vast Coma Super-cluster (roughly 600 kilometers per second), or the speed in which the Coma Super-cluster is moving toward the Great Attractor (between 500 and 600 kilometers per second), and so on.⁹

This dizzying process highlights the fact that spatial positions are relational. It is impossible to have an *actual*, *ultimate*, or *fixed* position in the vacuum, because in a quantized vacuum the mixing about of those quanta means that the vacuum state, its geometry and connectivity, is something that completely evaporates as one moment passes, reemerging the next moment slightly rearranged.¹⁰ Positions are defined in reference to the current vacuum state, but the vacuum state evolves. This is why x, y, z position is a relational parameter.¹¹

In summary, the hypothesis of vacuum quantization allows us to: simultaneously and intuitively access more than three dimensions of space, map state evolution and curvature without using tensors, precisely define spatial distances and amounts of space, and explain why spatial positions are inherently relational. The next step in our attempt to immerse ourselves into the knowable great beyond is to consider how to interpret time in our map.

> *"[The call of space] inspires me, like the call of a distant horizon, to seek what lies beyond. That it is knowable moves me to immerse myself in it [to discover] its grandeur and vastness and endless possibilities..."*
>
> *Carolyn Porco*¹²

[1] Pleiades is Greek for "to sail." The constellation lies approximately 440 light-years away from Earth. That means that the light we are seeing today from the stars of Pleiades has been traveling through space since Shakespeare and Galileo were toddlers. (2006, January-February). *Night Sky*. The star within the Pleiades group that appears the brightest from Earth is named Maia—the oldest of the seven sisters who, in Greek mythology became the mother of Hermes by Zeus. The month of May is named after the star Maia. (May is the only month named after a star.) The other stars that make up the group are: Electra, Atlas—the namesake of both "Atlantis" (where he was born) and the ocean that surrounded Atlantis which is now known as the Atlantic ocean, Poseidon—father of Atlas and the seven sisters and also the Greek God of the sea (the Romans called him Neptune), Merope, and Alcone. Then there are some other stars that are just a little too dim for most people to see without optical assistance: Pleione—the ocean nymph mother of the seven sisters, Asterope, Taygeta, and Celaeno. Over a hundred additional stars can be seen in the Pleiades cluster with just a small telescope or even large hunting binoculars.

[2] Neil deGrasse Tyson. (2007), pp.124-125.

[3] To give us a little perspective on just how small a quantum is, let's use a comparison. The volume of one quantum compared to the volume of one hydrogen atom is roughly equivalent to the difference in volume between one hydrogen atom and a sphere with a radius 100,000 times the distance from the Sun to Pluto (which is approximately one eighth of a light-year)! What is amazing about this is that even though we are enormously withdrawn from the quantum scale we can easily visualize and internalize higher-dimensional quantum geometries through the quantization of spacetime. This, in turn, gives us a richer understanding of distance.

[4] Douglas J. Soccio. (2004). *Archetypes of Wisdom: Introduction to Philosophy*, 5th Edition Wadsworth, p. 66.

[5] To calculate how many meters you move per second at your latitude, multiply the cosine of your latitude by 464 meters/second.

[6] Technically, because Earth follows an elliptical path, this velocity is not constant. In addition to its constant change in direction, the Earth moves faster when it is closer to the Sun (early January) than when it's further away (July).

[7] Out of the bright stars that can be seen by the naked eye from the surface of earth, Deneb is intrinsically the brightest. Absolute magnitude is a value that represents how bright a star would appear from a standard distance of **32.6** light-years. On this scale Deneb has a magnitude of **−7.2**, which makes it almost the brightest star in the Milky Way. (The brightest star is Cygnus 0B2#12 which has an absolute magnitude of **−9.9**.) The word "Deneb" is derived from the Arabic word for "tail" and, of course, it marks the tail of Cygnus the Swan.

[8] Our entire galaxy is rushing toward the Andromeda Galaxy at nearly 129 kilometers/s (80 miles/s).

[9] The "Great Attractor" was discovered by a group of seven astronomers that are collectively nicknamed the Seven Samurai. It is located in the direction of the Perseus—Pisces cluster some 200 mega-parsecs away. Lawrence Krauss. (2000). *quintessence—The Mystery of Missing Mass In The Universe*. Basic Books, pp. 84–85, 218.

[10] Those interested in mathematically representing a map consisting of points that fit together to form a connected whole wherein there is no pointwise identification between

one location to the next may consider using fiber bundles, which uses a base space E^1 and a fiber E^3. This allows each spacetime event to have an assigned time, but there is no natural assignment of a spatial location. (See Chapter 15.2 of Roger Penrose's *The Road to Reality* for more on fiber bundles.) This representation falls short of the goal because it requires a notion of time that is absolute—passing identically for each location. Because of this, I haven't introduced fiber bundles in the main text.

[11] In a chapter about space, it is worth noting that when we zoom in to the Planck scale, space and time show up in discrete increments. Eventually further division becomes nonsensical. This means that the vacuum cannot be adequately represented by the smoothness of the Real numbers. It should, instead, be represented by a system of integers $\mathbb{Z}$. Roger Penrose observes that, "It is a very striking fact, according to the state of our present physical knowledge, that all known additive quantum numbers are indeed quantified in terms of the system of integers, not general real numbers, and not simply natural numbers." Roger Penrose. (2004). *The Road to Reality: A Complete Guide To The Laws Of The Universe*. Alfred A Knopf, p. 66.

[12] Carolyn Porco, leader of the imaging science team on the Cassini mission to Saturn, in response to the question: "What is it about Space that inspires you?" *New Scientist*, September 8, 2007, p. 61.

Chapter 7 — **Time**

> *"If I had my time over I would do the same again.*
> *So would any man who dares call himself a man."*
>
> Nelson Mandela

> *"Henceforth space by itself, and time by itself, are doomed to fade away into mere shadows, and only a kind of union of the two will preserve an independent reality."*
>
> Herman Minkowski[1]

> *"Time is throned, men say, in the loftiest realm."*
>
> Atharva Veda XIX.53

In a secret location, searching for clues.

At age eleven I played a secret game—sneaking books from the shelves of the library, scanning them for clues, and returning them without being noticed. My goal was to acquire all the information I could about time travel. I was to be the world's best time traveler, but first I needed to figure out how to invent my own time machine. I had no idea if time machines already existed, but I did know that if anyone had one they probably wouldn't share it with me. So, it was up to me to make my own.

Before I figured out how to master the secrets of time, I needed to define time. When I listened to people speak loosely about *time* the concept seemed simple, but when I tried to definitively articulate it I was stumped. Out of all the mysteries I had cast my spotlight on (light, space, dimension, mass and so on), time felt special—as if it were more embryonic, but I couldn't put my finger on why. Turning to several dictionaries for guidance, I found only nebulous terms and circular references.

To me this meant that only two possibilities existed. Either no one had ever really understood the concept of time intuitively enough to define it in a clear way, or time travelers had gone back in time to protect their secrets by hiding even the basic definitions that would encourage discovery of their technology. Intrigued about these possibilities my senses became piqued. Searching for any clues that might have been overlooked by the potential protectors of this secret, I became a detective in hiding.

In the corner of that library, under low lights, I began uncovering a rich dialogue on time that had been developing for thousands of years. The more I explored the topic, the more I became intrigued by the possibility that the very foundation of what I believed might be wrong, but I never solved the mystery. *Fortunately*, I was given enough time as an adult to get back on the trail.

In the fifth century, Saint Augustine asked, "what then, is time? ... If no one asks me, I know; if I want to explain it to someone who does ask me, I do not know."[2] Augustine's words gave validity to the mystery I was investigating and they allowed me to take the next step—to ask why this is the case. Why is it so difficult for us to intuitively grasp or even define time in more than just vague terms?

As I searched for the answer to this question, I stumbled upon the suggestion that our nebulous understanding of time may have something to do with the way we use our brains to make sense of the world around us. Modern studies have extensively documented that our brains are constantly employing fancy editing tricks when it comes to the interpretation of time. Evidently, it does this to make events it believes to be simultaneous, feel simultaneous—overriding the fact that our different senses process information about those events at different speeds.

For example, our brains react more quickly to sounds than they do to flashes of light. This is why hundred-yard dashes begin with a gunshot rather than a strobe light. When we snap our fingers in front of us, our auditory system processes information about that snap approximately 30 milliseconds faster that our visual system does. But every time we snap our fingers, the sound of the snap and the visual cue of the snap seem simultaneous. Our brains go to a tremendous amount of trouble to convince us that the world is as it believes it to be.[3] In other words, our brains are always playing tricks on us—tricks with time.

When we become aware of this, we begin to notice many previously overlooked inconsistencies. To discover one of those tricks, look in the mirror directly at your left eye and then shift your gaze to your right eye. Obviously, our eye movements take time, but we do not see our eyes move from side to side as we switch our gaze. It is as if the world instantly transitions from one view to the next. What happened to that little gap in time? For that matter, what happens to the 80 milliseconds of darkness we should see every time we blink our eyes?

Our state of awareness comes with the conviction that time is something continuous and smooth, but when we look beneath the surface, we discover that our notion of continuous and smoothly passing time may be a mere construction of our brain—a lie that we subconsciously tell ourselves at every waking moment.[4]

When I became aware of the temporal tricks that our brains employ, I was sure that this knowledge would greatly benefit my quest to explore the nature of time. I thought that being aware of these tricks would automatically enable me to investigate time in a way that transcends the illusions that our brains are busily projecting. I soon discovered that this wasn't necessarily the case. Scientists have long been aware of the tricks our brains play on us, but they still haven't made much progress in coming up with a definition of time. Even though we know that our brains are lying to us, we still haven't learned to let go of the lie.

Humans have been keeping time for more than 20,000 years. Ice age hunters notched holes in sticks and bones to track the days between different phases of the moon. The early Babylonians, Egyptians, and Mayans devised calendars that told them when to plant their crops, and when heavenly events such as eclipses would occur.[5] Eons passed, as the Sun made its daily trek through the sky, drifting through 13 constellations year by year, and the moon waxed and waned in regular succession. This heavenly dance seemed to suggest that time was a function of celestial motions. The heavens, it seemed, were made up of regular intervals that controlled the passage of time.

Our modern fundamental unit of time (the second) was originally defined in relation to the most practical of these celestial motions—the solar day. The solar day was divided into 24 hours, the hour was divided into 60 minutes, and the minute was divided into 60 seconds. These divisions had navigational significance. For example, 4 minutes roughly corresponds with a rotation of 1 degree, which means the Earth rotates at a rate of 15 degrees per hour, or 360 degrees per 24 hours. Under this scheme the second was defined as 1/86,400th of the mean solar day.

In the nineteenth and twentieth centuries, scientists discovered that the mean solar day is getting longer. This forced them to redefine the second based on something that had a much more reliable period. Today the passage of time is regularly measured by the number of harmonic resonations piezoelectric quartz crystals undergo in common watches,[6] or (according to the new time standard) to the number of electron spin-flip transitions that occur in a cesium-133 atom when probed by microwaves of a specific wavelength.[7]

The ability to accurately measure time can make the difference between life and death. As a historically poignant example, consider the plight of eighteenth century Great Britain. In 1707, under Vice Admiral Sir Clowdesley Shovell, the British fleet ran aground on the Scilly Isles. This tragic disaster claimed the lives of 2,000 men and wrecked four ships. The culprit was something that plagued all seafarers attempting to determine their longitude—the inability to accurately tell what time it was. Had Sir Shovell possessed an accurate chronometer synchronized with Greenwich time, he could have precisely determined his longitude with relative ease. All he had to do was determine his local time (by observing the position of the Sun or stars) and subtract it from the chronometer's time—the difference is a direct measurement of the longitudinal difference between the measurer's position and where his chronometer was set.

For example, if Shovell was lost in the Atlantic, but knew that in Greenwich the sunset occurred exactly at 6:00 p.m. that day, he could determine how far west of Greenwich he was by recording what time the sunset occurred at his current location. If his Greenwich-synchronized chronometer read 7:00 p.m. as the Sun completed its descent from his vantage, then he would know that he was exactly one hour west of Greenwich which translates to exactly 15° west of Greenwich. But in order to make such measurements Shovell needed an accurate chronometer—something that did not yet exist.

In response to this tragedy, the British Parliament commissioned a Board of Longitude in 1714, which publicly announced substantial prizes for practical solutions to finding longitude at sea. The largest prize, £20,000 (which is equivalent to about $12 million today), was pledged to the inventor of an instrument that could determine a ship's longitude to within half a degree, or 30 nautical miles, when reckoned at the end of a voyage to a port in the West Indies, whose longitude could be verified using proven land-based methods.[8] Any instrument capable of keeping time within two minutes or better over the duration of the voyage would qualify to win the prize. This means that the cheapest watch you can buy today would have been worth the equivalent of 12 million dollars in 1714!

This prize money lured several groups across Europe to wrestle with the problem. Among them was a group led by Sir Isaac Newton, who ruthlessly and routinely attempted to sabotage rival inventors. Newton was attempting to perfect Galileo's method for telling time, which was based on tracking the orbital positions of Jupiter's four largest moons (Io, Europa, Ganymede and Callisto), but he never achieved the precision needed to fetch the prize.

In 1735, after enduring several of Newton's raids, a brilliant English mechanic named John Harrison met the Board of Longitude's challenge. He had designed a portable, palm-sized clock that quickly attained the reputation as being as valuable to a navigator as the keen eyes of the man standing on watch at the ship's bow. From that point on, this handy timekeeper would carry a reflection of its original use in the name "watch."[9]

Questioning Time

Our ability to measure smaller and smaller increments of time has improved, but we still haven't been able to resolve time's fundamental mystery. Basic questions about time still baffle us. What is it that we are measuring? Is there a limit to how accurately time can be measured? Is there a natural fundamental unit of time—a discrete "chronon"—underlying all larger increments? Or, is time infinitely divisible?[10] Does time flow ineluctably?

> *"Is time associated with every motion or just one special motion?"*
>
> *Averroes*

Our goal of defining time is complicated by the fact that some of our deepest intuitions about time change from one culture to the next. For example, today, the western world perceives time as progressing linearly and continuously[11]—an idea that is often represented by the timeline (Figure 7-1). Everything about the Judeo-Christian worldview implicitly presumes the linearity of time. Conceptions of excitement, remorse, nostalgia, anticipation, worry, hope, fear, and so on, are uniquely informed by this assumption. By contrast, the Babylonians, Ancient Greeks, Hindus, Buddhists, Janists, Incas, Maya, Hopi, and other Native American Tribes see time as something that proceeds in a cyclical sequence and loops back on itself.

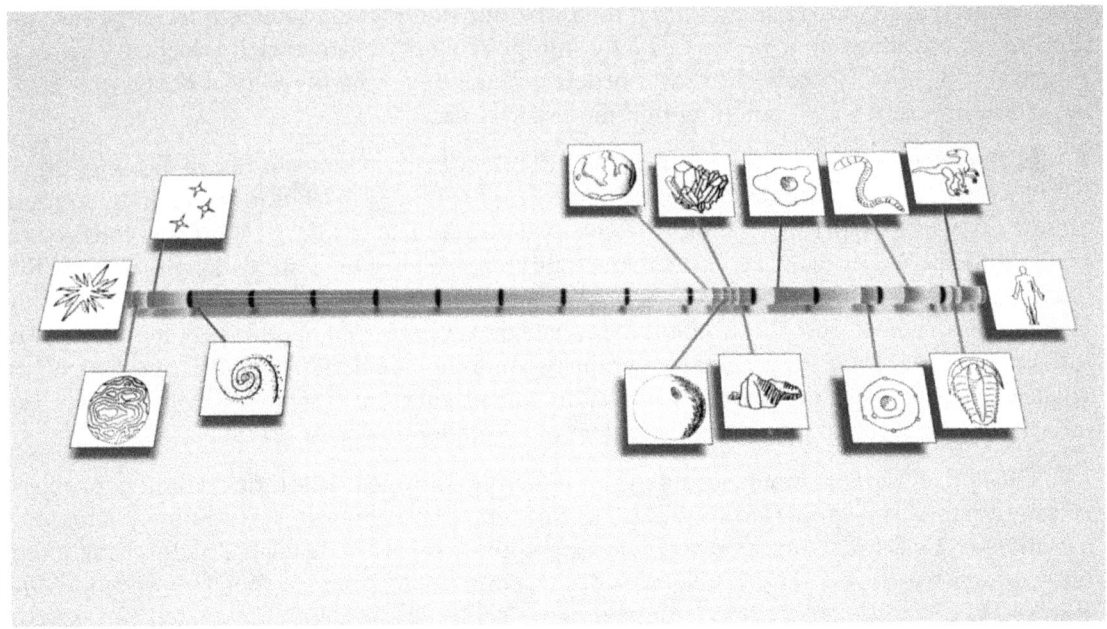

Figure 7-1 The time line

Other possibilities exist. For example, there might be something to the idea of thinking of time as something that has only two categories—*now*, and *other than now*. From this perspective, it might not be considered edifying to distinguish the past from the future because both of them, in a sense, exist only in our imaginations. In this view, it wouldn't matter how distant an event is from the present—if it is not *now*, it's just *other than now*.

In the Andean highlands of western Bolivia, southeastern Peru, and northern Chile, the native speakers of Aymara think of the future as something that is behind them while the past is thought of as being in front of them. They recognize that the "future is unknown, so it is out of sight, behind them."[12] As time flows, the unknown future transitions into the known past as it comes into view. This act of looking forward into the past mimics what astronomers do when they peer into the cosmos—observing the beauty of the far-off heavens, whose light belongs to the distant past. Maybe the speakers of Aymara have a better handle on time than we do. Or then again, maybe we're all way off.[13]

Although time is something that feels familiar, we are still missing its fundamental definition. If our goal is to master time travel, this simply will not do. We're going to have to figure out what time really is, but to accomplish this we're going to have to start from scratch.

We've got some messy questions to deal with. For example, is time illusory? Is there something that separates the past from the present and the future? Or, are these categories illusory? Does time flow or tick? If time does possess some quality of movement, how fast does it go? One second per second? What does that mean? Relative to what does time move? Is time physically real? Does it exist independently of stars, galaxies and other objects (a position known as substantivalism), or is it merely an artificial device used to describe how physical objects are related (relationism)? Exactly how is time tied to space in Einstein's spacetime? Why would a time dimension depend on space dimensions? Can the rate of passing time be changed? Is time travel possible? Why does the "arrow of time" seem to always point in one direction? Where does this temporal asymmetry come from? Was there a beginning to time? Will time ever end? Are there regions in space that are void of time? Are there objects that are "outside" of time? Are there things that do not experience time? And finally, what is the source of time, and how has it come to be so deeply embedded into the fabric of physical reality?

To answer those questions we need to know whether or not there are any properties that fundamentally belong to time. Is there a property of time that exists in all frames of reference? At first glance we might feel that the rate of time's passage is a universal property. While ten minutes pass in Hong Kong, ten minutes pass on the moon—right? Well, technically that's wrong. Our experiences may lead us to believe that time passes at equal rates in all places, but this is not the case.

We now know that the rate of time's passage can vary. The reason we are intuitively unaware of this is that, under circumstances that are normal to us, the differences are so slight that they cannot be detected by human senses. As long as we stay in weak gravitational fields, and consider only slow speeds relative to c, there are no easily detectable changes in the rate of time's passage.

Nevertheless, machines capable of measuring time far more accurately than our biological senses have recorded variable rates of time that depend upon the strength of the gravitational field and velocity.[14] The passage of time is slightly slower in London than in Athens because the gravitational field is stronger in London.

Due to effects of "mountain-valley systems resulting from old tectonic clashes,"[15] and differences in the thicknesses of the various rock formations beneath the surface of the earth, and variations in the densities of those layers, gravity's pull is "strongest in the southwestern Pacific and weakest just off the southern tip of India. So the fastest way to lose weight is a direct flight from Singapore to Sri Lanka."[16] But you also age faster in Sri Lanka. How much faster? Not much. Even if you were on the International Space Station, where this effect is more pronounced, it would take 700 years for you to age one second more than you would have if you had remained on the surface of the Earth.

The ability to travel through time at different rates "is a proven fact, even if it has so far been in rather unexciting amounts."[17] If we visited some of Nature's more exotic locales, where the passage of time is quite different from Earth, we could travel significant distances into the future. For example, "At the surface of a neutron star, gravity is so strong that time is slowed by about 30 percent relative to Earth time."[18] This means that if we spent seven years on the surface of a neutron star, everything on Earth would have aged approximately ten years during our seven-year stay. Effectively, we would have jumped into Earth's future by three years.

Closer to home, if we take an 8-hour airline flight at approximately 920 kilometers per hour, at the end of that flight we will be about 10 nanoseconds younger than if we had stayed on the ground. If we rode that flight nonstop for our entire lives, we would age 1/10,000th of a second less than we would with our feet on stationary rock the entire time. Likewise, spending 8 months in a submarine at 300 meters below the surface would shoot us into the future by 500 nanoseconds relative to someone at sea level. These amounts may not be all that impressive, but if we found a way to travel as fast as a Cosmic Ray Neutrino, 15 minutes of time for us would translate to 30,000 years for everyone left behind on earth.[19] The point here has less to do with the amount of time travel that can be practically experienced and more to do with the fact that time can move forward at different rates. The question is, does the fact that time has no universal rate mean that it has no fundamental, nonrelational properties?

Making Time Real

If time possesses inherent properties, what are they? Without any physical properties to point to, which don't depend on the observer's reference frame, it could always be argued that time is merely a measuring device, or a relational abstract, used only to compare events. If time is something more than this, then what is it? Einstein's theory of general relativity helps us answer this question by revealing the property of curvature, which belongs to both space and time. When space warps, the rate at which time passes in that region also warps (inversely). Also, according to relativity, each point in space must evolve through time uniquely. Both of these qualities of time are inseparable from space.

Quantum mechanics has something to say about time too. It suggests that time cannot be infinitely divided, that on the Planck scale further division yields nonsensical results. To obtain a definition of time that captures these three properties, we need to escape the fog that our traditional nebulous notion of time has matured in. We need to conceptualize what lies beyond the dimensionally tenuous brume, and open our mind's eye to new geometric possibilities.

If the vacuum is an inviscid fluid, then time is a marker of independent quantum resonations. This definition reveals why time unfolds uniquely at each location in the vacuum. Time is a measure of unique location persistence—a measure of location independence. The more independent quanta are—the more they exist without bumping into other quanta—the greater their time signature is. Locations in space move forward through time by one Planck time (one chronon) for every collision-free resonation they complete.[20] Each quantum evolves through time uniquely because time is a measure of this uninterrupted *quivering*.

During collisions, quanta lose their geometric uniqueness. Their evolutions are conjoined such that they represent a single location in the map. This lack of independence forbids evolution through time. When energy passes through space—for example, kinetic energy transferring from one quantum to the next—it goes through different stages. At first it is located at one quantum, its location is unique and independent, and it experiences time. Then the collision occurs, two unique locations in the map of space become one unique location, and no time is experienced. Finally the quanta rebound, regaining their status of being unique locations, and the kinetic energy now persists at its new location in space, independently resonating and therefore experiencing time. This process sacrifices evolution through time for evolution through space.

According to this definition, time is a manifestation of the dynamic properties of the vacuum. Its rate of passage is a function of the independence each quantum of space experiences. As curvature (density) of a spatial region increases, the average signature of time in that region decreases, because the independence of the quanta in that region is reduced. Less independence means that free resonations become less and less likely, which ebbs the flow of time.

This definition gives us the ability to answer the question of how fast, and with respect to what, time moves. The answer is that for each location, time progresses at a rate of one Planck time per quantum resonation. Four-dimensional descriptions forbid the exposure of quantum resonations, which makes it impossible to describe how fast, and with respect to what, time moves (not to mention why time flows at different rates in regions with different amounts of curvature). Quantization allows us to resolve those resonations.

The amount of time that elapses between two events that occur at the same location is equal to the number of resonations experienced at that location between events, multiplied by the value of the Planck time. Just as the amount of space, or distance, between two events is found as the sum of space quanta between those events, the amount of time between two events at one location is the sum of whole free quantum resonations experienced at that location between events.

This definition inherently captures some of time's most bizarre properties. It reveals how and why time evolves independently at each location in space. It explains why the passage of time depends upon the curvature of that region. And it automatically sets a minimum limit for time's resolution—a fundamental chronon that cannot be further divided without yielding nonsensical results. Does this definition do more for us than that? Can it explain why time always seems to flow in one direction? Does it tell us whether or not it is possible to reverse time and travel back to an earlier moment? Does it allow for regions without time—places where time has stopped?

The Temperament of Time

> *"Time is just another name for the motions of the heavens."*
>
> *Al-Ash'ari*

If the passage of time depends upon the number of whole resonations experienced at each quantum, then the directionality of the geometric contortions that make up those resonations will not affect their absolute measure. Because temporal measures only relate to the number of complete resonations each quantum undergoes, opposing resonate phases make identical contributions to the flow of time. From this it explicitly follows that time can only be described in terms of absolute change. Time can only be a positive quantity—just as length can only be positive. (There is no such thing as a rope with a negative length, despite what my high school physics teacher believed.) Time is an absolute value quantity.[21]

On the ultramicroscopic scales, time is not continuous. It manifests as a series of punctuated increments and progresses in discrete ticks. Each tick is equal to the Planck time, which is $5.39124(27) \times 10^{-44}$ seconds. Over longer durations, the passage of time smooths out to approximate a continuous flow.

Space and time are the two parameters through which energy (distortions/objects within the vacuum), carried by the fundamental constituents of Nature, can evolve. Between collisions, energy evolves through time and occupies the same position (the same quantum). During collisions, energy transfers to a new position in space but doesn't evolve through time. In this sense, space and time are mutually exclusive; the experience of one inhibits the other.

Zooming in on a region of the vacuum, the signature of time begins to wildly vary as we approach the Planck scale. At any given instant some quanta will be colliding while others are not. One quantum may undergo many collisions while another remains collision-free. Quanta progress through time only between collisions—one Planck time for every collision-free resonation it completes.

At extremely high resolutions the number of free resonations experienced by each quantum becomes highly variable, but the collision probability over longer periods evens out in regions with equal density (zero curvature). This leads to the statistical expectation that, in regions of equal spatial density, all locations will age at equal rates on average.

On scales much much larger than the quanta themselves, we can reduce time's rate of passage by moving to a region of greater spacetime curvature, or by increasing our speed. Either method increases our quantum collision rate. This effect is unfamiliar because we live out our lives at fantastically slow speeds compared to c, and have only very humble exposures to spacetime curvature. Enough curvature or enough speed can cause time to stop. But even those extremes do not alter our measurements of the speed of light. All (macroscopic) observers internal to the vacuum will measure the speed of light (c) as a constant.[22]

To illustrate how the rate at which time passes depends on speed, let's imagine that we have a very tall ship rigged with mirrors at the top and bottom of the main mast. When we introduce a beam of light into this set up, bouncing that beam up and down the mast, our ship becomes a light clock. Knowing that the speed of light (c) is exactly 299,792,458 meters per second from all reference frames, we can measure the passage of time by counting the number of reflections between the mirrors ($time = distance/c$). If our mast were an enormous 30 meters then this would enable us to register approximately ten million measurements every second. To an observer on this ship the light will take ~$2.00138457118891229 \times 10^{-7}$ seconds to travel up and down the mast.

However, if we describe the path of that light from another reference frame (e.g., as an observer on land), we get a different result. From the ship's perspective, the light travels parallel to the mast, but from any perspective in which there is a relative velocity between the ship and the observer, the light follows the hypotenuse of two triangles whose sides are equal to 30 meters and the distance the ship travels between reflections (Figure 7-2). If the ship is moving at 11 knots from the land's point of view (the fastest speed I achieved during my journey across the Atlantic), then between each reflection the ship will move ~$5.662917644 \times 10^{-7}$ meters forward. This means that instead of traveling a total of 60 meters, the light must now travel 60.0000000000000106896 meters between mirrors, which corresponds to a travel time of ~$2.00138457118891265 \times 10^{-7}$ seconds. That's an increase of ~3.6×10^{-23} seconds.

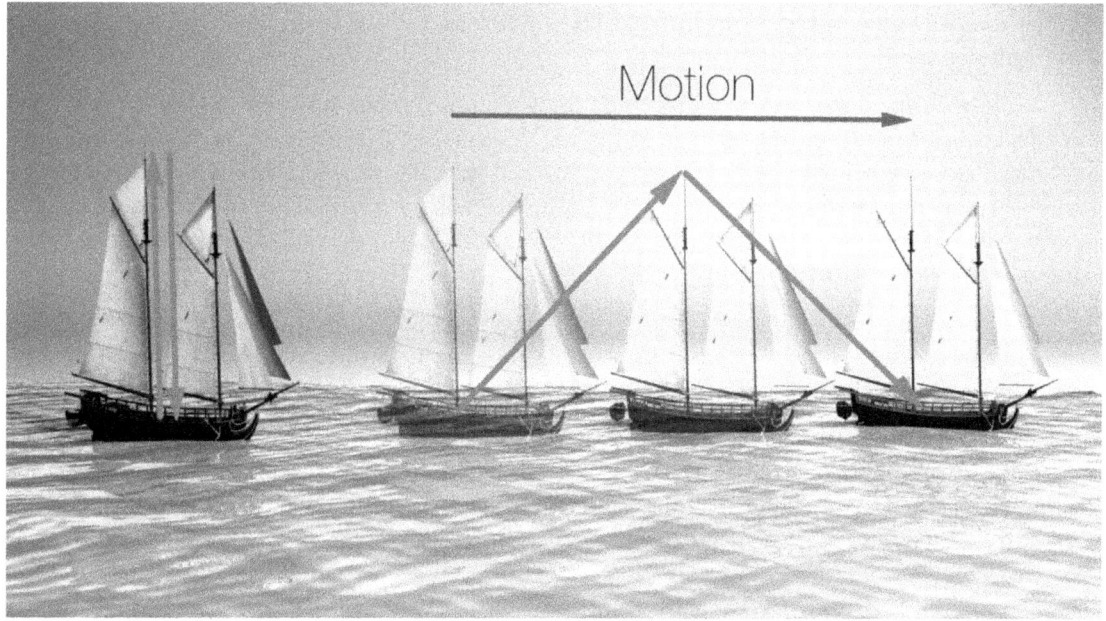

Figure 7-2 Motion slows time

What is important here is not the magnitude of the difference in this example, but that a difference exists at all—that events are separated by different amounts of time in different reference frames. If light travels at a constant speed, then it must take more time to follow the triangular path than the straight up-and-down path. The shape of that path depends on your reference frame. The observers on the ship measure the minimum amount of time, because to them the light only goes straight up and down the mast. However, from the external observer's point of view, the ship's clocks appear to be running slowly because the light has farther to travel. In general, any observer in an inertial reference frame will find the clocks of a moving reference frame to be advancing at a slower rate.[23] Consequently, events that are simultaneous in one reference frame are not simultaneous in another.

In Newton's map, motion through time was considered entirely unrelated to motion through space. In Einstein's map, the two motions became intimately linked.[24] Our new map gives us a rich explanation for this linkage. As motion through space increases, motion through time decreases because spatial independence diminishes. Waves propagate through the vacuum because the vacuum's parts, the quanta, elastically collide. The faster the wave, the more its parts are colliding, and the less they freely resonate. With increasing speed, travel through time is swapped out for travel through space. As this swapping continues, more and more time is drained out. At the maximum limit, the speed of light (c), the object (the wave) experiences no time regardless of how far it travels.

This explains time as an absolute value quantity and makes the notion of reversing the flow of time logically incoherent. This means that it is not possible to travel back in time and change the events that follow. We can alter the rate that time flows, but there is no way to reverse the flow of time. Because the passage of time is solely measured as a function of the number of free resonations completed by each quantum, a time machine's task is to take advantage of differences in the flow rate of time that exist in Nature, or to alter the flow rate by manipulating the environment. Simple time machines might carry us to places with greater spatial densities or increase our speed, and more advanced time machines might alter the flow rate of time by uniformly increasing the spatial density inside the machine. If the spatial density is increased, then the quanta will interact more frequently and the passage of time will decrease.

Time machines may carry us forward through time either slower or faster than the rate we are accustomed to on the surface of the earth, but they cannot carry us backward in time to a time existing prior to the moment we entered into the time machine. Traveling back in time is ruled out because each moment in time (each event) exists—just as each point in space exists. It does not change and it is not stochastically connected to the past or the future; rather, it is deterministically connected.

Budding time travelers might find it disappointing to learn that, according to this model, traveling backwards through time is impossible, but the true time traveler will also recognize that there is some exciting news in all of this. We now have a fundamental definition of time that, if correct, gives us intuitive access to what's going on behind the curtain.

Instead of ruling out time travel, our model clearly reveals mechanisms (in agreement with general relativity) for traveling forward in time at different rates. To master these methods, we have to allow our intuitions to absorb the implications of vacuum quantization. To assist us in that task, let's talk about the dimension of supertime and then test our understanding of this new system by examining how to interpret velocity, acceleration, and warped spacetime in the model.

Supertime

Researchers, such as Itzhak Bars at the University of Southern California, have shown that it is possible to formulate theories with a second time dimension while remaining physically reasonable.[25] Here "physically reasonable" means mathematically sound and empirically consistent. But what exactly does it mean to have more than one time dimension? What might another time dimension be?

Until now we have been focused on defining the familiar dimension of time (t) and understanding the origin of its peculiar properties. Now we shall turn our attention to the other dimension of time that exists in our map—supertime ($\matht{t}$).

To get an idea of the role this other time dimension plays, notice that quantizing the vacuum resolves more than just the number of resonations completed by each quantum. The undulatory motions of the quanta expose temporal information that is richer than just a count of completed free resonations. Those undulations make use of a temporal resolution that is much finer than the familiar dimension of time. They reveal a temporal evolution that is completely independent and separate from time.

For example, vacuum quantization resolves the superspatial motions of the quanta. A superspatial velocity is defined as a change in superspatial position divided by a change in supertime. The dimension of time is strictly tied to the number of resonations each quantum completes, but supertime is needed to resolve those resonations and to map superspatial motion.

We cannot use information about how many resonations a quantum of space has completed to determine its motion through superspace. This means that time is orthogonal to supertime. To back this up, note that it is entirely possible for a quantum in space to be experiencing supertime without experiencing time—just like it is possible for something to be moving in x without moving in y.[26]

If we assume a fractal structure to vacuum quantization, then superspace and supertime are also quantized. However, the resolution of supertime is much finer than the resolution of time because the *quanta of superspace* are much smaller than the *quanta of space*—they are the same size as the *subquanta* that make up the *quanta of space*, both of which are unresolved on the eleven-dimensional level. (We will explore the full fractal structure that resolves them in Chapter 11.) Without the presence of supertime, we would only be able to envisage snapshot images of the orientations and positions of the quanta as they move about through the superspatial dimensions. We would have only a strobe light depiction of reality that lacked a wealth of information and was incapable of restoring determinism.[27]

Even with two time dimensions, it is physically impossible to climb into time's second dimension for the purposes of traveling backward in time's first dimension. Temporal dimensions do not intermingle in the way that spatial groups do (like x, y, z or σ, μ, δ). They are not interchangeable descriptions of a metric. Temporal dimensions are naturally constrained by the geometric symmetry of dimensional hierarchy. Time (t) strictly belongs to the dimensions of familiar space (x, y and z). Supertime ($\matht{t}$) strictly belongs to the dimensions of superspace (σ, μ and δ).

Nevertheless, regions do exist where familiar time does not apply—regions void of quanta, or regions packed so tightly with quanta that they are unable to independently resonate. Whether or not we are caught in its rhythmic flow, time remains an absolute value quantity.

Supertime helps us understand the flow of time by giving us a temporal comparison. Having a clear definition for time allows us to solve its mysteries. Time is not illusory; it is a physical process. The past, present, and future are all fixed components of one deterministic timescape. Time proceeds in integer multiples at each location in the vacuum, having a base chronon to its structure. Its speed is one Planck time per resonation. Time is not an artificial measuring device; it is an expression of the number of uninterrupted elastic cycles that each quantum of space undergoes. Because this expression of time is *environment dependent* its rate can vary, and variable rates of time allow for time travel. The "arrow of time" that we observe in Nature is merely an artifact of the universe's present state. The asymmetry between past and future directions exists because our universe has not yet reached a state of maximum entropy.

Time ultimately originates from the physical properties of the individual quanta of space. It is ingrained as a potential that manifests itself in relation to its surroundings. It belongs to space. Every single moment that is experienced at every single location in x, y, z space is the completion of a harmonic resonation in the fabric of physical reality.

[1] From *Space and Time*, an address to the 80th Assembly of German Natural Scientists and Physicians, Cologne, Germany, September 21, 1908; Reprinted in *The Principles of Relativity*, by A. Lorentz, A. Einstein, H. Minkowski, & H. Weyle. (1052). New York, Dover, p. 75.

[2] Gary Stix. (2006). Real Time. *Scientific American*, Special Edition—A Matter of Time, p. 5.

[3] Technically, what our brains are doing is "fixing" the data to better match what it believes is happening. The expectations are usually Euclidean. In other words, our brains are altering our out-of-sync perceptions to match the intuitive model by which our brains comprehend physical reality—a model that inherently assumes simultaneity. Our brains also routinely change our memory of events so that they coincide—changing the "time stamp" on a cluster of sensory memories so that they are all remembered together.

[4] David Eagleman. (2007, August). 10 Unsolved Mysteries of the Brain. *Discover*, p. 59.

[5] Gary Stix. (2006). Real Time. *Scientific American*, Special Edition—A Matter of Time, p. 4.

[6] Piezoelectric crystals change their shape when a current is passed through them. This metamorphosis is a uniform process and can easily be adapted as a reliable time standard. For example, when a current flows through the crystal, it shrinks (contracts) and therefore breaks the connection. With the connection broken, the current no longer flows through the crystal and the crystal rebounds to its original shape—reestablishing the electrical connection. This harmonic process sets up a consistent ticking based on periodic electrical pulses. This is how the common quartz watch measures time.

[7] 9,192,631,770 of these transitions defines one atomic second. Guy Gugliotta, 'Catching Up With the Atomic Clocks—Talk About Second Guessing,' Washington Post, reported in The News Tribune, December 27, 2005, p. A6. One second is also defined as the amount of time it takes light to propagate 299,792,458 meters through a vacuum.

[8] William J. H. Andrews. (2006). A Chronicle of Timekeeping. *Scientific American, Special Edition—A Matter of Time*, p. 52.

[9] Neil DeGrasse Tyson. (2007). *Death By Black Hole and Other Cosmic Quandaries*. W. W. Norton, p. 314.

[10] The shortest time interval observed to date was measured by Ferenc Krausz at the Max Planck Institute of Quantum Optics in Garching, Germany. Krausz used "ultraviolet laser pulses to track the absurdly brief quantum leaps of electrons within atoms. The events [lasted] for about 100 attoseconds, or 100 quintillionths of a second." For a little perspective, 100 attoseconds is to one second as one second is to 300 million years. Tim Folger. (2007, June). In No Time. *Discover*, p. 78.

[11] In ceremonial awe of that mysterious time, when the end of the year somehow looped around on itself to become the beginning, the Romans used to celebrate Saturnalia, an upside-down holy week when masters served their servants, and slaves held the great offices of state. The mystery of this "transition" is an expected result of a strong belief in linear time. Jared Diamond. (2005), p. 383.

[12] Stephen Pincock. (2006, July 15). Back to the Future. *Financial Times*.

[13] When Islamic natural philosophers led the world in science they debated whether or not time was infinite and what its "topography" was. For example, Avicenna advocated a rectilinear infinite conception of time while Averroes supported a cyclical conception of time that also required time to be infinite but only in the way that the path of a circle is

infinite. For more about how the concepts of time vary throughout human cultures, see: Nunez and Sweetser, *With the Future Behind Them.*

[14] This was precisely confirmed in 1976, "when a Scout D rocket carried a hydrogen-maser clock to an altitude of 10,000 kilometers, or about 6,000 miles. In the two hours before it plunged in to the Atlantic a thousand miles east of Bermuda, the rocket's time-keeping was compared with that of an identical clock on Earth. A hydrogen-maser emits microwaves at 1.42 gigahertz; at the rocket's apogee the clock on board was about one hertz faster than its terrestrial twin." Robert Kunzig. (2004). Testing the Limits of Einstein's Theories. *Discover,* p. 59.

For an overview of the subsequent experiments that have verified time-dilation see Hans C. Ohanian and Remo Ruffini (1994) "Gravitation and Spacetime," W.W. Norton and Company, table 4.1, p. 186.

[15] Bjorn Carey. (2006, August). Mapping Earth's Fourth Dimension. *Discover,* p. 18.

[16] Ibid.

[17] Paul Davies. (2006). How to Build a Time Machine. *Scientific American,* Special Edition—*A Matter of Time,* p. 16.

[18] Ibid.

[19] Ibid., p. 18.

[20] Compare this with ar-Razi's postulation (1939) of "an unchanging duration."

[21] Although this measure of change gives time a positive or absolute value interpretation, there is no preferred origin of time in this model. This means that we cannot claim that the dynamical laws of Nature change as time proceeds away from that "origin".

[22] This maximum speed is only macroscopically fixed. Macroscopic measurements made by observers internal to the vacuum will always give a constant speed for c. But the propagation speed of spacetime (c) is not necessarily constant from an eleven-dimensional perspective. Varying quantum densities alter the speed in which connectivity waves of quanta move through superspace.

[23] The observers on the ship also consider themselves to be in an inertial reference frame. Therefore, they will conclude that the land-based clocks are slow.

[24] This is an extremely well established effect of special relativity. If physicists did not account for such effects then modern GPS satellites would yield errors that would increase by a measure of nearly one kilometer every two hours. See also: Neil Ashby. (2002). Relativity and the Global Positioning System. *Physics Today,* 55(5), pp. 41–47; (2003). Relativity in the Global Positioning System. Living Reviews in Relativity 6: http://relativity.livingreviews.org/Articels /Irr-2003-1/index.html.

[25] Brian Greene. (2004). *The Fabric Of the Cosmos,* Random House, New York.,p. 529. More recently, Itzhak Bars of the University of Southern California in Los Angeles has developed a theory he calls "two-time" physics that appears internally consistent. See 'Marcus Chown. (2007, October 13). The Hypertime Trap. *New Scientist,* pp. 36–39.

[26] This orthogonal relationship appears to be one directional. Spatial dimensions seem to be symmetrically orthogonally related. For example, an object can move through x and y without moving through z, it can move through y and z without moving through x, or it can move through z and x without moving through y. Each of these is an example of an orthogonal relationship. Temporal dimensions, however, are slightly different. For

example, time and supertime are orthogonally related in that it is possible for something to move through supertime without moving through time. However, according to our map, moving through time without moving through supertime does not seem to be a valid option. This relates to the idea that single temporal dimensions are internal to other single temporal dimensions, while spatial dimensions are internally related to each other only as groups.

[27] This is not the first time the possibility of an extra time dimension has been postulated. A theory called F–theory also specifically suggests the existence of two temporal dimensions.

Chapter 8 — **The Speed of Spacetime**

"From a wild weird clime that lieth sublime, Out of Space—Out of time"
Edgar Allan Poe, Dreamland

"You don't see what you're seeing until you see it, but when you do see it, it lets you see many other things"
William Thurston

Three weeks before harvest, Syracuse, Utah.

Stepping into the cornfield always made my heart quicken—as if I was disappearing from the rest of the world. Entering from random points along the edge of the field I followed each row to its end. One day, as I walked with my arms out in front of me, pushing the tapered leaves away from my face, I came upon a small irrigation pond.

I began visiting this place regularly, protecting the secret of its location by making sure that no one saw me duck into the row that lead to it. In this secret place there was no reason to feel guilty about freely imagining possibilities, asking questions, wondering, or wishing.

I gave every frog in the pond its own name. Sometimes the pond was too deep for catching frogs. When this happened, I sat on the concrete wall and turned the big metal wheel to crack open the irrigation dam. I had to drain the pond slowly because I didn't want any of the frogs to get sucked into the maelstrom that would be created if I opened it too much. The wait gave me plenty of time for daydreaming.

As I was waiting for the water level to drop, I began tossing small rocks into the pond. I found it soothing to watch the waves of concentric circles traveling outward. When I ran out of small rocks I began tossing larger and larger ones. I wondered why those concentric waves traveled at the speed they did. What determined their speed?

In memory, this was the first time my curiosity raised questions about wave speeds. I wondered if the speed of those waves depended upon the depth of the pond. I also wondered if waves from smaller rocks travel slower than waves from larger rocks? And if there are differences, how precisely would I have to measure the waves in order to detect them?

Later, I learned that there are two categories of waves: p-waves (pressure waves, or primary waves), which propagate throughout the body of a medium, and s-waves (surface waves, or secondary waves), which occur at the interface of two media—for example, where water meets the air (Figure 8-1). Every medium has its own characteristic p-waves and s-waves. The difference between these two wave speeds enables geologists to determine the distance to the epicenter of an earthquake. I was familiar with s-waves, as the concentric circles that moved outward on the surface of that pond, but I remained confused about p-waves until I learned about sound waves.

My junior high school science teacher explained that every note has its own frequency. Later, I learned that each frequency travels at a slightly different speed because air is a dispersive medium—it has a nonzero viscosity. When I heard this, I remembered the observations I had made in that pond. I tried to visualize these waves traveling through the air around me and realized that the difference between *s*-waves and *p*-waves was that *p*-waves are not restricted to surfaces. Instead *p*-waves propagate through the bulk of a medium. Suddenly it was easy to imagine all of the invisible sound waves traveling through the air as compressions of the small particles of air emanating from a source.

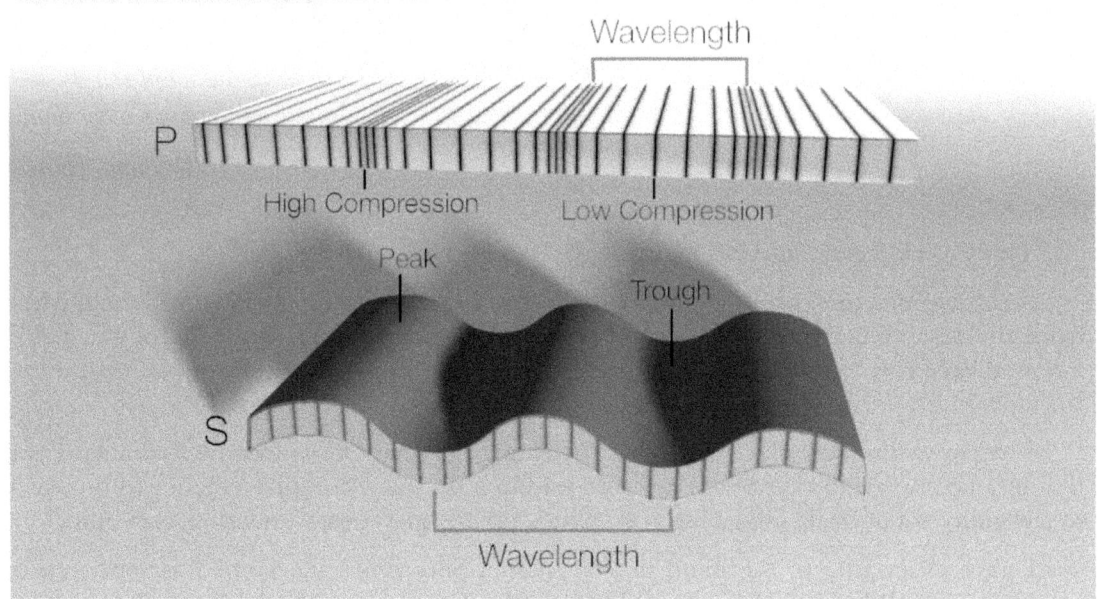

Figure 8-1 p-waves and s-waves

S-waves can be reasonably represented when we assume continuity and smoothness of the waving medium, but to picture *p*-waves we need a description that either reveals the individual molecules of the medium or invents some sort of magical "curvature". For example, if we didn't believe that air was made up of molecules, we would have a very difficult time explaining *p*-waves in air (sound waves).

This is an important realization. Particulate media inherently carry the ability to adjust their emergent properties. If a medium is held at a constant pressure and temperature, then a particular frequency will have a fixed *p*-wave and *s*-wave speed. However, if a medium supports a range of pressures and temperatures, then the wave speeds of that medium will adjust as those parameters change in a way that has nothing to do with dispersion. For example, because the atmosphere's pressure and temperature varies with height, the speed of sound in our atmosphere varies with altitude.[1]

To understand why wave speeds depend upon those parameters, let's imagine a fluid from a scale that enables us to identify its internal molecules. When we do this, it becomes readily evident that if the molecules are moving faster (higher temperature) they will be, on average, communicating with each other more rapidly, which in turn increases the speed that they pass distortions (phonons) through the medium. It also makes sense that if these molecules are packed closer together (higher pressure) they will communicate with each other, on average, more frequently. This condition also increases the internal propagation speed of that medium.

When we quantize the vacuum, variable vacuum densities produce variable propagation speeds in relation to the deeper background—the superspatial volume that the quanta are arranged in. Perspectives internal to the vacuum are incapable of registering these variations in propagation speed. As the density increases, the propagation speed increases, but the amount of space in that region also increases. The wave moves faster through denser space, but compared to the number of quanta the wave traverses, its speed has remained constant. Consequently, the natural propagation speed of the vacuum (c) is constant for all perspectives that are internal to the vacuum (perspectives that do not resolve vacuum quantization).

All waves in the vacuum are confined by this maximum propagation speed, as long as their wavelength is defined on scales much larger than the size of the quantized pieces that make up the medium. This implies that the vacuum is a fluid with zero viscosity. Otherwise, different wavelengths of light would have different propagation speeds.

This confinement is loosened as we approach the Planck scale, where wave speeds begin to exhibit a propensity for going faster and slower than the macroscopically observed 'fixed' wave speed. Over long distances, these two propensities will cancel each other out, but when the distances are short, these effects become vitally important. In other words, as we near the quantum scale, we find that light reveals the molecular structure of the vacuum fluid in two critical ways. One, it doesn't always go in straight lines, and two, it doesn't only go at the speed of light![2]

Cosmic Ripples

In college, I discovered a place where I could wade in a pond of cosmic secrets. At night, I would rush up to the observatory, where I worked, and spend my fantasy hours trying to find as many cosmic 'frogs' as I could. With fingerless gloves, under the red glow of the computer screen, I began each night with an attempt to focus the ST9 CCD camera. As I waited for the old computer to download each adjusted frame, I ignored the cold by scanning my mental list of unanswered questions.

One night I focused in on the topic of *speed* and was able to identify four critical questions:

1. Why are all sublight speeds purely relational?

2. Why is there a limiting speed in the vacuum? In other words, why can't anything move *through space* (macroscopically) faster than the speed of light?

3. How does c manage to be nonrelational when all other speeds are relational? and

4. What natural properties, or parameters, set the speed of light at exactly 299,792,458 meters per second instead of some other value?

Just as quantization allows us to understand *p*-waves, vacuum quantization enables us to answer these questions. For example, consider the relational quality of all sublight speeds. The term *relational* means that these speeds only gain definition once a specific reference is chosen for comparison. All sublight speeds are comparative measures only. Consequently, it is entirely meaningless to say that something has a speed of 100 miles per hour unless we mention what this speed is being compared to. But why are there no unique reference frames?

A unique reference frame requires the existence of a rigid, static, and geometrically fixed spacetime metric. The absence of a unique reference frame suggests that the metric of spacetime is not rigid—that it does not retain its exact identity from instant to instant. The picture we have been exploring captures this condition because its base constituents (the quanta) are continuously interacting and rearranging their geometric orientations. By definition, this metric is geometrically nonstatic, nonrigid, and background independent. This is why position and all sublight speeds are relational.

Why does this relational property not extend to Nature's one special speed, which is known as the limiting speed of spacetime, the natural speed of spacetime, the speed of light, or c? Why does this limiting speed even exist? Why does that speed remain constant as we change reference frames when we have clearly established that speed is something that gains meaning only when we compare it to something else?

To answer these questions, we need to make sure that we don't misunderstand what it means to say that nothing can move through space faster than the speed of light.[3] If we did not know that space has non-Euclidean characteristics—if we thought space was purely Euclidean in form—then the suggestion that a limiting speed exists would force us to a contradiction. In Euclidean geometry, it makes no sense to have an ultimate finite speed. Under the assumption that space is Euclidean, it is perfectly reasonable to conclude that faster-than-light speeds exist. If light travels at 299,792,458 meters/second then 299,792,459 meters/second exists as a greater speed. Under Euclidean assumptions that claim is valid.

People who use Euclidean intuitions to frame their comprehension of physical reality often interpret the physicist's statement that "nothing can move through space faster than the speed of light" to mean that humans have not yet found a way to boost something beyond this speed. They tend to respond with the suggestion that advanced technology will eventually enable mankind to overcome this restriction, but they have missed the point entirely.

Physicists are not making a claim about human potential. What we are saying is that within the familiar four dimensions of Nature, there does not exist a definable speed that exceeds the speed of light. The option for going faster than the speed of light entirely disappears, because the meaning of speed itself is maxed out at the speed of light. There is no definable speed beyond that point because Nature is not geometrically Euclidean.

When we learn to comprehend Nature through its true geometric form, this fact becomes no more fantastic than pointing out that you cannot go farther north than the North Pole, and you cannot have a color more red than the exact color defined as red. These statements are true by definition. They are tautologies. For the same reason, by geometric definition, Nature possesses a limiting speed. Let's explore this in greater detail.

An object's speed through space is equal to the amount of space that it traverses (from the observer's point of view) divided by the amount of time experienced *by the observer* during that interval. This definition critically sets a finite limit for the maximum speed in space.

To explore how, let's imagine that we deploy a powerful rocket, or a giant intergalactic miracle machine that possesses the ability to constantly accelerate with the force of one g for a period of 10,000 years. During the entire span of the rocket's journey its speed will be increasing each second by 9.8 meters per second (from the point of view of those on the rocket). Due to the constant thrust of the rocket's engines, those aboard will feel a uniform constant acceleration. As it accelerates, the rocket's speed increases. As a consequence, the rocket's experience of time begins to decrease relative to Earth's experience of time. The significance of this is that, although everyone aboard the rocket will continue to feel a constant acceleration of one *g*, observers from Earth will see the acceleration of the rocket diminishing asymptotically toward zero, as the rocket's speed increases asymptotically toward the speed of light.

This asymptotic speed limit remains exactly the same (approaches the same limiting value) independent of the magnitude of acceleration we choose for our rocket. This tells us that the limiting speed in Nature has something to do with the way time is swapped for space as speed increases. Because this limit represents the point at which the rocket's clocks have entirely stopped, it possesses an infinite association. If the ship reached the speed of light, it would move through space without experiencing any time. If *speed* were defined as the distance an object travels, divided by the amount of time *the object experiences* during that trip, then the speed of light would give us an infinite value.

$$\frac{nonzero\ measure\ of\ distance}{zero\ measure\ of\ time} \rightarrow \infty\ velocity$$

This limiting infinite value is one reason that c is nonrelational. Infinity is equidistant from all locations. As we change our reference frame, we change the value of the numerator in this equation, but the denominator remains zero. A positive number divided by zero yields infinity (∞). This means that, in some sense, to reach the speed of light is to touch infinity.

In any reference frame we choose, our description of the speed of an object not experiencing time must be identical. This is why c is the only nonrelational speed. It does not change when we change our perspective for the same reason that infinity remains identically distant when we change position.

If we chose to define speed as a measure of the distance an object travels (compared to the observer) divided by the time *experienced by the object* during that translation, then infinite speeds would be at least theoretically attainable. But, because we have specifically defined the speed of an object to be the distance it travels (compared to the observer) divided by the time *experienced by the observer* during that translation, the maximum value allowed for speed is a finite value known as c instead of ∞.[4]

This brings us to our fourth question. Why is the limiting speed exactly 299,792,458 meters per second, instead of some other value? What natural parameters precisely determine this value? To answer this question let's recall that the definition of speed becomes maxed out when the object being described has traded all of its motion through time for motion through space. An object traveling at speed c experiences zero time. Any sublight speed observer will experience a certain amount of time as the object is displaced from point A to point B. If the object has a velocity of c, then the number of Planck lengths the object travels (n_1) will be identical to the number of Planck times the observer ages (n_2). When we divide these two values we get the following ratio:

$$\frac{n_1 l_P}{n_2 t_P} = \frac{l_P}{t_P}$$

Because $n_1 = n_2$ whenever the traveling object is experiencing no time, this ratio reduces to the Planck length divided by the Planck time, which is equal to 299,792,458 meters per second *exactly*.

An object traveling through space must interact with the number of quanta that define the distance between its starting location and ending location; otherwise, it doesn't travel through that *amount of space*. Only two types of interactions are available—spatial or temporal. Because the object traveling at speed c is not experiencing any time, it travels only through space. An observer determines the object's speed by dividing the amount of space the object travels (compared to the observer) by the amount of time traveled by the observer during that translation.

As the observer speeds up in the direction of the object, the difference in spatial experience between the object and the observer decreases, but the observer also begins to experience less time. Any object experiencing no time (any object going the speed of light) will always appear to move one Planck unit of distance for every Planck unit of time experienced by any sublight observer. Therefore, the speed of light is a value that directly reflects the intrinsic properties that define the quanta of spacetime.

In our familiar (but arbitrary) units of meters and seconds, the speed of light is equal to 299,792,458 m/s, but in the natural (non-arbitrary) units of the vacuum's geometry the speed of light is equal to one natural unit of distance divided by one natural unit of time.

$$\frac{l_P}{t_P} = c$$

Here is another way to understand this limiting vacuum speed. Quantization sets a limit for how fast long-wavelength phonons can travel through a medium—any medium. Phonons are collective excitations in the elastic arrangements of the atoms, molecules, or quanta that make up a medium. When one or more quanta are displaced from the equilibrium position, that displacement creates a set of vibration waves that propagate through the medium. The propagation speed of those waves is given by the slope of the acoustic dispersion relation. For longer wavelengths, this dispersion relation asymptotically becomes linear. Consequently, the slope of that line, which represents the wave propagation speed, becomes fixed for all long wavelengths. The value of that macroscopic limiting speed is equal to the size of the elastic constituents multiplied by their vibrational frequency.

In the vacuum, this limiting phonon speed is equal to one quantum length (the Planck length) multiplied by the quantum frequency, which is equal to one divided by the quantum time (the Planck time). In other words, the speed of light is a material parameter of the superfluid vacuum. In the low-energy limit—the limit in which phonons (metric distortions) are capable of naturally propagating through the medium—this is the maximum attainable speed. Perspectives internal to the vacuum observe this maximum speed as a fundamental constant, because they are blind to changes in vacuum density.

On small scales, acoustic waves cease to exist,[5] because the concept of a wave is valid only when the wavelength is much larger than the distance between the constituents that make up the *fluid*. Therefore, in a quantized vacuum, where acoustic waves constitute light, wavelengths of light are inherently cut off as we approach the Planck scale. This condition does a lot of explanatory work. Let's see why.

The observation that blackbody radiation falls off on the shortest wavelengths (highest energies) led to the blackbody ultraviolet catastrophe. Eventually this *catastrophe* was subdued with the conclusion that light must only come in discrete quantum packets—now called photons. Although nobody had any idea why light would come in discrete packets, it was clear that if this assumption was made, then predictions matched observations. With our new insight we can shed light on this mystery. Wavelengths of light are cut off because the vacuum is quantized. In a quantized framework the shortest wavelengths are not allowed because acoustic waves cease to exist on small scales.

Furthermore, if the medium that light propagates through is a fluid, then non-uniformities in this fluid (changes in pressure from region to region) naturally account for alterations in the wavelengths and direction of the waves in motion. To assay whether or not these changes are analogous to what occurs within other non-uniform fluids, Theodore A. Jacobson and Renaud Parentani performed experiments on acoustic waves traveling through non-uniform media and found that the wave's direction and wavelength can become stretched, just like photons in curved spacetime. In fact, they found that when sound waves in a river enter a narrow canyon, or move through water that is swirling down a drain, they become distorted and follow a bent path, just like the light that comes close to a star or a black hole.

In order to create what we might call an "acoustic black hole," Jacobson and Parentani used a device that hydrodynamicists call a de Laval nozzle. The nozzle is designed so that the fluid reaches and exceeds the speed of sound at the narrowest point without producing a shock wave (an abrupt change in fluid properties). The effective acoustic geometry is very similar to the spacetime geometry of a black hole. The supersonic region corresponds to the hole's interior: sound waves propagating against the direction of the flow are swept downstream, like light pulled toward the center of a hole. The subsonic region is the exterior of the hole. Sound waves can propagate upstream but only at the expense of being stretched, like light being redshifted. The boundary between the two regions behaves exactly like a black hole horizon.[6]

Jacobson and Parentani discovered that sound waves respond to distortions in water in the same way that light responds to curved spacetime around a black hole. This discovery implies that spacetime has fluid properties.

Speed and Resolution

The meaning of *speed* slightly differs when we switch from a continuous to a quantized vacuum. In continuous geometries it makes sense to speak of speeds based on continuous numbers, where an infinite amount of possible speeds exist between any two specific speeds. In a quantized vacuum, however, speed is discrete and the range of possible speeds depends on temporal resolution.

When we resolve events to an accuracy of one Planck time, there are two possible associated speeds, 0 and c, which correspond to moving 0 or 1 unit of length during this interval. An event whose duration is resolved to an accuracy of two Planck units of time has three possible speeds, $0, \frac{1}{2}c$, and c, which correspond to moving *0*, *1*, or *2* units of length during that duration. Over longer durations more and more speeds become available.

Technologically, we have a long way to go before we start resolving the Planck scale. The shortest-period events that we can measure to date are quick bursts of laser light, which can be as short as a femtosecond $(10^{-15}\ seconds)$, or even a few attoseconds $(10^{-18}\ seconds)$. Nevertheless, we have been able to indirectly infer even shorter transitory episodes—on the order of 10^{-25} seconds (using extremely energetic collisions in particle accelerators). For example, "the mean life time of the top quark, the most massive elementary particle so far observed, has been inferred to be about 0.4 yoctoseconds (0.4×10^{-24} seconds)."[7] On these scales the discreteness of speed is significantly blurred.

Another thing to note about velocity is that we get different formulas for the addition of velocities when we assume that the vacuum is Euclidean, Minkowskian, or a quantized fluid. To consider those differences lets imagine an astronaut moving with constant velocity u compared to his spaceship, and the spaceship moving at constant velocity v compared to earth (Figure 8-2). The question we now ask is, how fast is the astronaut moving from earth's point of view?

Under Euclidean assumptions the answer is given by the velocity addition formula $v' = u + v$. This equation captures Euclidean intuitions about how velocities should add, but at fast speeds it gives us completely wrong answers.

Under Minkowskian assumptions spacetime warps in response to speed, morphing measures of time and distance. Therefore, according to relativity the velocity addition formula becomes

$$v' = \frac{(u+v)c^2}{c^2 + uv}$$

This formula shows how the addition of velocities in Nature differs from the way velocities are added in Euclidean geometry. Those differences, however, only become significant as we approach relativistic speeds.[8]

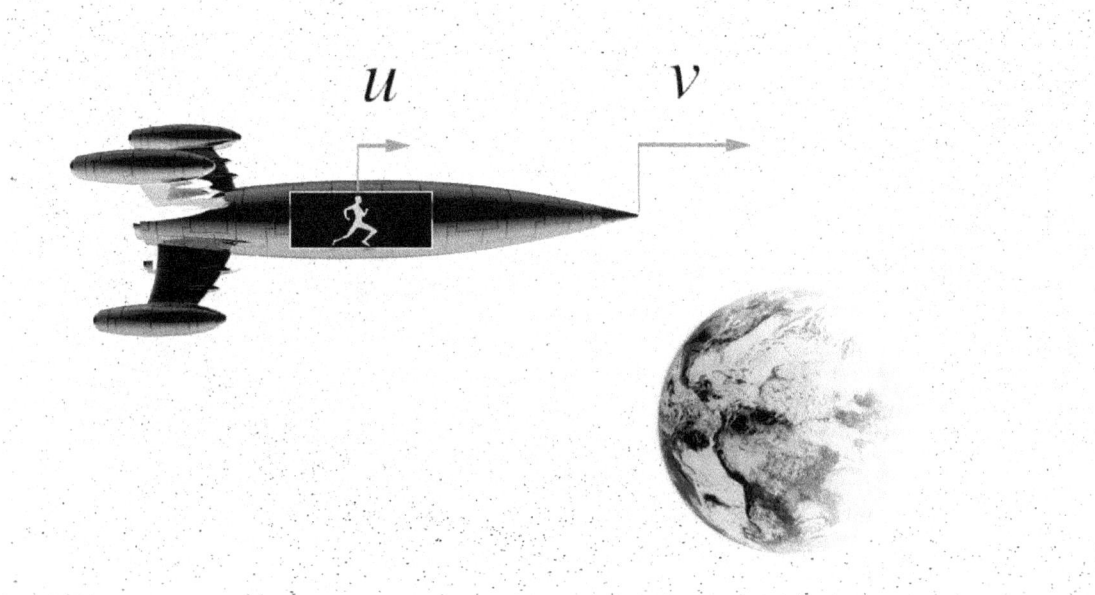

Figure 8-2 Addition of velocities

Under the assumption that the vacuum is a quantized fluid, measures of time and distance also warp in response to changes in speed but, as previously mentioned, quantization also limits the available speeds.[9] To correctly compute the addition of relativistic velocities for any temporal resolution we need a quantized version of Einstein's velocity addition formula, which can be expressed as:[10]

$$v' = \frac{c(ae + bd)}{(be + ad)}$$

Here u represents the speed of the astronaut compared to his spaceship, defined as the number of Planck units of space (a) he moves, compared to the number of Planck units of time (b) observers on the spaceship experience while the astronaut is running ($u = \frac{al_P}{bt_P}$). And v represents the speed of the spaceship compared to earth, defined as the number of Planck units of space the astronaut moves through (d), compared to the number of Planck units of time (e) observers on earth experience during that translation ($v = \frac{dl_P}{et_P}$).

If the astronaut is running at a speed of $\frac{2}{3}c$ compared to the ship, and the ship is moving at a speed of $\frac{4}{5}c$ compared to earth, what speed will the observers on earth measure for the astronaut? The Euclidean formula spits out a completely wrong answer → $1.4\overline{66}\ c$.

According to the other two equations, the observers on earth observe the astronaut to be moving at a speed of $0.956521739\ c$. Although both equations agree, our new velocity addition equation is different from Einstein's velocity addition equation in that it contains a quantization limit (a, b, d, & e are always integers and $a \leq b, d \leq e$). It depicts all possible solutions as a fraction of the speed of light $(be + ad) \geq (ae + bd)$.

Lorentz Contraction

Vacuum quantization also implicitly envelops FitzGerald-Lorentz contraction, which characterizes the tendency of an object's length to shrink toward zero (in the direction of its motion) as its speed approaches the speed of light. In a quantized vacuum this effect can be explained in the following manner.

When an object is at rest, the number of independent quanta between its ends determines its spatial length. A sphere at rest, with a diameter equal to one centimeter, consists of $\sim 10^{33}$ space quanta along its orthogonal diameters of length, width, and height, each of which independently resonate $\sim 10^{44}$ times per second. As the sphere begins to move (in the direction of its length), the number of independent quanta defining its width and height remains the same, but the number of quanta in its length decreases because the quanta that make up its length begin to lose their independence.

As the sphere increases speed, the set of quanta that span its length lose their ability to uniquely express position. Those quanta become more and more connected, contributing less and less unique positions to the sphere's length. At the speed of light, the set of quanta defining the sphere's length become geometrically conjoined—contributing only a single unique location to the map (Figure 8-3).

Figure 8-3 Shrinking length

At the speed of light, the sphere's entire length is effectively located at one space quantum per chronon of the observer's time. Its measurable size reduces to one quantum unit of length (one Planck length). However, the magnitude of the sphere's width and depth (the directions perpendicular to its velocity) do not change because the total number of independent space quanta defining those measures does not change. This is the origin of the effect known as FitzGerald-Lorentz contraction.

Quantization puts our representation of FitzGerald-Lorentz contraction in slight disagreement with the traditional representation, which allows an object's length to shrink to zero as it reaches the speed of light. This difference, though very slight, may provide falsifiable predictions.

Speed Outside of Space

Both Galileo and Newton suspected that our understanding of light would play a central role in our quest to unveil Nature's true character. In 1638, Galileo attempted to measure the speed of light by stretching lanterns from hill to hill. Although his method was far too crude to actually measure light's enormous speed, Galileo was the first person to empirically test for a finite speed of light.[11]

Galileo and Newton were held back in their investigation of light because they lacked an understanding of quantum properties. Building on Michael Faraday's brilliant insights, James Clerk Maxwell eloquently laid down the modern theory of electromagnetism in 1865.

Today light is seen as so intimately connected with the force that binds atoms together, that our modern theory of electromagnetism *is* our theory of light.[12]

Vacuum quantization allows us to reproduce that theory with an ontologically stirring, background dependent, model. According to that model, objects traveling *through* all of the locations between two points in x, y, z space, cannot do so faster than light can travel between those points. Nevertheless, it is possible for objects to arrive at a distant location in space before light is able to travel to that location—if the object travels outside of space, or in a superspatial path. Anything that allows the object to avoid some of the *space* between two spatial locations makes it possible for the object to arrive at a distant location in less time than it takes light to reach that location (e.g. quantum tunneling).

It is also possible for two locations to separate faster than light can travel through the space between them. This is the case because the collision frequency between quanta characterizes the average portion of quanta that are in contact at any moment. During collision, quanta lose their status as unique and independent positions. When they separate, uniqueness of location is restored. In a cooling system the collision frequency decreases, causing the number of unique locations in the map to increase. Therefore, the number of total independent locations between two distant locations in the vacuum can increase over time, making it possible for two locations in space to separate faster than light can travel between them.

Vacuum quantization explains why all sublight speeds are relational, reveals the origin of a limiting speed in the vacuum (on all macroscopic scales), elucidates why that special speed is nonrelational, and precisely dictates the value of that speed. It also enhances our ontological access to warped spatial geometries—the acme of Einstein's general relativity. This shall be our next topic.

[1] To simplify the system, we can pretend that the atmosphere is composed only of nitrogen (N_2) molecules. In reality there are many other components, each of which have their own distributions.

[2] Richard Feynman. *QED—The Strange Theory of Light and Matter*, pp. 89–90.

[3] This is a macroscopic condition. To explore why see Richard Feynman, *QED—The Strange Theory of Light and Matter*, pp. 89–90.

[4] For a more technical explanation of why c is a limiting velocity see: Mermin. (1968). Chapter 15; Nahin. (1999), pp. 342–353 and Tech Note 7.

[5] Theodore A Jacobsen & Renaud Parentani. (2005, December). An Echo of Black Holes. *Scientific American*, p. 71.

[6] Ibid., p. 72.

[7] Scott A. Diddams & Thomas R. O'Brian of the Time and Frequency Division of the National Institute of Standards and Technology. *Scientific American*. (2005, April), p.108.

[8] For nonrelativistic velocities, hyperbolic and Euclidean geometries have practically indistinguishable velocity addition results.

[9] Measures of distance and time relate according to the following equation:

$$dl^2 = dx^2 + dy^2 + dz^2 = c^2 dt^2 \qquad ds^2 = -dx^2 = dy^2 - dz^2 + c^2 dt^2$$

(dl^2 expresses space-like displacements as a positive quantity, and ds^2 yields a positive quantity for time-like curves—paths composed of more time than space.)

[10] Both of these equations are simplifications of the full velocity addition formula—restricted by the assumption that the velocities being considered are parallel or antiparallel, and by the assumption of zero curvature. Removing the first restriction requires us to break each velocity into parallel and anti-parallel components, combine those components according to the equations above, and then recombine the resultant components via the Pythagorean theorem. Removing the second simplification takes us from special relativity to general relativity.

[11] Seventy years after Galileo's experiments, the Danish astronomer Olaf Romer used observations of the eclipses of Jupiter's Galilean moons (Io, Europa, Ganymede, & Callisto) to calculate the speed of light. Given the era, his answer was surprisingly accurate. The first person on record to consider whether or not light's speed was finite was Empedocles of Acragas, in the mid-fifth century BCE.

[12] Maxwell figured out how to codify Michael Faraday's electromagnetic insights—how to represent them with clear mathematical equations. As a symbol of how important astronomers and physicists regard this scientific contribution, he has been given the distinguished honor of being the only man to have a feature on Venus named after him. All of the other features on Venus are named after women (either real or fictional). The feature bearing his name is Venus' most prominent mountain chain (approximately 500 miles long) whose peaks are coated with very reflective minerals, possibly pyrite and chalcopyrite, named Mount Maxwell. Metaphorically, Maxwell is spending eternity as the only man on a planet populated by women like Aphrodite and Cleopatra.

Chapter 9 — **Warped Spacetime**

"When a blind beetle crawls over the surface of a curved branch, it doesn't notice that the track it has covered is indeed curved. I was lucky to notice what the beetle didn't notice."

Albert Einstein[1]

"The laws of Nature in an accelerating frame are equivalent to the laws in a gravitational field."

Einstein's Equivalence Principle

Mine Shaft, La Oroya, Peru.

The butterfly feeling we get when we experience the rapid drop of a roller coaster, swiftly drive over the top of a hill, get pulled over the edge of a plunging cascade, or cliff dive off a canyon wall, can teach us a lot. As the effects of free-fall become familiar, we begin to explore questions that can expose us to a radically new way of seeing the world.

Einstein framed his exploration of free-fall with his famous elevator thought experiments. His investigation led to an insight that gave birth to his equivalence principle, which serves as the foundation for general relativity. What exactly was the connection that Einstein made? What profound insight did he grasp? The answer is that he came to terms with the fact that gravity and acceleration are the same thing. He recognized that it is incorrect to think of objects on the surface of the Earth as *at rest*. If an object is in a gravitational field, then it is accelerating (unless it is free-falling).

When I was first introduced to the connection between acceleration and gravity, I was confused. I imagined floating in space inside a large vessel. Then I imagined that the external rockets of that ship were ignited. I understood that as soon as one of the walls of my chamber caught up with me and started pushing me—at which point I could justifiably call the wall a floor—I would start to have weight. As long as I was being pushed through space faster and faster, I would feel my weight. The magnitude of that weight depended only on the amount of acceleration provided by the rockets. With the right acceleration I could have the same weight that I have on earth (Figure 9-1). However, if the wall/floor beneath me were to stop accelerating "upward" at any point in time, then my experience of weight would immediately cease. It was easy for me to accept that in this sense acceleration can give me weight, and in that way it was like gravity, but to me this relationship seemed artificial, or at least a bit one-sided.

If gravity was exactly the same as acceleration then wouldn't that mean that the ground is always pushing up on me just like the wall/floor of the space vehicle? If this were the case, then shouldn't the surface of the Earth always be expanding outward (Figure 9-2)? When I came upon this question, I thought that there was no way the surface of the Earth could be accelerating away from its center because that would require exponential growth in its size. The Earth wasn't exponentially getting larger, so this couldn't be true. For the most part, this is where my state of confusion hovered until I started to explore zero gravity weightlessness.

One way to stimulate curiosity about the connection between acceleration and gravity, or the question of what gravity is, is to change our gravity experience. Reduced gravity, and zero gravity experiences, have just the right mixture of adrenaline and fascination to facilitate a breakthrough. Of course, it helps to couple those experiences with a little reading on general relativity.

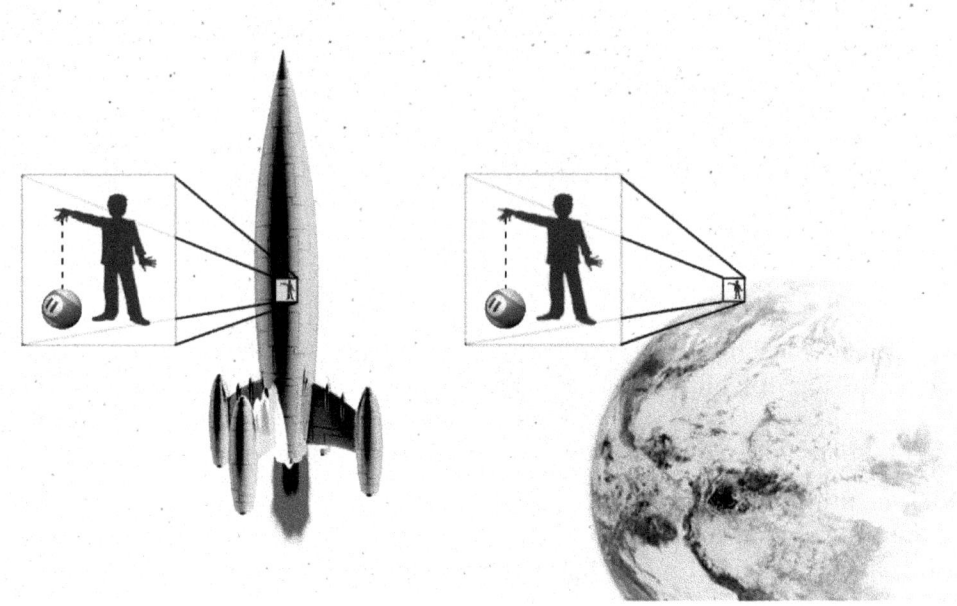

Figure 9-1 Weight from acceleration

My initial taste of something that was baked from this complete recipe occurred during my first solo flight. Leaving Salt Lake City's controlled airspace in a 1970 Cessna 172, I had one thing on my mind. Climbing higher and higher, I announced my position over the local channel and then prepared to turn Tooele's class G airspace into zero-G airspace.

I had practiced stall recovery techniques many times before, pulling up until an airflow stall began to announce itself as an annoying high-pitched squawk in my headset, but this time, instead of waiting for the full stall, I pulled the throttle back all the way and pushed the yoke away from me. Not knowing exactly how far to go, I overshot it a bit. As a result, I went from a normal positive gravity experience to a slightly negative one. My body activated the slack in my seat belt and everything that wasn't fastened down leapt into the air. Unfortunately, this included a rather thick layer of dust and dirt that had been hiding on the floor. I tried to balance out to zero-G, but within a few seconds, it was time to pull up. When I did, I was unexpectedly rewarded with a rather violent thud from my flight bag (which was stuffed full of textbooks on flying and physics). Although I had stowed the bag in the tail section, it had floated its way directly above my head. Obviously it needed its own seat belt.

In sinusoidal fashion, I practiced falling out of the sky, collecting a few more seconds of weightlessness with every cycle. After I learned how to hold the plane exactly at zero gravity, I discovered that the butterfly feeling I mentally associated with free fall actually only occurs when changing from one value of G to another. Once I reached zero-G, I was weightless, but my stomach wasn't churning. However, as soon as I pulled up, the butterflies resumed flapping their wings in my belly until I leveled off again.

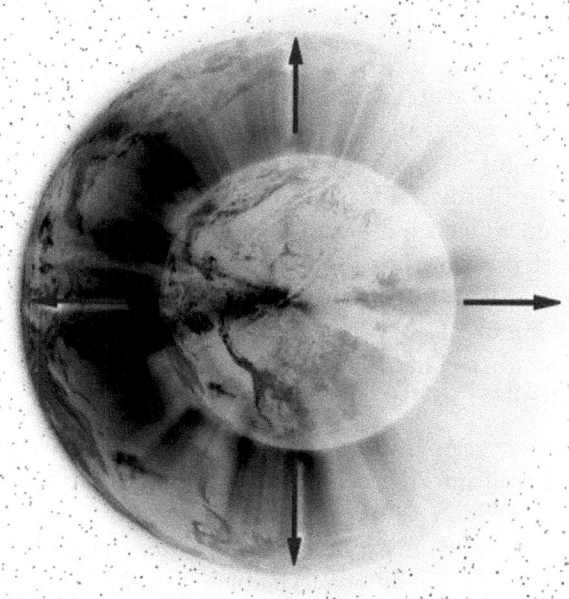
Figure 9-2 Earth's radius growing with an accelerated pace

This experience gave me the personal motivation to investigate this thing we call gravity. A few months later, I had another experience that brought more light to my investigation. This experience took place in a more historically appropriate setting—an elevator. I was on a Peruvian research mining expedition in conjunction with the University of Utah and the University of Texas, Austin. The entrance to the mine system that hosted this particular elevator was at such a high altitude that a brisk walk would result in a blackout in less than five seconds.

Most modern elevators aren't ideal laboratories for Einstein's thought experiment because they never really have an extended period of free-fall. They accelerate briefly, hold a constant velocity for the majority of the ascent or descent, and then make an abrupt stop at the end, preceded by another short acceleration period. That's a normal elevator. What I stepped into in the high Andes mountain mines of Doe Run, Peru, was definitely not a normal elevator. This contraption was truly worthy of Einstein's thought experiment.

The rickety structure I entered wasn't designed to transport people back and forth between a few floors of an office building. It was intended to function as a portal between Earth's surface and its mysterious depths. The chamber held twenty of us between its opposing scissor gates. Its frame was held in place by brakes that clamped onto metallic I-beams, which served as the backbone of this bottomless shaft. There was no cable involved. A cable long enough for this plunge was presumably impractical. An old three-button control device hung on the wall. There was no light source in the elevator itself. Instead a single bulb shone into the elevator from the dynamite blasted tunnel we had just traversed.

Half of the occupants were from our research expedition. The majority of them were either world-renowned precious metal geologists, or operators/owners of substantial mines around the world. The other half of the occupants consisted of Peruvian miners who were all wearing knee-high water boots, Speedos, and hard hats adorned with oversized lamps. The tallest among them was roughly five feet. For their proportions, they all had extremely wide chests—no doubt an evolutionary advantage in these high mountains, where the oxygen was noticeably scarce.

Our guide lifted the control device off its hook and unceremoniously pressed the bottom button. What happened next caught me completely off guard. The elevator just dropped. For all practical purposes, we were falling like a stone into a bottomless pit. The light disappeared, leaving us blindly falling into the abyss. Another light appeared and then, just as quickly, disappeared as we sped past a deeper horizontal level. Each level was separated by over ten meters of metamorphosed rock, and each announced itself with a single fast-moving light bulb. As we continued to fall, the blips of light morphed into streaks of light that pulsated with decreasing breaks between them. For a brief period, it looked like a strobe light was showering the chamber. That strobe light quickly transformed into a solid smear of light that stretched from the floor to the ceiling.

We weren't at perfect zero-G. Some friction somewhere barely kept us on the floor, but I could easily launch myself into the ceiling by twitching my calves. The miners were calmly looking at their feet like this was normal—which was the only thing that was keeping me from screaming. Slowly I made my way from the center of the elevator towards its walls. The imminent crash seemed completely unsurvivable, even if I used everyone else as cushioning, but I had to do something.

As time went on, I noticed something very important. Even though we were plunging into the womb of the Earth, the only manifest difference between this reference frame and the more familiar reference frame (standing on the Earth's surface) was my weight. If I were unable to see outside the elevator, it would have been impossible to distinguish whether I was falling or gravity had been turned down to a very small fraction of its regular strength. The only thing that had changed was that the Earth beneath me had reduced how much it was pushing on me—that little bit of friction was the only push remaining. Einstein was right. The Earth was pushing on me and causing me to accelerate.

Thankfully, the operator did eventually activate the brakes, and we ended up a mile and a half below the surface in the tempestuously heat-laden air of the underworld, where I discovered that the rocks were hot enough to burn holes in my jeans—but that is another story. It was time to rethink my earlier analysis of acceleration and gravity.

Rethinking Gravity

Although I had accepted Einstein's equivalence principle as true, I struggled to understand how the acceleration of a rocket was equivalent to acceleration associated with the gravitational field. What exactly is a gravitational field? How do I describe it geometrically? How does an accelerating rocket mimic that same field? Are there any differences between these two kinds of acceleration?

Modeling the vacuum as a compressible fluid resolves the vacuum's particulate nature and allows a simple and complete description of spatial density. This is advantageous, because gradients express how densities change with respect to either position or time, and the experience of a density gradient is an experience of acceleration.

To unravel that explanation, let's back up a little, and pay attention to how the density of space responds to changes in velocity. From our eleven-dimensional perspective we discover that when a spaceship is at rest in the vacuum fluid, the density of space is, on average, uniform in all directions—so long as we are referencing a region that has no measurable curvature. By contrast, when a ship is moving through the vacuum fluid with a constant velocity, it interacts with quanta more frequently in its direction of travel. With increasing speed, the ship's experience of space becomes less and less uniform (Figure 9-3).

Except for effects that show up when making comparisons between different reference frames, this variable experience of space, which also changes the ship's experience of time, is entirely undetectable four-dimensionally. This is the case, because seven Planck lengths of space is equal to seven Planck lengths of space, no matter how they are separated through superspace. In other words, space travelers restricted to four-dimensional observations are incapable of distinguishing between different uniform spatial dispersions. This is why all sublight constant velocity reference frames appear identical. Without a way to distinguish zero velocity from constant non-zero velocities all inertial frames are on equal footing.

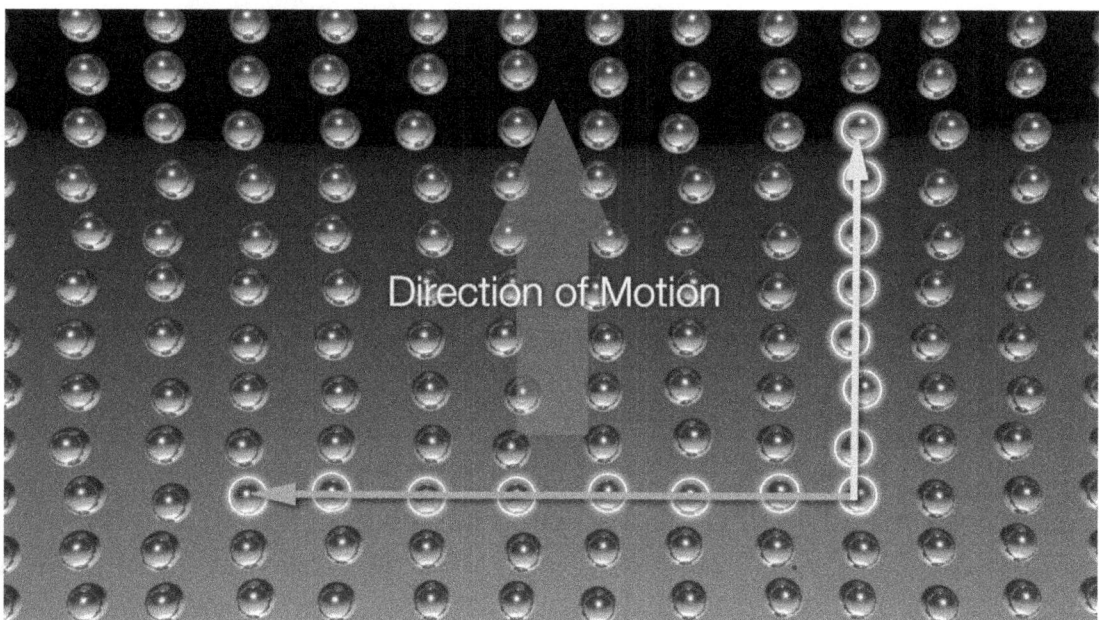

Figure 9-3 Four-dimensional measurements cannot distinguish different uniform spatial densities

Acceleration is different from velocity—you can feel it. After experiencing a truly inertial reference frame (like floating in zero-G), no one in an accelerated reference frame would confuse his or her reference frame for an inertial one, that is, a nonaccelerated frame (constant or zero velocity). The reason acceleration is different from constant velocity, is that it references non-uniform spatial dispersions, or spatial density gradients. To understand this, we need to explore the nuances that come with the claim that the density of space can macroscopically change from one region to another.

To that end, let's imagine that the relative velocity between two spaceships (Atlantis and Columbia) is half the speed of light. If neither of the spaceships is accelerating, then both reference frames are inertial. To the astronauts aboard Atlantis, Columbia is moving at half the speed of light, its length is shortened (in the direction of its motion), and its clocks are running slow (Figure 9-4). Likewise, the astronauts on Columbia see Atlantis moving at half the speed of light, its length is shortened, and its clocks are running slow.

These observations directly reflect the difference between the spatial distributions between the two ships. As their difference in velocity increases, the relative vacuum density between them increases, along the axis of motion. It makes no difference which ship is actually stationary from the vantage of absolute volume (if either of them are) because only relative velocities can have any measurable meaning in the four dimensions of spacetime.

Figure 9-4 Switching perspectives

How does all of this help us understand acceleration or gravity? What does a change in spatial density have to do with acceleration? To answer that question, consider what would happen if one of the pilots of those ships decided to change the velocity of their ship, until there was no remaining velocity difference between the two ships. Matching velocities means that both of these ships will, once again, have lengths that match and clocks that tick at equal rates. However, in order to accomplish this synchronicity, a change has to occur. Specifically, the difference in spatial density between the two ships, along their direction of motion, must go from something to zero. This means that a *change* in spatial density must be experienced; which means that a change in speed (acceleration) has something to do with a change in spatial density.

Acceleration is literally the experience of a change in spatial density. If all of an object's parts experience space equally over time, then the object is not accelerating. Acceleration can be accomplished by firing your rockets, so that you experience a greater density of space from moment to moment. It can be accomplished by spinning something around a central axis, so that its distant parts experience more space than its proximate parts. Or it can be accomplished simply by standing still within a radial spatial density gradient.[2]

The difference between linear acceleration, and the kind of acceleration that we associate with gravity, is that the density gradient associated with linear acceleration is linear and time dependent, while the field associated with Earth's mass is radial and position dependent. During a period of linear acceleration, from one velocity to another, the object being accelerated experiences a spatial density of a particular value at one moment and a different value the next moment. Its experience of space is not equal through time. This means that it is experiencing a change in spatial density—a gradient over time.

Gravity's spatial density gradient is a different kind of gradient—one that references a density distortion that can be statically defined (without time). Just in case such gradients are unfamiliar, let's spend a moment discussing exactly what these kinds of gradients look like.

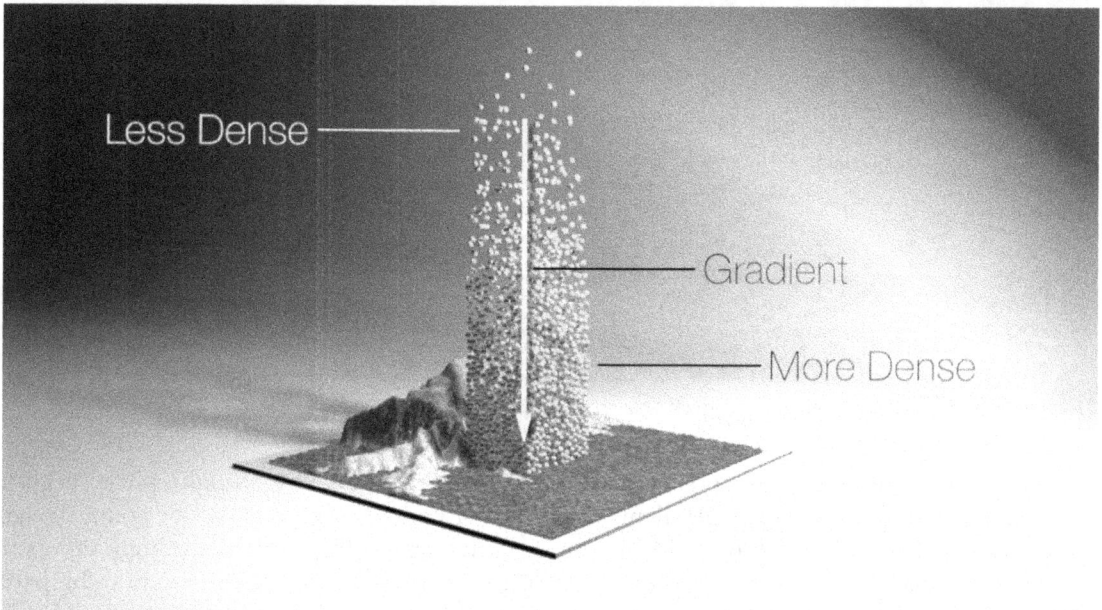

Figure 9-5 Atmospheric density gradient

To use a simple example, let's imagine the molecules that make up the atmosphere of our Earth. When we fly in an unpressurized aircraft, or drive up mountainous terrain, we discover that air pressure, or density, decreases as we increase in altitude.[3] An open container at sea level contains more air molecules than it does at a higher altitude. The same amount of volume contains a different number of particles or molecules. Therefore, our atmosphere is a good example of a density gradient that varies with height. A spatial density gradient is very similar to this. It is defined by a change in the number of space quanta occupying equal regions of superspatial volume.[4]

Gradients are depicted using arrows. Each arrow points toward the greatest local increase in density with a length that signifies the magnitude of local density increase. In our atmosphere, the gradient arrows all point towards the center of the earth. If we looked at a portion of our atmosphere above just one square mile of land, its gradient arrows would point straight down (Figure 9-5). But this scale of observation is slightly misleading. The gradient arrows for the adjacent square miles may be close to parallel, but they are not perfectly parallel. As we zoom out we begin to resolve the curvature of the earth and the radial structure of the gradient.

For straight-line acceleration, the gradient arrows always point in the direction parallel to the accelerated motion (Figure 9-6), and gravitational gradients are always radial. These gradients are important because they define how *straight* paths through the vacuum differ from Euclidean projections.[5]

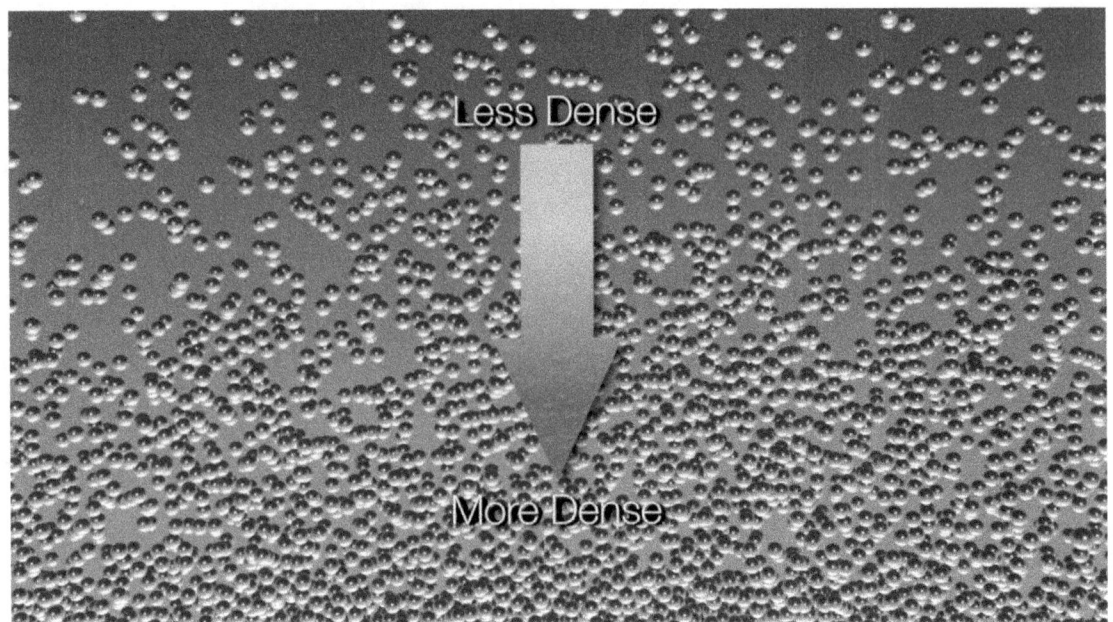

Figure 9-6 The density gradient of linear acceleration

Our eleven-dimensional perspective allows us to explain why straight paths become distorted within spatial density gradients (Figure 9-7). Following a geodesic, or going straight through space, means that one side of an acoustic wave must travel through the same amount of space as the other. When the density of space is non-uniform, straight paths through space differ from Euclidean expectations. The side nearest the gradient must travel a greater spatial distance in order to continue on the Euclidean straight path. But, if it followed that path—if the near side covered more spatial distance than the distant side—then the path of the acoustic wave would not be a geodesic, or a straight path in x, y, z *space*.

In order to follow a straight path through superspace the wave would need to get some kind of boost, something would have to push the left side through more space than the right side during the same amount of time. Without this extra boost the object will follow the natural straight path—the path in which both sides propagate through the same amount of space and cover the same spatial distance per unit time. In the absence of external forces all objects will follow these straight trajectories through the vacuum.

The insight is that when we assume a flat continuous metric for space, we interpret the paths of objects that move through radial spatial density gradients to be curved in space.[6] Our ignorance of the true geometry of spacetime leads us to assume that paths, like orbits, are curved. Once we account for the spatial density gradients in Nature, all objects in orbit are understood to be following straight paths—spacetime geodesics.

The fact that such paths appear curved to us is an unavoidable consequence of our attempt to map the eleven dimensions of Nature using only four dimensions. Our dimensional tenuity leaves us unable to chart or comprehend the density gradients of spacetime that define the true straight paths.

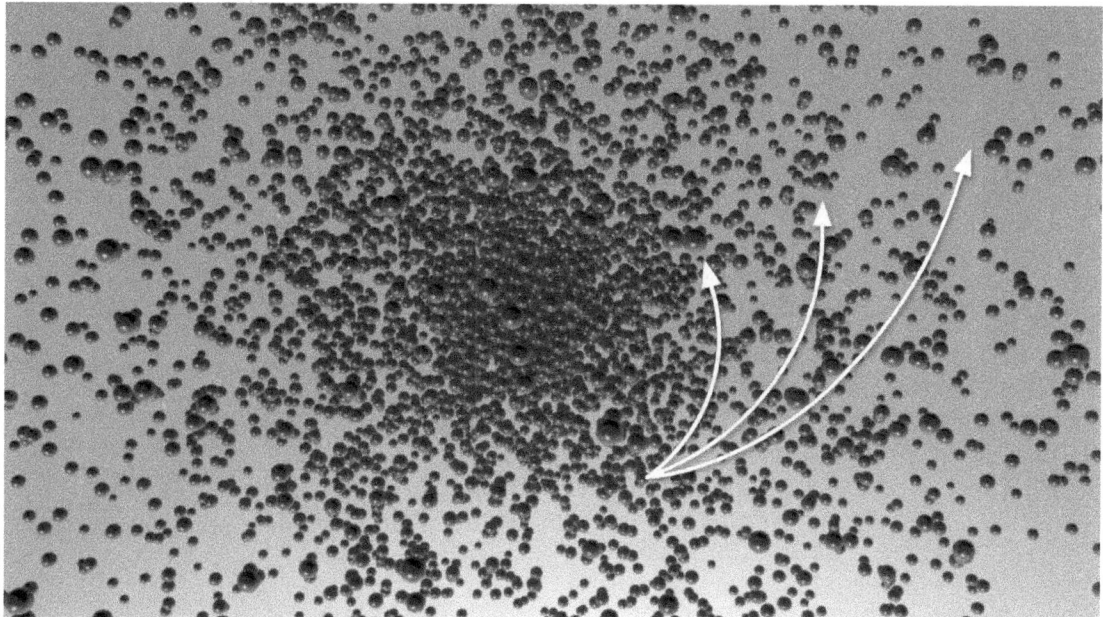

Figure 9-7 The straight path through radial spatial density gradients

Because we don't see these density gradients with our eyes, we make up *forces* (gravity, the weak nuclear force, electromagnetism and the strong nuclear force) to explain why objects follow paths that appear curved to us, or behave in ways that are not expected by our four-dimensional intuitions.

Taking a different route, vacuum quantization allows us to conceptualize the dynamic geometry of space without restricting us to partial representation. For example, instead of representing the curvature of a slice of space using a rubber sheet diagram (Figure 9-8a), it allows us to represent it in terms of density (Figure 9-8b). This representation is easily extended beyond that slice. As a consequence, modeling the vacuum as a compressible fluid allows us to depict the complexities of spacetime curvature in a superior way.[7]

Acceleration is a manifestation of our unequal experience of spacetime. We feel our weight on Earth because we are not following a natural or straight path through space: we are continuously subjected to an uneven spacetime experience. The spacetime around Earth takes the form of a radial density gradient, and the ground beneath us forces us to experience the unevenness of that gradient.[8] Boarding a rocket with a constant linear acceleration of 9.81 meters/second^2 will reproduce the same local effect, but globally the shape of the gradients will differ.

Given that spatial density gradients require no external energy—no chemical rockets—how are we to understand their origin? Where do radial density gradients come from, and how are they sustained? You might guess that mass has something to do with it. From Einstein's $E = mc^2$ equation, we learn that mass is a form of energy, and because energy denotes a distortion in the geometry of spacetime, it is natural to be curious about how these distortions (spatial density gradients) come about and gain dynamic stability.

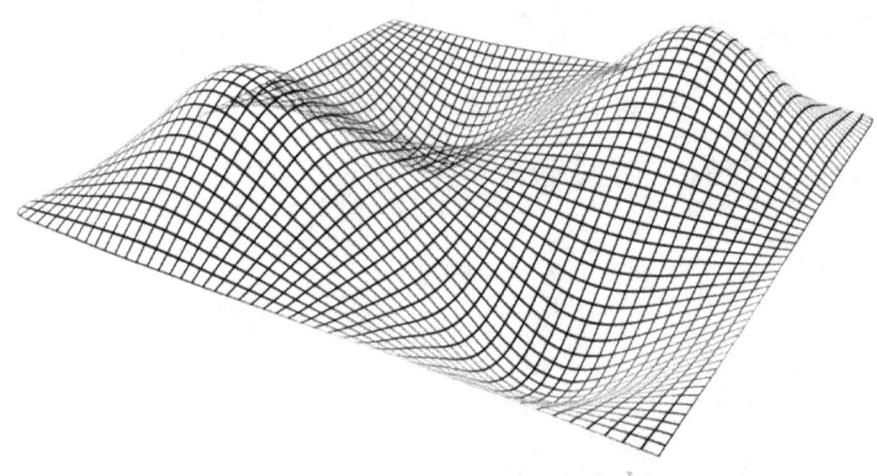

Figure 9-8a Complex curvature of a plane as traditionally represented

Figure 9-8b Complex curvature of a plane represented by vacuum quantization.

According to our map, the solution has to do with the fact that the vacuum is a superfluid. In regular fluids, vortices cannot maintain themselves indefinitely. When a vortex is not being actively created, its energy begins to dissipate throughout the medium. In a superfluid, however, vortices can be stably maintained—indefinitely. If the vacuum is a superfluid, then specific quantum vortices that form in the vacuum will persist (as well as conglomerates of vortices). In the absence of vortices, quantum interactions are expected to be randomly oriented, bouncing off each other such that the average density is uniform. In a spatial vortex, the average motions of the quanta begin to align as they swirl about a common center. This alignment allows the quanta to become densely packed, which is why the density of the medium radially increases near the center of the vortex ring (Figure 9-9).

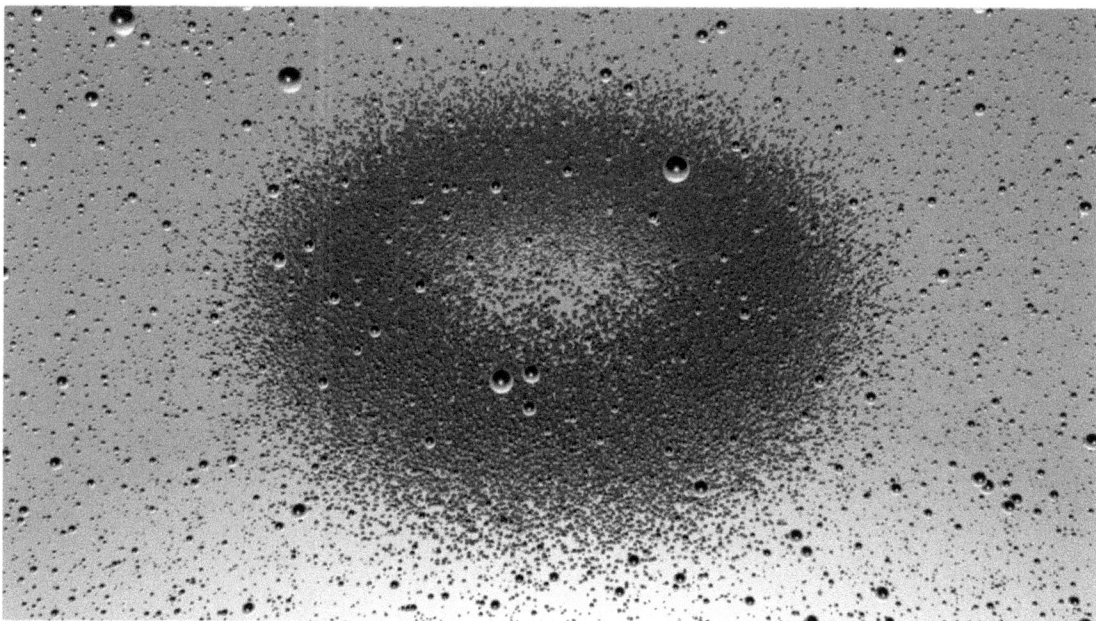

Figure 9-9 Stable vortices

This explanation jives with the projection that matter formed as the temperature of space dropped below a critical limit—the critical superfluid temperature.[9] It treats matter as a geometric distortion—a vortex in the superfluid vacuum that initiates a spatial density gradient.

Beyond Newton's Apple and Einstein's Rubber Sheet

Although Newton predicted the effects of gravity with an accuracy of one part in ten million, he failed to explain the origin of gravity. The idea that gravity is an invisible magical force that instantaneously tugged between all bodies of mass was just as distasteful to his philosophical senses as it is to ours today, but he never figured out how to transcend this description.

That 'little' problem hung in the air until Einstein revealed gravity as a geometric effect. Einstein's geometric interpretation of gravity explained Mercury's anomalous perihelion advance, which deviates from Newtonian gravity by a mere 43 seconds of arc per century (an amount that translates to being off by one whole orbit every 3 million years).[10] It also predicted that starlight would bend by about 1.75 arc seconds as it grazed the Sun, which was famously observed during the solar eclipse of May 29, 1919, by Arthur Eddington's expedition to the island of Principe off the coast of Africa.[11] And, it predicted that clocks should slow in increased gravitational fields, in a manner that agrees with today's most precise measurements. Tying all of these effects to the geometry of spacetime was a revolution, but the full power of that revolution has not yet been realized. The trouble is that Einstein's notion of curvature, and how the geometry of spacetime evolves, has been restrained by the limitations of rubber sheet diagrams.

Rubber sheet diagrams fail to reveal the full underlying structure of spacetime's geometry. They map only a single slice of space at a time, and they make use of some *other* dimension, but do not explain what this other dimension is (Figure 9-10).

When it comes to explaining gravity, rubber sheet diagrams only push the question of gravity's cause back to, "What causes the rubber sheet to stretch?" We get no sense of why, or how the presence of mass stretches the space around it, or why acceleration would do the same. If anything, we may be tempted to think that the weight of the bowling ball causes the rubber sheet to stretch "down." This isn't allowed because weight is a function of gravity—we cannot presuppose its existence. In other words, we cannot use gravity to explain gravity. Doing so wouldn't illuminate anything. Furthermore, the meaning of the word "down" is lost in the rubber sheet diagrams because the dimension used to graph curvature—in this case σ—replaces the familiar third spatial dimension. Standard images of warped space may help us understand that the shape of space changes around a massive object (space having an increased presence near a massive body), but it doesn't help us answer why or how the shape of space actually changes around mass to begin with.

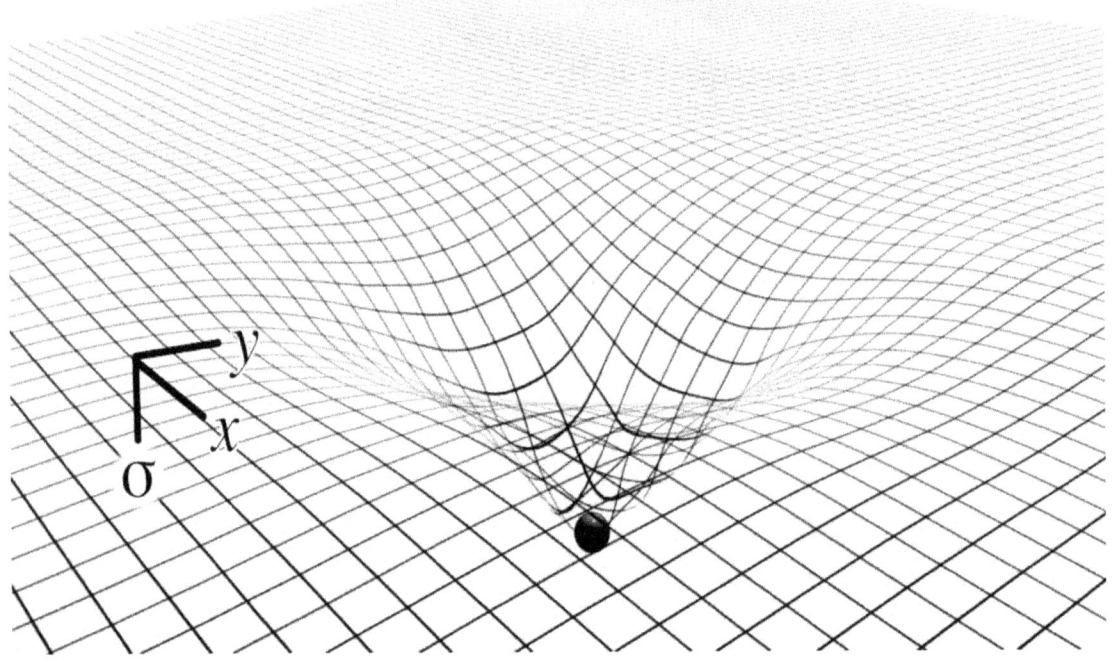

Figure 9-10 The limitations of rubber sheet diagrams
If space is continuous, then this may be the best visual model, but if space is quantized, a better model can be introduced.

When we replace rubber sheet diagrams with radial spatial density gradients, we gain the ability to simultaneously graph all three spatial dimensions and the curvature of those dimensions (Figure 9-11). This is quite an improvement over graphing just the curvature of a single slice of space at a time. By substituting radial density gradients for the traditional rubber sheets, we avoid the elimination of one of the familiar spatial dimensions, gain the ability to comprehend warped time, see exactly what the other seven dimensions of physical reality are, and become able to explain the source of gravitational fields.

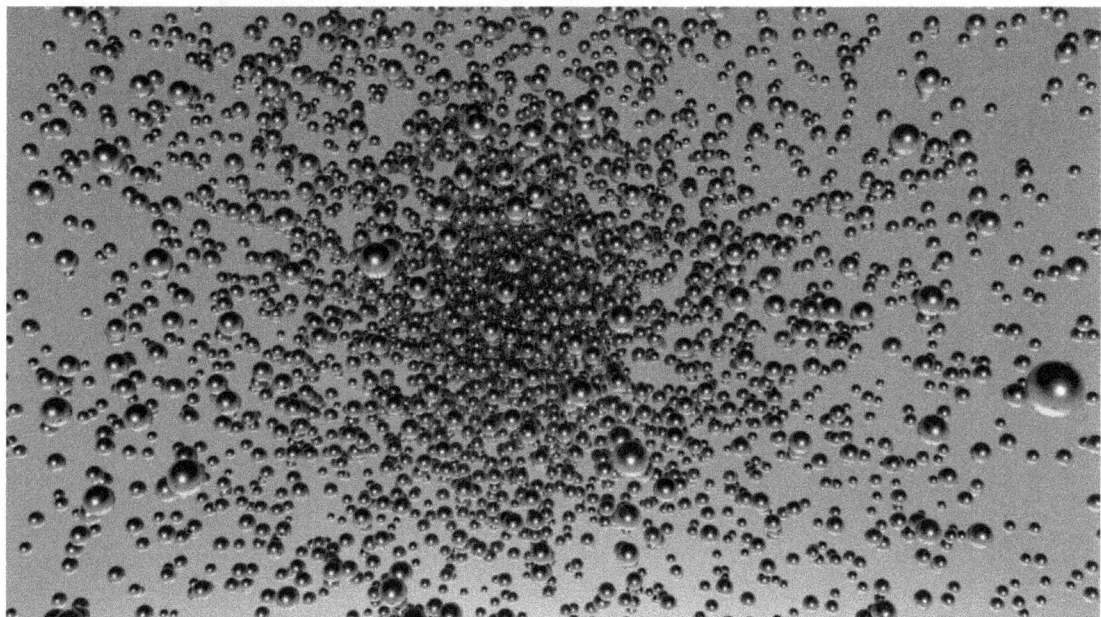

Figure 9-11 Modeling the curvature of space without dimensional reduction

Depicting Warped Time

> *"Time is warped if its rate of passage differs from one location to another."*
>
> *Brian Greene*

Traditional rubber sheet diagrams also don't do much to help us understand warped time. This shortcoming is inherent in the presupposition that spacetime is smooth and continuous. A model framed by that presupposition retains, at best, a very nebulous definition of time. But a quantized model of the vacuum enables a straightforward definition of time. In such a model, the illustration of warped time becomes inherently tied to the illustration of warped space. That is to say, a density gradient of space quanta not only helps us visualize warped space, where the geometric relations of space are distorted, but the same graphic representation conveys the meaning of warped time.

Because time is simply a measure of the free resonations experienced at each quantum of space, the ability for these space quanta to undergo collision-free resonations inversely depends on density. Wherever a spatial density gradient exists, time automatically warps.

141

This means that time-like geodesics, which maximize "proper" time, must follow the antigradient.

This insight reveals that the amount of time an object experiences (or inversely, the rate of spatial interactions it experiences), can vary in two ways. We can decrease the amount of time an object experiences by moving the object to a region with a greater quantum density, or by increasing the object's speed through space.

Figure 9-12 Time near a black hole

If two identical clocks were placed in regions of different spatial densities, they would tick at different rates (Figure 9-12). The clock in the denser space would have a reduced time signature—it would run slower. Likewise, time runs slower on the surface of a planet that is more massive than earth because the density of space quanta is greater at that surface than on the surface of the earth (Figure 9-13). Standing on that planet, all the quanta within the astronaut's body will statistically experience fewer free resonations before interacting with the next quanta. In this manner, every quark, atom, and molecule in the astronaut's body trades travel through time for travel through space. She is moving through space and time, but because she occupies a region of higher spatial density, she is trading more of her movement through time for movement through space than someone standing on Earth.

When the astronaut returns home, she notices that she has experienced less time than her counterparts. Because her family and friends have been traveling through time at the rate that is characteristic for the spatial density of the surface of Earth, everyone on Earth has aged more than she has during her voyage around the black hole.

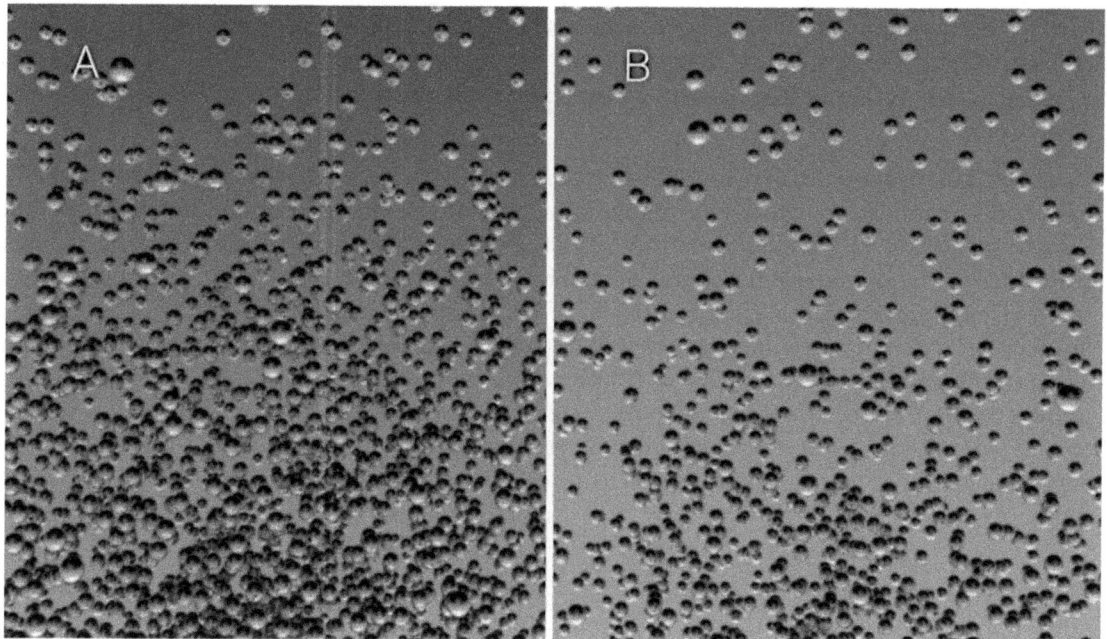

Figure 9-13 Spatial density on a massive planet (A) versus Earth (B)

Gravity's Rainbow

> *"As ever when we unweave a rainbow, it will not become less wonderful."*
>
> *Richard Dawkins*

The assumption that space is a superfluid leads us to the expectation of compressibility (spacetime curvature), and gives us a rich way to geometrically depict gravitational fields. This picture resolves both the similarities and the differences between radial density gradients (gravitational fields) and densities that change in time (fields associated with linear acceleration). It allows us to visualize warped spacetime without eliminating one of the three familiar dimensions of space, and it gives us an intuitive explanation for why time slows down as one approaches a region with a greater spatial density.

Vacuum quantization allows us to resolve another problem that has plagued higher-dimensional models since they were first introduced—the problem of how there can be more than three spatial dimensions that compose physical reality when gravity clearly depends upon only three dimensions. The inverse square law describes the strength of gravity as getting weaker by a factor of four as distance increases by a factor of two $f(g) = G\frac{m_1 m_2}{r^2}$. This equation implies that gravity is dispersed throughout three dimensions only.

To understand why the inverse square law implies a three-dimensional relationship, imagine a star shining in space. Now imagine a transparent sphere positioned around that star with a radius of one light-year. The light emanating from the star and penetrating the surface of that sphere has a specific intensity. How would that intensity change if we doubled the radius of our sphere? Because the amount of light coming out of the star remains constant, we end up with the same amount of light falling on a surface of four times the area (surface area of a sphere = $4\pi r^2$). So when r is doubled, the intensity decreases by a factor of four. This result explicitly depends upon the geometry of three-dimensional volumes. Therefore, any force that obeys an inverse square law must also implicitly depend on three dimensions.

No previous model has been able to satisfactorily comply with this gravity restraint. But under the assumption of vacuum quantization this three-dimensional condition is inherent. Gravity fields are radial spatial density gradients, which can only be defined and depicted within the three-dimensional volume of superspace. Therefore, it is the three dimensions of superspace, and only these dimensions, that the inverse law makes reference to. Gravity is limited to a three-dimensional relationship because it is a manifestation of a density gradient. We cannot talk about the density of space quanta within the volume of a single quantum, and we cannot define a density of space within the dimensions of x, y, z space. Out of nine spatial dimensions, gravity can only be represented within the three dimensions of superspace—the dimensions that allow us to depict and resolve spatial densities and density gradients.

[1] Albert Einstein explaining to his son Edward why he was so famous in; Einstein to Hendrik Lorentz, Feb. 15, 1911; Walter Isaacson. *Einstein: His Life and Universe*, p. 196.

[2] When we illustrate a density gradient we are depicting a field. Gravity is not alone in being a "force" that arises from a field. Magnetism and electricity are also examples. A changing electric field creates a magnetic field and a changing magnetic field creates an electric field. It is the gradients inherent in fields that give rise to what we call forces. Maxwell's equations relate the rate of change of the electromagnetic fields to the presence of electric charges and currents.

These equations imply a speed for electromagnetic waves equal to $\frac{1}{\sqrt{\varepsilon_0 \mu_0}}$ which is equal to c.

$$\nabla \cdot \boldsymbol{E} = \frac{\rho}{\varepsilon_0}$$

$$\nabla \cdot \boldsymbol{B} = 0$$

$$\nabla \times \boldsymbol{E} = -\frac{\partial \boldsymbol{B}}{\partial t}$$

$$\nabla \times \boldsymbol{B} = \mu_0 \boldsymbol{j} + \mu_0 \varepsilon_0 \frac{\partial \boldsymbol{E}}{\partial t}$$

Where: $\boldsymbol{E}$ = the electric field, $\boldsymbol{B}$ = the magnetic field, ρ = the electric density, j = the electric current density, ε_0 = the permittivity of free space and μ_0 = the permeability of free space, $\boldsymbol{\nabla}$ = "gradient of," and ∂ = "partial derivative of."

Because fields are geometric distortions of spacetime we may argue that all "forces" are nothing more than our way of accounting for the difference between what we expect to see, based on our assumptions about the geometry of Nature, and what we actually see. Once we model Nature's true geometry we no longer need to evoke mysterious forces to explain what occurs (Chapter 21).

[3] Although it is well established that matter is composed of atoms, and molecules, it is quite remarkable that humans have reached a level of comprehension that includes these miniscule ingredients when we consider that, for the most part, humans only experience the macroscopic properties of most media.

As an example of how small the molecules of air are—consider that the volume of an average sized breath of air is about one liter. Near sea level, there are roughly ten sextillion (1×10^{22}) air molecules in a liter. Coincidentally, there are also roughly 1×10^{22} breaths (liters) of air in the entire atmosphere. This means that there are as many molecules in a single breath as there are breaths in the atmosphere—a condition that is particularly interesting in terms of the interconnectivity of life on Earth.

Once we allow for sufficient mixing in the atmosphere (a few years is plenty) we can statistically say that, on average, every breath we breathe contains one molecule from every breath ever taken by any past human. Einstein lived roughly 76 years, which translates to about 500 million breaths. Therefore, we can say that, on average, each and every breath we take contains roughly 500 million molecules that were taken in by Einstein during his life. With every breath we share particles that have been a part of nearly every person that has breathed on this planet. That's interconnectedness.

It is also interesting to note that because the visible universe contains roughly 70 septillion stars (7×10^{25}), it only takes us roughly 8 hours and 20 minutes to take in one molecule of air for every star in the visible universe! Donald Ahrens. (2003). An Introduction to Weather, Climate, and the Environment. *Meteorology Today*, 7th Edition, p. 4.

[4] Within the gradient the manifestation of time and the geometrical relationships within space are altered, or warped, in direct correlation to this changing spatial density. However, the quantity $E^2 - p^2c^2 = m_0c^4$ remains equal in every region.

[5] Mathematically we can express a spacetime density as the Lagrangian $\mathcal{L}$. For example, if we have: $\mathcal{L} = \phi^a \phi^b \nabla_a \Psi_b$, then $\frac{\partial \mathcal{L}}{\partial \phi^c} = \phi^b \nabla_c \Psi_b + \phi^a \nabla_a \Psi_c$ where $\frac{\partial \mathcal{L}}{\partial \nabla_c \phi^d} = 0, \frac{\partial \mathcal{L}}{\partial \Psi_c} = 0$, and $\frac{\partial \mathcal{L}}{\partial \nabla_c \Psi_d} = \phi^c \phi^d$.

Therefore, a spatial density gradient can be expressed as a change of the Langrangian ($\mathcal{L}$) from one region to the next.

[6] Except when the motion is exactly parallel to the gradient. In this situation, the path is observed as curved only through time. When it is exactly perpendicular to the gradient arrow it is curved only through space. All other motions appear curved through variable degrees of both space and time.

[7] This gives us a richer depiction of Einstein's discovery that no one geometry is intrinsic to space—that regions of elliptic, hyperbolic and Euclidean (flat) geometries can all exist within the universe.

[8] A change in position with respect to time $\frac{dx}{dt}$ is defined as a velocity. A change in velocity, or the derivative of velocity with respect to time $\frac{dv}{dt}$ is called acceleration. The derivative of acceleration $\frac{da}{dt}$ is called jerk. Interestingly enough, the next three consecutive derivatives with respect to time are called snap $\frac{dj}{dt}$, crackle $\frac{ds}{dt}$ and pop $\frac{dc}{dt}$.

[9] When we treat the medium as a gas of non-interacting particles with no apparent internal degrees of freedom the critical transition temperature is given by: $T_c = \left(\frac{n}{\zeta(3/2)}\right)^{\frac{2}{3}} \frac{2\pi\hbar}{mk_B} \approx 3.3125 \frac{\hbar^2 n^{\frac{2}{3}}}{mk_B}$ where T_c is the critical temperature, n is the particle density, m is the mass per boson, $\hbar$ is the reduced Planck constant, k_B is the Boltzmann constant, and ζ is the Riemann zeta function; $\zeta(3/2) \approx 2.6124$.

[10] Newton calculated that Mercury's perihelion, its closest approach to the Sun, should advance by 531 arc seconds per century. That number was based on the gravitational influences of Venus (277"), Jupiter (153"), Earth (90"), and the other planets (10"). The problem was that Mercury's perihelion actually advances by 574" per century.

According to Einstein's general relativity, Mercury gets a tiny additional push because the high–speed orbital motion causes its "effective mass" to increase a little, an effect that is exacerbated because the planet moves fastest near its perihelion, where the Sun's gravity is strongest. This adjustment produces an advance of perihelion that is 43" per century more than Newton predicted and in perfect accord with observation. Richard Talcott. (2006). *Astronomy*—Collector's Edition, Cosmos, p. 26.

More recently, smaller deviations of the orbits of the other planets have been measured. All of these measurements support Einstein's theory of general relativity.

[11] Technically, when the photographs from that expedition were examined more closely it was found that errors inherent in the measuring process were as large as the effect they were trying to resolve. The reported measurements were, therefore, as much of a reflection of the fact that the experimenters wanted Einstein to be right, as they were true measurements. Nevertheless, later observations have accurately confirmed this effect.

Chapter 10 — **The Bucket**

"Where ignorance lurks, so too do the frontiers of discovery and imagination."

Neil deGrasse Tyson[1]

"What had been the purview of the gods, or God, came now within the comprehension of mortals."

Gary Zukav[2]

President's Circle, University of Utah.

Walking to my *Extra Terrestrial Intelligence* class—taught by astrophysicist and master storyteller George Cassiday—I noticed a metallic dome on the roof of the South Physics building. After class I approached the professor and asked what that dome was used for. He told me that he thought it looked like an observatory, but he wasn't sure because he hadn't heard anything about an observatory in use on campus. All of his observations were done in collaboration with the FLY's Eye II project on the Dugway Proving Grounds. So he directed me to the physics department's counselor Lynn Higgs.

Lynn's door was wide open. He was on the telephone in a swivel chair in front of his computer. I waited in the doorway until his conversation was over. He seemed to be a jolly person. I introduced myself briefly and asked my question. "Oh," he said with a surprised expression. "That's our observatory." "We have an observatory?" I asked. I wondered why I had never heard of it before. "Well, kind of. No one has been up there in seven years." "Why?" "Well there hasn't been much public interest in it, and if I remember correctly the electronic drive doesn't work."

He told me about the telescope's old days for a minute then he turned to his desk and pulled out two large keys. "Here," he said as he handed each key to me. "The square one will get you on the roof, and the round one will get you into the dome. Take them, and bring them back to me this time tomorrow." I didn't know how to respond. No one had ever given me private access to an observatory before.

Immediately, I went to the top floor of the South Physics building and found a long staircase that led to the roof. At the top of the stairs I inserted the sturdy square key and opened the door. Stepping into the bright sunlight I unknowingly crossed into a new chapter in my life.

The aluminum colored dome was impressive. As I unlocked its rust brown heavy door, I imagined the telescope inside and felt a thrill of exhilaration. But when I stepped in I couldn't see anything—it was completely dark. I felt around for a light switch and ended up running into a mat of spider webs.

I found the light switch, turned it on and saw the massive cylinder. It was much bigger than I had expected, about six feet long and two feet in diameter. Then I noticed that the top half of the dome was designed to rotate and that a section making up about a fifth of the roof opened up to give the telescope access to the sky. Standing on a chair, I unlatched the roof and rotated the old turnstile crank until I heard the banging of metal on metal. On the wall an old electric control box was hanging with two large buttons. I discovered that one rotated the dome clockwise and the other rotated it counterclockwise. As I spun the roof around to look at different parts of the sky, I lost track of where the exit was. I instantly knew that I liked this place.

That evening my sweetheart and I spent a couple of hours cleaning, vacuuming, and wiping all the surfaces down in order to transform this forgotten relic into our celestial palace for the night. We removed the cap from the end of the telescope, and began practicing the art of loosening the big metal wheels that unlocked its axes, moving it into a new position, and then retightening the wheels. The first thing we pointed it at was the moon.

Because the electric motor didn't work, every image we focused on drifted slowly across the view, due to the rotation of the earth. The primary mirror turned out to be warped, keeping us from achieving a crisp, clean focus on anything. Still, it was spectacular. The brightest star in the nighttime sky, which I later learned was Sirius, drifted toward the western horizon and appeared to be twinkling different colors. Through the telescope it looked like a distant rotating prism sending out all the colors of the rainbow. It was strange to realize that somehow I had gone my entire life without seeing the heavens this way.

On the following day the key to that magical place burned a hole in my pocket as I tried to work through the math of my physics homework in the rotunda of the main physics building. After staring at my paper for about twenty minutes I announced to myself that I couldn't concentrate with all these distractions. I needed to go somewhere where I could be alone. It was the perfect excuse. I excitedly made my way up to the forgotten observatory, opened the dome to let the sunlight in, and spread my homework out on one of the freshly cleaned tables. Then I just stared at my assignment.

After some time, I began to wonder who I was trying to fool. I put my mechanical pencil down and admitted that I was up here because I was going to try to fix the motor drive. I reassured myself with the fact that no one had been up here in seven years, so if I screwed it up worse it wouldn't really matter. A small cabinet inside the observatory contained a quaint tool set. I found a screwdriver, electrical tape, and wire cutters and got to work. After removing the casing I saw some really old fuses and frayed wires. I removed everything that looked bad and remembered where they came from.

In the basement of the same building there was an old shop. I'd never been in, just walked by, but I knew it was there. I took the parts that I had removed downstairs and approached the elderly man stationing the shop. I handed him the parts and told him I needed replacements. He quickly went about the aisles and returned with the replacements. He rang it up and then said, "On the physics account?" I noticed that the register had no money in it and realized that everything here ran on account only. I didn't have an account so I nodded as if this was a normal process for me and went on my way. It didn't take long to put everything back together. Before I put the casing back on I plugged it in. I crossed my fingers as I flipped the small toggle that activated the motor drive. Bzzzzzzzz—it was working! The motor was rotating the telescope at just the right speed to counter the earth's rotation, keeping the field of view stationary in the telescope.

All of my other commitments were easily postponed as I played with this monstrous metallic toy. Even though it was midday, I busied myself with the process of loosening the wheels that allowed me to swing the telescope on its two axes and then tightening them down when I had a new object in view. I would climb up the stepladder, look into the eyepiece, focus it, and examine my target closely. After a few seconds of absorbing each image, it was back down the ladder and off to a new point of interest. I discovered a radar tower up on top of the mountains to my east that looked like a whitewashed billboard—except someone had painted an eye on it that looked like the Egyptian symbol for the god Ra. I smiled as I thought of the Egyptian god watching in secret over the Salt Lake Valley. As soon as I thought to look at my watch it was already time to return the key to Dr. Higgs.

With a bit of reluctance I went to Lynn's office, handed the keys to him and thanked him for the opportunity. As I was leaving I remembered the money I had charged to the physics account and asked him if he wanted me to reimburse the department. "What was it for?" he asked. I told him and his eyes widened. We went back up to the observatory. He noticed how clean it was and asked about it. I told him about my experience and watched his face light up as he heard the buzzing sound of the motor. Then he stood silent for a moment, looked at me and said, "How would you like to run this observatory and offer public star parties once a week?" I was hesitant. "I don't know anything about astronomy," I said. He laughed and with his hand on my shoulder he said, "When I was your age I didn't either. All you have to do is stay one lesson ahead of the crowd." I really couldn't believe what was happening. "And after you do it for one semester," he added, "I'll make it a paid position."

There was no going back after that point. Lynn gave the key back to me. Soon after, I installed a brand-new telescope, and founded the University of Utah Astronomical Society (UUAS). Two years later funding was secured to put a new surface on the roof, and to install an elevator that went to the roof making the observatory handicap accessible. Part of this money also kickstarted an undergraduate observational astronomy class, complete with six new permanent mount telescopes with CCD cameras, laptops, and wheeled locking covers. I could see the difference from an airplane.

This experience helped me understand that questions are the most honest and powerful things we have. Questions initiate profound investigations; they help us escape our debilitating preconceptions and pry us loose from our convictions. Questions are more than just expressions of our imaginative wonder—the art of asking questions is what guides humanity's moral and intellectual progress.

Newton's Question

As discussed in Chapter 2, a significant question in the history of physics has been—is there an ultimate reference frame in Nature? Isaac Newton brought this question to center stage by exploring the philosophical implications of the fact that the surface of water in a spinning bucket takes on a concave shape.

Newton's claim was that the water inside a bucket possesses a concave shape because it is truly spinning—in contrast to the idea that is was simply spinning *compared to an arbitrary reference frame*. In other words, Newton was proposing the physical existence of an ultimate reference frame. Without an ultimate reference frame Newton would not be able to defend the idea that the bucket had true accelerated motion. Newton picked a particular candidate for this ultimate reference frame and called it *absolute space*.

Despite the advantages of Newton's concept of absolute space, relationists (such as Leibniz and Mach) found it unsettling to accept the existence of an ultimate reference frame. Following their intuition they posed questions that eventually eroded the possibility that Newton's particular reference frame (absolute space) could be the ultimate reference frame in

Nature. Their questions, however, did not rule out the possibility that Nature possesses some other ultimate reference frame. The crux of the relationist argument is that constant velocity motion can only be described in a relative sense; likewise position and time can only be defined in a relative sense. So, why would accelerated motion be nonrelational?

To be clear, this argument does not deny that accelerated motion is different from constant velocity motion. Acceleration has noticeable effects—we can feel it, and it changes the shape of the surface of the water in a spinning bucket. Instead, the argument is focused on Newton's inability to explain how and why acceleration is any different from the other measures of motion. Assuming that absolute space exists may give us a way to identify true accelerated motion, but it offers no explanation for why accelerated motion is different from constant velocity motion. It doesn't explain why constant velocity motion cannot be defined in relation to that unique reference frame. It also doesn't explain why all positions and measures of time don't seem to require the existence of an ultimate reference frame. If an ultimate reference frame exists, the relationists pointed out, it appears to manifest only for acceleration. Should such an ultimate reference frame be taken seriously?

Newton's absolute space failed to reveal why accelerated motion appears so different from constant velocity motion—why position and time are fundamentally relational, but acceleration is not. An accurate map of physical reality should be able to explain how a unique reference frame is capable of giving meaning to acceleration, without rigidly fixing the notions of position, time, and velocity within that map.

Because velocity is purely a measure of space divided by a measure of time ($\Delta x/\Delta t$), measures that are both purely relational,[3] it comes as no surprise that velocity is also a relational measure. But acceleration is also made up of component measures that are relational (space and time) $(\Delta x/\Delta t)/\Delta t$. So how is it that the combination of relational measures results in a relational measure in one case, but a nonrelational measure in another case?

Leibniz and Mach answered this question by saying that they don't. They argued that, despite how it appears, accelerated motion is also relational. This belief forced them to claim that in an otherwise empty universe, an object could experience no measure of acceleration. It couldn't even be spinning because, in this view, there would be no reference by which the motion of the object could be compared. According to their view *acceleration* loses all meaning without a comparison.

This argument puts accelerated motion on equal footing with constant velocity motion— an act that is intended to give us a conceptual (or at least theoretical) advantage. But when it comes to making sense of what motion has to say about Nature, this argument gives us no real advantage. This proposal falls short in the same way that Newton's concept of absolute space did. It does not explain why accelerated motion seems so different from constant velocity motion—why we can feel acceleration, but cannot feel constant velocity motion. More importantly, it does not explain why the surface of the water in Newton's bucket does not change shape if we change our reference to one that is accelerating with the bucket. If acceleration is purely relational, and we are spinning with the bucket, shouldn't we see a flat surface of water because there is no relative acceleration (spinning) between us and the bucket?

Both of these views leave us with unanswered questions. Neither Newton's map of absolute space, nor Leibniz's/Mach's purely relational conception, tell us why the basic measures of distance and time are relational, while acceleration is not. We need a map of physical reality that is capable of revealing the geometric magic behind this mystery.

If a purely relational view could offer an explanation for why acceleration appears so separate from constant velocity motion, then it would be palatable. On the other hand, if a framework that possessed an ultimate spatial reference frame could explain why position, time, and velocity are purely relational measures, while simultaneously exposing a reason for why accelerated motion is uniquely defined in reference to that frame, then perhaps Newton's intuition would be vindicated. But which vision is right?

Einstein's Question

Although Einstein started out trying to vindicate the relationist position championed by Mach, he ended up fortifying the other camp. Einstein realized that space and time were inversely proportionate expressions of a single entity called spacetime, which is replete with warps and ripples. Einstein noticed that in order to make sense of these warps, ripples and curves, it was necessary to assume the existence of an unwarped state. This unwarped state became Einstein's ultimate reference frame. He called it *absolute spacetime*.

Today, Einstein's view is widely accepted, but rarely understood. According to this view, accelerated motion gains its definition in comparison to a real, tangible, and changeable vacuum field—instead of being relative to material objects like planets, stars, or ships. The existence of this field means that there is an underlying, inherently encoded, ultimate reference frame in physical reality. That reference frame is the zero field—the unwarped vacuum state of general relativity (Figure 10-1). Even where there is no mass present—where the curvature is zero—the shape of space still defines a field. A zero-field is still a field, because it can be measured and changed. It also provides a reference by which acceleration can be defined.

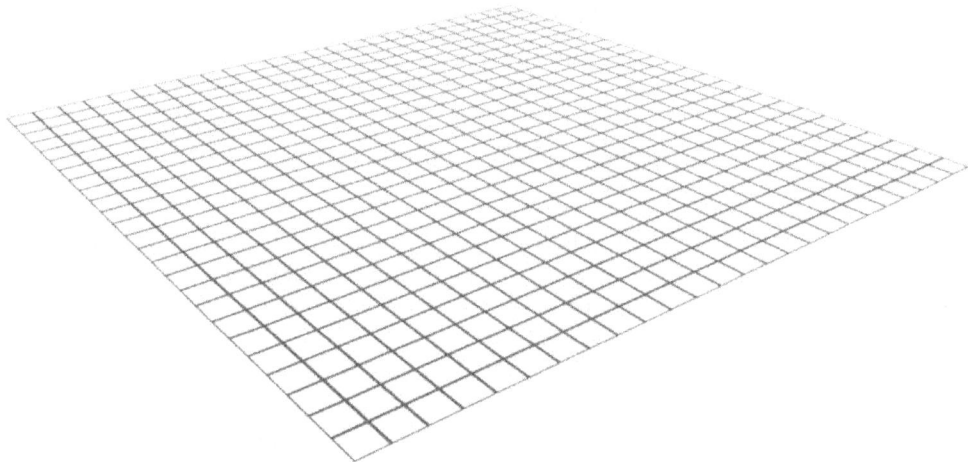

Figure 10-1 The zero field reference frame

Technically, Nature's zero field only has an average value of zero on macroscopic scales. On the ultramicroscopic scales there are tiny moment-to-moment changes in density, fluctuating gradients, and undulating warps. Quantizing the vacuum allows us to uniquely resolve these warps in terms of the geometric arrangements of the quanta (Figure 10-2). The degree to which the vacuum's geometry is warped is meaningful only in reference to its hypothetical unwarped state. In other words, warps reference deviations from a lattice spatial geometry, in which all of the quanta of space are equally arranged throughout superspace. The smallest of those deviations are resolved only when we quantize the vacuum.

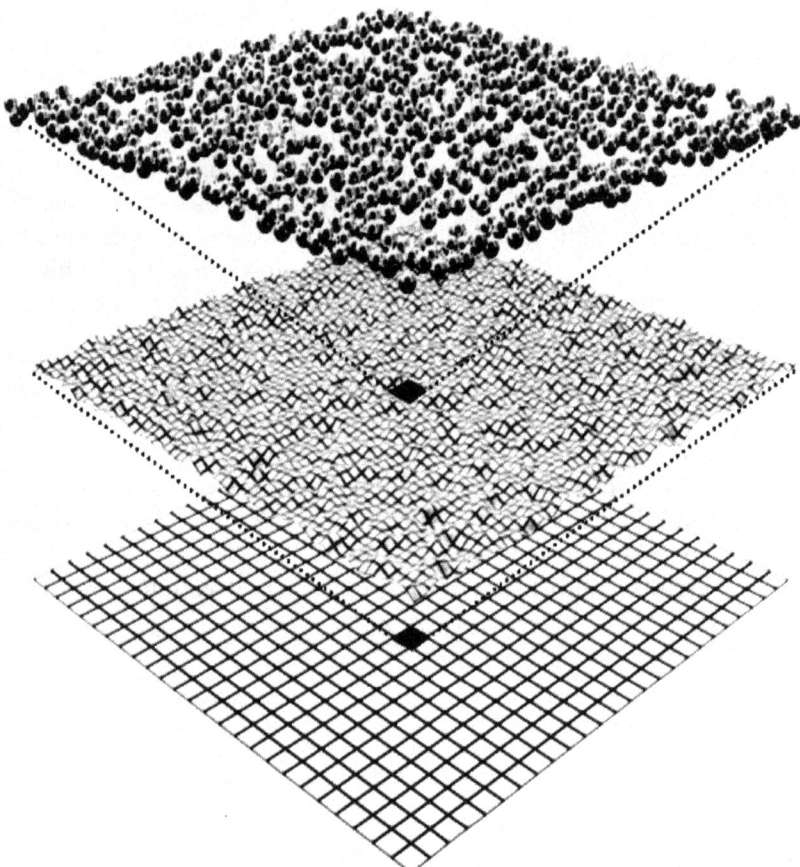

Figure 10-2 An infinitely smooth 'rubber sheet' masks microscopic details

Like Newton, Einstein came to the conclusion that there is an underlying ultimate reference frame in Nature. The existence of that unique reference frame—the zero field—allows us to uniquely define acceleration, explaining why it is nonrelational. But Einstein's depiction of this zero field fails to explain why distance, time, and velocity are relational measures. Why aren't these measures also defined in relation to that reference frame? How is it that position, time, and velocity manage to remain purely relational if an ultimate reference frame exists? Einstein's conception of the vacuum does not allow him to answer these questions. Will a different picture bring us the answers?

Transforming the Question

In Einstein's continuum, accelerated motion is correlated with curvature (an uneven experience of the vacuum), and the ultimate reference frame is defined in relation to the minimum theoretical value of the curvature field—the zero-field. In a quantized vacuum this zero-field, the ultimate reference frame, only has macroscopic existential cohesion.

Acceleration directly references the magnitude of the gradient, which is defined in relation to the zero-field. It is to be understood macroscopically because the field from which it is derived only comes into focus on macroscopic scales. Position, time, and velocity cannot be accurately captured by this zero-field. Here's why.

As the quanta of space interactively mix they take on new geometric relationships. Because positions in space ultimately derive their claim through these relationships, position is sensitive to these changes. The zero-field is neither sensitive to these changes, nor capable of resolving them. This is why it is incapable of precisely capturing the notion of position.

The zero-field cannot uniquely capture the advancement of time for two reasons. Microscopically it does not resolve the variability of time, and macroscopically it references only a zero density gradient. Knowledge about the gradient of a field does not constrain the magnitude of the density of that field. A zero-field is incapable of distinguishing two regions (separated by distance or time) with different densities. Any density that is uniformly distributed will be characterized by a zero-field. The zero-field is incapable of precisely capturing the notion of time because time's passage critically depends upon the density of space, which it cannot resolve.

Velocity is not uniquely set by the zero-field either. In the vacuum, the more precisely we know an object's position the less we know about its momentum or velocity. As we approach the Planck scale we resolve position more accurately, but we also lose the averaging over effects that allow us to determine its x, y, z, t velocity. Once we pinpoint an object's position down to one specific quantum we lose all ability to claim anything about its velocity (as long as we are restricted to four-dimensional information).

If our perspective is internal to the vacuum, then precise knowledge about position forces complete ignorance about its velocity. Four-dimensionally, there is no way of knowing how the other locations in space are distributed and moving about. Therefore, there is no way of knowing where that point will be next, or where it was last. Knowledge of its velocity is completely absent.

Therefore, the hypothesis that the vacuum is a compressible inviscid fluid offers a natural solution to the ultimate reference frame mystery. Acceleration references the magnitude of a vacuum gradient, which is defined in terms of the zero density gradient—the zero field. This is why acceleration is nonrelational. By contrast, position, time, and velocity are relational because they represent details of the map that are tied to greater degrees of freedom than the zero field has access to. For example, we could swap out two quanta in that field, drastically changing their positions relative to the rest of the vacuum, without changing the field at all. Likewise, because the field only references changes in density of space, it is possible to have identical density gradients in regions with different densities, and therefore with different time signatures. Vacuum quantization offers us our first possible solution to this obstinate mystery.

[1] Neil DeGrasse Tyson. *Death By Black Hole—And Other Cosmic Quandaries*, p. 77.

[2] Gary Zukav. *The Dancing Wu Li Masters*, p. 24.

[3] The Greek letter delta (Δ) represents "change in." Therefore Δx is equal to $x_{final} - x_{initial}$, and Δt is equal to $t_{final} - t_{initial}$.

Chapter 11 — **Dimensional Analysis**

> *"To see a World in a Grain of Sand*
> *And a Heaven in a Wild Flower,*
> *Hold Infinity in the palm of your hand*
> *And Eternity in an hour."*
>
> William Blake, *"Auguries of Innocence"*

> *"Now that science has invented magic, anything is possible."*
>
> *Marge Simpson*

The Hole, Federal Prison, Florence, Colorado.

"Troublemaker," the guard said as he removed my handcuffs and passed me off to the FCI guards. They gave him a nod and watched him leave. I stood there wondering what would happen next. The unfamiliar guards looked me over, slowly shaking their heads as if to say, "Why won't you ever learn?" For a couple of minutes they sat there creating stories in their heads about what I must have done, fueling their need to punish others for the things that live in their imaginations. "Let's put this one in the cooler," one guard suggested. "Yes, yes," another replied as he rubbed his hands together. I couldn't tell if they were playing one of their mind games, but they seemed to be fairly open to suggestions, so I mimicked their tone and said, "No, let's don't put this one in the cooler." As it turned out, they were only open to first suggestions.

They tossed me a jumpsuit and plastic sandals and instructed me to change out of my compound clothes. Then they grabbed me by the arm and directed me down the hallway. The cooler was a large cement holding cell with a vent blowing cold air into it from the middle of the ceiling. A raised cement block, big enough to support a twin mattress, rose from the center of the room. Instead of a mattress, pillow, or blanket, there were iron loops sticking out from each corner of the block—places for the guards to tie people down.

I searched for the place with the least direct air, crouched down and tried to warm up by minimizing my exposed surface area. It was roughly 3:00 in the morning. I was tired, but it was too cold to fall asleep. I wondered how long I would have to freeze in here before they were satisfied. I drew comfort from the fact that there was no toilet in the room. They couldn't expect someone to stay in here for long without a bathroom. Surely they would take me to the actual "hole" soon.

I spent the first few hours hunched over on my feet, trying to keep my back from touching the cold wall. I had to shift my weight back and forth in an attempt to keep my legs from falling asleep. It was getting difficult to ignore the effect the temperature was having on my bladder. A couple more hours passed and my bladder was about to erupt. Then I heard the sound of jingling keys. For a moment I thought they were coming to let me out, but then the keys moved on and I realized it was only shift change. I knew the next crew would be told some grandiose story about why this "troublemaker" needed to be put in here, and their need for punishment would be fresh. I also knew that my bladder wasn't going to survive this change so I began to bang on the door, demanding to use a bathroom.

This was a risky thing to do. In county jail I'd seen inmates beaten half to death for less than banging on a door. In Atlanta's transfer center, my cell mates and I were greeted with twenty cops in full combat gear, shocking shields, beating batons and all, just for banging on the door and yelling "there's a dead one in here." When that door opened, we were backed up on the far wall. The body was on the floor right next to the door. The cops rushed in and proceeded to bash the corpse's face in with their batons. You might think that one or two hits would be enough to verify our story, but these cops were thorough. You couldn't accuse them of not doing their job.

The response I got was unexpected. One guard unlocked the door, smiled, and handed me a milk carton with the top cut off. As the door closed, laughter burst out in the hallway. I smiled. I had seen through their plot. If they could get away with leaving me in the cooler for more than two days, they would have given me a five gallon bucket—like in Atlanta. Instead, they gave me this carton, which would be full in about two days. This told me that they weren't allowed to keep people in here, that their superiors would eventually be making rounds. I would be in the official hole by then, which compared to this was an improvement. I started the countdown in my head. I only had to keep myself warm for up to two more days.

Thirty hours later, I was escorted down another hallway and locked away in the hole. After a couple of days in this cell another inmate from the camp was confined with me. "What's your name?" I asked. "Rico," he responded. "I'm…" I began to reply. "I know who you are Moon Rock, I've seen you around camp," he said. Rico's dreadlocks and perfectly chiseled muscles advertised a man's man, but his diction and pensiveness told me that there was much more to the man than he usually gave away. "What'd they throw you in here for," he asked. "I was in the bathroom when they came through for count," I said. "Tarnaski?" he asked. "Yup." He leaned back and put his hands behind his head. "Figures."

This wasn't the first time that Tarnaski had flown off the handle because someone was using the bathroom during count. Instead of just adding 1 to 31 when this occurred, he would make a scene, wait for that person to walk back down the wing, and then recount the entire wing. He felt that this inconvenience was a serious kind of disrespect. It was almost as if the idea of inmates needing to go to the bathroom at night shattered his worldview. Inmates weren't human, and even mimicking human functions was disrespecting him.

"Why did they throw you in here?" I asked. Rico didn't respond. I could tell this was a very personal matter. I didn't want to intrude, so I left it alone. Several hours later Rico decided to share. He told me about his beautiful wife, and their two kids. He told me how he sold drugs until he had enough money to start up his own legitimate business. His eyes lit up when he talked about being a legitimate businessman. Then his eyes went dark as he told me how the Feds busted the circle he used to be involved with, got his name, and sentenced him to several years for his past dealings. They seized his money, his business, and left his family with nothing to survive on. His brother was helping out with his wife's rent and food, but they were struggling.

Rico told me that he got visits two weekends a year. He told me about last weekend, when his brother flew into Denver and drove his wife and kids to the prison camp in a rental car. They were there for the full visiting hours on Friday, and Saturday, but on Sunday they had to leave a little early so his brother could catch his plane. Believing this to be his last visit, he gave his kids a long hug telling them, "Daddy is going to be home in just a couple more months." Then he kissed his wife and watched his family leave.

The next day, Rico was called into the Chaplin's office and was informed that there had been an accident. His brother had hit black ice, and his wife was thrown from the vehicle. She died in the hospital. His 2-year-old boy was in critical care, and died 3 days later, and his 4-year-old boy was still in the hospital, but was expected to make a full recovery. Rico's brother survived with only minor injuries. As he told me this story he had tears in his eyes and unmistakable anger in his throat. He was angry that he couldn't do anything to help, he was angry that he wasn't there, and he was angry at himself for being here.

"So why did they throw you in the hole?" I asked. "Because I requested a furlough, for the funeral." Technically this was an option for camp prisoners. There were forms for both escorted and non-escorted furloughs for extenuating circumstances. The only problem was that, as far as anyone on the compound could remember, the current decision makers had never approved a furlough. "I had to try," Rico said. "I went to Sliger's office, filled out the paperwork for both kinds of furloughs, and she said they would consider it."

"So what happened?" I asked. "Two days later they called me into their office. Lyde, Porco, and Sliger were in there looking at me in that way, you know, like they are going to enjoy this. They told me they looked over my request and decided to turn it down. I pleaded with them, but then Sliger put the paper down, looked me right in the eyes, and said, "Well, if you hadn't done what you did to get in here, your family would still be alive—wouldn't they!""

My heart sank. I had finally glimpsed the depth of his pain, and I was not prepared for it. "What did you do?" I asked. "I walked out of there with my fists clenched. Later they came to throw me into the hole as a "precautionary measure," until the funeral is over."

We sat in silence for nearly a day, until Rico finally said, "Give me something else to think about. I need to put something else on my mind." I thought for a moment and then asked, "Have you ever heard of complex numbers?" "No," he replied. "Well," I said, "let me introduce you to one of Nature's grand mysteries."

Complex Magic

> *"Perhaps we should be seeking a role for discrete combinatorial principles somehow emerging out of complex magic, so 'spacetime' should have a discrete underlying structure rather than a real number based one."*
>
> *Roger Penrose*

One of the most unexpected discoveries in science is that, on the microscopic scales, Nature is controlled by what is known as *complex magic*. Vacuum quantization plays a key role in explaining this complex magic—revealing additional *internal* dimensions that are responsible for its presence.

To unravel that explanation, we note that a system characterized by the familiar x, y, z, t dimensions is expected to have properties that map to the set of "real" numbers. The discovery that some properties conform to the rules of complex magic means that the "imaginary" numbers also play a role, which suggests that the normal characterization is incomplete.

One way to represent a system that follows the rules of complex magic, is to expand our expression of possible events from a four-dimensional representation (x, y, z, t) to a seven-dimensional representation, wherein three of the new variables remain attached to the imaginary realm, for example, x, y, z, t, iu, jv and kw. Note that i, j and k are all independent square roots of negative one. Doing this as a mathematical exercise is trivial, but if we mean to suggest that this set represents physical reality, we find ourselves facing a lot of questions that need answers.

Does our discovery of complex magic in Nature suggest that imaginary parameters actually exist? Or are these imaginary numbers some sort of mathematical smoke and mirror trick? Exactly how does Nature exhibit imaginary properties? How can additional dimensions reveal the origins of those imaginary properties? And how can the square root of negative one represent anything real?

Before we explore these questions, it is important to recognize that the term *imaginary*, in the mathematical sense, does not make reference to "unreal" or "flight of fancy". It is a historical term that, unfortunately, has become as misleading as the public's repetitive misuse of the scientific word *theory*. Imaginary parameters are simply parameters that follow the rules of the imaginary set of numbers. This, in and of itself, may be remarkable, but it is not a flight of fancy.

Despite the questions that arise when we encounter complex magic, the literal inclusion of imaginary properties is unavoidable. When it comes to the microscopic realm, no theory is capable of accounting for our observations without making room for complex numbers. According to quantum mechanics, the microscopic realm is so opulently laced with complex magic that "ordinary solid matter could not exist without its consequences."[1] This is the case because electrons, protons, and neutrons all behave in a manner that defies the commutative property of real numbers, in favor of the commutative property of imaginary numbers.

Mathematicians have long recognized that complex numbers are suggestive of additional dimensions. This connection between complex numbers and additional dimensions is so straightforward that complex functions are routinely analyzed and graphed within what is called the complex plane[2] (Figure 3-1). This plane combines a familiar spatial dimension (x) with an unfamiliar spatial dimension (u).[3] Mimicking how the union of real and imaginary numbers gives us complex numbers, the union of a familiar and unfamiliar dimension forges the complex plane. The presence of this unfamiliar dimension is responsible for the imaginary properties of objects in that plane.

This representation incorporates an unwanted limitation—the same limitation that impedes the usefulness of rubber sheet diagrams—because it requires us to suppress a real (familiar) dimension for every imaginary (unfamiliar) dimension we include. As we have seen, this kind of dimensional suppression forces us away from an intuitive grasp of a complete solution. In an attempt to escape the confines of this conceptual barrier, let's take a closer look at the imaginary realm—let's examine the role of imaginary numbers.

Imagine a Number

Imaginary numbers can be used to construct an algebraic system that satisfies all of the normal laws of algebra except for the commutative law of multiplication. The commutative law states that $ab = ba$. Within complex space this relationship no longer holds. Instead, the equality becomes $ab = -ba$.

To explore the geometric implications of this new relationship, consider the following. When we select a macroscopic object at random (a sword, Space Ship One, a dog, etc.), we can say with confidence that it possesses the geometric property of returning to its original configuration upon being rotated by 360° through any randomly chosen axis. If the object is only rotated by 180° then it will be geometrically "backward" compared to its starting arrangement. In other words, a rotation of 180° leaves the object in its geometric negative (Figure 11-1). This is all very straightforward and familiar, but it is worth mentioning because it turns out that not all of Nature's objects behave this way!

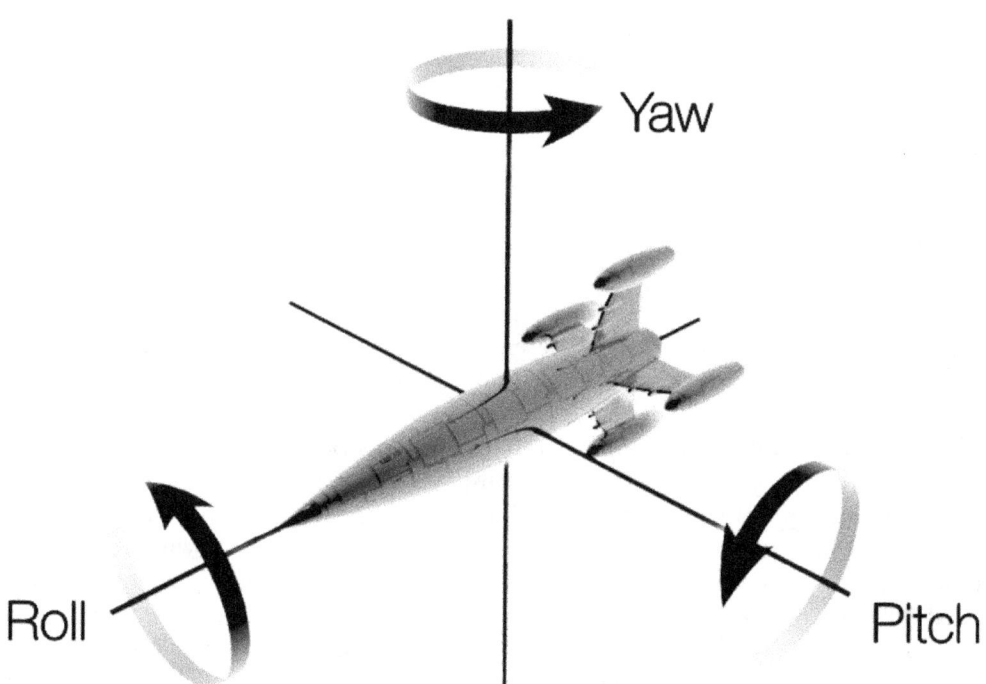

Figure 11-1 Familiar rotations

Physicists have found that the objects that live in Nature's smallest realms require a rotation of 720° to return to their original configuration. We call these objects *spinors*. They are, quite literally, the physical manifestations of complex magic.[4]

161

Spinors first entered the realm of physics in 1928 when Dirac formulated his famous equation for the electron, which describes the electron's state as a spinorial quantity. In an attempt to explain this discovery, it was postulated that time itself might be some sort of spinorial entity.[5] But it turns out that time isn't physically allowed to be a spinorial entity because the notion of spinorial time clashes with relativity—it does not allow rigid interpretation of dimensional parameters. Within frames of different velocity the real and imaginary coordinates get all mixed up, but the differences between these different kinds of dimensions do not allow for that. This rules out the possibility that time is a "quaternion" or spinorial entity, and it leaves us with one option—spinors must have a geometric (spatial) explanation.

The fact that spinors seem to be scale dependent also suggests that spinors acquire their unique characteristics from spatial parameters. Supporting this, Roger Penrose said, it is best to think of each dimension that we need to describe spinors as "referring to a kind of *spatial* dimension that is additional to those of ordinary space and time. These extra *spatial dimensions* are frequently referred to as internal dimensions, so that moving along in such an *internal direction* does not actually carry us away from the spacetime point at which we are situated."[6]

Can our eleven-dimensional model reveal what these internal dimensions are? Can it provide us with an intuitive explanation for how spinors end up in their geometric negative orientation after being rotated by 360°? Can it explain why they must undergo a rotation of 720° to return to their original geometric configuration?

To fully appreciate the answers to these questions, let's explore the quantum mechanical representation of spinors and compare it to the one offered by our eleven-dimensional model.

Spinor Visuals Compared

In quantum mechanics, we can attempt to explain spinors by joining a complex plane to a real plane. We end up with a representation of three spatial dimensions—two *real* or familiar dimensions (x and y), and one *imaginary* or unfamiliar dimension (u) (Figure 11-2a). The inclusion of this imaginary dimension allows us to graphically account for spinorial qualities because it allows us to split up 360° rotations between the two different planes. For example, if an object, like a teapot (Figure 11-2b), undergoes a rotation of 180° in the xy plane, it ends up in the configuration shown in Figure 11-2c. After another rotation of 180° in the ux plane, the teapot ends up in the configuration shown in Figure 11-2d. Within the real dimensions, the object ends up in its geometric negative after being rotated through a total of 360°, because half of those rotations take place in an additional dimension. If these rotations are repeated, then the object returns to its original geometric configuration once it has completed a total rotation of 720°.

If we allow for the literal existence of the additional dimension that this representation depicts, then this explanation is fairly satisfactory. But there is a problem with doing that. The problem is that the additional dimensions required by spinors come with the geometric condition that they are *internal*. Because the representation in Figure 11-2 depicts the additional dimension (u) as an expansive dimension—treating it as physically identical, but perpendicular to x and y—it cannot be an accurate depiction. An accurate representation must depict the dimensions responsible for imaginary properties as *internal* instead of expansive. Moving along in those dimensions should not actually carry us away from where we started.

One of the problems with this traditional explanation is that it is dimensionally reduced. Full descriptions of spinors should include at least six spatial dimensions.[7] A three-dimensional depiction cannot be considered a complete representation of complex space.

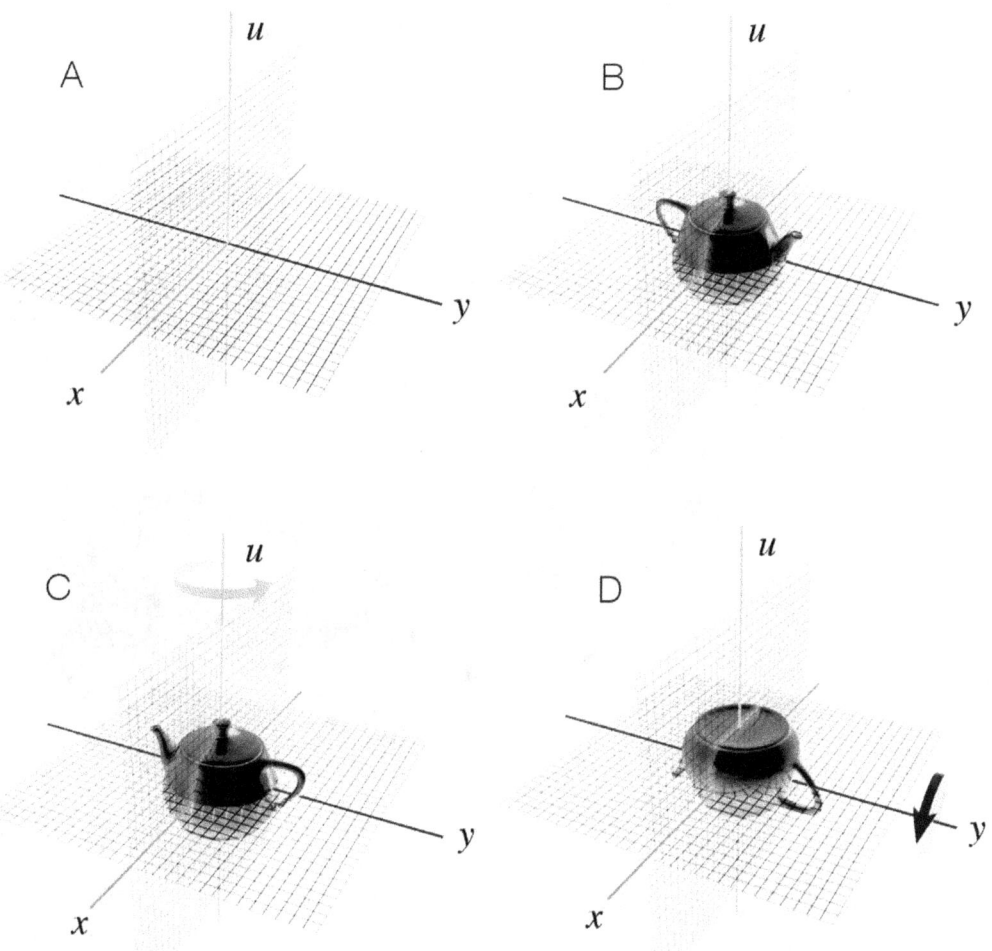

Figure 11-2 a-d Rotation in a complex-real plane
(a) Union of complex plane and real plane (b) Object in combined planes (c) 180° rotation in the xy plane (d) rotation of 180° in the ux plane—leaving the object geometrically reversed

Qst identifies the internal spatial dimensions of complex magic as the intraspatial dimensions—the volume of the individual quanta and the degrees of freedom they introduce. As we can clearly see, moving along in these internal dimensions does not actually "carry us away from the spacetime point at which we are situated." And rotating quanta does not alter the state of space—the arrangements of the quanta (Figure 11-3). [8]

Spinorial objects return to their original geometric configuration when they are struck with the amount of energy we would expect to rotate them by 720° because half of that energy goes to rotating the quanta that make up those objects. In other words, half of the rotations are intraspatial rotations.

Our model gets another boost from the quantum mechanical discovery that there are many equivalent paths that represent the minimum amount of space between two points (*a*) and (*b*). In four-dimensional representations of spacetime there is only one minimum path that connects point *a* and *b*, called a line. On large scales, the shortest distance between two points in space approximates a line when curvature approaches zero, but on smaller and smaller scales this approximation evaporates. Once again, we must turn to complex space, or higher dimensional representations of space, to explain this. When we do, we find that there can be many different spatially equivalent paths (in the familiar dimensions of space) that represent the shortest distance between points *a* and *b*.[9]

Figure 11-3 Internal dimensions and internal rotations

Vacuum quantization removes the smoke and mirrors from complex magic. When we include all nine spatial dimensions, and both temporal dimensions, it is no longer necessary to attach i, j or k to any of our variables. Rotations through the intraspatial dimensions, in relation to the spatial dimensions, automatically result in the geometric configurations that yield "quaternion" or spinorial effects. Instead of the reduced dimensional signature of (x, y, z, t, iu, jv, kw) we end up with the complete form $(x, y, z, t, \sigma, \mu, \delta, t, u, v, w)$. This signature is far more intuitive because it does not require ad hoc attachments of independent square roots of negative one.

All objects defined on quantum scales are spinors—collections of quanta whose connectivity (arrangement pattern) defines its form. In addition to the rotations that correlate to changes in x, y, z connectivity for macroscopic objects, the rotation of a spinor also involves the rotation of the quanta that make it up. In other words, in order to geometrically reverse an object—like an arrow arranged from quanta (Figure 11-4a)—we must rotate all of the quanta within the object and rotate its connectivity pattern (making the arrow point backwards). Rotating just the quanta (Figure 11-4b), or just the connectivity (Figure 11-4c), doesn't geometrically reverse the object. To accomplish that, we have to preserve the original relationship by rotating the connectivity of the object and the quanta that compose it (Figure 11-4d).

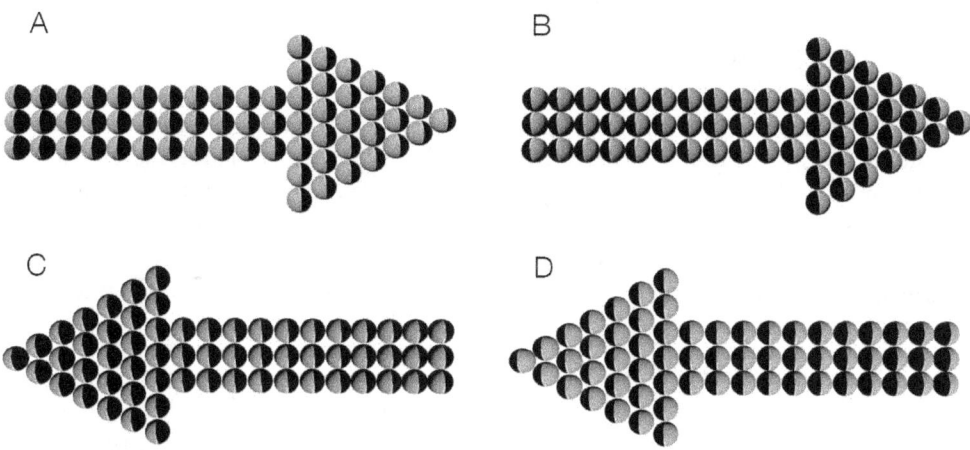

Figure 11-4 Internal rotations and spinors

Rotating the quanta may not affect our macroscopic x, y, z description of an object, which depends upon the geometric connectivity of the quanta that make it up, but it adds complexity to our inventory of the world and accounts for why twice as much energy is required to rotate a spinor than naïvely expected.

When we try to rotate a spinor, half of the energy we apply goes to distorting the geometric connectivity of space—altering its x, y, z configuration—while the other half contributes to intraspatial rotations—rotating the quanta that make up the object. If this occurred macroscopically, for example, if half of the energy we put into rotating a line of ants was absorbed by rotations of the ants, then a line of ants would also be a spinorial object.

Infinite Dimensional Cascades

When we quantized the fabric of spacetime we jump from a description composed of one kind of volume (three spatial dimensions) and a single time dimension, to a description that possesses three kinds of volume (three spatial groups each possessing three dimensions) and two temporal dimensions. This specific result was not an accident, or a coincidence—it was a geometric requirement of quantization. A deeper pattern emerges from the assumption that the quanta of space are composed of subquanta, which are in turn composed of sub-subquanta and so on. Extending the assumption of quantization transforms the vacuum into a perfect fractal, which amplifies the intellectual reach of our map. The additional texture revealed by this pattern exposes levels of self-symmetry that give rise to cascading infinities.

Probing the texture of Nature on smaller and smaller scales means increasing our dimensional resolution and resolving the internal structure of the individual quanta—the subquanta that they are made up of. The next step is to resolve the sub-subquanta that the subquanta are made up of. As we continue this process, a pattern emerges—tying the number of dimensions in the map to its resolution in a very specific way. That pattern impacts a mystery that has particularly perplexed string theorists for decades.

As theorists brought string theory into maturity, five independent ten-dimensional solutions emerged. Eventually, Edward Witten discovered a symmetry that came from an overlooked dimension that allowed those five solutions to be combined into one eleven-dimensional theory, which he called M-theory. Before M-theory came along, theorists had noticed something very strange that they could not explain. For some unknown reason, the equations of string theory were found to be compatible in ten and twenty-six dimensions. This discovery left theorists puzzled. They wanted to be able to explain what mechanism, or physical property, was responsible for linking ten and twenty-six dimensions.

> *"One of the deepest secrets of string theory, which is still not well understood, is why it is defined in only ten and twenty-six dimensions."*
>
> *Michio Kaku*

The connection between ten and twenty-six dimensions still needs to be explained, because if we subtract the additional symmetry introduced by the structure of M-theory, then the old pattern still persists. To solve this mystery we need the dimensional hierarchy equation, which emerges from our quantization process $f(n) = 3^n + n$, where n = order of perspective and $f(n)$ = the total number of dimensions within that perspective. This equation dictates how the total number of dimensions included in our map of physical reality depends on the number of times we quantize spacetime. It reveals that the number of dimensions in our map reflects the level of detail included in our perspective.

A first order perspective ($n = 1$) recreates the Euclidean perspective, framing three continuous spatial dimensions and one temporal dimension. This order of perspective does not resolve any vacuum quantization. A second order perspective ($n = 2$), quantizes the vacuum, distinguishing three different groups of spatial volumes and two temporal dimensions. This produces an eleven-dimensional map. As we ascend through more detailed representations of the vacuum, the number of dimensions included in our map increase according to the dimensional hierarchy equation.

The right side of that equation has two components. The first component is an exponential term that represents the number of spatial dimensions in the map. It comes in groups of three because it reflects the volumetric nature of all spatial dimensions. The second term is linear, denoting the fact that every increase in dimensional perspective resolves another independent dimension of time. This extended relationship produces the following pattern:

$$f(n) = 3^n + n$$

Although *n* can continue indefinitely, dimensionally complete maps (self-consistent geometric constructs) must have a number of dimensions that conforms to $f(n)$. Can you see the numbers ten and twenty-six in this pattern? Look closely. Remember that the ten and twenty-six dimensions were maintained by string theory when it was missing certain symmetries that came from missing dimension(s). The ten-dimensional solutions were missing one dimension, and the twenty-six dimensional solutions were missing four dimensions.

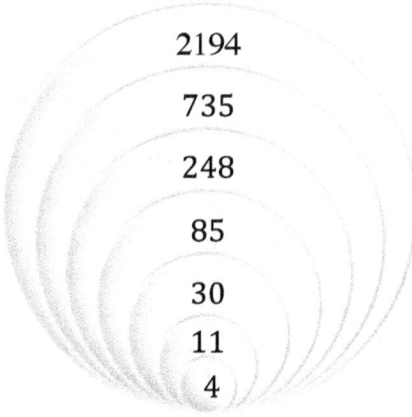

$f(n)$ = number of dimensions in perspective n = order of perspective

According to our new pattern, a ten-dimensional solution is equivalent to a second order description minus the dimensional contributions of a zero order description (11 − 1 = 10). Continuing this pattern by increasing the order of each part by one, we find that a twenty-six-dimensional solution is equivalent to a third order description minus the dimensional contributions of a first order description (30 − 4 = 26).

This pattern suggests something that may be testable as the processing speed of our supercomputers increases. It predicts that the old equations of string theory will also find compatibility in 74, 218, 650 dimensions and so on (85 − 11 = 74, 248 − 30 = 218, 735 − 85 = 650 ...). It also suggests that the eleven-dimensional equations of M-theory (or supersymmetric sets) will find compatibility in 30, 85, 248, 735 dimensions, etc. By dedicating supercomputers to the verification of this claim we could, for the first time, reach into the dimensional symmetries of cascading infinities. If we find any supersymmetric set that is not predicted by this equation we could falsify this pattern.[10]

> *"...any real body must have extension in four directions: it must have Length, Breadth, Thickness, and Duration."*
>
> H. G. Wells

Before we explore the extended implications of dimensional hierarchy, let's discuss the role the number three plays in this hierarchical equation. Spatial objects cannot be void of volume. In order for something to physically exist, it must have non-zero extension. Spatially, real objects must be at least three-dimensional. Two-dimensional Euclidean objects contain exactly zero space and therefore have no spatial existence. A 'plane' must have nonzero thickness in order to spatially exist. It may be very thin, but it cannot have zero thickness. It cannot be a true Euclidean plane. In accordance with this rule, our eleven-dimensional model depicts the smallest x, y, z objects in Nature as three-dimensional space quanta.

A Euclidean line, or point, does not exist in Nature. Every point (every smallest divisible location) must represent, in and of itself, a spatial measure. It must, therefore, be three-dimensional. Three is the minimum number of allowable spatial dimensions because it enables a nonzero volume—a nonzero measure of space.

One or two-dimensional approximations can provide valuable simplifications, but they can also result in ontological disaster. In Nature nothing exists, in a spatial sense, that possesses less than three dimensions. Extension is *a priori* to form and must, at minimum, reference a volume (an intraspatial volume, a spatial volume, a superspatial volume, etc., or a combination of these). A first order dimensional description ignores this distinction, and this oversimplification has been the source of many conceptual struggles.

Architectural Detail

> *"Everything is in everything at all times."*
>
> Anaxagoras[11]

> *"Behind every atom of this world, hides an infinite universe."*
>
> Rumi

A second order dimensional description (an eleven-dimensional map) is also an approximation. It is a much finer approximation, but an approximation nonetheless. One of the limitations of an eleven-dimensional approximation is that it cannot resolve the mechanism of the Big Bang. To capture that causal story we must expand to a third order perspective.

Of course, if a description composed of thirty dimensions (a multiverse map) describes the origins of the eleven-dimensional universe—the mechanism of the Big Bang—then the origins of that multiverse find their explanation in an eighty-five-dimensional map, and so on. This relationship means that true genesis solutions come from the structure of cascading infinities—from the pattern of dimensional hierarchy.

So far we have only dealt with dimensional perspectives of increasing resolution—allowing us to resolve finer and finer dimensional details within our picture at each step. The dimensional hierarchy equation equally applies in the direction of increasing expanse. Each expansive step in perspective resolves the entire previous map into a discrete collection. If we increase our dimensional perspective, from our eleven-dimensional picture, by one expansive leap, we find that our entire eleven-dimensional universe is reduced to a small part of a greater picture—a multiverse, which is filled with a vast number of universes floating about and interacting in a manner quite similar to the way quanta interact in the eleven-dimensional picture. Here the symmetries of cascading infinities come into focus. The levels are self-referential. In perfect fractal form, each order of perspective repeats.[12]

This means that our particular dimensional experience is non-unique, making our map of physical reality a wholly "invertible map."[13] This is philosophically very satisfying. Quantization not only allows us to derive the rules of our formal system, providing us with a useful and self-consistent theory with which to map that system; it also automatically derives the rules of the larger, exterior system, and the system that lies outside that system, *ad infinitum*. It provides us with a theory, and a meta-theory. If we project ourselves throughout this infinite hierarchy, we find that the same theory (or identical copies of one theory) govern all the rules of any immediate formal system. All of those theoretical constructs collapse into a hierarchical monism, which gives us intuitive access to all levels.

If we imagine the cascading infinities as an "infinite dimensional tree,"[14] or an infinite ladder of perspective, it is easy to see that every starting point on that tree, or ladder, appears entirely identical to the rest. This suggests that *dimensional invariant equality* exists in Nature. Wherever we choose our primary reference point along the dimensional ladder, we consequently end up with four directly observable (extended) dimensions.

From each of these starting points, a four-dimensional map is unable to map relativistic and quantum mechanical effects—a fact that will prompt an increase in dimensional resolution. Eventually, the desire to understand ultimate origins will necessitate an expansive increase in dimensional perspective. As this process unfolds the geometric pattern of dimensional hierarchy will be revealed. Observers from any universe will likely begin the endeavor to map their universe from a first-order perspective. Because all first-order dimensional descriptions are identical, the path of discovery will be very similar in all cases. The noble art of curiosity will lead explorers of each universe to discover the structure of the cosmos—a geometry whose dimensional skeleton cascades into the infinities.

"The macrocosm is the microcosm."

Twelfth Century Sufis[15]

The explanatory power we are after dictates the level of description we must use. In the everyday world that we are accustomed to, a first-order description will generally do, even though that description leaves out the seemingly most important feature of the map—the quanta. From this point of view, the fundamental building blocks of the map are, to a certain degree, dispensable. If we wanted to describe a book to someone we would run into the same situation. To describe the book we might convey details about the central characters in the book, or the plot, but we probably wouldn't mention the letters composing the words, or the words composing the sentences. Clearly, the story would not exist without the words, or the primary building blocks—the letters, but those building blocks are the medium, not the message.

By analogy, if our concern involved a desire to understand how the book was put together, its structure, origin, or how it relates messages to us, then we could no longer afford to ignore its fundamental building blocks. The same applies to our dimensional map of Nature. If we want to understand precisely how a single quantum quivers about, then we need to include information about the internal subquanta that compose that quantum. If we want to understand the origin of quantum mechanical and relativistic effects, then we need to jump to a second-order (eleven-dimensional) perspective. And, if we want to understand the conditions that caused the Big Bang, we need to expand our perspective until our entire universe takes on the depiction of a single superquantum within the multiverse.

The fractal structure of our map allows us to easily expand our perspective. Hierarchical quantization resolves a pattern that repeats on each level. In other words, a single quanta may make just a small contribution to the vacuum of our universe, but that same quanta holds within it another entire universe, made of subquanta, which all resolve to be their own universes internally, and so on. Going the other way, a more expansive evaluation reveals an entire universe contributing just a single quantum to a universe defined on a different scale.

Perspectives of both increasing resolution and increasing expanse produce the repeating pattern of self-similarity. If the dimensions that compose three universes (A, B, and C) are offset by one level each, then it is possible that universe A is just a single quantum in Universe B, and Universe B contributes just a single quantum of space to Universe C (Figure 11-5). If observers in each universe map their reality as eleven-dimensional, by quantizing their vacuum once, then the dimensions between their scales overlap in a very specific way—the dimensions of A's map and C's map are bridged by B's map, which shares dimensions with both of them.

The extended spatial dimensions of universe A become the intraspatial dimensions of one of B's quanta, and B's extended spatial dimensions represent C's intraspatial dimensions. Each resonation of A represents a single tick in time at a single location within universe B, yet inside universe A those evolutions trace out all of history—possibly extending from one Big Bang to the next. Time in universe A is supertime for universe B, and time for universe B is supertime for universe C.

In effect, by raising our expansive perspective by one rung on the dimensional ladder, we reduce B's entire universe to one quantum, which minutely participates in the construction of C's universe. Within one tick of C's time, the evolution of that quantum (B's whole universe) goes through eons. This suggests that Big Bangs are cyclical, and the mechanism responsible for that cyclical process can be modeled by higher-dimensional dynamics.

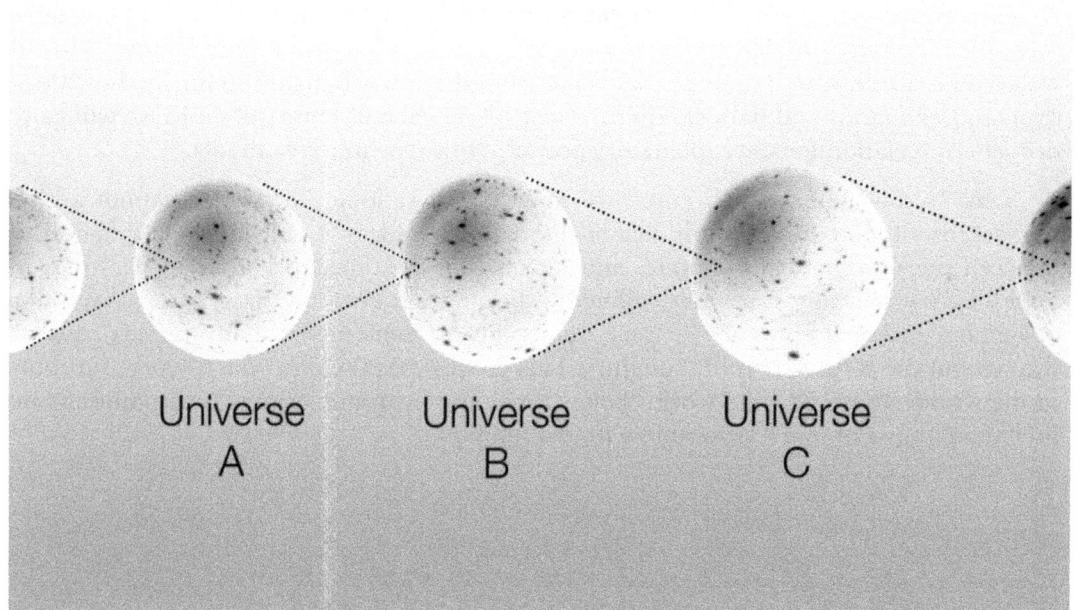

Figure 11-5 Hierarchical perspective.

"It is not impossible that to some infinitely superior being the whole universe may be as one plain, the distance between planet and planet being only as the pores in a grain of sand, and the spaces between system and system no greater than the intervals between one grain and the grain adjacent."

Samuel Taylor Coleridge 'Omniania'

The equation $f(n) = 3^n + n$ describes dimensional perspectives of increasing resolution and increasing expanse in a symmetrical sense. It includes an infinite number of perspectives (orders of perspective) in both directions. It is reflective and invariant. Observers at any level will macroscopically see themselves occupying a universe whose geometry is based on a first order description ($n = 1$). Observers that find themselves aware of only four extended dimensions will be able to deduce, from close examination, the necessary presence of eleven dimensions regardless of their position on the dimensional ladder. Eventually, inquiries into low entropy origins (the cause of their Big Bang) will illuminate the need for third order perspectives (thirty-dimensional explanatory maps) and will further expose the cascading dimensional structure that is built into the nonlocal physical reality we share.

With this understanding we are now ready to venture into the perplexing mysteries of quantum mechanics. Using our new map we will try to make sense of: the state vector, wave-particle duality, the mysterious origins of matter and dark matter, the mechanism of quantum tunneling, teleportation and the formation of worm holes, the connection between black holes and elementary particles, what black holes are like inside their event horizons, and the origin of Nature's universal constants. Our new map offers us some very interesting insights into the origins of these effects.

After exploring consequences of our new geometry that relate to the aforementioned effects, we will take on questions like: is physical reality deterministic or stochastic? What is the emergent process of reality? Do all of the forces have geometric origins? How do we solve the hierarchy problem? Why is symmetry a congenital property of Nature? How do we account for entropy's time arrow? What caused the Big Bang? And finally, how do we explain dark matter and dark energy? As we explore each of these questions we will gain a deeper appreciation for the explanatory power of our new intuitive model.

Leo Tolstoy realized that empirical science and rational knowledge cannot address the depths of humanity's most personal questions unless it introduces a relationship between the finite and the infinite, and makes that relationship known to us.[16] It is my hope that by following Einstein's willingness to question even the most basic assumptions of our day, we will continue to hone our ability to explore the many islands scattered throughout the strange seas of thought.[17] This process has the potential to open our minds to the worlds above and the worlds below, exposing beautiful surprises, and offering us a precious glimpse of our relationship with the infinite.

[1] Roger Penrose. *The Road to Reality*, p. 204.

[2] This also applies for higher dimensional versions called complex spaces.

[3] Traditionally mathematicians represent the complex plane as $z = x + iy$. However, because y can easily be confused with the familiar spatial dimensions that we commonly label y, I am choosing to use a variable that won't be a source of confusion.

[4] All fermions (electrons, protons, neutrons, quarks, neutrinos, muons, etc.) have wave functions that can be described as spinorial objects. The property of spin is also a spinor entity—an entity that makes no intuitive sense in four dimensions. In eleven dimensions, however, an understanding of how a particle, or even every location in space, can possess the property of spin automatically emerges. Furthermore, in eleven dimensions, it becomes expected that spin will be a spinorial measure because it expresses the quaternionic properties of spacetime resolved by the picture.

[5] If this were the case then we could represent time as $\tau = it$. The advantage of this would be that the dl^2 for the Minkowskian metric would look exactly like the ds^2 of E^4, $dl^2 = dt^2 + dx^2 + dy^2 + dz^2$. From a reductionist point of view this would have been considered aesthetically pleasing.

[6] Roger Penrose. *The Road to Reality*, pp. 325–326, original emphasis was in single quotes.

[7] The reason that spinors do not necessarily require the full eleven-dimensional description to produce complex magic is that they are independent of relativistic effects. Because relativistic effects result from spatial density changes within absolute volume, they emerge only when we include superspatial dimensions. Intraspatial dimensions, and therefore the quaternion properties of spacetime, do not come into play for relativistic effects. This is equivalent to saying that the intraspatial dimensions, which possess the quaternion properties, remain unaffected as spatial densities vary.

[8] Roger Penrose. *The Road to Reality*,. pp. 325–326.

[9] On the complex plane, the Cauchy-Riemann equation expresses this spatial characteristic as:

$$\frac{\partial \alpha}{\partial x} = \frac{\partial \beta}{\partial y}, \quad \frac{\partial \beta}{\partial y} = -\frac{\partial \beta}{\partial x} \quad \text{for } \phi = \alpha + i\beta$$

In part, this equation represents the idea that within complex space a single line is a nonsmooth manifold. It is a description that involves dimensions that are separate from the measure that tells us how much space is between two points. Because of these additional dimensions it is possible for several paths through those dimensions to yield the same spatial value for distance between points a and b.

This geometric adjustment also allows us to visually resolve the mysteries of what physicists call *holomorphic* functions, which are geometric descriptions of a region that we call *nonsmooth*, in the three dimensions of familiar space, but becomes *smooth* in complex space. In other words, these functions are what we call *complex smooth*. (A smooth map has no disconformities or disjunctions.)

In a quantized map of space, the nonsmooth character of familiar space is explicit, but the smoothness of complex space is also explicit. In our eleven-dimensional depiction, our framework is defined by dimensional consistency—every region can be immediately associated with a volume. Because of this, the complete picture is smooth—there are no regions that aren't immediately taken up by either the intraspatial dimensions or the superspatial dimensions. Therefore, holomorphic functions are natural inclusions of our model and they are revealed symbiotically with complex magic.

[10] This prediction was made in 2006. In 2007, Jeffery Adams from the University of Maryland, College Park, and 17 other mathematicians verified part of this pattern. They did this by tasking a supercomputer to analyze what is known as an E8 Lie group. The structures that we call Lie groups were first conceived in 1887 by the Norwegian math genius Sophus Lie. The E_8 Lie group was first formulated by German mathematician Wihelm Killing shortly thereafter. The analysis used almost 100 times as much data as the Human Genome Project and took four years to complete. The supercomputer tasked with analyzing this E_8 table of integers with more than 450,000 rows and columns eventually discovered supersymmetries that allowed it to mathematically map a 248-dimensional structure. That specific result aligns perfectly with the prediction of our dimensional hierarchy equation because a 248-dimensional map is equivalent to a fifth order perspective in that equation. Alex Stone. (2007, June). Fearful Symmetry—Mathematicians Triumph over 248 Dimensions. *Discover*, p. 18; (2007, March 24). The Taming of the Symmetries. *New Scientist*, p. 6; Graham P. Collins. (2008, April). Wipeout? *Scientific American*, pp. 30–32.

[11] J. Burnet. (1930). *Early Greek Philosophy*, 4th ed. London. Fragment 11, p. 259; Douglas J. Soccio, *Archetypes of Wisdom*, p. 65; Sider D. (2005) The Fragments of Anaxagoras: Edited with an Introduction and Commentary. Second Edition, Sankt Augustin: Academia Verlag.

[12] In music, this sort of phenomenon, whereby one moves up or down through the levels of some hierarchical system only to end up right back where they started, is called a "strange loop."

[13] Manfred Requardt. (2003, March 25). A Geometric Renormalization Group in Discrete Quantum Space-Time. arXiv:gr-qc/0110077v3. p. 27.

[14] Ibid.

[15] "The Greek philosophers Anaximenes of Miletus, Pythagoras, Heraclitus, and Plato; the ancient Gnostics; the pre-Christian Jewish philosopher Philo Judaeus; and the medieval Jewish philosopher Maimonides—all embraced the macrocosm-microcosm idea." Henry Corbin. (1969). *Creative Imagination in the Sufism of Ibn 'Arabi*, trans. Ralph Manheim, Princeton, NJ: Princeton University Press, p. 259; Michael Talbot, *The Holographic Universe*, p. 290.

[16] Leo Tolstoy. *Confession*, translated by David Patterson. Chapter IX.

[17] In 1921, Einstein received an honorary degree from Princeton University "for voyaging through strange seas of thought." Walter Isaacson, *Einstein, His Life and Universe*, p. 297.

Part Three—Physical Reality in Eleven Dimensions

"Looking at the stars always makes me dream, as simply as I dream over the black dots representing towns and villages on a map. Why, I ask myself, shouldn't the shining dots of the sky be as accessible as the black dots on the map of France?"

Vincent van Gogh

"One peek is worth a hundred finesses."

E. O. Wilson

Chapter 12 — The Questions of Quantum Mechanics

"The way we have to describe Nature is generally incomprehensible to us... because the more you see how strangely Nature behaves, the harder it is to make a model that explains how even the simplest phenomena actually work."

Richard Feynman[1]

"The non-intuitive nature of quantum mechanics—or, rather, of Nature herself at the level of quantum-mechanical activity—leads many people to despair of finding any kind of trustworthy picture of quantum-level phenomena."

Roger Penrose[2]

"I just found out there's no such thing as the real world, just a lie you got to rise above."

John Mayer

Federal Prison Camp, Florence, Colorado.

By the time I became a triple-digit-midget (someone with less than 1000 days left) my life had become quite monastic. I would go to the chow hall for breakfast, tend to my remedial job, and then return to my cell to read physics and astronomy books. After lunch I returned to my cell, and busily worked on researching and writing this book. From 3:00 until the 4:00 recall I ran around a small dirt track outside with a fellow named Joseph Smith (his real name). After count I stood in line for an hour for dinner, and then returned to my writing until it was time to teach Astronomy, Basic Science, or Beginning Art.

One day, in the middle of the routine, a new inmate came to me with a question that I will never forget. He approached me and said "Hey, you dat astronaut guy right... the one that stole the moon?" I was expecting some kind of set up, or to be the butt of some joke, and I didn't feel like explaining to him that I hadn't actually been to space yet, or that it wasn't the entire moon, so I assumed that he meant "the guy that worked for NASA", and replied "Yeah." Then he said, "Let me axe you a question." I suspiciously waited for the question.

"They say the world is round right?" For a few seconds of awkward silence I stared into his face trying to determine whether or not this was a serious question. Eventually I apprehensively said, "Yeah..." To which he said, "So... are we on the inside or the outside?"

My mind began to race to figure out how this question could have come about. He must have never seen a globe, a weather report, or a picture of earth from space. How could he have missed all of this?

As I was contemplating this mysterious state of confusion, he revealed the deeper root of his question. "Because if the world is spinning and we are on the inside, then that would explain why we stay on the ground." I didn't expect this turn of events. He was trying to make sense of gravity. He was also trying to use a geometric explanation to get there. In a way, this completely naïve question was brilliant. He had recognized that gravity posed some rather big questions and was apparently unsettled by the fact that he didn't have answers to those questions. He wanted to understand why things worked as they did. He may have had a lot to learn about scientific observations, and how to tie them together, but he also had an advantage—his imagination was free of the preconceptions that reinforce our worldview and keep us from progressing. He wasn't even aware of the current "truth" we cling to. As a consequence, he was free to explore all options.

Upsetting Our Worldview

"The voyage of discovery lies not in seeking new horizons, but in seeing with new eyes."

Marcel Proust

Making sense of the world, mapping its full character, means finding a way to include the enigmas that are not addressed by our old maps—the effects in Nature that have managed to evade all our attempts to explain them based on our Euclidean assumptions. In the next chapter we shall discuss our eleven-dimensional solutions to some of these mysteries. In this chapter our goal is to acquire a taste for the quagmires of quantum mechanics. An examination of the double-slit experiment will give us a good introduction to these mysteries. However, in order to make that examination worthwhile, we need to make sure that we are familiar with an important effect known as interference.[3]

Interference applies universally to all interacting waves. A water wave, for instance, can be described as a disturbance in the shape of the water's surface. This disturbance produces regions where the water level is higher and regions where it is lower than the undisturbed value. The highest part of each ripple is called a peak and the lowest part is called a trough (Figure 12-1). Typically waves involve periodic succession, peak followed by trough followed by peak and so on. In general, we can define a wavelength as the distance between identical parts of adjacent waves. Measurements from peak to peak, or trough to trough, for example, give the same value for wavelength.

When waves interact in a medium, they interfere. For example, if we drop two rocks into spatially separated parts of a pond, their waves will interfere when they cross (Figure 12-2). When a peak of one wave and a peak of the other wave come together, the height of the water rises to a height equal to the sum of the two peaks. Similarly, when a trough of one wave and a trough of another wave cross, the depression of the water's surface dips to the sum of the two depressions. And when a peak of one wave crosses with a trough of another, they (at least partially) cancel each other out. The peak of one wave contributes a positive displacement while the trough of the other wave contributes a negative displacement. If the two waves have equal magnitude, then there will be perfect cancellation and the water's surface will be flat, just as it was before any wave existed.

Keeping these rules of interference in mind, let's turn our attention to light. If we take a laser emitting a single wavelength—a single color, and shine it on a screen that has a slit etched into it (Figure 12-3), what image should we expect to see on the wall behind the screen?[4] Classically speaking, we would expect to see a stripe of light on the wall. (Classically means according to our four-dimensional intuition, or the rules of Euclidean geometry.) It turns out that this *is* what we see. In this sense light's behavior correlates perfectly with our Euclidean intuition.

Figure 12-1 Peaks and troughs of waves

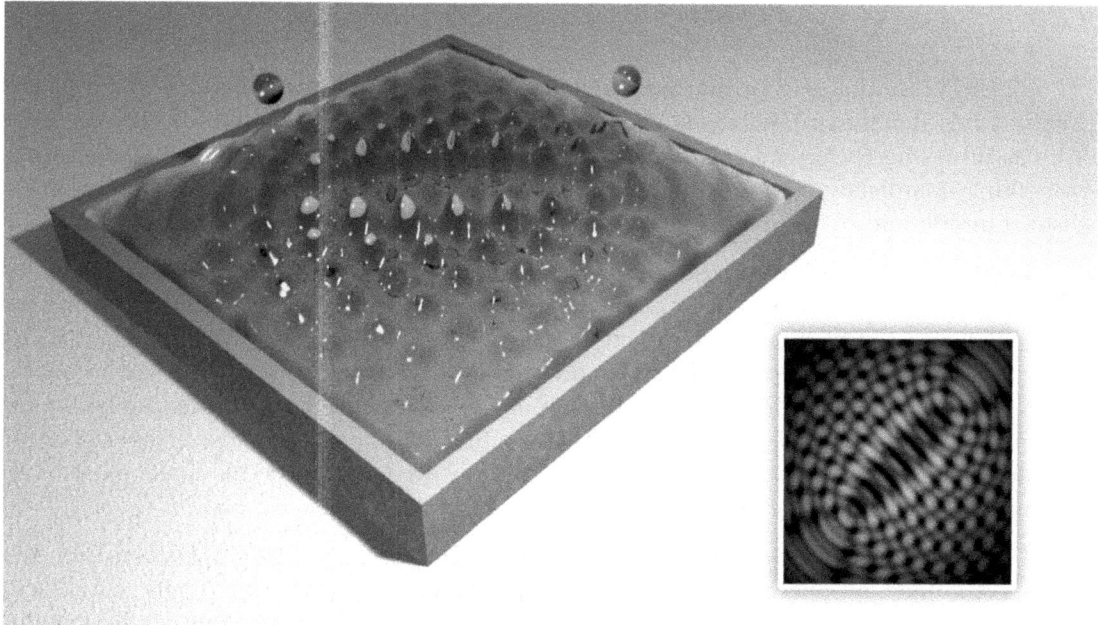

Figure 12-2 Constructive and destructive interference

What image should we expect to see on the wall if we etch a second slit on our screen and cover the first slit with a black piece of tape? Well, our classical intuitions tell us to expect

179

a line of light projected on the wall, just like we did before, except this line of light should be offset from the first. Again, this is exactly what we see when we perform the experiment. So far all of this is straightforward and conceptually trivial. But as it turns out, we are only one step away from a profound mystery. We discover this mystery by removing the piece of tape. To understand the impact of this mystery, ask yourself: What sort of projection do we expect to see on the wall when both slits are open?

Figure 12-3 Expected single slit projection

Classical intuition tells us that we should see two parallel bands of light on the wall as in Figure 12-4. But this is where our classical training (our Euclidean intuition) lets us down. This is also where classical mechanics breaks down. When we perform this experiment, something completely counterintuitive happens, contradicting our Euclidean intuitions. A distinct interference pattern is projected on the wall (Figure 12-5).

The bright and dark bands produced in this double-slit experiment are telltale signs that light propagates as a wave.[5] Interference patterns are key signatures of waves. The problem is that this wavelike characteristic directly clashes with our observations of light's particulate behavior. After all, photons are always found in point-like regions rather than spread out like a wave, and individual photons are always found to have very discrete amounts of energy. When measuring a wave, you would expect to find its energy spread out over a region instead of being concentrated in one location. So how are we supposed to make sense of this observation? What is going on?

These diametrically opposed properties of light are verified facts. Contradictory as they may seem, they are here to stay. They have forced us to the seemingly paradoxical conclusion that light is both a wave *and* a particle. But how can this be? How can it be both?

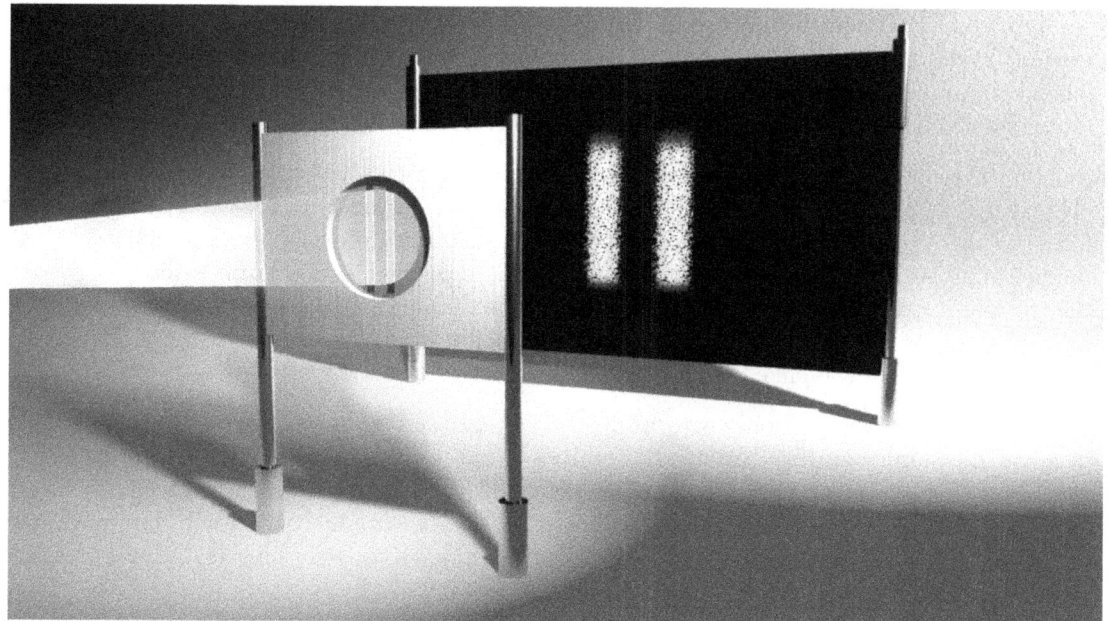

Figure 12-4 Expected double slit projection

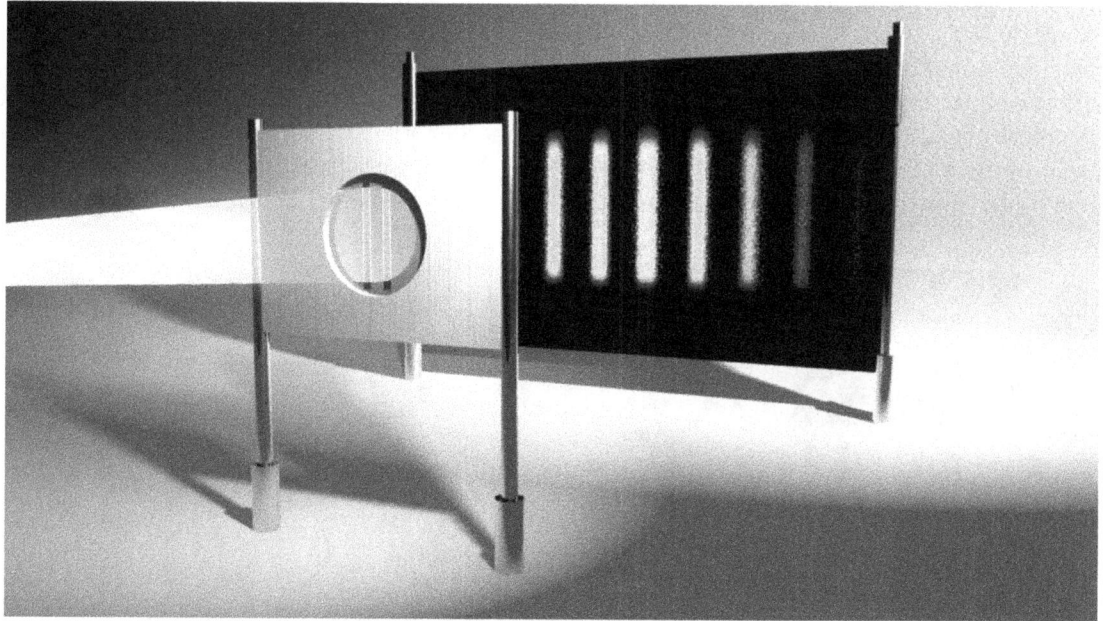

Figure 12-5 Actual double slit projection

The Dual Nature of Light and Matter

Although many scientists have found the *wave-particle duality* of light to be conceptually vague and schizophrenic, this description has persisted. In fact, after the wave-particle concept was adopted as an accurate description of light, it was extended to describe electrons and, eventually, all of matter. This transition was nothing short of a revolution.

Up until 1910, atoms were simplistically viewed as miniature solar systems with the nucleus making up the "central star" and orbiting electrons being "planets".[6] The wave-particle duality of light and matter rejected this view and pointed to a significantly different architecture for atoms. Of course, this conceptual transition did not take hold over night.

In 1924, Prince Louis de Broglie found that in addition to their particle like character,[7] electrons also possessed a wavelike character. In 1927, Clinton Davisson and Lester Germer followed this up by firing a beam of electrons at a piece of nickel crystal, which acted as a barrier analogous to the one used in the double-slit experiment. A phosphor screen recorded the resultant pattern of electrons.[8] When they examined the screen, they observed an interference pattern just like the one produced in the double-slit experiment, showing that even electrons have wavelike properties.

These experiments shook the foundation of physics by threatening the structure of classical mechanics and destroying humanity's intuitive framework of reality. But it didn't stop there. The next step was to tune the beam of electrons down so that the electron gun fired just a single electron at a time. Similar experiments were later used with lasers wherein individual photons were fired seconds apart from each other. The results were mind-bending.

Completely against expectation these experiments also produced interference patterns over time as the collection of electrons (or photons) continued to build (Figure 12-6). These observations only added to the confusion. Waves are supposed to be a collective property—something that has no meaning when applied to separate, particulate ingredients. (A water wave, for example, involves a large number of water molecules.) So how can a single electron, or a single photon, be a wave? Furthermore, wave interference requires a wave from one place to interact with a wave from another place. So how can interference be relevantly applied to a single electron or photon? While we are considering such questions, we should also ask, if a single electron or photon is a wave, then what is it that is "waving"?[9]

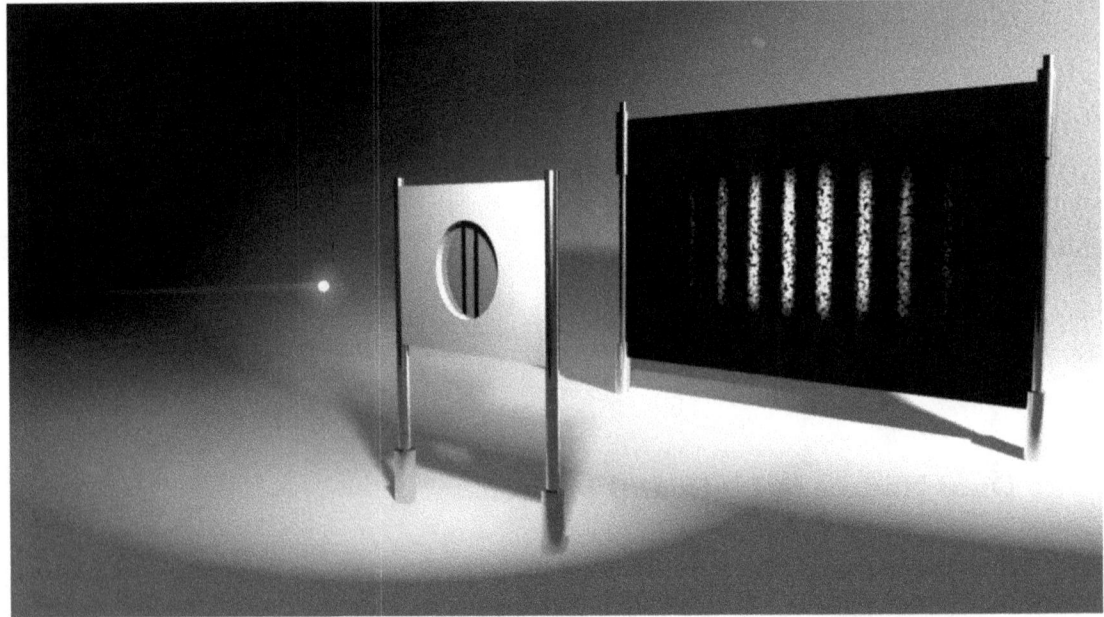

Figure 12-6 Over time individual photons construct an interference pattern

To answer these questions, Erwin Schrödinger proposed that the stuff that makes up electrons might be smeared out in space and that this smeared electron essence might be what waves. If this idea was correct then we would expect to find all of the electron's properties, spread out over a distance, but we never do. Every time we locate an electron, we find all of its mass and all of its charge concentrated in one tiny, point-like region.

Max Born came up with a different idea. He suggested that the wave is actually a probability wave.[10] Einstein tinkered with a similar idea when he hypothesized that these waves were optical observations that refer to time averages rather than instantaneous values.

Inserting a probability wave (also called a state vector, or a wave function) as a fundamental aspect of Nature delivers another blow to our common-sense ideas about how things truly operate. It suggests that experiments with identical starting conditions do not necessarily lead to identical results because it claims that you can never predict exactly where an electron will be in a single instant. You can only define a probability that we will find it over here, or over there, at any given moment. Two situations with the same probabilistic starting conditions, say of a single particle, might not produce the same results, because the particle can be anywhere within that probability distribution. From a classical perspective, the discovery that the microscopic universe behaves this way is absolutely baffling. Nevertheless, it is how we have observed Nature to be.

This leads us to a rather interesting precipice. It seems that the map we have been using to chart physical reality somehow dissolves when we look closely at it. The rules of four-dimensional geometry simply fail to accurately map Nature when we examine the smallest scales. Nature doesn't strictly behave as our old Euclidean map dictates. Stumbling upon this discovery forces us to face a vital question. Is Nature ultimately and fundamentally probabilistic in a way that we may never understand, as many modern physicists have chosen to believe; or, is this probabilistic quality a byproduct of our reduced dimensional representation of Nature?

After pondering these questions long and hard, some physicists have come to believe that the tapestry of spacetime is analogous to water: that the smooth appearance of space and time is only an approximation that must yield to a more fundamental framework when considering ultramicroscopic scales. As far as I can tell, however, up until now this point has only been entertained abstractly. Geometrically resolving a molecular structure for space might resolve our greatest quantum mechanical mysteries, but as of yet, no one has taken that final step. No one has developed a self-consistent picture from this geometric insight. No one has moved beyond the mathematical suggestion that spacetime is analogous to water, or interpreted the theoretical quanta of space as being physically real. Consequently, a framework that enables conceptualization of what is meant by the "molecules" or "atoms" of spacetime has not been developed.

Eight decades of meticulous experiments have confirmed the predictions of quantum mechanics based on this wave function, or probability wave, description with amazing precision. "Yet there is still no agreed-upon way to envision what quantum mechanical probability waves actually are. Whether we should say that an electron's probability wave is the electron, or that it's associated with the electron, or that it's a mathematical device for describing the electron's motion, or that it's the embodiment of what we can know about the electron is still debated."[11]

Although quantum mechanics describes the universe as having an inherently probabilistic character, we don't experience the effects of this character in our day-to-day lives. Why is this? The answer, according to quantum mechanics, is that we don't see quantum events like a chair being here now and then across the room in the next instant, because the probability of that occurring, although not zero, is absurdly miniscule. But what exactly makes the probability for large things to act, as electrons do, so small? At what scales do such effects become important? And, why should the macroscopic universe be so different from the microscopic universe?

As if these newly uncovered characteristics of reality weren't obscure enough, quantum physicists conceptually fuddle things further by suggesting that without observation things have no reality. They claim that until the position of an electron is actually measured the electron has no definite position. Before it is measured, the position exists only as a probability, and then suddenly, through the act of measuring, the electron miraculously acquires the property of position.

Einstein acutely recognized the absurdity of this claim. When approached with this conjecture, he famously quipped, "Do you really believe that the moon is not there unless we are looking at it?"[12] To him everything in the physical world had a reality independent of our observations. Measurements that suggested otherwise were mere reflections of the incompleteness by which we currently map and comprehend physical reality. To many quantum physicists, however, the unobserved Moon's existence became a matter of probability. To them, a discoverable, complete map of physical reality, with the ability to resolve an underlying determinism, became nothing more than a myth—a romantic dream.

The mathematical projection of quantum mechanics can be statistically matched with our four-dimensional observations, but when it comes to a conceptual explanation of those observations, it completely lets us down. Intuitive explanations cannot be gleaned from a framework of physical reality that is assumed to be fundamentally probabilistic. By definition, randomness blurs causality. This vague description of physical reality keeps us from grasping a deeper truth by allowing what should be the most basic of concepts to drip into a realm of nonsense. When we restore determinism with higher-dimensional depictions, we no longer have this problem.

As an example of the confusion that stems from swallowing the standard quantum mechanical interpretation "guts, feathers, and all," consider the fact that a probabilistic treatment of quantum mechanics leads us to the conclusion that the double-slit experiment can be explained by assuming that a photon actually takes both paths. We can combine the two probability waves emerging from both slits to statistically determine where a photon will land on a screen. The result mimics an interference pattern.

According to this, we can explain interference patterns by assuming that one photon somehow always manages to go through both slits, but is this really what is going on? Does a photon really travel along both paths? Can this count as an *explanation* if we have no coherent sense of what it means? You might notice that if we were to design our experiment with three slits, then we would have to consider whether or not the photon really travels all three routes. This question can be extended for as many slits as you like, but the fundamental conceptual problem remains the same.

In order to solve this mystery, you may suggest that we place detectors in front of the slits to determine if the photons are actually going through both slits, or just one. When we do this, we always find that individual photons pass through one slit or the other—never both. But, when we measure the position of individual photons we no longer get an interference pattern and so the question retains its ambiguity. Some have taken this to mean that the act of observation forces wave properties to collapse into a particle, but how and why this theoretical collapse occurs still lacks explanation.

Because probability waves are not directly observable and because photons (and electrons) are always found in one place or another when measured, we might be tempted to think that probability waves might not be real—that they were never really there. If that is true, then how are the interference patterns created? Surely these probability waves exist, but in what sense? What are they referencing? Why is it that whenever we know which path the photon takes, we get a classical image instead of an interference pattern? How does the detection of a photon, or an electron, change its behavior?

To date, these questions have yet to be resolved. In fact, more clever experiments designed to solve these questions have only deepened the mystery. For example, let's perform the double-slit experiment again, but this time let's place devices in front of the slits, which mark (but do not stop or detect) the photons before they pass through the slits. This marking allows us to examine the photons that strike the screen and subsequently determine which slit they passed through. Thus we only gain knowledge of which path the photon takes after the path has been completed. For some reason, however, when we do this we find that the photons do not build up an interference pattern. They form a classical image (Figure 12-4).

Once again, it seems that "which-path" information inhibits us from probing these ghostly waves. But is it really the fact that we gain the ability to determine which path a photon goes through—independent of when we gain that information—that disrupts the interference pattern? Or does our marking of the photon somehow disrupt its interference potential?

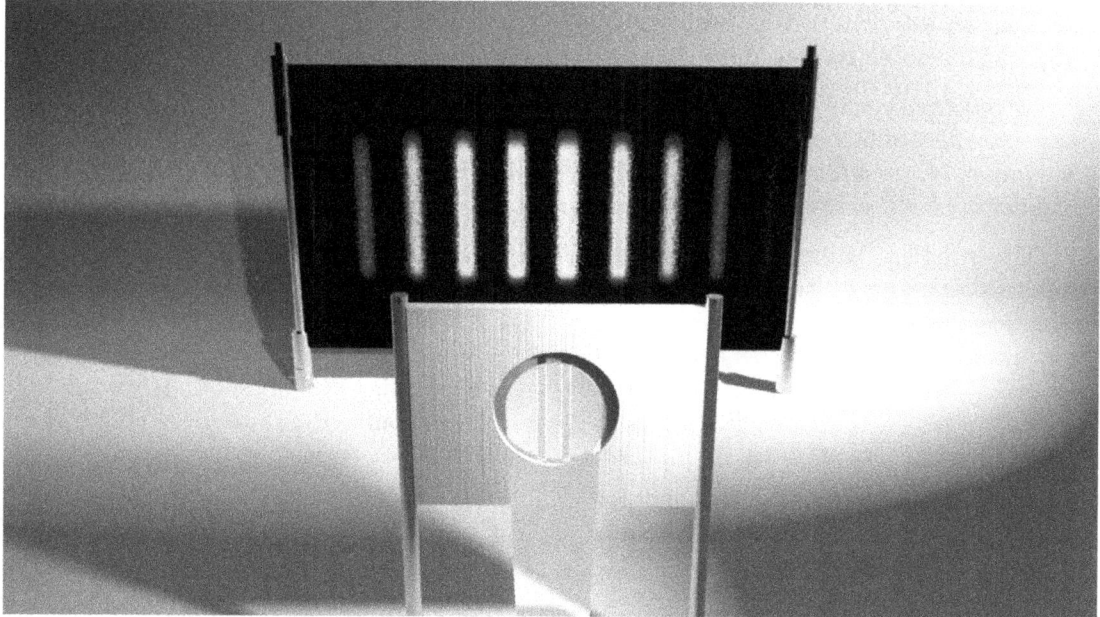

Figure 12-7 A classical pattern emerges

To explore this question, we perform what's known as the quantum eraser experiment. We start with the same set up we just described. Then we place another device between each slit and the screen, which completely removes the mark from the photon. We already know that the marked photons project a classical image. Will an interference pattern reemerge if we remove the effects of this mark—if we lose the ability to extract the which-path information?

When we perform this experiment the interference pattern does return (Figure 12-7). Does this mean that photons somehow choose how to act, based on our knowledge of them? Or does it imply something even stranger—that the photons are always both particles and waves simultaneously? How are we to understand either conclusion?

The Photoelectric Effect and the Uncertainty Principle

Another curiosity of Nature is known as the photoelectric effect. Philipp Lenard first discovered this effect through controlled experiments in 1900. When light shines on a metal surface, it causes electrons to be knocked loose and emitted. Knowing this, Lenard designed an experiment that allowed him to control the *frequency* of the incoming light. During the experiment, he increased the frequency of the light—moving from infrared heat and red light to violet and ultraviolet. Greater frequencies caused the emitted electrons to speed away with more kinetic energy. After discovering this, Lenard reconfigured his experiment to allow him to control the *intensity* of the incoming light. He used a carbon arc light that could be made brighter by a factor of 1,000.

Because both experiments involved increasing the amount of incoming light energy he expected to have identical results. In other words, because the brighter, more intense light had more energy, Lenard expected that the electrons emitted would have more energy and speed away faster. But that's not what happened. Instead, the more intense light produced more electrons, but the energy of each electron remained the same.[13]

In response to these experiments Einstein suggested that light is composed of discrete packets called photons. Under this assumption, light with higher frequency would cause electrons to be emitted with more energy, and light with higher intensity, that is, a higher quantity of photons, would result in emission of more electrons—just as we observe.

The problem with this solution (a solution that is now universally accepted among physicists) is that it doesn't provide us with a clear description for what the light quanta are. Why does light come in quantized packets? Near the end of his life Einstein lamented over this problem in a letter to his dear friend Michele Besso. He wrote, "All these fifty years of pondering have not brought me any closer to answering the question, what are light quanta?"[14] It's been another fifty years and we seem as confused as ever over how it is that light is quantized into little discrete packets called photons.

In the midst of these enigmas lies the uncertainty principle, which states that knowledge of certain properties inhibits knowledge of other complimentary properties. For example, the more accurately we determine the position of an electron, the less we can determine its momentum, and vise versa.

Heisenberg tried to explain the uncertainty principle by appealing to the observer effect; claiming that it was simply an observational effect of the fact that measurements of quantum systems cannot be made without affecting those systems.[15] Since then, the uncertainty principle has regularly been confused with the observer effect.[16] But the uncertainty principle is not a statement about the observational success of current technology. It has nothing to do with the observer effect. It highlights a fundamental property of quantum systems, a property that turns out to be inherent in all wave-like systems.[17] Uncertainty is an aspect of quantum mechanics because of the wave nature it ascribes to all quantum objects.

If our current description of quantum mechanics is fundamental, if there is nothing beneath the state vector—a claim that defines the heart of the standard interpretation of quantum mechanics—then this uncertainty principle may be a sharp enough dagger to kill our quest for an intuitive understanding of physical reality. The corrosive power of the uncertainty principle, when mixed with our current paradigm, is poignantly illustrated by an old story involving Niels Bohr. According to the story, Bohr was once asked what the complementary quality to truth is. After some thought he answered—"clarity."[18]

The State Vector

Unlike classical mechanics, which describes systems by specifying the positions and velocities of its components, quantum mechanics uses a complex mathematical object called a state vector $|\psi\rangle$ (also called the wave function[19]) to map physical systems. Interjecting this state vector into the theory enables us to match its predictions to our observations of the microscopic world, but it also generates a relatively indirect description that is open to many equally valid interpretations. This creates a sticky situation, because to "really understand" quantum mechanics we need to be able to specify the exact status of $|\psi\rangle$ and to have some sort of justification for that specification. At the present, we only have questions. Does the state vector describe physical reality itself, or only some (partial) knowledge that we have of reality? "Does it describe ensembles of systems only (statistical description), or one single system as well (single events)? Assume that indeed, $|\psi\rangle$ is affected by an imperfect knowledge of the system, is it then not natural to expect that a better description should exist, at least in principle?"[20] If so, what would this deeper and more precise description of reality be?

To explore the role of the state vector, consider a physical system made of N particles with mass, each propagating in ordinary three-dimensional space. In classical mechanics we would use N positions and N velocities to describe the state of the system. For convenience we might also group together the positions and velocities of those particles into a single vector **V**, which belongs to a real vector space with $6N$ dimensions, called *phase space*.[21]

The state vector $|\psi\rangle$ can be thought of as the quantum equivalent of this classical vector **V**. The primary difference is that, as a complex vector, it belongs to something called *complex vector space*, also known as *space of states*, or *Hilbert space*. In other words, instead of being encoded by regular vectors whose positions and velocities are defined in *phase space*, the state of a quantum system is encoded by complex vectors whose positions and velocities live in a *space of states*.[22]

The transition from classical physics to quantum physics *is* the transition from phase space to space of states to describe the system. In the quantum formalism each physical observable of the system (position, momentum, energy, angular momentum, etc.) has an associated linear operator acting in the space of states. (Vectors belonging to the space of states are called "kets.") The question is, is it possible to understand space of states in a classical manner? Could the evolution of the state vector be understood classically (under a projection of local realism) if, for example, there were additional variables associated with the system that were ignored completely by our current description/understanding of it?

While that question hangs in the air, let's note that if the state vector is fundamental, if there really isn't a deeper-level description beneath the state vector, then the probabilities postulated by quantum mechanics must also be fundamental. This would be a strange anomaly in physics. Statistical classical mechanics makes constant use of probabilities, but those probabilistic claims relate to statistical ensembles. They come into play when the system under study is known to be one of many similar systems that share common properties, but differ on a level that has not been probed (for any reason). Without knowing the exact state of the system we can group all the similar systems together into an ensemble and assign that ensemble state to our system. This is done as a matter of convenience. Of course, the blurred average state of the ensemble is not as clear as any of the specific states the system might actually have. Beneath that ensemble there is a more complete description of the system's state (at least in principle), but we don't need to distinguish the exact state in order to make predictions. Statistical ensembles allow us to make predictions without probing the exact state of the system. But our ignorance of that exact state forces those predictions to be probabilistic.

Can the same be said about quantum mechanics? Does quantum theory describe an ensemble of possible states? Or does the state vector provide the most accurate possible description of a single system?[23]

How we answer that question impacts how we explain unique outcomes. If we treat the state vector as fundamental, then we should expect reality to always present itself in some sort of smeared out sense. If the state vector were the whole story, then our measurements should always record smeared out properties, instead of unique outcomes. But they don't. We always measure well-defined properties that correspond to specific states.

Sticking with the idea that the state vector is fundamental, von Neumann suggested a solution called state vector reduction (also called wave function collapse).[24] The idea was that when we aren't looking, the state of a system is defined as a superposition of all its possible states (characterized by the state vector) and evolves according to the Schrödinger equation. But as soon as we look (or take a measurement) all but one of those possibilities collapse. How does this happen? What mechanism is responsible for selecting one of those states over the rest? To date there is no answer. Despite this, von Neumann's idea has been taken seriously because his approach allows for unique outcomes.

The evolution of the state vector $|\psi(t)\rangle$ between t_0 and t_1 is given by the Schrödinger equation:

$$i\hbar \frac{d}{dt}|\psi(t)\rangle = H(t)|\psi(t)\rangle$$

$H(t)$ characterizes the Hamiltonian evolution of the system—it expresses how energy in the system can swap from one kind to another (between potential and kinetic forms).

The problem that von Neumann was trying to address is that the Schrödinger equation itself does not select single outcomes. It cannot explain why unique outcomes are observed. According to it, if a fuzzy mix of properties comes in (coded by the state vector), a fuzzy mix of properties comes out. To fix this, von Neumann conjured up the idea that the state vector jumps discontinuously (and randomly) to a single value.[25] He suggested that unique outcomes occur because the state vector retains only the "component corresponding to the observed outcome while all components of the state vector associated with the other results are put to zero, hence the name *reduction*."[26]

The fact that this reduction process is discontinuous makes it incompatible with general relativity. It is also irreversible, which makes it stand out as the only equation in all of physics that introduces time-asymmetry into the world. If we think that the problem of explaining uniqueness of outcome eclipses these problems, then we might be willing to take them in stride. But to make this trade worthwhile we need to have a good story for how state vector collapse occurs. We don't. The absence of this explanation is referred to as the *quantum measurement problem*.

Many people are surprised to discover that the quantum measurement problem still stands. It has become popular to explain state vector reduction (wave function collapse) by appealing to the observer effect, asserting that measurements of quantum systems cannot be made without affecting those systems, and that state vector reduction is somehow initiated by those measurements.[27] This may sound plausible, but it doesn't work. Even if we ignore the fact that this 'explanation' doesn't elucidate *how* a disturbance could initiate state vector reduction, this isn't an allowed answer because "state vector reduction can take place even when the interactions play no role in the process."[28] This is illustrated by *negative measurements* or *interaction free measurements* in quantum mechanics.

To explore this point, consider a source, S, that emits a particle with a spherical wave function, which means its values are independent of the direction in space.[29] In other words, it emits photons in random directions, each direction having equal probability. Let's surround the source by two detectors with perfect efficiency. The first detector D_1 should be set up to capture the particle emitted in almost all directions, except a small solid angle θ, and the second detector D_2 should be set up to capture the particle if it goes through this solid angle (Figure 12-8).

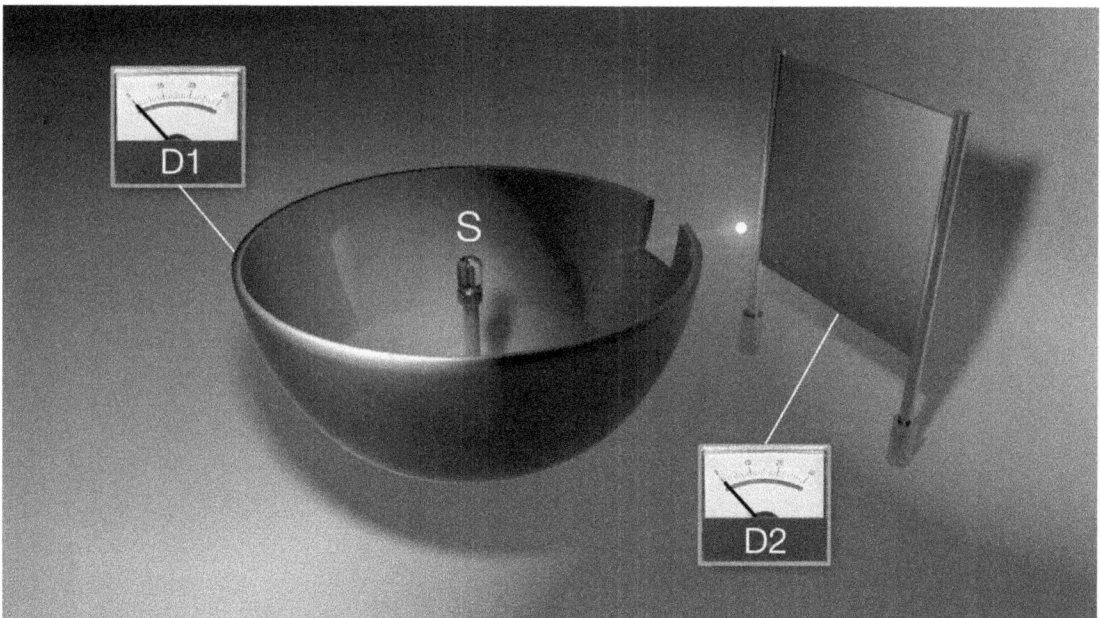

Figure 12-8 An interaction-free measurement

When the wave packet describing the wave function of the particle reaches the first detector, it may or may not be detected. (The probability of detection depends on the ratio of the subtended angles of the detectors.) If the particle is detected by D_1 it disappears, which means that its state vector is projected onto a state containing no particle and an excited detector. In this case, the second detector D_2 will never record a particle. If the particle isn't detected by D_1 then D_2 will detect the particle later. Therefore, the fact that the first detector has not recorded the particle implies a reduction of the wave function to its component contained within θ, implying that the second detector will always detect the particle later. In other words, the probability of detection by D_2 has been greatly enhanced by a sort of "non-event" at D_1. In short, the wave function has been reduced without any interaction between the particle and the first measurement apparatus.

Franck Laloë notes that this illustrates that "the essence of quantum measurement is something much more subtle than the often invoked 'unavoidable perturbations of the measurement apparatus' (Heisenberg microscope, etc.)."[30] If state vector reduction really takes place, then it takes place even when the interactions play no role in the process, which means that we are completely in the dark about how this reduction is initiated or how it unfolds. Why then is state vector reduction still taken seriously? Why would any thinking physicist uphold the claim that state vector reduction occurs, when there is no plausible story for how or why it occurs, and when the assertion that it does occur creates other monstrous problems that contradict central tenets of physics? The answer may be that generations of tradition have largely erased the fact that there is another way to solve the quantum measurement problem.

Returning to the other option at hand, we note that if we assume that the state vector is a statistical ensemble, if we assume that the system does have a more exact state, then the interpretation of this thought experiment becomes straightforward; initially the particle has a well-defined direction of emission, and D_2 records only the fraction of the particles that were emitted in its direction.

Standard quantum mechanics postulates that this well-defined direction of emission does not exist before any measurement. Assuming that there is something beneath the state vector, that a more accurate state exists, is tantamount to introducing additional variables to quantum mechanics. It takes a departure from tradition, but as T. S. Eliot said in *The Sacred Wood*, "tradition should be positively discouraged."[31] The scientific heart must search for the best possible answer. It cannot flourish if it is constantly held back by tradition, nor can it allow itself to ignore valid options. Intellectual journeys are obliged to forge new paths.

The Bell Theorem: Blurring Local Realism

Violations of the Bell theorem have conclusively shown that local realism does not hold in the vacuum.[32] On the microscopic scales, superluminal effects arise—violating the speed of light restriction of causality. (For a review of local realism see Chapter 4: The Case For Quanta.)

Locality, which is short for *local relativistic causality*, encodes the idea that the mutual influence of events decreases when their distance increases; that the mutual influence of events directly depends upon their spatial separation. Realism is the view that the world described by science represents the real world in a mind-independent way.

As a scientific presupposition, local realism is absolutely fundamental, yet violations of local realism pop up in even the simplest quantum systems—like two spins in a singlet state.

Do these violations undermine the possibility that Nature has a causal structure? To explore that question we need to understand Bell's theorem.

Any time the properties of two sub-systems (two spins in this case) are correlated inside the state vector itself, we say that they are in an "entangled state." If two particles with spin ½ are emitted by a source S in a single spin state and propagate in opposite directions, their combined spin state is described by:

$$|\psi\rangle = \frac{1}{\sqrt{2}}[|+,-\rangle - |-,+\rangle]$$

The first component of this two-spin state $|\pm,\mp\rangle$ has a spin that is in an eigenstate with an eigenvalue of $\pm\frac{\hbar}{2}$ along the O_z component, and the second has a spin that is in an eigenstate with eigenvalue $\mp\frac{\hbar}{2}$. When the two particles reach distant locations, the first particle is submitted to a spin measurement with a Stern-Gerlach apparatus oriented along the direction defined by angle a, while the other particle undergoes a similar measurement along the direction defined by angle b.

If $\theta_{ab} = a - b$ is the angle between the directions defined by a and b, then quantum mechanics predicts that the probability for detecting identical results (+1, +1) or (-1, -1) is:

$$P(+,+) = P(-,-) = \frac{1}{2}\sin^2\frac{\theta_{ab}}{2}$$

and the probability of detecting opposing results $(+1,-1)$ or $(-1,+1)$ is:

$$P(+,-) = P(-,+) = \frac{1}{2}\cos^2\frac{\theta_{ab}}{2}$$

If the orientations of the measurement apparatuses are parallel, the probability of detecting identical results vanishes, while the probability of detecting opposite results goes to ½ and ½.

If $\theta_{ab} = 0$, then $P(+,+) = P(-,-) = 0$ and $P(+,-) = P(-,+) = \frac{1}{2}$

Using these probability equations we can calculate the average value of the product of the two results obtained by measuring the components of the two spins along directions a and b with relative angle θ_{ab}. The result is:

$$P(+,+) + P(-,-) - P(+,-) - P(-,+) = \cos\theta_{ab}$$

In 1969 Bell, Clauser, Horne, Shimony, and Holt showed that, independent of the mechanism that is creating the correlations between these entangled particles, if locality is obeyed, then this equation must hold.[33] Eberhand later noticed that under this restraint the sum of products becomes:[34]

$$M = AB - AB' + A'B + A'B' = A(B - B') + A'(B + B')$$

Because one of the brackets in the right-hand side of this equation always equals ± 2 while the other vanishes, this sum is always equal to $+2$ or -2. And the average value $\langle M \rangle$ of M is:

$$-2 \leq \langle M \rangle \leq +2$$

This is the BCHSH form of the Bell theorem.[35] The predictions of any local realist theory must obey this inequality. To check whether or not quantum mechanics adheres to local realism we take the quantum equivalent $\langle Q \rangle$ of the combination of four products in our expression for M. From this we calculate the same combination of averages of products of results:[36]

$$\langle Q \rangle = -\cos\theta_{ab} + \cos\theta_{ab'} - \cos\theta_{a'b} - \cos\theta_{a'b'}$$

If quantum mechanics is a local realist theory, then the confinement of $\langle Q \rangle$ should mirror that of $\langle M \rangle$. The surprise is that it does not. For some appropriate choices for the angles of a, a', b, b', $\langle Q \rangle$ reaches values of $\pm 2\sqrt{2}$.

$$-2\sqrt{2} \leq \langle Q \rangle \leq 2\sqrt{2}$$

This result violates the BCHSH inequality, and therefore contradicts the Bell theorem, by a factor of $\sqrt{2}$. This means that quantum mechanics quantitatively and significantly violates local realism. Quantum mechanics simply cannot be a local realist theory.

If quantum mechanics is not a local realist theory, then what is it? How could nonlocality be inscribed by Nature? If a deeper-level theory ends up accounting for the predictions of quantum mechanics, would it too have to violate local realism? Or could the deeper-level theory conform to local realism?

Einstein was very uncomfortable with the implications of a probabilistic universe. To him, things possessed an absolute quality at all times rather than a smeared essence until observed. Despite the precise manner in which quantum mechanics statistically predicts experimental observations of the microscopic world, Einstein remained convinced that the universe is not fundamentally probabilistic. His intuition told him that there must be a deeper and less bizarre description. He held the firm belief that rational inquiry can speak to the hunger for truth. Like John Wheeler, he felt that "when we come to the final laws of Nature, we will wonder why they were not obvious from the beginning."[37]

As we chase Einstein's dream, let's explore the possibility that, instead of being exact, quantum mechanics is a very accurate approximation of a deeper-level deterministic theory—that the properties it describes are emergent. Let us consider whether the assumption of vacuum quantization can expose this deeper-level deterministic theory by determining whether or not the enigmas of quantum mechanics are projected by that assumption.

We seek to gain access to the causal story behind quantum mechanics and general relativity; to intuitively grasp what's going on behind the curtain. Whether or not our current assumptions lead us to success in this task, we must let these mysteries draw us toward new intellectual horizons, because "facts so bizarre cry out for a deeper explanation."[38]

[1] Richard Feynman. (1988). *QED, The Strange Theory of Light and Matter*. Princeton University Press. pp. 77, 82.

[2] Roger Penrose. The Road To Reality, p. 527.

[3] The discussion on interference and the double-slit experiment that follows is further developed by Brian Greene, (2004). *The Fabric of the Cosmos: Space, Time and the Texture of Reality*. New York: Knopf, pp. 84–84. Greene's discussion was used as a general guide here.

[4] In order to show diffraction (a fuzzy border of light on the projected image) the slit must have a width that does not greatly exceed the wavelength of the color of the light that we have chosen.

[5] Light's wave nature was first revealed in the mid-seventeenth century through experiments performed by the Italian scientist Francesco Maria Grimaldi, and was later expanded upon by experiments performed in 1803 by the physician and physicist Thomas Young. (1807). *Interference of Light*; Alan Lightman. *A Sense Of The Mysterious*. pp. 51–52, 71.

[6] Before the "planetary model" of the atom, physicists pictured the atom being a plum-shaped blob (the nucleus) with tiny protruding springs that each had an electron stuck to its end. When the atom absorbed energy it was thought that these electrons would jiggle (oscillate) on the ends of their springs. Consequently, any atom that was above its ground state of energy was understood to be an "excited atomic oscillator," This depiction of the atom wasn't overthrown until 1900. At that point in history the physical existence of atoms was still controversial. It was replaced by the planetary model, which in turn was replaced by the electron cloud model we use today—a model that was initiated in 1910 and was secured by 1930. Gary Zukav. *The Dancing Wu Li Masters*, pp. 49–50.

[7] Electrons can be individually counted and you can individually place them on a drop of oil and measure their electric charge. Richard Feynman. (1988). *QED, The Strange Theory of Light and Matter*. Princeton University Press, p. 84.

[8] According to de Broglie's doctoral thesis all matter has corresponding waves. The wavelength of the "matter waves" that "correspond" to matter depends upon the momentum of the particle. Specifically, $\lambda = \frac{\hbar}{mv}$, which falls into an important group of equations along with Planck's equation $E = h\nu$) and the ever famous $E = mc^2$. (λ, pronounced "lambda," stands for wavelength, h is Planck's constant, and ν pronounced 'nu' represents the frequency of a photon) From this equation we are told to expect that when we send a beam of electrons (something we might traditionally think of as a stream of particles) through tiny openings, like the spacing between atoms in a piece of nickel crystal, the beam will diffract, just like light diffracts. The only requirement here is that the spacing between the atoms of the material must be as small, or smaller, than the electron's corresponding wavelength—just like the slits in our double-slit experiment. When we perform the experiment, diffraction and therefore interference, occurs exactly as wave mechanics predicts.

[9] Part of the problem here is that in keeping with our four-dimensional intuition we tend to assume a particle aspect in the double-slit experiment without accounting for nonlocality. By doing this we are technically violating Heisenberg's uncertainty principle and missing the bigger picture.

[10] M. Born. (1926). Quantenmechanik der Stossvorgänge. *Zeitschrift für Physik* **38**, 803–827; (1926). Zur Wellenmechanik der Stossvorgänge. *Göttingen Nachrichten* 146–160.

[11] Brian Greene. (2004), p. 91.

[12] Albert Einstein quoted in *Einstein* by Walter Isaacson.

[13] Walter Isaacson. *Einstein*, pp. 96–97.

[14] Ibid.

[15] Werner Heisenberg. *The Physical Principles of the Quantum Theory*, p. 20.

[16] Masano Ozawa. (2003). Universally valid reformulation of the Heisenberg uncertainty principle on noise and disturbance in measurement. *Physical Review A* **67** (4), arXiv:quant-ph/0207121; Aya Furuta. (2012). One Thing Is Certain: Heisenberg's Uncertainty Principle Is Not Dead. *Scientific American*.

[17] L. A. Rozema, A. Darabi, D. H. Mahler, A. Hayat, Y, Soudagar, & A. M. Steinberg. (2012). Violation of Heisenberg's Measurement—Disturbance Relationship by Weak Measurements. *Physical Review Letters* **109** (10).

[18] Steven Weinberg. *Dreams Of A Final Theory*, p. 74.

[19] For a system of spinless particles with masses, the state vector $|\psi\rangle$ is equivalent to a wave function, but for more complicated systems this is not the case. Nevertheless, conceptually they play the same role and are used in the same way in the theory, so that we do not need to make a distinction here. Franck Laloë. *Do We Really Understand Quantum Mechanics?*, p. 7.

[20] Franck Laloë. *Do We Really Understand Quantum Mechanics?*, p. xxi.

[21] There are $6N$ dimensions in this phase space because there are N particles in the system and each particle comes with 6 data points (3 for its spatial position (x, y, z) and 3 for its velocity, which has x, y, z components also).

[22] The space of states (complex vector space or Hilbert space) is linear, and therefore, conforms to the superposition principle. Any combination of two arbitrary state vectors $|\psi_1\rangle$ and $|\psi_2\rangle$ within the space of states is also a possible state for the system. Mathematically we write $|\psi\rangle = \alpha|\psi_1\rangle + \beta|\psi_2\rangle$ where α & β are arbitrary complex numbers.

[23] Franck Laloë. *Do We Really Understand Quantum Mechanics?*, p. 19.

[24] Chapter VI of J. von Neumann. (1932). *Mathematische Grundlagen der Quantenmechanik*, Springer, Berlin; (1955). *Mathematical Foundations of Quantum Mechanics*, Princeton University Press.

[25] I challenge the logical validity of the claim that something can "cause a random occurrence." By definition, causal relationships drive results, while "random" implies that there is no causal relationship. Deeper than this, I challenge the coherence of the idea that genuine random occurrences can happen. We cannot coherently claim that there are occurrences that are completely void of any causal relationship. To do so is to wisk away what we mean by "occurrences." Every occurrence is intimately connected to the whole, and ignorance of what is driving a system is no reason to assume that it is randomly driven. Things cannot be randomly driven. Cause cannot be random.

[26] Franck Laloë. *Do We Really Understand Quantum Mechanics?*, p. 11.

[27] Bohr preferred another point of view where state vector reduction is not used. D. Howard. (2004). Who invented the Copenhagen interpretation? A study in mythology. *Philos. Sci.* **71**, 669–682.

[28] Franck Laloë. *Do We Really Understand Quantum Mechanics?*, p. 28.

[29] This example was inspired by section 2.4 of Franck Laloë's book, *Do We Really Understand Quantum Mechanics?*, p. 27–31.

[30] Franck Laloë. *Do We Really Understand Quantum Mechanics?*, p. 28.

[31] T. S. Eliot. (1921). *The Sacred Wood*. Tradition and the Individual Talent.

[32] For an in depth analysis of nature's nonlocal character see Tim Maudlin. *Quantum Non-Locality and Relativity*, Second Edition, Blackwell Publishing.

[33] Ibid, p. 58–61; J.F. Clauser, M.A. Horne, A. Shimony, & R.A. Holt. Proposed experiment to test local hidden-variables theories. *Phys. Rev. Lett.* **23**, 880–884.

[34] P. Eberhard. (1977). Bell's theorem without hidden variables. *Nuov. Cim.* **38 B**, 75–79; (1978). Bell's theorem and the different concepts of locality. *Nuov. Cim.* **46 B**, 392–419.

[35] J. F. Clauser, M.A. Horne, A. Shimony, & R. A. Holt. (1969). Proposed experiment to test local hidden—variables theories. *Phys. Rev. Lett.* **23**, 880–884.

[36] For an in depth analysis of this discussion, see Franck Laloë. *Do We Really Understand Quantum Mechanics?* Chapter 4: Bell theorem.

[37] Quoted by Steven Weinberg. *Dreams of a Final Theory*, p. 236.

[38] Lisa Randall, p. 454.

Chapter 13 — **Beneath Quantum Mechanics**

"I believe, indeed, that a new perspective is certainly needed, and that this change in viewpoint will have to address the profound issues raised by the measurement paradox of quantum mechanics and the related non-locality that is inherent in EPR effects..."

Roger Penrose

"There must be a fuller explanation of how the universe operates, one that would incorporate both relativity theory and quantum mechanics."

Walter Isaacson paraphrasing Einstein's deepest belief [1]

"Phenomena are a view of that which is unseen."

Anaxagoras[2]

Before the Etymologiæ of St. Isidore of Seville was popularized,[3] it was widely believed that the Earth was shaped like a flat disk. The Etymologiæ presented careful arguments for why the Earth must be spherical, which initiated a slow, but monumental, intellectual metamorphosis. As people let go of their previous assumptions they began to see the world, and their place in it, differently. They slowly overcame the misconceptions and fears that were woven into the fabric of the old model.

To get a feel for how dramatic such a change in perspective can be, let's try to peer into the mind of someone that believed the world was flat. Let's try to make this practice real by connecting to an ancient terror that once dominated the high seas. This task is complicated by the fact that we have lost track of what it was like to live in a world where a simple voyage into the distant horizon was only attempted by the intrepid. We might relate to a fear of monsters ruling the deep, and take pause in the thought of deadly storms that can swallow an entire ship, and we still connect to the fear of being lost at sea, and dread the idea of being abandoned or the suffering that would accompany the wait to die of hunger and thirst; but being from the twenty-first century, we are completely withdrawn from an ancient fear of the sea that used to dwarf all others—a fear inscribed by a level of dimensional tenuity we have overcome.

Our new perspective leaves no room for some of the fears that echoed through our past. We are no longer haunted by dreams of the distant horizon, and we no longer set sail, as Tennyson's Ulysses did, and "follow knowledge like a sinking star, beyond the utmost bounds of human thought," into the unknown Atlantic.[4]

Conceiving the world as a three-dimensional oblated sphere tames the imposing Atlantic. A flat disk world necessitates the existence of a vast and dangerous rim that we might call the "ends of the Earth." Any oceans extending into this forbidden realm would escape the terrestrial domain and pour into the heavens in an endless cascade. A ship that got too close to this rim would be swept up by powerful currents and dragged right off the edge of the world. To sail into this territory would be to suffer a horrendous fate. Get too close to the edge and you will spend eternity falling farther and farther from everything you know—from the place you call home—never being afforded the chance to rest in peace. This fear haunted the minds of our ancestors. Stemming from their comprehension of the world, it was a fear that grew into a dark terror of exploring the horizon.

Before the Etymologiæ the flat disk model of the earth was so deeply accepted that educated philosophers would debate its implications. They would ask questions like: What do the creatures look like who inhabit the underbelly of the Earth? Where do the clouds go when they drift beyond the horizon? What holds the Earth up in the heavens?[5] And, exactly how far away are the edges of the Earth?

Today it may seem easy to overlook the conceptual framework that our ancestors used to understand reality, and dismiss the notion of a flat Earth as absurd or ignorant. This is only because our conceptual roots are in a higher-dimensional model. These roots give us the ability to identify the misconceptions and confusions that are inherent in the lower-dimensional model. Looking back, we discover that when our questions and beliefs live in a framework that misinterprets Nature's true geometry, our comprehension is stifled. Looking back is easy. The trick is to learn how to look forward. To do this we must consider the possibility that today's unanswered questions are signaling that our map of physical reality is still dimensionally incomplete. Perhaps today's most serious unanswered questions are also nothing more than naïve extensions of our oversimplified and incomplete map.

A New Scene of Thought

"Problems cannot be solved by thinking within the framework in which the problems were created."
Albert Einstein

Throughout the great human adventure, our grandest intellectual leaps have come from transitions in perspective. Every time observations have been made that cannot be made to fit the map currently in use, a newer, higher-dimensional map eventually replaces it.

Conceptually this process can be divided into harmonic and inharmonic periods of comprehension. Harmonic periods occur when the model in use accounts for all observed phenomena. At some point in history, each of our two-dimensional, three-dimensional, and four-dimensional models were, in turn, considered harmonic.[6] Inharmonic periods reference a state of confusion. These periods occur when observed phenomena resist explanation within our current perspective. They are limbo phases wherein the model used is known to be incomplete, but a model capable of encapsulating the new observations does not yet exist.

Unexplained observations are the initiators of inharmonic eras. When the Greeks first observed that the sails of an approaching ship are seen before its hull[7], the flat Earth map became inharmonic. When scientists observed the bending of starlight during the eclipse of the Sun and measured how the passage of time can differ between one location and another, the Newtonian map became inharmonic. Then, when the double-slit experiment was observed to produce an interference pattern, general relativity also became an inharmonic explanation of the cosmos. Observations that dethrone our maps are often fingerprints of extra dimensions.

Our task is to explore that which appears irrational from the perspective of our last map, and to naturally integrate those occurrences into an expanded description. The goal is to come into possession of a theory (a map) that is self-consistent and explains all observable phenomena. Einstein called this harmonic state the "ideal limit of knowledge."[8]

As we examine the evolutionary history of our attempts to model Nature, we find that, in agreement with our exponential ascertainment of knowledge, changes to the parameters in our model have not been linear in time. Instead, it seems that every time we improve our model, the new model endures for less time than its predecessor.

Our two-dimensional map reigned for thousands of years. Then, in 340 BCE, when Aristotle wrote his book *On the Heavens*, the authority of that map was challenged. In 1632 CE, Galileo noted in his *Dialogue Concerning the Two Chief World Systems*, that the laws of motion and mechanics were the same in all constant velocity reference frames (the laws of electromagnetism had not yet been discovered).[9] However, it wasn't until 1687 CE that a new map was formulated with the ability to incorporate this principle. That was the year Sir Isaac Newton published his *Philosophiae Naturalis Principia Mathematica*. This work ushered in a new three-dimensional map of physical reality with the ability to explain everything known at the time.

By the end of the nineteenth century, observations were made (like the advancement of Mercury's perihelion) that could not be explained by this new map. Slowly more and more fissures continued to fray the Newtonian map until it lost its stature. In 1905 Einstein initiated construction of a new map (special relativity), but did not complete it (general relativity) until 1916. By that point, many quantum mechanical observations had been made that could not be described by the theory. Consequently, Einstein never considered general relativity to be a complete map of physical reality.

If we complete Einstein's task, if we come upon a theory that captures the effects of general relativity and quantum mechanics, how will we determine whether that map is a final map or just a better approximation? Are all maps just approximations? If they are, is it possible to identify a symmetry between maps of increasing resolution that will allow us to extend the explanatory power of our map infinitely without being stifled by the task of depicting an infinite number of dimensions? Is a final map of physical reality accessible?

Paradigm Shift

> *"In our endeavor to understand reality we are somewhat like a man trying to understand the mechanism of a closed watch. He sees the face and the moving hands, even hears it ticking, but he has no way of opening the case. If he is ingenious he may form some picture of a mechanism which could be responsible for all the things he observes, but he may never be quite sure his picture is the only one which could explain his observations."*
>
> *Albert Einstein*[10]

In part, modern theorists have failed to construct a map that unifies general relativity and quantum mechanics because they have overlooked the axiom "a picture is worth a thousand calculations." They have not grounded themselves in a strong, intuitive picture.

They stopped believing that it was possible. Lisa Randall echoes this state of affairs with the words, "One reason that quantum mechanics seems so bizarre is that we are not physiologically equipped to perceive the quantum nature of matter and light."[11]

Among physicists, belief in this limitation is almost universal. Nearly everyone has assumed that we are incapable of simultaneously visualizing or intuitively grasping more than three spatial dimensions. They often cite physiological reasons for this limitation—claiming that we cannot comprehend more than we physiologically see, but they also tend to overlook the glaring fact that we only physiologically see two-dimensional pictures.

When we visualize the world around us, we conceptually place everything into a three-dimensional construct—a model that exists in our minds. But we never actually see three spatial dimensions. Every image our eyes receive is nothing more than a flat (two-dimensional) image—like a photograph. We take those two-dimensional images and process them through the three-dimensional model that we use to make sense of the world around us.

We do this in two ways. First, we have learned to interpret smaller objects as farther away than larger objects. When we see a familiar object we use its apparent size to accurately arrange it within the three-dimensional framework in our minds. When we see an unfamiliar object this method alone is incapable of telling us how far away the object is. This is where the second mode comes in—parallax. Our eyes have overlapping fields of view. Each eye sees a two-dimensional image, but a slightly different image. When we compare these two images we find that the amount of relative displacement each object possesses depends upon how distant it is. The closer an object is, the more it is displaced from one image to the next. Our brains use the differences between these two images to construct a three-dimensional interpretation of the exterior world.

Even though our eyes only take in two-dimensional inputs, we are completely capable of comprehending a world with more dimensions. There are no physiological limitations that are blocking our imaginations from processing higher-dimensional frameworks.

Beneath Quantum Mechanics

In the previous chapter we discussed some of the most pressing quantum mechanical questions. Our task now is to answer those questions through the lens of vacuum quantization. Let us begin by addressing wave-particle duality.

The mere assumption that the vacuum is a superfluid, or more generally that it is a medium made of elastically interacting parts, leads to the expectation that disturbances will ripple through that medium in the form of compression waves. More formally, any disturbance necessitates collective excitations in the periodic arrangements of the vacuum quanta. When one or more quanta is displaced, that displacement creates a set of vibration waves that propagate through the medium as other quanta "fill in the gap" behind it and "get pushed together" in front of it.

The propagation speed of these waves is given by the slope of the acoustic dispersion relation, which is why all long wavelengths have a propagation speed that is equal to the speed of light (c) in the vacuum (see Chapter 8 for the derivation).

Any time a metric displacement occurs, a wave propagates through the vacuum to adjust for that displacement. The displacement itself must involve a whole number of quanta, which is why photons (in the particle sense) are quantized. The collective excitations that ripple through the vacuum in response to that metric displacement also come in discrete wavelengths because they too are composed of the same discrete entities. The periodic distortion they reference can involve 2000 quanta, 1999 quanta, etc., but they cannot be composed of a non-whole number of quanta. On top of this, the shortest wavelengths are not allowed because acoustic waves cease to exist on scales that approach the size of the medium's constituents. The meaning of "wave" dissolves as we approach that scale.

From this description alone, we fully expect the results of the double-slit experiment. A particle (the quanta that make up the metric displacement) has an ever-present wave function (the "guiding wave" that ripples through the vacuum in response to that displacement) that suffers interference when subjected to the double-slit experiment—just like an acoustic wave. The way that this wave alters the arrangements of the quanta guides the particle's motion through the vacuum, such that the associated particle is likely to land where the wave function is large and unlikely to land where it is small (Figure 13-1).

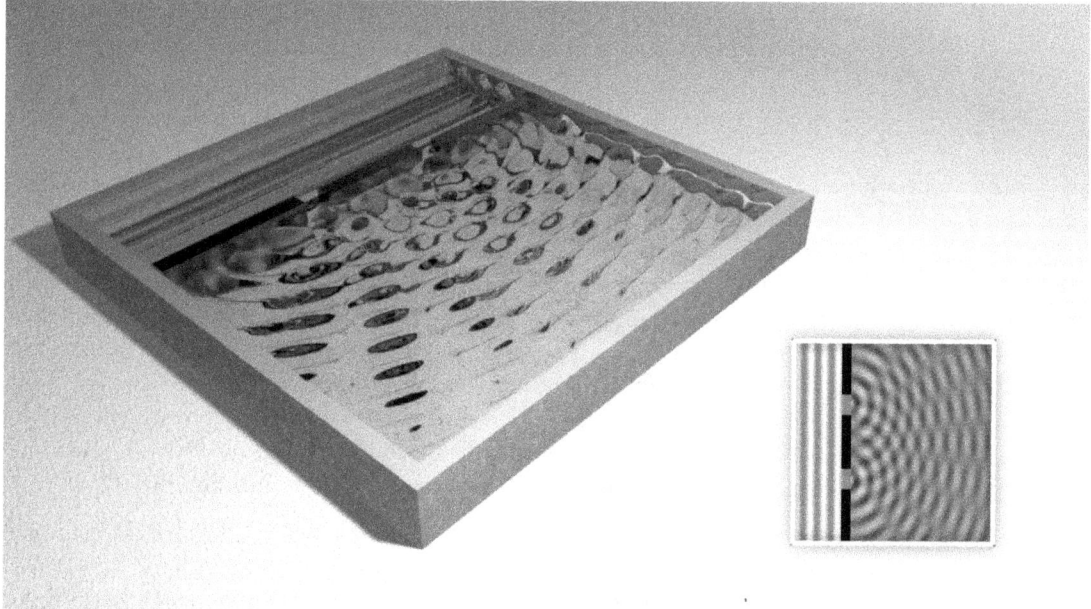

Figure 13-1 The wave function suffers interference and guides the particles

The thing that is waving in this case is the connectivity of space itself.[12] For every momentary state of space, photons have precise x, y, z locations in the vacuum, but they move through the vacuum with a compression wave in front of them, one that is capable of suffering interference. This is how light manages to behave as both a particle and a wave.

Under the assumption of vacuum quantization even single photons are expected to be guided by the wave function. The wave function associated with that photon goes through both slits, suffers interference, and guides the photon in such a way that the interference pattern is revealed. It doesn't matter how many slits there are, the wave function passes through all of the open slits, and the particle (photon, electron, etc.) goes through only one slit. Geometrically, the distortions that we call bosons (i.e. photons) are plane waves or impulse waves, whereas fermions (i.e. electrons) are best described as sonons—like smoke rings—in the superfluid vacuum.

This makes it clear why we get a classical image, instead of an interference pattern, whenever we know which path the photon takes. To learn which path the particle takes, we have to either interact with the particle as it goes through one slit (or the portion of the wave function that goes through the slit with it), or we have to interact with the portion of its wave function that goes through the other slit. If we absorb the particle as it goes through one slit, it will never reach the screen. If we mark the particle as it goes through the slit, instead of absorbing it, it will follow a classical trajectory because the marking process necessarily interrupts the portion of the wave function that went through the slit with the particle, eliminating interference. Likewise, if we absorb or reflect the portion of the wave function that passes through the slit that the particle doesn't go through, then interference will no longer occur and the particle will strike the screen in a manner that reproduces a classical pattern. However, if we alter the particle, or the portion of the wavefunction that went through the slit with the particle, but then cleverly reverse that effect, as we do in the quantum eraser experiment, then the interference pattern will reemerge.

> *"I thus believe that the next phase of theoretical physics will bring us a theory of light that can be interpreted as a kind of fusion of the wave and the emission theories of light."*
>
> *Albert Einstein*[13]

While Einstein was pondering the mysteries of light, he received a letter from Theodor Kaluza who had reformulated Einstein's equations in five dimensions instead of four (one dimension of time and four dimensions of space). Mathematically, this was no problem, because Einstein's equations can be trivially extended to any number of dimensions. But Kaluza's letter contained a startling observation. If you assume that the extra spatial dimension is small and circular, and then manually separate out the four-dimensional pieces contained within the five-dimensional equations, you "automatically find, almost by magic, Maxwell's theory of light! In other words, Maxwell's theory of the electromagnetic force tumbles right out of Einstein's equations for gravity if we simply add a fifth dimension."[14] The inference was that light could be characterized as ripples in the fifth dimension.

When Einstein read this he was intrigued, but he was also shaken. Where is this higher dimension? Exactly how does space ripple through this dimension making light waves? And, what are these light quanta? After mulling over Kaluza's paper for two years Einstein finally agreed to have it published.

With our newfound intuitive grasp of the vacuum we are no longer held back by such concerns. The "higher" dimensions come into focus as we quantize spacetime, allowing us to depict propagating ripples in spacetime—light. These ripples are simply plane wave distortions in the connectivity of the acoustic medium we call the vacuum.[15] (Figure 13-2) To get beneath the mysteries of quantum mechanics, to bring wave-particle duality within our intellectual horizon, we need only to resolve the vacuum's underlying structure (the state of space) and the evolution of that structure.[16]

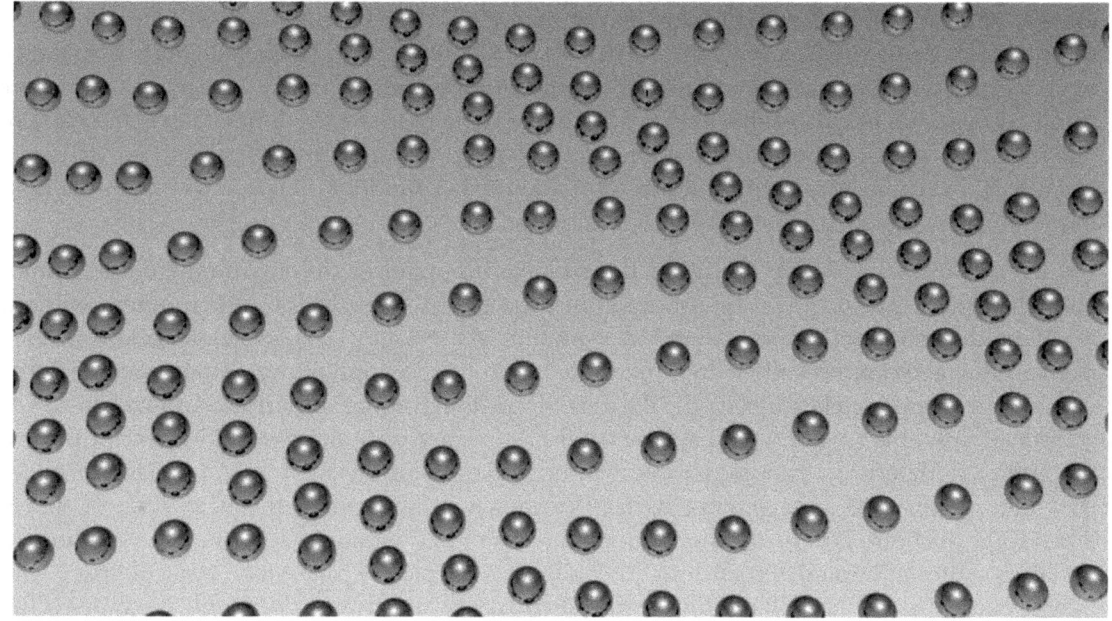
Figure 13-2   Connectivity ripples

The State of Space

> *The state vector "is only an ensemble of potentialities to which probabilities are assigned, it is only an entity manifesting itself to us in a fugitive way..."*
>
> Louis de Broglie[17]

Several other mysteries dissolve when we get beneath the state vector, resolving the state of space and how that state evolves. For example, in a quantized vacuum it is natural to characterize the vacuum's geometric flux, to represent the tendency of the quanta to mix about causing the state of space to change, via a wave equation. The character of that wave equation depends on the fluid properties of the medium. The specific assumption that the vacuum is a superfluid leads to the expectation that the guiding wave equation is the Schrödinger equation. In Chapter 21 we will devote a section for this derivation. For now, we need only to recognize that instead of being an extra ad hoc postulate, or a brute assertion, the Schrödinger equation is a natural consequence of the assumption that the vacuum is a superfluid.

The idea that the state of space is always in flux also necessitates the uncertainty principle. The uncertainty principle calls to our attention that features of the microscopic realm (particle positions and velocities, energies and angular momenta, and so on) cannot be precisely ascertained simultaneously. For example, the more precisely we know the position of a particle the less we can know about its momentum. This complementarity is a consequence of geometric flux. In a quantized vacuum, knowing a particle's precise position at a precise moment, locating the exact quantum that carries its energy, forbids the extraction of its momentum. To nail down the particle's momentum we need to know its (x, y, z, t) velocity. The most precise way to do this would be to compare where the particle is now to where it will be located in the next moment (the next Planck time), or where it was in the previous moment. If space had a static construction, if its geometric connectivity were unchanging, then these two comparisons would be four-dimensionally obtainable, and they would provide identical results. But in a quantized vacuum we no longer have a means of making these comparisons. Because the state of space changes from moment to moment, because the vacuum stirs about, information about what the state of space will be in the next moment is completely irretrievable four-dimensionally.

Even if we knew which quantum the particle occupied in the last moment, we cannot know how that quantum will be arranged in the state of space in the next moment without resolving the actual state of space and its evolution. As the state of space evolves (in the space of states) the geometry is sufficiently mixed such that precise four-dimensional expectations of velocity are completely dissolved. Because four-dimensional descriptions incorporate a statistical ensemble of states of space, instead of an exact state of space, they do not possess the ability to determine how a particle will be arranged in the next state of space, nor can they determine what arrangement its last location will take on in the next state of space. Therefore, without access to the underlying structure, without depicting the additional variables—the additional dimensions that allow us to resolve the exact state of space and how it changes—the complementarity inscribed by the uncertainty principle becomes a full expectation.

From a four-dimensional perspective, the momentum state is infinitely spread out when the position state is concentrated to its maximum value. On the other hand, when we take advantage of the averaging effects that take place on larger and longer scales, we gain the ability to accurately determine a particle's (averaged) momentum, but we lose the ability to precisely describe its position.[18] The same logic applies to the complementary relationship between time and energy.

The state of space at any given moment is defined by the positions of all of its quanta relative to each other and their superspatial velocities. That state evolves according to the Schrödinger equation. Because we cannot resolve the state of space in four dimensions, our best option is to represent the vacuum state in terms of a superposition of all of its possible states, a statistical ensemble called the state vector $|\psi\rangle$, which evolves according to the Schrödinger equation. In other words, the set of all possible states of space (at any given time) forms a statistical ensemble that can be represented with a single complex vector $|\psi\rangle$, which lives in a complex vector space (also called space of states, or Hilbert space).

The quantum measurement problem is cast by the tendency to use this statistical ensemble as a fundamental descriptor. The state vector $|\psi\rangle$ is *not* a fundamental descriptor. Once we recognize this, the 'problem' dissolves. After measurement we always find properties that map to a well-defined state of space because for every moment the vacuum has a well-defined state. Before measurement we describe those properties in reference to a statistical ensemble of states because we are restricted to descriptions that rely on the state vector, which blurs the specific actual state of space. This gives us only a probabilistic projection of possible states of space. When we make a measurement we gain access to details of the specific state of space in a strobe light manner. We momentarily glimpse the underlying state of space. Observations do not collapse many possible states of space to one specific state of space. There is no state vector reduction; the wave function does not collapse. There is always only one specific state of space.

In classical probability theory, distributions of probabilities that undergo sudden jumps are inherently observer-dependent and subjective. An observer with more information about the state of the system will describe future possible states of that system with a distribution that is narrower than that of another less informed observer. If the observer has perfect knowledge of the state of the system, then the probability of the future state of the system (at any given time) will be perfectly peaked at one value. For this observer only one possibility exists, and the process is deterministic.

It would be a strange thing to conclude that the observer's imperfect knowledge of the system plays some sort of essential role in determining the state of the system. Classical distributions of the system's state undergo sudden "jumps" to more precise distributions because our measurements reveal more precise values of the variables that already exist in that state. These variables existed before the jump. The jump merely references our acquisition of knowledge.

This sort of observer-dependence characterizes our access to quantum mechanical systems. To expose that dependence consider the following Wigner-like thought experiment.[19] Inside a closed laboratory Chris is performing an experiment, a Stern-Gerlach measurement for instance. Audrey is outside of the laboratory and cannot access the experiment in any way until the door is opened. She naturally considers the whole ensemble of the closed laboratory, containing the experiment and Chris, as the "system" to be described by a big state vector. This state vector will continue to contain a superposition of the experiment's two possible results until the door of the laboratory is opened and the result of the measurement is shared. At this point, and at this point only, state vector reduction should be applied to her description of the system. Clearly Chris will take a different position. From her perspective state vector reduction is to be applied as soon as the particle has emerged from the analyzer. Does this contradiction mean that we should consider two state vectors, one reduced, and one not reduced, between the points in time when Chris and Audrey learn of the result?[20]

This furtive contradiction plays off of the assumption that the state vector is a fundamental descriptor. Once we recognize that the state vector is not a fundamental descriptor—that it is a statistical ensemble—the question dissolves and the observer dependence of quantum mechanics becomes enveloped by classical probability theory.

In classical systems the probability of the system's state can be peaked differently for many different people. By contrast, quantum systems are generally characterized by only two kinds of distributions. Until a measurement is made, our best description of the state of space is coded by the state vector. Before the measurement there can be no variability in the probability distribution between observers. We all use the same probability distribution, the one set by the state vector because, as observers that are internal to the vacuum, we have no way of narrowing down the real state of space without looking (taking a measurement). Therefore, before a measurement is made, no observer can claim special knowledge about the state of a quantum system. Only after an experiment has been performed can observers vary in their probability distribution of the state of space. That variance depends solely on whether or not observers have access to the experiment's results.

In quantum systems, we either have access to a strobe light glimpse of the state of space, or we don't have access at all. Observer dependence may be less flamboyant in quantum systems than in classical systems, but the classical claim ultimately carries through. An observer external to the vacuum, one that could observe the positions and velocities of the quanta of space, would find the stochastic events prescribed by quantum mechanics to be deterministic. All quantum mechanical events are determinate, but the best description available to observers with access to only four spacetime dimensions remains indeterminate.

The state vector represents only the information to which we have access about the physical system (four-dimensional information).

In contrast to the standard interpretation, this positivistic interpretation of the state vector is both contextual and objective. It remains in the genre of ordinary classical distributions of probabilities, and it does not assert any asymmetry in time. One specific state of space evolves forward in time, changing from moment to moment in a way that is guided by the Schrödinger equation. If we reversed time, the system would evolve back to the original state. When a particle is measured, its associated wave function disperses into the connectivity of the medium, and therefore remains present. If we run the clock backward the process will be mirrored. Nature is deterministic, time-reverse symmetric, and conforms to local realism. The vacuum itself is manifestly non-local. It mixes about in a way that will always be unpredictable to internal observers. Nevertheless, beneath the vacuum we find a structure that gives rise to a specific state of space. This state of space evolves according to the rules of local realism.

> *"A theory is more impressive the greater the simplicity of its premises, the more different things it relates, and the more expanded its area of applicability."*
>
> *Albert Einstein* [21]

Unexplained mysteries are often artifacts of narrow and incomplete perspectives. The ultimate goal in physics is to unite quantum mechanics and general relativity under one rubric, and to gain ontological access to all of the phenomena in physics. As Einstein put it, scientific exploration is about the possibility of achieving "a cosmic connection based on an intellectual comprehension of the rules of reality."[22] Vacuum quantization fits well with that goal because it introduces additional variables that enrich our perspective—elucidating wave-particle duality, the Schrödinger equation, the uncertainty principle, and the state vector, while eliminating state vector reduction, and restoring local realism. Just as the mystery of the "ends of the Earth" disappeared when we expanded the dimensions of our terrestrial map, our higher-dimensional perspective eliminates state vector reduction.

[1] Walter Isaacson. (2007). *Einstein*, p. 335.

[2] Quoted by Dana Mackenzie. (2008, June 14). Don't blame it on the gods. *NewScientist*, pp. 50–51.

[3] The Etymologiæ of St. Isidore of Seville gave sound logic supporting the idea that the earth was round instead of flat. It was produced in Spain in the early 7th century.

[4] Weinberg, p. 240.

[5] In response to this question consider the following anecdote: "A well-known scientist (some say it was Bertrand Russell) once gave a public lecture on astronomy. He described how the earth orbits around the sun and how the sun, in turn, orbits around the center of a vast collection of stars called our galaxy. At the end of the lecture, a little old lady at the back of the room got up and said: 'What you have told us is rubbish. The world is really a flat plate supported on the back of a giant tortoise'. The scientist gave a superior smile before replying, 'What is the tortoise standing on?' 'You're very clever, young man, very clever', said the old lady, 'but it's turtles all the way down!'" Stephen Hawking. *A Brief History of Time*, p. 1.

[6] Inharmonic periods are initiated as observations of paradoxical occurrences in Nature are observed. If these occurrences do not fit within the dimensional parameters allowed by the map in use, then the map is incomplete.

[7] This was only one of the many observations that led to the failure of a flat earth model. Many of the other features are dealt with in greater detail in Chapter 1.

[8] Gary Zukav. *The Dancing Wu Li Masters*, p. 18: Einstein quoted from, Albert Einstein & Leopold Infeld (1938). *The Evolution of Physics*. New York, Simon and Schuster, p. 31.

[9] Walter Isaacson (2007). *Einstein*, p. 108 [chart information, pp. 92 and 108].

[10] Albert Einstein & Leopold Infeld (1938). *The Evolution of Physics*, New York, Simon and Schuster, p. 31; Gary Zukav, p. 8.

[11] Lisa Randall. *Warped Passages*, p. 118.

[12] This waving medium is a little different from the œther previously posited and rejected. While it requires a preferred reference frame, it allows no preferred frame within the dimensions of spacetime.

[13] Lecture in Salzburg, "On the Development of Our Views Concerning the Nature and Construction of Radiation," Sept. 21, 1909, CPAE 2:60; Schlipp, 154; Armin Hermann, *The genesis of the Quantum Theory* (Cambridge, MA: MIT Press, 1971), pp. 66–69; Walter Isaacson, 2007, *Einstein*, p. 156.

[14] Michio Kaku. *Parallel Worlds*, p. 199.

[15] Five dimensions make it possible to depict, and mathematically describe, plane wave ripples associated with photons traveling in one familiar spatial dimension. In order to extend this description so that we can depict ripples in all three familiar space dimensions simultaneously, we need the full eleven dimensions (nine space dimensions and two time dimensions) of our quantized map.

[16] Einstein preferred to interpret the state vector as a statistical ensemble too. He said the state vector "does not in any way describe a condition which could be that of a single system; it relates rather to many systems, to an 'ensemble of systems' in the sense of statistical mechanics.... If the function $|\psi\rangle$ furnishes only statistical data concerning

measurable magnitudes... the reason lies... in the fact that the function $|\psi\rangle$ does not, in any sense, describe the condition of one single system." Albert Einstein. (1936). Physik und Realität. *Journal of the Franklin Institute* **221**, 313–347.

[17] L. de Broglie. (1952). La physique quantique restera-t-elle indéterministe? *Revue des sciences et de leurs applications* **5**, 289–311. French Academy of Sciences, 25 April 1953 session, http://www.sofrphilo.fr/telecharger.php?id=74

[18] This is codified by Heisenberg's uncertainty relation which tells us that the product of two of these quantities cannot be smaller than the order of Planck's constant: $\Delta p \times \Delta x \geq \frac{1}{2}\hbar$. Remember $\hbar$ is equal to the Planck mass, multiplied by the Planck length squared, divided by the Planck time (p stands for momentum, which equals mass multiplied by velocity).

[19] E. P. Wigner. (1961). Remarks on the mind-body question. *The Scientist Speculates*, I. J. Good editor, Heinemann, London, pp. 284–302; Reprinted by E. P. Wigner (1967). *Symmetries and Reflections*, Indiana University Press, pp. 171–184.

[20] Hartle has actually proposed that we should say yes to this question. J. B. Hartle. (1968). Quantum mechanics of individual systems. *Am. J. Phys.* **36**, 704–712.

[21] Albert Einstein; Isaacson. (2007), p. 512.

[22] Corey S. Powell. (2006, October). My Three Einsteins. *Discover*.

Chapter 14 — **Quantum Tunneling & Entanglement**

"The best scientist is open to experience and begins with romance—the idea that anything is possible."

Ray Bradbury

Huntington Beach, California.

There was something pleasing about this tumultuous, rhythmically restless boundary. Sauntering along the beach, under the midday sun, I found myself wrestling with the fact that electrons regularly disappear from one place in space and then appear at some other place, without ever being located between those locations. The foamy water surged around my legs, churning about in an effort to transform seashells into beach sand as it actively erased my footprints, and I slowly lost track of time.

Beckoned by the horizon to take a break from my thoughts, I dove into the cool water and began scanning the shelf below me as I swam. Just past the cresting waves I spotted a small cache of pristine sand dollars, causing my heart to quicken with the thrill of discovery.

Holding my breath as I dove, I began scouring the treasure site. During my third ascent I saw something strange out of the corner of my eye—some kind of amorphous shape, perhaps 8 feet long. A quick burst of fear struck me as my imagination proposed "shark," but then my rational voice reminded me that there were no known human-eating sharks in these waters.

After clearing my goggles, I dipped my head back below the surface and began scanning. Nothing, nothing… there, in the distance, another shadowed impression. This time it looked like it was coming towards me. Instinctively I put my hands out in front of me, but before it got close it turned and faded into the distance. Its tail was moving up and down instead of side to side. It was a dolphin.

Before I knew it, a pod of dolphins were circling me. Two of the dolphins were particularly playful, circling in closer than the rest, darting away, and then returning from a different direction. In the middle of this game, one of the dolphins turned right for me, curiously bumped me with a deliberate nudge, and then scurried away.

The next time they approached I extended my arm. One of them slid under my hand, smooth and firm, and the other brushed my leg with its side. I tried to grab onto a dorsal fin, but didn't grip hard enough. They circled around again. This time I held tighter and felt a jolt of acceleration, immediately causing me to slip off. The dolphin I had tried to hold onto circled around and gave me another chance. For a brief moment we danced to a melodic flurry of clicks and trills. Then the dolphin pulled away, rejoined the group, and vanished.

The appearance and disappearance of these dolphins delectably echoed through my senses. As I caught my breath, I watched the surface pattern of their curiosity—popping up here and there. That's when it occurred to me that miraculous and disconnected events are nothing more than persistent illusions. Mismatches between our perspective and Nature's true form give rise to the illusion of "miracles", and misplaced belief in the accuracy of that worldview reinforces the illusion's persistence—supporting the view that the phenomena in question are unexplainable.

Euclidean Misdirection

"The real voyage of discovery consists not in seeking new landscapes, but in having new eyes."

Marcel Proust

According to the rules of Euclidean geometry, objects that move from one location to another must successively occupy the locations between those points. There are many possible paths between two points, but in Euclidean space there is only one minimum path—defined by the line segment that connects them.

On macroscopic scales, the conclusion that Nature is not Euclidean follows from the fact that there is no room for spacetime curvature in Euclidean geometry. On microscopic scales this conclusion is much more richly encoded. For example, on small scales, the minimum path between two locations in space doesn't always look like a line, it's not always fixed in magnitude, and more than one path can represent the spatial minimum between two positions. More importantly, in the microscopic realm, objects are able to go from here to there without occupying all of the space between the end points—an event we refer to as quantum tunneling.

More than two thousand years before we knew about curvature and quantum tunneling, the Greek philosopher Zeno of Elea discovered that the assumptions of Euclidean geometry engender a paradox—preemptively undermining the logical soundness of the idea that space is Euclidean. Zeno reasoned that if space is continuous and smooth, then the distance across a river can be subdivided into an infinite number of points.[1] He then realized that anything crossing the river must experience each one of the possible locations along that path for a nonzero duration of time. It must, at some point in time, be officially located at each of those positions. This implies that it will take an infinite amount of time to cross a river.

Zeno's proof, known as *Zeno's paradox of extension*, concludes that it is impossible for anything to move at all. In a continuous metric of space, it will always take an infinite amount of time to move across any finite distance (which, confusingly, is made up of an infinite number of points). In a non-quantized vacuum, motion itself appears to be a contradiction.

With the advent of calculus, we can attempt to overcome Zeno's conundrum by using a mathematical trick called diminishing limits. The basic idea is that, as we consider smaller and smaller regions of space, the length of time each region needs to be occupied becomes less and less. As the region of space shrinks to zero length, the amount of time assigned to the occupancy of that region also diminishes toward zero.

The goal is to use these diminishing limits to claim that an infinite number of points can be crossed in a finite amount of time. But these limits only allow this extrapolation if we align the diminishing functions so that the limit of the smallest region of space (a point) corresponds with the limit in which a zero amount of time references the occupancy of that point—instead of a positive but very small length of time. This condition does not actually solve Zeno's paradox. Instead, in a self-defeating way, it only approximates the limit of the argument—requiring that all the locations along a path end up being experienced for exactly zero time. Conceptually, this complicates the mystery. If an object moves from one point in space to another, but spends zero time at each point along its path, then it cannot possess position along that journey. If an object spends zero time at each location, then it never occupied those locations. How can an object exist if it never has a location?

Modern discussions on this conundrum tend to claim that Zeno's questions are resolved by the fact that Cauchy's definition of infinite sums only applies to "countable" infinite sums while Zeno's paradoxes deal with what are called "uncountable" infinite sums. Therefore, because we cannot apply the Cauchy definition of infinite sums, there is no paradox. This response avoids the truly interesting aspects of Zeno's questions, and does nothing to resolve the mystery. The fact remains that Zeno's questions expose a problem with our tendency to map space as a continuous metric.[2]

The interesting thing is that these so-called 'solutions' to Zeno's paradox continue to assume continuity in spacetime. They rely on the ability to subdivide space and time into smaller and smaller portions until we end up with an infinite collection of zero-sized points. Quantum mechanics has shown that beyond the Planck scale, measurements of x, y, z, t spacetime become meaningless—a fact that eliminates the use of infinite diminishing limits, and the use of either countable or uncountable infinite sums, to represent the space between any points in the universe. Still, Zeno's paradox is often considered solved based on diminishing limits, or the cop out of uncountable infinite sums; solutions that are, at best, ontologically silent.

We do not gain a valuable solution by suggesting that an object traversing any non-zero distance never actually experiences, or spends any time whatsoever located at any of the points of space between the starting and finishing points. If objects cannot, even for a brief moment, possess a location, then they cannot ever be located exactly half way across the river. Worse yet, if they cannot have position, then they cannot change position. This solution is absurd. Nature simply is not composed of zero-sized points.

> *"Zeno's difficulty demands an explanation; for if everything that exists has a place, place too will have a place, and so on ad infinitum."*
>
> *Aristotle*

A reasonable way out of this quagmire is to acknowledge that the familiar dimensions of spacetime cannot be infinitely subdivided. There is a natural restriction when it comes to dividing measures of space (and time). That restriction manifests on the Planck scale.

When we map the vacuum as quantized, we no longer face the philosophical contradictions that Zeno wrestled with.[3] From this dimensionally richer perspective, an object traveling from point A to point B in space can do so in a finite amount of time, even though it spends a nonzero amount of time at each possible position along the path between the two points. This is the case because each smallest possible position manifests as a non-zero measure of space—one quantum length, or one Planck length.

In addition to resolving Zeno's paradox, vacuum quantization also enables us to penetrate a more recently discovered mystery—the mystery of quantum tunneling, or movement via wormholes. Like Zeno's paradox, this mystery is a geometric one. It suggests that in the vacuum you don't always have to interact with a predetermined minimum amount of space to get from one point to another.

Quantum Tunnels

"Spatial propinquity is not required for direct causal connection."

Tim Maudlin

The mystery of quantum tunneling was first tied to quantum uncertainty in the 1920s when Russian physicist George Gamow discovered how radioactive decay was possible. At the time, scientists knew (from the work of Madame Curie and others) that the uranium atom was unstable and emitted radiation in the form of alpha particles.[4] What they didn't know was why. The strong nuclear force was more than a sufficient barrier to prevent alpha particles from ever escaping the nucleus, so the fact that alpha particles did escape presented a true mystery.

To explain his insight, Gamow famously used an analogy of a prisoner sealed in a jail, surrounded by huge prison walls. He noted that in a classical Newtonian world, escape is impossible. But in our strange non-Newtonian world, where it is impossible to simultaneously know the prisoner's precise position and velocity, there is an ever-present scale-dependent probability that any object might "tunnel" from one four-dimensional location to another. If the prisoner bangs against the prison walls often enough, there is a finite, calculable probability that he will pass right through them.

This conclusion is in direct violation of Newtonian common sense, but it represents what happens in the real universe. Large objects like humans have to wait several times longer than the lifetime of the universe for this miraculous event to have the statistical chance of happening once, but for subatomic particles, like alpha particles and electrons, it happens all the time.

Gamow (and R.W. Gurney and E.U. Condon) discovered that radioactive decay occurs because the uncertainty principle makes room for quantum tunnels. Recall from Chapter 13 that the uncertainty principle stems from the never-ending shuffling of the quanta—redefining the vacuum state from moment to moment. Radioactive decay occurs when an object tunnels from one location to another in familiar space—skipping at least some of the space that separates here from there.

Today, quantum tunnels, or microscopic wormholes, explain much more than radioactive uranium. Electrons regularly disappear from the fabric of spacetime and then rematerialize on the other side of walls inside the components of our electronic devices. And according to our best models, the temperature inside the core of the Sun is not high enough to fuse hydrogen into helium unless we take into account the effects of tunneling. In the words of Michio Kaku, these effects are consequences of that fact that the smallest scales are "constantly boiling with tiny bubbles of spacetime, which are actually tiny wormholes and baby universes."[5]

Vacuum quantization allows us to intuitively access the mechanism behind this effect, providing us with a picture in which a quantum tunnel—a wormhole—forms whenever the motions of the quanta collectively produce a region of superspace whose quantum density is less than the background (Figure 14-1). Large tunnels are increasingly less likely; because the choreography required to create a tunnel increases radically as the size of the tunnel increases.

When a particle is "tunneling" through the vacuum, it is moving through the superspatial dimensions, navigating through the dimensional voids that separate the quanta of space. During the tunneling event, its connectivity with the rest of the medium is suspended. This is what makes it possible for a particle to go from one position in space to a distant one, without being located at any of the points between those positions.

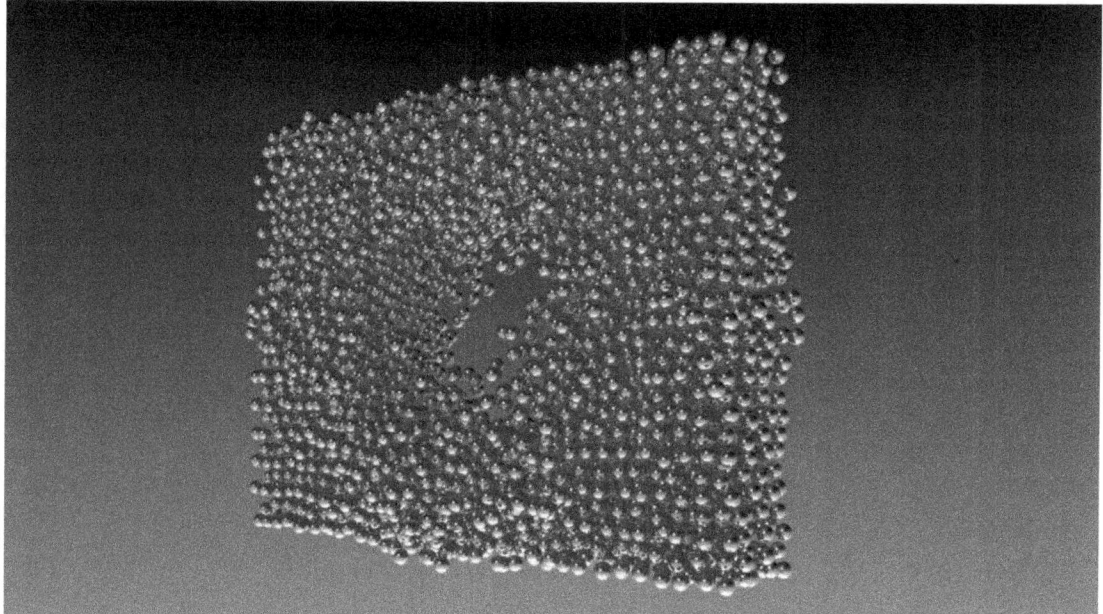

Figure 14-1 A quantum tunnel

Gaining lucid access to the mechanism of quantum tunneling allows us to make specific predictions about its behavior. For example, in addition to its scale dependence, the likelihood of tunneling should depend upon vacuum density, pressure, temperature, and the superspatial velocity of the subatomic particle. Where there is more spacetime curvature—higher vacuum density—the sea of vacuum quanta is less likely to provide an available tunnel for a particle to sail through. Therefore, higher spatial densities should make it more difficult for any object to move through the superspatial dimensions without interacting with any other quanta of space.

According to this model, the frequency of quantum tunneling should have increased as the temperature of space has decreased. In the past, when the temperature of empty space was higher, quantum tunneling was less frequent. Because quantum tunneling plays a role in nuclear fusion there should be observational differences between contemporary galaxies and those of the distant past. The practical difference between now, and say one million years ago, may be negligible to a first-order approximation, but when we trace our universe's history back toward the Big Bang, we may find measurable differences.

Tunneling particles should also have a greater time signature than non-tunneling particles. A particle that does not maintain spatial propinquity with the spatial geometry around it will experience more independence, and therefore more time.[6] Furthermore, tunneling objects are not restricted by the speed of light. Therefore, it should be possible for a tunneling object to arrive in a distant location in less time than it would take long-wavelength light to propagate to that location by moving through the vacuum.

The likelihood for a subatomic particle to pass through a quantum tunnel should also depend on its superspatial velocity. Quantum tunnels are highly unstable. The motions of the quanta that open up the tunnel are also responsible for closing it. If a particle is going too fast, or too slow, its chances of tunneling will decrease. The most favorable speed for tunneling will depend upon the temperature and spacing of the quanta.

While a quantum tunnel is open, its path represents a region of what we call negative energy. Positive energy (e.g. matter) is characterized by a local increase of vacuum density compared to the cosmic average. Negative energy is a manifestation of a decrease in vacuum density compared to the background average density.

The larger an object is, the less likely it is to pass through a quantum tunnel.[7] The likelihood that quantum motions will come together in just the right way to open up a tunnel large enough for a human to pass through is astronomically small. In fact, "the amount of negative energy needed to keep open a one-meter-wide wormhole is roughly equal in magnitude to the total energy produced by the sun over about 10 billion years."[8] This is why quantum tunneling doesn't occur on macroscopic scales.

Some scientists have suggested that large tunnels might occur naturally in the form of an Einstein-Rosen bridge (Figure 14-2). They envisage these macroscopic wormholes as shortcuts through space, created by the union of two black holes. But the geometry they imagine for these shortcuts is riddled with logical incongruities.

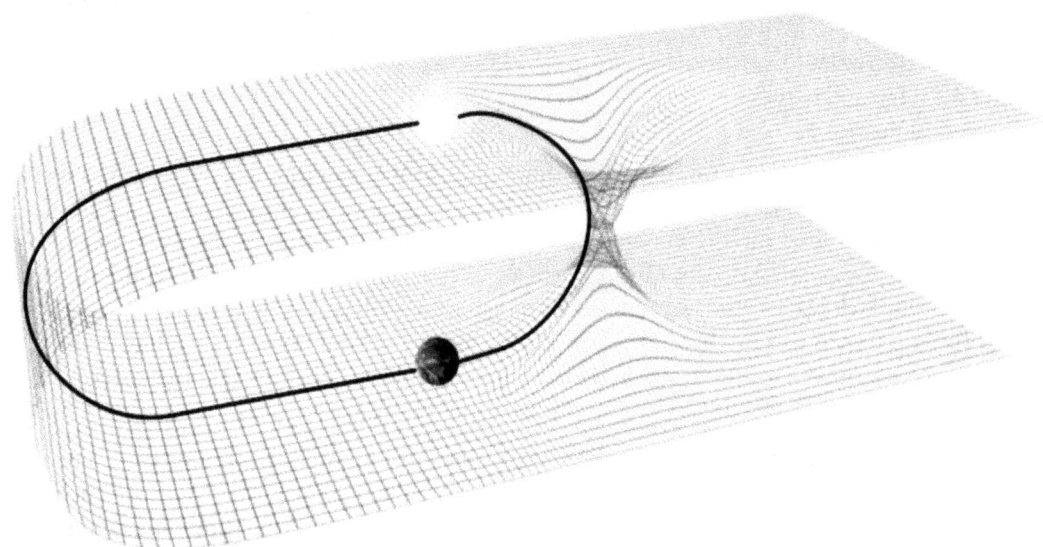

Figure 14-2 A common depiction of a wormhole

For example, in the traditional depiction of a wormhole, a spatial shortcut is created between otherwise distant regions, because the curvature of one black hole projects downward and the other projects upward. In rubber sheet models "up" and "down" can be arbitrarily chosen to reference curvature, but once we select the direction for positive curvature it is globally fixed. If this were not the case, if we could arbitrarily change the direction of positive curvature from region to region, then we could have equally assumed that the curvature from the top black hole stretches upward and the bottom downward. There would be no way to logically distinguish between these two options, yet they lead to drastically different conclusions. In one, a bridge is created, and in the other it is not.

To be logically consistent we must use one direction to represent positive spacetime curvature and leave the other to represent negative curvature. In this case, black holes will never connect to form bridges. Instead, bridges must represent the union of a black hole with a white hole.

A more significant problem comes from the assumption that the entire plane can fold back on itself, somewhere in the distance. This leads to a contradiction. Even if we pretend that far away there is enough matter to warp a large section of the plane, the limit for how matter can stretch the plane is represented by the slope of the plane being at 90 degrees to the flat plane, which is reached in the throat of the black holes. At this point space is maximally warped. Additional mass will not push the curvature past 90 degrees—it will simply expand the size of the curved region. The very notion of curvature, represented in the rubber sheet diagrams, forbids the possibility of warping a plane of space back on itself. Any warp past 90 degrees contradicts general relativity.

Vacuum quantization helps us understand that macroscopic wormholes do not exist as Einstein-Rosen bridges. Travel via wormholes is about—"harnessing the uncertainty principle to dart across the vastness of intergalactic space."[9] To accomplish this we must have access to vast amounts of negative energy, or restrict ourselves to the microscopic scales.[10]

As we progress toward such possibilities, let us remember that although it may seem completely natural to expect a particle to be located at, or at the very least, *near* the location it was a moment ago, Nature does not strictly make it so. The vacuum is nonlocal; it is made up of unconnected, interactive, quantized parts. In light of this, quantum tunnels, or microscopic wormholes, explain why "spatial propinquity is not required for direct causal connection."[11]

Stepping Out of Spaceland

> *"If there are many things, they must be both small and large; so small as not to have size, but so large as to be unlimited."*
>
> *Simplicius*[12]

A being with the ability to navigate within a higher-dimensional space would be able to disappear at will and walk through walls. She would be able to suddenly vanish from space and then rematerialize out of nowhere. No jail could hold her. No secret treasure could be kept hidden from her. She could perform surgery without ever cutting the skin. And if some priceless treasure, locked within a high security vault, were thought by her to be desirable, she could extract it without even opening the vault.

On the other side of that coin, beings that are intellectually restricted to fewer spatial dimensions will have stunted comprehension. For example, a Flatlander that is completely ignorant of the third spatial dimension will be unable to deterministically map any process that makes use of that dimension. They will find these processes to be baffling and magical, and their best models of them will be statistical.

Figure 14-3 The miracle of the orange dot-line

To explore a simple example, imagine that as a normal day begins a Flatlander is gazing into the familiar linear background. Then, out of nowhere, an orange line segment suddenly appears before him. The line segment widens and moves to the left. Still moving to the left, it becomes a green line segment for a brief period, and then disappears from the entire universe (Figure 14-3).

To the Flatlander this occurrence is utterly mystical. From their perspective, something appeared out of nowhere, and then disappeared right back to nowhere, an occurrence that is beyond comprehension or explanation. But from a higher-dimensional vantage this event is trivial—just a carrot dropping through the Flatlander's plane of view (Figure 14-4).

Like the Flatlanders, we have witnessed completely baffling phenomena in our world. Believing that there are only three spatial dimensions, we have modeled those occurrences as statistical, and have claimed that Nature's most fundamental description is stochastic.

A quantized map of space gives us the ability to correct that mistake, making it possible to understand how something disappears from one point in space and reappears somewhere else during quantum tunneling. Within a Euclidean perspective (or any perspective that maintains a smooth and connected spatial structure) this sort of movement (traveling through quantum tunnels) will be interpreted as "disappearing" from one location in space and "reappearing" in another distant location, without occupying any of the locations between them. This, of course, is not the whole story. Nevertheless, if we are trapped within a perspective of three dimensions of space and one dimension of time, then it is as close of a description as we can get.

Vacuum quantization allows us to expand that picture—to visualize and comprehend the additional dimensions of space. Whether or not this model stands the test of time, whether or not it perfectly maps Nature, it is a step in the right direction because it exposes our minds to a higher-dimensional representation—something long thought impossible. That act collapses previous mysteries, making it possible to understand, for example, how it is possible to go from one location to another in our four-dimensional realm without being located at all the points between them. Quantum tunneling is an expectation of a quantized vacuum.

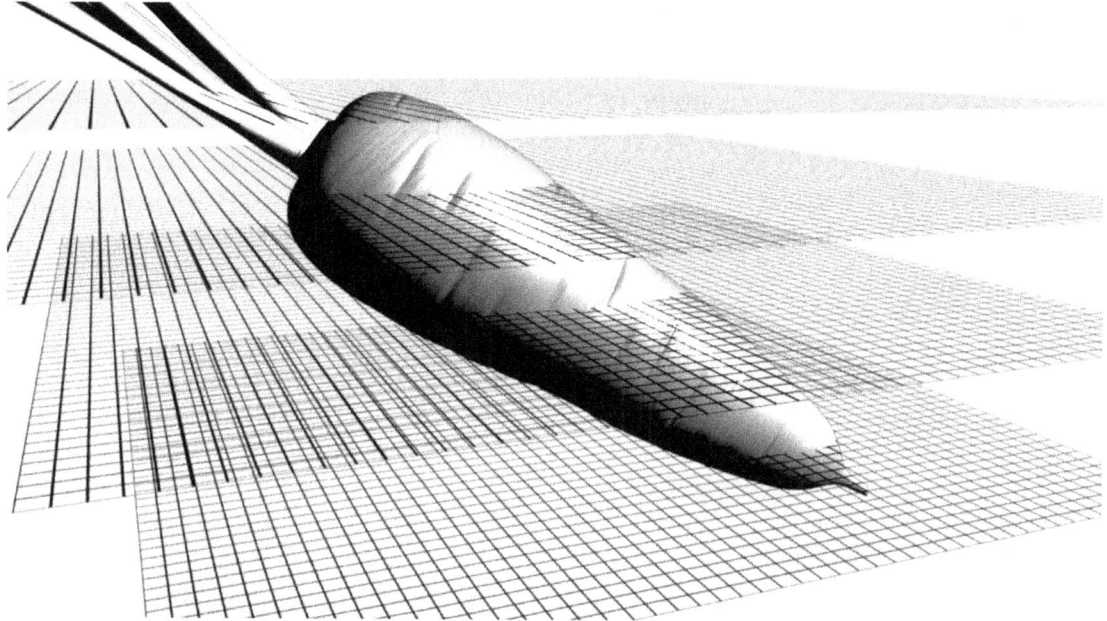

Figure 14-4 Explaining the "miracle" in higher dimensions

Schrödinger's Cat

Now it is time for us to talk about Schrödinger's poor cat—the unfortunate creature that has been kept in limbo between life and death for far too long. Qst offers a clear resolution to the mystery behind this cat's schizophrenic state, and helps us comprehend what's really going on.

The debate over Schrödinger's cat started with a thought experiment proposed by Erwin Schrödinger. He asked us to imagine a cat sealed in a soundproof, non-transparent box. Inside this box he imagined a bottle of poison that was rigged to a hammer, which was in turn connected to a Geiger counter placed near a pile of uranium. In quantum mechanics, the radioactive decay of uranium is modeled as purely a probabilistic quantum event. Therefore, as Schrödinger pointed out, once we put the cat in this kind of set up we cannot precisely predict ahead of time what state the cat is in—whether it is dead or alive.

This thought experiment drastically conflicts with our experiences of the world. As a consequence, it highlights some pretty significant philosophical problems that are inherent in a four dimensional view of the world. For example, let's say there is a 50 percent chance that an alpha particle will come zooming out of our uranium pile in the next 10 minutes. If the decay event does happen it will set off the Geiger counter, which will set off the hammer that breaks the glass, and kill the cat. Before we open the box, it is impossible to tell whether the cat is dead or alive because we cannot claim with certainty whether the uranium has decayed.

In reaction to this (trying to explain this thought experiment within the bounds of four dimensions) physicists have chosen to describe the "cat system" by adding up the wave function (the state vector) of both the live cat and the dead cat. Consequently, 10 minutes after the box is closed, physicists claim that the cat is in a nether world—being 50 percent dead and 50 percent alive simultaneously. But does this description actually represent the physical state of the cat inside the box?

What happens when we open the box? Some physicist's claim that by peering into the box (making an observation) we cause the wave function to collapse (somehow). After this collapse, the cat takes on a single state and we observe that the cat is either dead or alive. To Schrödinger, this interpretation was silly. His whole point was that it is absurd to assume that a cat can be both dead and alive at the same time, so long as we haven't looked at it.

The property of *alive* or *dead* cannot just spring into existence merely because we observe it. If it could, then we would have to carry this interpretation to its logical end and claim that any macroscopic system will be in a large-scale superposition in the absence of a *conscious observer*. This would require that before consciousness evolved, the entire universe existed in a giant state of superposition. Then, presumably when the first conscious creature came into existence, its first observation caused the state of the entire universe to suddenly collapse. This, of course, is ridiculous. It also isn't helpful. It doesn't explain how or why observations cause state vector reduction (wave function collapse). How exactly does observing the cat cause it to be alive instead of dead? How is the particular state selected?

Einstein was also very displeased with this interpretation. When guests came over to his house, he would say, "Look at the moon. Does it suddenly spring into existence when a mouse looks at it?" Einstein believed the answer was no.[13]

Did you spot the logical flaw in this thought experiment? The hinge point is the claim that the radioactive decay of an atom is purely a probabilistic quantum event. In a four-dimensional map that may be the best description we can formulate, but what about in Nature's complete map? What does our quantized geometry have to say about the decay of an atom? Is it still a probabilistic quantum event, or do events that appear to be probabilistic in four dimensions become deterministic in eleven dimensions?

According to our new map there are no random or probabilistic events in the vacuum. Four-dimensional maps do not have access to how the quanta are moving about, changing the vacuum state, which is why wormholes appear to open randomly or stochastically in those maps. Because radioactive decay makes use of quantum tunnels, specific decay events also appear to be entirely stochastic occurrences within a four-dimensional map. However, from an eleven-dimensional viewpoint the interactions and positions of all the quanta are entirely deterministic. The vacuum has an exact state that evolves deterministically. When and where a tunnel will open up, and enable an alpha particle to escape the nucleus of an atom, can be precisely predicted in an eleven-dimensional framework.

When we describe Schrödinger's cat as something that exists in a confusing nether state, we become similar to Abbot's Flatlanders, who could not understand the carrot that fell through their view. The additional variables that come from vacuum quantization make it clear that all cats are always either alive or dead, never both. Schrödinger and Einstein were correct—things do not spring into existence, or take on vital qualities, just because we observe them. Alive cats are alive—even when we are not observing them, and the moon still exists even when nobody is looking.

Because of this resolution we no longer need to entertain the outrageous conjecture that a cosmic consciousness watches over all of the universe, or that every quantum possibility is played out in an infinite number of ghostly linked quantum universes. The four-dimensional depiction of Nature that we are accustomed to is only a partial representation of the geometry of spacetime—a strobe-light version of the richer, more complete fabric of the cosmos. Each four-dimensional "frame" or "state" is a geometric snapshot of the system, but the system itself does not collapse to create that frame. State vector reduction never occurs; the wave function never collapses. Beneath the statistical ensemble known as the state vector $|\psi\rangle$ there is a specific state of space $\langle\varphi|$, whose evolution is guided by the Schrödinger equation.

Entanglement

"A perfect description of the whole does not contain a perfect description of the parts."

Franck Laloë[14]

With the assumption that the vacuum is quantized we become equipped to resolve the mysteries of quantum entanglement—or what Roger Penrose calls quanglement.[15] To work our way to that resolution let's take two particles, for example photons, and entangle them. We can do this via a process referred to as parametric down-conversion. Basically we shoot a photon from a laser into a non-linear crystal, which converts it into a pair of photons. The photons emitted from the crystal are entangled in various ways. "Their momenta must add up to the momentum of the incident photon, and their polarizations are also related to one another…"[16]

Once we have two entangled particles we slowly and carefully separate them; so they don't interact with their surroundings and, for example, get absorbed.[17] When they are sufficiently separated, our set up is complete. Explaining what happens next is a task that many of today's physicists are still tangled up over.

Let's say that one of our particles is located in a tent on a black sand beach in Hana, (on the island of Maui, Hawaii), and the other is in a lab at the University of Utah. The time it will take light to travel between these two locations is several orders longer than the minimum increment of time that our finest chronometers can measure. This will allow us to trust our measurements. These measurements will focus the problem of *quantum entangled information*.

Because we know the total state of the two particles, knowing the exact state of one particle at any precise moment will tell us the exact state of the other particle at that same moment. But until we measure one of these particles, we have no idea what state either of them is in because they (at least four-dimensionally) stochastically flip from one value to another. When we measure the spin of the particle in the physics lab at the U of U we instantaneously learn what the spin of the particle on Hana's black sand beach is.

Some have found this instant acquisition of information about an object at a distant location confusing. Some have even gone so far as to claim that this occurrence violates Einstein's principle of relativity, but this is not the case. First of all, no information has traveled through space; therefore, there cannot be a contradiction on that front. The confusion has nothing to do with information travel, and everything to do with the '*stochastic*' process that controls the evolution of the two autonomously connected particles.

To make this clear, consider the following example. Imagine that I take the Queen of Hearts and the Seven of Clubs from the deck, show them to you, and then mix them up behind my back. Then I place them face down in front of me. You are no longer able to determine which card is which, because you could not see how they were being mixed. Probabilities reflect what we are licensed to believe based on our knowledge of the world. When you are ignorant of a system's precise state, you often have to refer to its state probabilistically—in a way that reflects your knowledge of its possible states in light of your ignorance. For example, at this point each card is said to have a 50 percent chance of being the Queen of Hearts. This probabilistic claim reflects the fact that we are missing important information about the history of this system. It does not mean that each card is simultaneously 50 percent the Queen of Hearts and 50 percent the Seven of Clubs.

Next I place both cards into separate envelopes, seal them, and ask you to choose one. The envelope you don't select will be taken to Madagascar. Once it is there you will be instructed to open your envelope and look at your card. What happens?

Well, as soon as you make your observation, you learn that the card you have is, for example, the Seven of Clubs. Therefore, you also simultaneously learn that the card in Madagascar is the Queen of Hearts. Have any laws of physics been broken here? No. In fact, this isn't even a miraculous event. So why should it be so mysterious when it happens in the quantum realm? The reason people have projected an air of magic around this issue has to do with the fact that in the quantum realm the possible outcomes become more difficult to predict ahead of time and are, therefore, laced with more confusion. But the principles at work are fundamentally the same.

In our card example, we knew that one of two possible outcomes would occur. Both possible outcomes were easy to identify, and therefore easy to understand. We also knew that the end result was determined by the manner in which I shuffled the cards behind my back. This means that when the cards were separated they were already *programmed*, or determined—we just hadn't seen the result.

In order to better represent what quantum-entangled particles do, let's carry out our experiment again, but in a slightly different way. This time let us consider two boxes that contain a single die each. There is a hole on the top that we can peer into, and the die inside each box has been programmed to rattle about, flipping in a seemingly random pattern with different numbers facing up moment to moment.

The dice are connected in such a way that they always have opposite orientations. Whenever one die has the number 1 facing upward, the other die has the number 6 facing upward (2 is matched with 5, 3 with 4, 5 with 2, and 6 with 1) (Figure 14-5). Because we don't have access to the program that determines how the dice flip in sequence (like the shuffling behind my back), we have no idea what orientations the dice are going to have until we look at them. If we were to watch one die flip from number to number and record its sequence, it would seem to us to be an entirely random sequence (6, 4, 1, 5, 4, 3, ...). But the second die simultaneously mirrors that sequence, producing the exact same pattern except with opposing numbers (1, 3, 6, 2, 3, 4, ...). This connection enables us to look at one of the die and instantly learn the orientation of the other die. More importantly, the fact that these sequences are mirrored tells us that they cannot be truly *random*.

Figure 14-5 Two dice with a shared 'random' history

With this adjustment, our experiment more adequately approximates a quantum system. When we separate the two boxes by a significant distance, we gain the ability to "teleport" information. However, the kind of information we can receive is restricted because teleported information is not sent information, it is only revealed information. Revealing information is completely different from sending it.

> *"Entangled particles are somewhat like twins still joined by an umbilical cord..."*
>
> *Michio Kaku*

Entangled particles remain autonomously connected, because, as long as they remain unperturbed, they share the same "umbilical cord", or program. The states the photons took on when we entangled them, and how those states evolve as the quanta mix about, determines how the macroscopically accessible parameters of the two particles change—determining the evolution of their states. From a four-dimensional perspective the individual state sequences may appear stochastic (for the same reason that quantum tunnels seem to randomly appear in a four-dimensional map of spacetime), but from an eleven-dimensional perspective those state sequences are deterministically controlled. From this perspective the mystery of quantum tunneling, wormholes, and teleportation dissolves.

[1] Remember that points are defined in the Euclidean map as "zero-sized."

[2] Philip Ehrlich offers a wonderful review of Zeno's Paradox of Extension, and offers a solution that is superior to Adolf Grünbaum's solution, which is currently treated as the standard resolution, based on Lebesgue measure theory. See: Philip Ehrlich. (2014, October). Ninetieth Birthday: A Reexamination of Zeno's Paradox of Extension. Philosophy of Science, 81, pp. 654–675. 0031-8248/2014/8104-0005

[3] Of course Zeno predated Newton, but the metric he assumed for space was in many ways comparable to the Newtonian metric.

[4] An alpha particle is the nucleus of a helium atom (two protons and two neutrons bound together).

[5] Michio Kaku. *Parallel Worlds*, p. 135.

[6] Here "spatial geometry" refers to the averaged, macroscopic geometry that is formed by the collection of quanta. A quantum that is "tunneling" does not remain localized compared to that geometry.

[7] Fundamental particles of matter, like electrons, represent solutions to the acoustic superfluid's wave equation known as "sonons"—vacuum smoke rings. Different sonon sizes are allowed, but smaller sonons are more likely to pass through quantum tunnels.

When photons of light—localized plane waves—move through wormholes, they delocalize if the hole is larger than the localization boundary of the photon because Anderson localization is no longer able to hold them together. By contrast, fundamental particles of matter, like sonons, do not rely on Anderson localization and are able to maintain their form while passing through a wormhole.

[8] Brian Greene. *The Fabric of the Cosmos*, p. 467; Matt Visser. (1996). *Lorentzian Wormholes: From Einstein to Hawking*. New York: American Institute of Physics Press.

[9] Michio Kaku, *Parallel Worlds* p. 146.

[10] If the wormhole is not empty, but instead contains exotic matter that can exert an outward push on its walls, then it might be possible to keep the wormhole open and stable. Although similar in its effect to a cosmological constant, exotic matter would generate outward pushing repulsive gravity by virtue of having negative energy (not just the negative pressure characteristic of a cosmological constant). Brian Greene, p. 467. To date there are no signs that any forms of exotic matter actually exist.

[11] Tim Maudlin. *Quantum Non-Locality and Relativity*, p. 236.

[12] Simplicus(a) *On Aristotle's Physics*, 141.2.

[13] Michio Kaku. *Parallel Worlds*, p. 158.

[14] Franck Laloë. *Do We Really Understand Quantum Mechanics?*, p. 53.

[15] Roger Penrose. *The Road to Reality*, p. 603.

[16] Ibid., p. 604.

[17] "The slightest contamination with the environment will destroy quantum teleportation."—Michio Kaku. *Parallel Worlds*, p. 178.

Chapter 15 — Black Holes

"One of the greatest things in the world is climbing to the edge of human knowledge and poking around."

William Prince

"Black holes, quarks, and gluons really do have a big thing in common: They can be described by equations that govern the behavior of fluids."

Dam Thanh Son[1]

Physics Classroom, University of Utah.

My fascination with exotic physics was kick started by Dr. Richard Price (whose doctoral advisor was Kip Thorne). Along with Hawking, Penrose, Kerr, Robinson, Carter, and others, Price noted that Nature's most extreme opposites—black holes and elementary particles—share the same defining traits. This led him to the audacious idea that black holes might actually be gigantic elementary particles.[2]

Price's classes were lessons in thinking—transitional exercises between the burden of memorization and the honesty of questioning. Like clockwork, he would arrive at the old painted cinder block room within a few seconds of the scheduled hour, say something unexpected, and force our minds to dance in new ways. His goal was to spark our hibernating curiosity—to help us see beyond force-fed solutions. He saw no point in memorizing definitions just to regurgitate them on exams. The rules were just echoes of comprehension—and it was the underlying comprehension we were supposed to be after.

To help wake us up to this fact, Price purposefully over-exaggerated his body motions, gestures and speech patterns. Watching him contort his face and lanky figure into awkward forms was both a source of amusement and clarity. One day he stood in the front of the class, with one hand on his chin, switching his gaze from the front of the room to the back. After a long moment of silence he said, "That's odd. Someone stole the projector, and put a new one back there." In his class the illusion of rigid answers began to dissolve and the mysteries of the physical world received direct attention.

New Solutions

"What looks like a singularity in four dimensions may not be one in five dimensions."

Paul Steinhardt[3]

How does modeling the vacuum as a superfluid, impact our characterization of black holes? Does it give us any special insights? Does it enable us to characterize the geometry of a black hole inside its event horizon? Can vacuum superfluidity explain why the laws of black hole mechanics are "simply the ordinary laws of thermodynamics applied to a system containing a black hole"?[4]

Black holes represent a unique junction in physics—a crossroad between the two most prominent descriptions of physical reality. Classically, black holes are treated as perfect absorbers, which means they emit nothing and have a physical temperature of absolute zero. But in quantum theory, black holes emit Hawking radiation with a perfect thermal spectrum, and therefore, have a temperature that is determined by the size of the black hole. This thermal spectrum allows us to consistently interpret the laws of black hole mechanics as "physically corresponding to the ordinary laws of thermodynamics."[5]

When we combine the classical laws of black hole mechanics with the formula for the temperature of Hawking radiation, we discover a quantity that plays the mathematical role of a black hole's entropy.[6] Does this quantity truly represent the entropy of a black hole? If so, can it help us better define entropy in general? Can it help us understand the degrees of freedom referenced by entropy?

To answer these questions, we need to understand the complete geometry of a black hole. Specifically, we need to know how to geometrically characterize a black hole at and around its singularity—inside its event horizon. This task has stymied theoretical physicists ever since black holes were first theorized.[7] Here's why.

In terms of mass and gravitation, black holes are extremely massive, which means that the map of general relativity must be applied to them. At the same time, singularities are extremely small, which means that quantum mechanics also claims dominion over them. General relativity and quantum mechanics make completely different claims about what a black hole is like inside the event horizon. To settle this disagreement, we need to reconcile these two theories—we need a quantum theory of gravity.

Until this discordance is resolved, physicists have chosen to limit themselves to describing the exterior traits of black holes—from its event horizon, or its Schwarzchild radius, outward. In this region, general relativity and quantum mechanics refrain from engaging in turf wars.[8]

Physically, a black hole defines a region where gravity is so strong that nothing can escape. This notion necessitates the specification of an event horizon encircling a region of spacetime that, once entered, is impossible to escape. In flat spacetime, that horizon has an infinite radius, because it is impossible to go further than infinity. But when curvature is introduced the region that is impossible to escape can take on a finite boundary.

Wherever spacetime is curved, familiar geometric relations are altered. For example, in Euclidean geometry a triangle is a shape that is defined by having three angles that sum to 180 degrees. This property changes in curved space. For positively curved space, the sum of angles in a triangle increases. In negatively curved space, that sum decreases. The degree to which a region of space deviates from traditional Euclidean expectations (for example, how much the sum of angles in a triangle in that region varies from a total of 180°), reflects the magnitude of curvature in that region.

This effect is traditionally illustrated by placing two-dimensional shapes on three-dimensional objects, like a triangle on a sphere or a saddle (Figure 15-1). The sum of angles in a triangle becomes greater than 180 degrees on a sphere, and less than 180 degrees on a saddle.

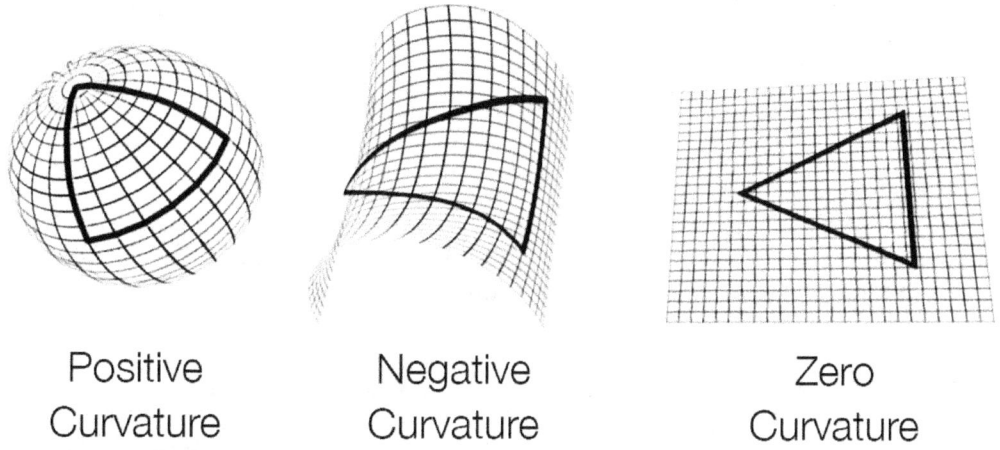

Figure 15-1 Geometric properties are altered in regions with positive or negative curvature

Another geometric expression that depends upon the curvature of space is the ratio of a circle's circumference to its diameter. In flat space, this value of this ratio is 3.14159265358979323846264338...—a number we refer to as π. In curved space, the numeric value of this ratio is altered. We can depict this alteration by placing a circle onto a sphere or saddle. On the sphere, which has positive curvature, the circumference to diameter ratio is less than the familiar numeric value of π, and on the saddle (negative curvature) the circumference to diameter ratio is greater than π. In other words, in positively curved spaces the diameter of a circle proportionally increases (compared to its circumference), and in negatively curved spaces the diameter proportionally decreases.

These traditional depictions are useful because they help us explain how geometric relationships change when two-dimensional objects are placed in curved space. To go further, to conceptualize how three-dimensional objects respond to warps, ripples or curves in spacetime, we need vacuum quantization.

Understanding the curvature of spacetime in terms of spatial density gradients, gives us the ability to understand how warps in spacetime alter three-dimensional objects, but it also sheds light on the geometric interior of a black hole. A medium that is composed of distinct parts (molecules, atoms, or quanta) possesses a maximum density boundary. Nitrogen gas, for example, is a specific type of medium with distinct traits. If we increase the density of N_2 molecules in a given region there is a limit to how much it can be compressed while retaining its gas-like characteristics. This barrier marks a phase change for the medium. Nitrogen gas can only bear so much change before it ceases to be nitrogen gas. Beyond that point the properties of the medium transform.

If the medium of spacetime is quantized, than it too can undergo phase changes. At low densities the vacuum dynamically acts like a gas. Approaching the center of a black hole, the connectivity of space undergoes alterations and the vacuum becomes less and less like a gas. Singularities exist wherever the quanta of space have been packed to their limiting maximum density.

Once this occurs, adding more quanta to the group does not increase the density any further. Instead, when additional quanta join the singularity, the event horizon extends outward. Technically these additions don't increase the size (in the spatial sense) of the singularity because the singularity represents a state in which a conglomeration of quanta acts as one unique point in x, y, z space (no matter how many quanta make it up). The entire conglomeration shares a unique history. This lack of independence means that, no time is experienced in the singularity, but it also means that a singularity contributes just a single unique location to the x, y, z map.

Approaching the singularity of a black hole, the spatial density increases and time is bled out (Figure 15-2). When the quanta become maximally packed, they cease to behave independently. At this point it is no longer possible for energy—phonons—to interact with a quantum in the conglomerate without interacting with the entire conglomerate.

Figure 15-2 As the spatial density increases the signature of time decreases

Because the passage of time requires free resonations (independence of position), and because this condition is forbidden within this region, the singularity of a black hole is without time (Figure 15-3). Therefore, a black hole's singularity can be characterized as a region where the individual quanta of spacetime are packed with maximum density—where they are all stuck together. Such a region, is purely spacelike, acts as a single x, y, z location, and is without the passage of time.

Figure 15-3 A singularity is a group of quanta representing a single location void of time

Entropy and Radiation

"Dare to reason! Have the courage to use your own minds!"

Immanuel Kant

In 1973, Jacob Bekenstein wondered what would happen if a box filled with hot gas fell into a black hole. According to the second law of thermodynamics, the entropy of a closed system never decreases. The random disordered arrangements of molecules in a box full of gas contains a lot of entropy, so what happens to that entropy when it falls into the black hole? Does it just disappear? Bekenstein realized that if the second law of thermodynamics were to hold, then black holes themselves must have entropy. The entropy of the black hole must increase, just as the size of its event horizon and its mass, increases when the box of gas falls into it. Working out the math, Bekenstein found that entropy in a black hole must be proportional to the area of its event horizon.

This insight led to a new question. If entropy is a measure of randomness, and random motion is heat, shouldn't black holes have temperature? In 1974, Stephen Hawking addressed this question and showed that, indeed, black holes have temperature. He also fixed the precise proportionality between the area of a black hole's event horizon and its entropy.[9] His result showed the temperature of a black hole to be inversely proportional to its mass. Unlike familiar objects, which heat up when you fuel them, feeding a black hole—putting energy, or mass, into it—cools it.

When it comes to black holes, the challenge of a quantum theory of gravity is to explain these insights of entropy and temperature from first principles. The goal is to gain a deeper understanding of Bekenstein's entropy and Hawking's temperature.

Entropy, which can be thought of as a measure of the number of rearrangements an object's internal constituents can undergo, without affecting its appearance or any of its measurable properties, is a significant marker of evolution. All systems with low entropy (high order) tend to evolve toward a state of high entropy (low order).

A quantized metric leads to the expectation that there is a maximum bound of entropy within the medium of spacetime—where the constituents of the vacuum state play perfectly symmetric roles. This condition is met inside a black hole, because rearranging the parts of a black hole's interior, swapping out any of the quanta that make it up, leaves it unchanged. This is what it means to say that a black hole has maximal entropy—or, as Brian Greene puts it, entropy-sated.

Turning this around, we note that the amount of entropy contained within a black hole tells us something fundamental about space itself—that space retains a structure that is finitely bounded. This bound depends on the surface area of its minimum constituents. Therefore, the fact that entropy has a maximum possible bound suggests that the vacuum is quantized.

Once we embrace quantization, as we do when we model the vacuum as a superfluid, it is no longer surprising that "there is strong evidence that the laws of black hole mechanics are indeed a special case of the laws of thermodynamics, and that the black hole area does indeed denote its entropy."[10] Furthermore, if the singularity of a black hole is nothing more than an amalgamated gigantic quantum, then the surface boundary of that amalgamation should determine its evolution.

Our everyday experiences lead us to believe that entropy depends upon volume. For example, the entropy of a one-gallon container of carbon dioxide contains half the entropy of a two-gallon container filled with the same gas under the same temperature and pressure conditions. Technically, entropy is defined as the logarithm of the number of rearrangements, or states, the constituents can possess. Therefore, the number of CO_2 molecular arrangements for the two-gallon container is the square of the number of arrangements in the one-gallon container. When we take the logarithm we find that twice the volume yields twice the entropy.

In a black hole, entropy is proportional to surface area instead of volume. This limiting condition paves the road to ontological clarity, because it suggests that our universe is composed of discrete elemental entities.[11] If the number of constituents within a black hole of a given size is finite, then entropy itself becomes intelligibly bounded. In other words, if the entropy of a black hole depends upon the number of discrete Planck areas that can be arranged on the surface of its event horizon, then each quantum of space carries a single unit of entropy.[12]

Vacuum superfluidity further penetrates the notion of entropy because it allows us to understand the degrees of freedom responsible for entropy—the superspatial dimensions. This collapses the general notion of entropy—revealing it to be a reflection of the different ways the vacuum quanta can be arranged relative to each other.

This description enables us to define an equilibrium state for a black hole, and to expose what drives its evolution. On one hand, the growth of a black hole's singularity, through the addition of more quanta, is favored because compressing the quanta together decreases the total surface area of the quantum system, reducing the pressure of the vacuum.[13] On the other hand, the singularity is part of a thermodynamically interactive medium, which means that collisions from incoming quanta with sufficient energy can cause quanta to escape the surface of the conglomerate, or evaporate.

Because the collisions that lead to this evaporation lose their potency as the quantum amalgamations become larger, we expect small groups of maximally packed quanta (microscopic singularities) to evaporate more readily than large singularities.[14] The magnitude of this expectation, however, should also depend upon the background temperature of space. As the universe cools, the Schwarzschild radius of the minimally stable black hole should decrease. Consequently, the rate of black hole formation should slightly increase over time.

This enriches our grasp of Bekenstein's entropy and Hawking's temperature, but it also offers an additional insight. When we use this picture to examine microscopic black holes, we discover that miniature black holes are not perfect correlates of elementary particles. The thermal agitation that causes miniature black holes to evaporate is far less effective at disrupting fundamental particles—like electrons.

Externally, black holes and elementary particles are both completely characterized by their masses, charges, and spins. For this reason, it may seem natural to think of black holes as gigantic elementary particles. But there is a difference between the two—elementary particles are far more stable than a black hole with its same mass would be. This means that they must represent different geometric configurations. Black holes and elementary particles must be dynamically stabilized via different processes—they must be geometrically distinguishable.

To understand the process that might be responsible for the formation of elementary particles, let's talk for a moment about fluid properties. If we create a flow in a classical fluid, little waves will stir up, carry the energy away, and dissipate the flow. Things change, however, when we talk about quantum fluids: superfluids, or Bose–Einstein condensates (BEC's). Quantum fluids support flows that can persist unchanged forever. This is why we call them superfluids.

Applying the expectations of a superfluid to the vacuum allows us to forecast a variety of stable vortices—solutions to the vacuum's wave equation. If these vortices correlate to elementary particles, then the mass of each particle represents the overall geometric distortion maintained by each vortex, while charge and spin, are characterized by the divergence sustained in the vortex flow, and the braided twists that define the flow of its central toroid—imagine these vortices as sonons, or smoke rings. Therefore, in addition to offering us a geometric characterization of black holes inside their event horizons, vacuum quantization may also enable us to unveil the origins of elementary particles.

[1] Dam Thanh Son reported by Tim Folger. (2007, February). The Big Bang Machine. *Discover*, pp. 32–38.

[2] Technically, this idea can be traced all the way back to 1935 when Einstein proposed "a unified field theory in which matter, made of subatomic particles, could be viewed as some sort of distortion in the fabric of spacetime. To him, subatomic particles like the electron were actually "kinks" or wormholes in curved space that, from a distance, looked like a particle. Einstein, with Nathan Rosen, toyed with the idea that an electron may actually be a mini-black hole in disguise." Michio Kaku. *Parallel Worlds*, pp. 226–227.

[3] Paul Steinhardt, quoted in Kaku. (2006), p. 224.

[4] Robert M. Wald. (2000, September 30). The Thermodynamics of Black Holes. arXiv:gr-qc/9912119v2.

[5] Ibid.

[6] Entropy (S) is an expression of how disordered a system is, of the number of ways the system may be rearranged without changing its physical parameters. Once we specify the thermodynamic state of the system, entropy is the amount of additional information needed to specify the exact physical state of a system. Reaching maximum entropy means being in thermodynamic equilibrium. Entropy has units of energy/temperature $\left(\frac{kg\ m^2}{s^2\ K}\right)$.

[7] John Wheeler coined the term "black hole" in the 1960s.

[8] If our Sun collapsed into a black hole, its event horizon would be less than two miles in diameter.

[9] More precisely, the entropy of a black hole is equal to the Boltzmann constant k_B multiplied by the area of its event horizon expressed in Planck units A, and the speed of light cubed c^3, divided by the product of 4, the universal gravitation constant G, and the reduced Planck constant $\hbar$

$$S_{BH} = \frac{k_B A c^3}{4G\hbar}$$

[10] Bekenstein. (1973), (1974); Wikipedia, "General relativity."

[11] See also: James Owen Weatherall. (2008, May). The Tabletop Universe. *Popular Science*, pp. 72–76.

[12] Further support of this picture comes from the fact that the exact bound depends on a constant that guides molecular theory—the Boltzmann constant. This constant represents the ratio of the Universal Gas Constant $\left(R = 8.3144621\ \frac{m^2\ kg\ mol}{s^2\ K}\right)$ to Avogadro's number ($N_A = 6.02214129 \times 10^{23}\ mol^{-1}$) and originates from the Planck parameters

$$k = \frac{R}{N_A} = 1.3808 \times 10^{23}\ \frac{kg\ m^2}{s^2\ °C} = \frac{l_P^2 m_P^2}{t_P^2 T_P}$$

Translating the Boltzmann constant in terms of the fundamental Planck parameters allows us to rewrite the entropy of a black hole. If we represent the surface area of a black hole as

$4\pi r^2$, where r is equal to nl_P, and n is the number of space quanta that make up the surface of the black hole's event horizon, then our expression becomes

$$S_{BH} = \frac{\pi n^2 l_P^2 m_P}{t_P^2 T_P}$$

(See Chapter 16 for the conversions of k, G, and $\hbar$.)

Because $\frac{l_P^2}{t_P^2}$ is equal to c^2, and $m_P c^2$ is equal to the Planck energy E_P, this expression is also equal to

$$S_{BH} = \frac{\pi n^2 E_P}{T_P}$$

This expression is a reflection of molecular assumptions. It dictates that the entropy bound of a black hole is nothing other than the ratio of the Planck energy to the Planck temperature multiplied by π and the number of quanta that construct the black hole squared.

If we want to retain the Boltzmann constant in our equation, then the expression is

$$S_{BH} = \frac{4\pi k_B M^2}{m_P^2}$$

This is the case because the ratio of M^2 to m_P^2 is identical to the ratio of surface area divided by the Planck area l_P^2. In both cases we end up with a critical dependence on the number of quanta involved squared n^2.

For those unfamiliar with the molecular relations of R and N_A, they are as follows. The Universal Gas Constant is equal to the product of the pressure and the volume of one gram-molecule of an ideal gas divided by the absolute temperature. Avogadro's number represents the number of atoms in a gram-atom or the number of molecules in a gram-molecule. Avogadro's Law states that equal volumes of gases at the same temperature and pressure contain the same number of molecules. Thus, the molar volume of all ideal gases (gases held at **0 °C** and a pressure of 1 atmosphere) is equal to 22.4 liters. This means that the number of molecules filling 22.4 liters under ideal conditions is exactly equal to Avogadro's number for any ideal gas.

Amedeo Avogadro never calculated Avogadro's number. Johann Loschmidt, an Austrian scientist, calculated the number in 1865. Years after Loschmidt's death some chemists campaigned to name it in Avogadro's honor. In Germany this historical blooper has been corrected. There $6.02214129 \times 10^{23}$ is called the Loschmidt number. *NewScientist*. (2007, October 6). Zeroth theorem, p. 60.

[13] As the volume of the singularity increases (in Planck volume multiples), the total surface area of the system decreases. The surface area of a sphere, defined as having n multiples

of the Planck volume, will always be less than the surface area found from the sum of n individual Planck spheres, provided that n is a whole number larger than one. In other words, two identical spheres that are separated always have a greater combined surface area than one sphere would have if its volume equaled to the combined volumes of the two spheres.

[14] The time it takes for black hole evaporation to occur is proportional to the cube of the black hole's initial mass m_0. "For a solar-mass hole, the lifetime is an unobservably long 10^{64} years. For a 10^{12} kilogram one, it is 10^{10} years—about the present age of the universe." Bernard J. Carr & Steven B. Giddings. (2005, May). Quantum Black Holes. *Scientific American*, p. 51.

A 1000-kilogram black hole would entirely evaporate in less than a billionth of a second. The general equation for this evaporation time is

$$t_{evaporation} = (2^{10})(5)\,\pi\, t_P\, \frac{m_0^3}{m_P^3}$$

where t_P is the Planck time $(5.39106(32) \times 10^{-44}s)$ and m_P is the Planck mass $(2.17651(13) \times 10^{-8} kg)$.

Chapter 16 — **The Constants of Nature**

"It is hardly justifiable to suppose that universal laws of nature have no reason for their special form."
Charles Pierce

"God himself could not have arranged these connections any other way than that which does exist, any more than it would have been in His power to make four a prime number."
Albert Einstein[1]

NASA co-op party, Clear Lake, Texas.

Seeking refuge from the pounding music, I slowly made my way through an array of tightly packed young scientists, and their delicately balanced cups of alcohol. Under the moonless sky, caressed by a warm breeze, my thoughts found cohesion. As my eyes adjusted to the dark I discovered a sandy playground, complete with swings and a slide. In the center of this playground a shadowed silhouette was gently rocking back and forth. As I approached, the form became familiar. I sat in the swing next to my friend and shared his heavenly gaze in silence.

After a minute or two, without moving his eyes, my friend somberly asked, "What is the point of science?" I continued looking up and attempted to fully absorb his question. Nate Miller was a NASA co-op, an aficionado of math and physics, and a very serious thinker. He was, by every measure, a scientist. As I was considering his question he added, "Since it is impossible to prove anything, and all we can do is disprove based on logical contradictions, or mismatches between observation and theory, what's the point?"

I sympathized with Nate's desire to possess certainty and clarity, but I also felt that the situation he was addressing spoke to science's shining strength. I began swinging in rhythm with him, joining his search for the next meteor. After a short spell I said, "Maybe the point is to find new questions." Nate stopped swinging and I continued. "We may never know that our axioms, or our assumptions are correct, but as we explore new ideas, and challenge our core assumptions, we risk finding new *dangerous* questions that threaten to expand the limits of our imaginations. Along the way we might even come upon the truth—without the arrogance of knowing it." Nate started swinging again.

With light hearts we started talking about how science can be used as a tool to help free us from our delusions and our sense of self-importance. We discussed the implications of the discovery that we live on the surface of an oblate sphere, which rotates about its axis and orbits an average star, making its way around a common spiral galaxy—a galaxy that is scattered among hundreds of billions of other galaxies in the visible universe. Then we discussed the indelible stamp of Darwin and Wallace—noting how unearthing our relationship to the forms of life on this planet has expanded our minds. Instead of being trapped by the idea that we are the pinnacle of creation, the knowledge that we represent a single branch on the self-assembling tree of life, which owes its very form to simple ancestral roots, the complex history of evolving ecosystems, and the laws of Nature that underlie those processes, has put us one step closer to understanding our magnificent insignificance.

Rocking back and forth in our swings and gazing at the stars, Nate and I imagined the potential of humanity's intellectual evolution. Years later I still found myself concerned about pushing that threshold.

The lens of science has sublimely shown that the fear of becoming insignificant pales in comparison to the awe of participating in the journey toward the palpable unknown. But this lens is a rare treasure to possess. Why? Why is it still the case that honest questions, which challenge traditions, or beliefs, are violently combated by those with a talent for magnifying irrationality? When the question is more honest than the answer, why can't we let go of the answer?

As a somewhat benign example of this, consider the following. In third grade, I asked my teacher why the Moon goes around the Earth instead of just passing it by. She replied, "Oh, because of gravity." Naturally, my next question was, "What is gravity?" She scolded me with her eyes and said, "Gravity is what pulls the Moon toward Earth." I guess I was supposed to just accept that the Moon is pulled toward the Earth because of the thing that pulls the Moon toward the Earth, but that didn't feel like an answer to me.

This was the first time I remember encountering an evasive response to one of my questions outside of Sunday school. My teacher had completely avoided my question by using what is known as circular reasoning. This little trick enables people to appear like they have an answer, even though that "answer" is entirely empty.

"To know what you do not know is best. To pretend to know what you do not know is a disease."

Lao-tzu

Circular reasoning is still very much alive today. As a cancer of insecurity, doubt, and confusion, circular reasoning preys on humanity's desire to have an answer, planting new seeds every time someone utters the phrase, "everything happens for a reason." It projects a compelling argument that a deceptive circular answer is better than no answer, or the honest, "I don't know." But this conclusion runs entirely against the grain of our scientific quest. It attempts to reverse our progress by smothering curiosity and by robbing doubt of its nobility.

An incarnation of circular reasoning in use today (among scientific circles) is called the anthropic principle. There are two common varieties of this principle: the weak anthropic principle, and the strong anthropic principle. Both of these versions give an illusory "explanation" for why the universe is the way we see it today.

The weak anthropic principle assumes that our universe is either infinite in space (and/or time), or at least very large compared to the size of our visible universe. It then assumes that the physical parameters within this universe are able to vary over extremely large distances, or over long durations of time. Building upon these conditions, the weak anthropic principle then says that the conditions necessary for the development of intelligent life "will be met only in certain regions that are limited in space and time. The intelligent beings in these regions should therefore not be surprised if they observe that their locality in the universe satisfies the conditions that are necessary for their existence."[2] The conclusion is, that we see the universe the way it is, because we are here to see it.

Does that feel like an answer? Isn't it a bit like saying that the Moon goes around Earth because of the thing that makes the Moon go around Earth? To be fair, if we accept the foundational assumptions within this anthropic principle, then it is true; just like it is true that the Moon goes around Earth because of the thing that makes the Moon go around Earth. This truth, however, is trivial. It is a tautology—true by definition. It does not bring us closer to a profound understanding any more than saying "a red thing is red" does. This anthropic principle may be routinely confused with a profound truth, but it is not profound. It is a hand waving trick used to get us to ignore the real question. Can you identify the question that this circular reasoning diverts our eyes from?

The strong anthropic principle is slightly different from the weak anthropic principle. It assumes that there are either many different universes, or many different regions within a single universe. Critically, it then assumes that each of these universes, or regions, somehow randomly attains its own set of physical parameters, which means that they each have their own unique physical laws. Building on these assumptions, the strong anthropic principle then claims that in "most of these universes the conditions would not be right for the development of complicated organisms; only in the few universes that are like ours would intelligent beings develop and ask the question, "Why is the universe the way we see it?" The answer is then simple: if it had been different, we would not be here!"[3]

The active function of this argument is to get us to ignore the most interesting relevant question. Namely, how did the universe actually obtain the precise physical parameters that it has? What determines the values of Nature's fundamental constants—what prescribes them? By diverting our attention from this central line of curiosity, the anthropic principle attempts to save us from having to be embarrassed about not having an answer.

Both versions of the anthropic principle assume that the physical parameters in our universe, or our corner of the universe, just came into existence randomly. They also assume that it is possible for all of the physical parameters of Nature to wildly vary either from universe to universe, or within different regions of the same universe. Yet there is no compelling evidence supporting either of these claims.[4]

Of course, it appears to be true that *if* the physical parameters of our universe had been different by just a few percent, then the development of life (as we know it) would not have been possible. But this statement overlooks the question of whether or not those physical parameters are able to vary. It is possible that the physical constants of Nature are identical in every universe. It just might be the case that an underlying physical law circumscribes them. In fact, without evidence against this possibility, it may be argued that this condition is even more likely.

"The whole history of science has been the gradual realization that events do not happen in an arbitrary manner, but that they reflect a certain underlying order..."

Stephen Hawking

The blind assumption that the values of the physical parameters of our universe just sprang into existence by some random process contradicts the main corpus of our scientific knowledge. This assumption is deceptively used to make us think that the question of how things ended up the way they are has been resolved. It creatively persuades us to ignore the question by embracing a logical contradiction.

The claim that the physical parameters of our universe were randomly selected, assumes that a selection process exists—that some kind of process is responsible for that exact assignment (like a random number generator). But if events are *selected*, then they are *caused*. They may appear random, but this is merely an illusion that presents itself to those that remain ignorant of the full causal process. Simply put, it is a contradiction to assume that something is *randomly* selected. Events, phenomena, occurrences, etc., cannot be randomly caused. They may *appear* to be randomly connected to any observer that remains sufficiently ignorant of the variables in play, but they cannot be ultimately random. Random number generators are deterministic programs—they aren't truly random. Causes are deterministic.

Because we have no reason to expect that our universe is special, it would be very satisfactory to discover a theory that reveals a mechanism that determines the physical parameters—requiring them to have the values we observe. If vacuum quantization can make general relativity and quantum mechanics commensurable, can it explain the constants of Nature? If vacuum quantization is responsible for the constants of Nature, then we no longer need to use the anthropic principle to avoid the real questions, or to hide our ignorance—we can have a real answer.

Geometric Origins of the Constants of Nature

"Reality is the mirror of the sacred language."

Touching the Timeless[5]

Our new multidimensional perspective offers to save us from the drain of circular logic. Instead of turning us away from the question of how the constants of Nature came to have the values they have, vacuum quantization licenses the possibility that the geometry of the vacuum is responsible for setting the constants of Nature.

If the axiomatic parameters of quantization turn out to dictate the constants of Nature, then the laws of physics are exquisitely inscribed by the axiomatic structure of the vacuum itself. The question is—can we actually link the constants of Nature to the axiomatic parameters of our map?

The easiest way to discover how vacuum quantization encodes specific limiting values—defining a natural scale—is to note that every unit of measurement (knot, curie, fortnight, calorie, kilometer, volt, bushel, parsec, milligram, light-year, Mach, astronomical unit, Pascal, Dalton, slug, kilohertz, ohm, carat, psi, newton, decade, candle, pound, weber, fathom, dyne, furlong, watt, township, liter, tesla, kilogram, joule, decibel, Galileo, ton, farad, second, coulomb, degree Celsius, gallon, femtogray, ampere, btu, millibar, electron-volt, horsepower, foot, gauss, pico-Henry, Kelvin, lux, erg, hour, langley, acre, attopoise, stokes, etc.), can be reduced to an expression of length, mass, time, charge, temperature, or a combination of these five kinds of measure.

In a quantized vacuum, these five fundamental quantities have natural limits. Length and time have discrete minimum values, while mass, charge, and temperature have discrete maximum values associated with those minimums. For example, quantization theoretically sets a maximum amount of geometric distortion (mass-energy) that a minimum amount of space (a single quantum) can contribute.

These discrete limits define a natural scale—revealing the natural unit of space to be one quantum of space, the natural unit of time to be one quantum resonation, and the natural units of mass, charge, and temperature to be the maximum respective values that one quantum can contribute to each. The discrete values of these quantities (in arbitrary units—of meters, kilograms, seconds, Coulombs, and Kelvins) are contrasted with their value in natural units in the following table.[6]

Natural unit	Symbol	Value (arbitrary units)	Natural value
Planck length	l_P	$1.616199(97) \times 10^{-35}$ m	1
Planck mass	m_P	$2.17651(13) \times 10^{-8}$ kg	1
Planck time	t_P	$5.39106(32) \times 10^{-44}$ s	1
Planck charge	q_P	$1.875545946(41) \times 10^{-18}$ C	1
Planck temperature	T_P	$1.416833(85) \times 10^{32}$ K	1

When it comes to encoding the geometric character of the vacuum, these limiting values are necessary, but insufficient. A full geometric representation of a quantized vacuum must include two additional unitless expressions—representing the maximum and minimum limits of curvature allowed in that medium.

When a particular region of space maintains a consistent macroscopic quantum density, the density change, or curvature, across that region of space is zero. Any large circle within this region will have a circumference to diameter ratio that approximates π. In this sense, π represents the minimum limit of curvature (zero curvature).

What number represents the maximum limit of curvature? To get at that answer symmetrically, we have to explore what happens to the ratio of a circle's circumference to its diameter in a region with non-zero curvature. For example, in a radial density gradient, the ratio of a circle's circumference to its diameter decreases, because its diameter proportionally increases.

If the fabric of spacetime were continuous, then the minimum limit of that ratio would be zero; because, in a continuous vacuum, a circle drawn around a region of maximum curvature—a black hole—has an infinite diameter (Figure 16-1). But if spacetime is quantized, then there is a limit on how tightly the quanta of space can be packed, which means that the diameter of the circle is not infinite, and the minimum limit of that ratio is greater than zero (Figure 16-2). In a quantized vacuum, the value of the number that represents the maximum limit of spacetime curvature must be greater than zero and less than π. The ratio anticipated by vacuum quantization—the minimum limit of a circle's circumference divided by its diameter—shall be represented with the Cyrillic script ж (pronounced "zhe" or "je").

This number completes our geometric characterization of the quantum vacuum. Therefore, if the constants of Nature are products of the axiomatic structure of spacetime, and if our map adequately represents that axiomatic structure, then we should expect the constants of Nature to be functions of the five Planck parameters (l_P, m_P, t_P, q_P, T_P) and the two dimensionless numbers that represent the limits of curvature (π and ж). Are they? Can we generate the constants of Nature from these seven numbers? Without knowing the precise numeric value of Ж, our question gets translated into the following: Is there one dimensionless number whose value lies between zero and π that can be combined with l_P, m_P, t_P, q_P, T_P, & π to yield the constants of Nature in a simple and non-arbitrary way—without depending on ugly or unexplained adjustment factors?

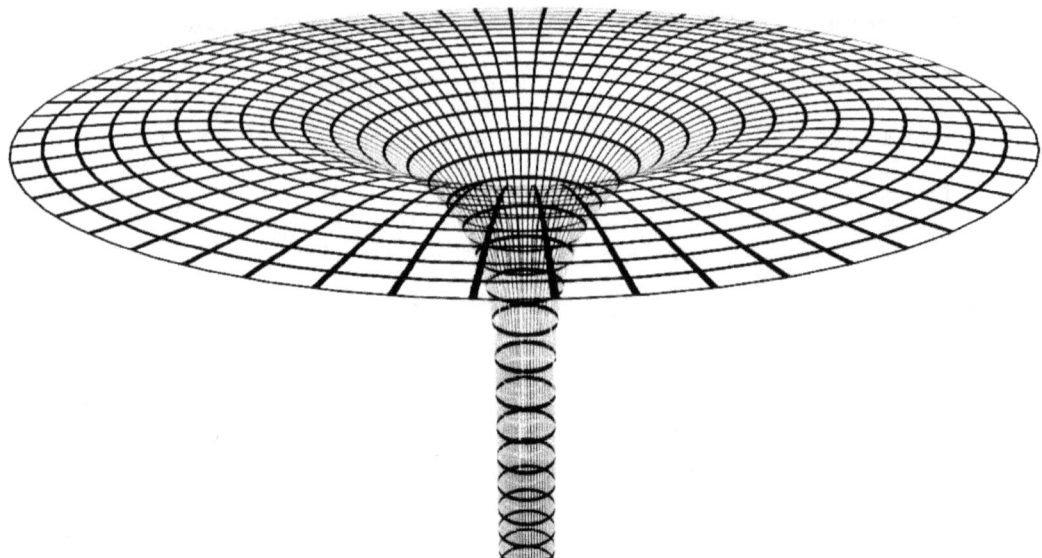

Figure 16-1 Does the tail of a black hole stretch to infinity as general relativity claims?

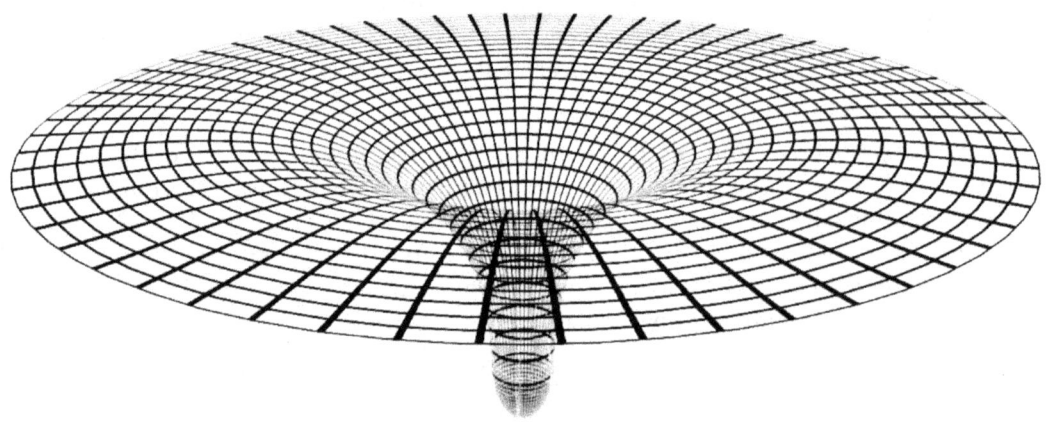

Figure 16-2 Quantization requires a cutoff

The constants of Nature we hope to explain in terms of the vacuum's geometry are:

Speed of light $= c$ *Planck's constant* $= \hbar$
Gravitational constant $= G$ *Fine-structure constant* $= \alpha$
Elementary charge $= e$ *Boltzmann constant* $= k_B$
Vacuum permeability $= \mu_0$ *Vacuum permittivity* $= \varepsilon_0$
Coulomb's constant $= k_e$ *Stefan-Boltzmann constant* $= \sigma$
von Klitzing constant $= R_K$ *Josephson constant* $= K_J$
Magnetic flux constant $= \Phi_0$ *Characteristic impedance* $= Z_0$
Conductance quantum $= G_0$ *Quantized Hall conductance* $= H_C$
First radiation constant $= c_1$ *Spectral radiance constant* $= c_{1L}$
Second radiation constant $= c_2$ *Molar gas constant* $= R$
Faraday constant $= F$ *Classical electron radius* $= r_e$
Compton wavelength $= \lambda_C$ *Bohr radius* $= a_0$
Hartree energy $= E_h$ *Rydberg constant* $= R_\infty$
Bohr magneton $= \mu_B$ *Nuclear magneton* $= \mu_N$
Compton angular frequency $= \omega_C$ *Schwinger magnetic induction* $= S_{mi}$
Gravitational coupling $= \alpha_G$

I began my search for this dimensionless number with very limited resources (in prison). Then, on July 24, 2008 a stroke of luck came my way when, against all odds, my bunky William Prince acquired a copy of the *Handbook of Chemistry and Physics*,[7] which contained an authoritative list of the values of the fundamental physical constants. After hours and hours of crunching numbers the old fashioned way, I discovered that there is one unitless number that fits this bill.[8]

In other words, I discovered that—if the hypothesis of vacuum quantization is correct, and if this number represents the maximum limit of curvature in the superfluid vacuum, then the constants of Nature listed above are all geometric expressions of the five Planck parameters and the two dimensionless numbers that set the boundaries of minimum and maximum curvature.

$$\pi = 3.141592653589\ldots \qquad ж = 0.085424543135(14)$$

A serious contender for a grand unification theory should offer a fundamental explanation of the constants of Nature. Here, we gain that possibility. To see how, note that in the arbitrary units used today (meters, kilograms, seconds, Coulombs, Kelvins and moles), the constants of Nature have the following numeric values:

$$c = 2.99792458 \times 10^8 \, \frac{m}{s} \qquad \hbar = 1.054571726(47) \times 10^{-34} \, \frac{m^2 \, kg}{s}$$

$$G = 6.67384(80) \times 10^{-11} \, \frac{m^3}{kg \, s^2} \qquad \alpha = 7.2973525698(24) \times 10^{-3}$$

$$e = 1.602176565(35) \times 10^{-19} \, C \qquad k_B = 1.3806488(13) \times 10^{-23} \, \frac{m^2 \, kg}{s^2 \, K}$$

$$\mu_0 = 1.25663706143592\ldots \times 10^{-6}\, \frac{m\, kg}{C^2} \qquad \varepsilon_0 = 8.854187817\ldots \times 10^{-12}\, \frac{s^2\, C^2}{m^3\, kg}$$

$$k_e = 8.98755178736821\ldots \times 10^9\, \frac{m^3\, kg}{s^2\, C^2} \qquad \sigma = 5.670373(21) \times 10^{-8}\, \frac{kg}{s^3\, K^4}$$

$$R_K = 2.58128074434(84) \times 10^4\, \frac{m^2\, kg}{s\, C^2} \qquad K_J = 4.83597870(11) \times 10^{14}\, \frac{s\, C}{m^2\, kg}$$

$$\Phi_0 = 2.067833758(46) \times 10^{-15}\, \frac{m^2\, kg}{s\, C} \qquad Z_0 = 3.767303134617\ldots \times 10^2\, \frac{m^2\, kg}{s\, C^2}$$

$$G_0 = 7.7480917346(25) \times 10^{-5}\, \frac{s\, C^2}{m^2\, kg} \qquad H_C = 3.87404614(17) \times 10^{-5}\, \frac{C^2}{m^2\, kg}$$

$$c_1 = 3.74177153(17) \times 10^{-16}\, \frac{m^4\, kg}{s^3} \qquad c_{1L} = 1.191042869(53) \times 10^{-16}\, \frac{m^4\, kg}{s^3}$$

$$c_2 = 1.4387770(13) \times 10^{-2}\, m\, K \qquad R = 8.3144621(75)\, \frac{m^2\, kg\, mol}{s^2\, K}$$

$$F = 9.64853365(21) \times 10^4\, \frac{C}{mol} \qquad r_e = 2.8179403267(27) \times 10^{-15}\, m$$

$$\lambda_C = 2.4263102389(16) \times 10^{-12}\, m \qquad a_0 = 5.2917721092(17) \times 10^{-11}\, m$$

$$E_h = 4.35974434(19) \times 10^{-18}\, \frac{m^2\, kg}{s^2} \qquad R_\infty = 1.0973731568539(55) \times 10^7\, m^{-1}$$

$$\mu_B = 9.27400968(20) \times 10^{-24}\, \frac{m^2\, C}{s} \qquad \mu_N = 5.05078353(11) \times 10^{-27}\, \frac{m^2\, C}{s}$$

$$\omega_C = 7.763441 \times 10^{20}\, s^{-1} \qquad S_{mi} = 4.419 \times 10^9\, \frac{kg}{s\, C}$$

$$\alpha_G = 1.7518(21) \times 10^{-45}$$

But in natural units—defined by the geometry of our quantized vacuum—the values of these constants are simply:

$$c = \frac{l_P}{t_P} \qquad \hbar = \frac{l_P^2 m_P}{t_P} \qquad G = \frac{l_P^3}{m_P t_P^2} \qquad \alpha = \text{Ж}^2$$

$$e = \text{Ж}\, q_P \qquad k_B = \frac{l_P^2 m_P}{t_P^2 T_P} \qquad \mu_0 = \frac{4\pi\, l_P m_P}{q_P^2} \qquad \varepsilon_0 = \frac{t_P^2 q_P^2}{4\pi\, l_P^3 m_P}$$

$$k_e = \frac{l_P^3 m_P}{t_P^2 q_P^2} \qquad \sigma = \frac{\pi^2\, m_P}{60\, t_P^3 T_P^4} \qquad R_k = \frac{2\pi\, l_P^2 m_P}{\text{Ж}^2\, t_P q_P^2} \qquad K_J = \frac{\text{Ж}\, t_P q_P}{\pi\, l_P^2 m_P}$$

$$\Phi_0 = \frac{\pi \, l_P^2 m_P}{\text{ж} \, t_P q_P} \qquad Z_0 = \frac{4\pi \, l_P^2 m_P}{t_P q_P^2} \qquad G_0 = \frac{\text{ж}^2 \, t_P q_P^2}{\pi \, l_P^2 m_P} \qquad H_C = \frac{\text{ж}^2 \, t_P q_P^2}{2\pi \, l_P^2 m_P}$$

$$c_1 = \frac{4\pi^2 \, l_P^4 m_P}{t_P^3} \qquad c_{1L} = \frac{4\pi \, l_P^4 m_P}{t_P^3} \qquad c_2 = 2\pi \, l_P T_P$$

And the constants that also depend on Avogadro's number N_A, the electron mass m_-, or the proton mass m_+ are:[9]

$$R = \frac{N_A \, l_P^2 m_P}{t_P^2 T_P} \qquad F = N_A \, \text{ж} \, q_P \qquad r_e = \frac{\text{ж}^2 l_P m_P}{m_-} \qquad \lambda_C = \frac{2\pi \, l_P m_P}{m_-}$$

$$a_0 = \frac{l_P m_P}{\text{ж}^2 \, m_-} \qquad E_h = \frac{m_- \, \text{ж}^4 \, l_P^2}{t_P^2} \qquad R_\infty = \frac{m_- \, \text{ж}^4}{4\pi \, l_P m_P} \qquad \mu_B = \frac{\text{ж} \, l_P^2 m_P q_P}{2 \, m_- \, t_P}$$

$$\mu_N = \frac{\text{ж} \, l_P^2 m_P q_P}{2 \, m_+ \, t_P} \qquad \omega_C = \frac{m_-}{t_P m_P} \qquad S_{ml} = \frac{m_-^2}{\text{ж} \, m_P t_P q_P} \qquad \alpha_G = \frac{m_-^2}{m_P^2}$$

This suggests that the geometry of the vacuum is responsible for setting the constants of Nature.

As this list fell together, I became overwhelmed with exhilaration. The fact that a number within the expected range actually completes the pattern means that it is possible that these constants of Nature originate from the axiomatic structure of spacetime—that they are all material dependent parameters. To secure this claim, it is necessary to show that this value ж = 0.085424543135(14) precisely follows from the structure of our quantized geometry—a task now underway.

Note that there are no ugly arbitrary correction factors in the previously listed expressions. These constants are derived directly from the parameters of vacuum quantization. The Boltzmann constant, for example, is simply *one* in natural units of length squared, multiplied by mass, divided by the product of time squared and temperature. Its value is automatically written into Nature as an inescapable result of quantization.[10]

Planck's constant $\hbar$, which represents the separation between allowed quantized values of the angular momentum of a particle, prescribes the wavelength of matter waves, and is central to Heisenberg's uncertainty principle, is also inherently encoded by the geometric descriptors of our quantized map.[11] In natural units of length squared, multiplied by mass, divided by time it is equal to *one*.

Einstein would have been thrilled to discover this connection. In a letter to his fellow physicist, Arnold Sommerfeld, he once wrote, "I have come to this pessimistic view mainly as a result of endless, vain efforts to interpret… Planck's constant in an intuitive way."[12] Now that an intuitive explanation for where the constants of Nature come from is on the horizon, the next step is to derive the value we used for ж from the limits of compressibility inscribed by vacuum superfluidity. Accomplishing this task will enable us to solidify the conclusion that the constants of Nature are derived from the vacuum's geometry, but it may also allow us to complete our quantum theory of gravity, giving rise to an equation for the curvature of spacetime near a black hole that doesn't contradict quantum mechanics.

All Universes Have the Same Constants

> *"In the particular is contained the universal."*
>
> *James Joyce*

The fractal pattern of quantization we have assumed dictates that the constants of Nature will be mirrored in all universes. Hierarchical quantization means that observers in every universe, regardless of hierarchical scale, will internally measure their constants to be identical to the values we now measure them to be—$c, \hbar, G, k_B, k_e, R, \omega_C,$ and α_G will be perfectly identical and $\alpha, e, \mu_0, \varepsilon_0, \sigma, R_k, K_J, \Phi_0, Z_0, G_0, H_C, c_1, c_{1L}, c_2, F, r_e, \lambda_C, a_0, E_h, R_\infty, \mu_B, \mu_N,$ and S_{mi} will be nearly identical throughout the evolution of each universe (with a slight dependence on local curvature). In arbitrary units (feet, millimeters, hours, or whatever else some alien might come up with), the values of these constants are arbitrary, but in natural units, which internally reference the geometric structure of the axiomatic system, their values will always be identically fixed.

To wrap our mind around this consequence, we note that in a quantized vacuum the speed of light is nothing more than a relationship between the elemental length of space and the elemental chronon of time. In today's arbitrary units for distance, the elemental measure of space—the Planck length—is approximately 1.616242×10^{-35} meters. However, in nonarbitrary, natural units it is just one quantum length. In arbitrary units, the elemental chronon of time—the Planck time—is approximately 5.391505×10^{-44} seconds. In natural units, it is just one chronon.

If we divide the value for the Planck length by the value for the Planck time we get $299,792,458 \ m/s$, which is the speed of light (c). However, in nonarbitrary, natural units, the speed of light is simply one Planck length per one Planck time. All universes will reference this limiting speed, because all universes are composed of elemental quanta. In other words, the speed of light in every universe will be one natural unit of length in the vacuum of that universe, divided by one natural unit of time in that universe.

The constants of Nature are reflections of a deeper symmetry hidden in Nature—the symmetry of quantization and dimensional hierarchy. The physical parameters of our universe are not randomly ascribed. They did not obtain their specific values by pure chance. On the contrary, the values of the physical parameters of our universe (and within all universes) are governed by the geometry of space itself. The constants of Nature supervene on the intrinsic spatiotemporal properties of the metric. They are written by the texture of the superfluid vacuum.

Lee Smolin writes that in science, "we aim for a picture of Nature as it really is, unencumbered by any philosophical or theological prejudice."[13] Part of the beauty of a quantized vacuum is that it offers us a picture unencumbered with arbitrary constants—it gives us a way to understand what it means to say that a universe only 13.7 billion years old is governed by laws that are eternally true. Vacuum quantization reveals these laws, telling us that throughout the countless universes, the conditions necessary for life are ubiquitous.

> *"No adequate theory or explanation can contain any brute, crude, unexplained facts."*
>
> Douglas J. Soccio

The ontological value of this claim artfully intertwines with the reductive process—the way in which we explain the phenomena of our world. Traditional (satisfactory) scientific explanations tend to heavily rely on the reductive process of explanation, where high-level or complex phenomena are explained in terms of more basic phenomena. For example, biological phenomena are explained in terms of cellular phenomena, which are explained in terms of biochemical phenomena, which are explained in terms of chemical phenomena, which are explained in terms of physical phenomena.[14]

Ultimately then, our explanation of biological phenomena relies on our understanding of the underlying physical mechanisms. Because our explanations assume a reliance on the reductive process, we cannot truly say that we have an understanding of those physical phenomena (the underlying physical laws, such as the physical constants of Nature) unless we can explain them in a reductive sense also. A consequence of this is that we will never gain an ultimate explanation of any phenomena unless we discover an echoing symmetry that enables us to reductively explain the emergence of all phenomena infinitely. As long as there is some lower level to our explanation, which has to be taken as brute, our ladder of explanation is truncated and, therefore, is incomplete.

The symmetries revealed by the dimensional hierarchy in our map remove this truncation. These symmetries connect us to an unencumbered, complete route of reductive explanation. Through this, for the first time, we gain access to infinitely echoing explanations of the phenomena in our world. Our answers no longer end with the statement, "Because of the physical constants of Nature, which just are as they are." With the help of dimensional symmetries (hierarchical quantization), we become able to explain those constants in terms of more basic phenomena and trace those phenomena through the infinite geometric cascade of a fractal. Through this we stand to gain a looking glass of determinism.

The philosophical impact of this perspective is quite potent. To get the flavor of what I mean, consider the following. Many people have come to believe that the "fine-tuning" argument can be taken as evidence for a theistic or deistic God. The "evidence" they refer to is based on the assumption that, in the absence of any way to explain the constants of Nature, we must assume that the constants could have had any random value. It is then assumed that the precise values for the constants of Nature that are found in our universe (which are clearly necessary for life as we understand it), were fine-tuned.

For some, this line of thought is used to necessitate a fine tuner—some process or entity to do the fine-tuning, who presumably has a personal interest in the evolution of life and consciousness. Put succinctly, a theistic or deistic God is postulated to explain fine-tuning. Those who are not persuaded by this reasoning tend to use the anthropic principle to escape it, but the fine-tuning argument and the anthropic principle are both inherently flawed. These popular arguments just distract people from discussing what is really responsible for encoding the constants of Nature.

Our new axioms overthrow the fine-tuning argument with the same insight by which they erode the anthropic principle. We are no longer in a situation wherein we have no way to explain the exact values of Nature's constants. Therefore, we can no longer reasonably assume that the values of the constants of Nature came about randomly. As a consequence, we can no longer logically rationalize an attempt to explain why things are as they are via theistic and deistic concepts.

> *"This day we rescue a world from mysticism and tyranny and usher in a future brighter than anything we can imagine."*
>
> *Dilios*[15]

Quantization automatically and naturally dictates the values of the constants of Nature, and these constants in turn fix the character of the laws of physics that have led to the evolution of life and what we call consciousness. No other postulate is needed. Indeed there is no room for any other postulate. The reduction of Nature's constants to combinations of the vacuum's geometric parameters is the simplest and most beautiful explanation we could seek.

[1] Einstein. (1929). Ueber den Gegenwertigen Stand der Feld-Teorie. AEA 4–38; Walter Isaacson. (2007). *Einstein*, p. 385.

[2] Stephen Hawking. *A Brief History of Time*. p. 128.

[3] Ibid., p. 129.

[4] In string theory we can trace the source of this anthropic problem to a confusion over higher-dimensional geometries. The mathematical equations of string theory do not account for the geometric differences between superspatial, spatial and intraspatial dimensions. They allow each spatial dimension to be folded or compacted (signified by its "Euler number") in many different ways producing geometric arrangements known as "the landscape." These possible configurations each correspond to different kinds of universes, each with its own specific type of vacuum, fundamental constants, and laws. The geometric flexibility that stems from this dimensional oversight leads to the prediction that there are 10^{500} possible ways to fold or 'compactify' these extra dimensions—leading to the claim that there are 10^{500} possible universes—each with its own arrangement of fundamental constants. One of the more popular tools mathematicians use to compactify these extra dimensions is a Calabi-Yau manifold. All of these possible geometries collapse into a single geometry under vacuum quantization and the fundamental constants reduce to geometric identities.

A Calabi-Yau manifold.

[5] From the video *Touching the Timeless*. GN 380M54 1992, Part 6.

[6] To derive the Planck parameters we can set the Compton wavelength λ_C, which is equal to $\frac{\hbar}{mc}$, equal to the Schwarzschild radius r_S, which is equal to $\frac{G}{mc^2}$ and then solve for m. The specific mass value we end up with is the Planck mass m_P and it is equal to $\sqrt{\frac{\hbar c}{G}}$. Once we

245

know the Planck mass, we can plug it into the equation for the Schwarzschild radius, or the equation for the Compton wavelength, to find the Planck length l_P. The length that we end up with is equal to $\sqrt{\frac{\hbar G}{c^3}}$.

[7] David R. Lide. (2004–2005). 85th Edition, CRC Press.

[8] This number turns out to be equal to the ratio of the electron's charge to the maximum quantum charge.

[9] Avogadro's number N_A is used in the molar gas constant and the Faraday constant. This number is the result of somewhat arbitrary historical conditions wherein the number of atoms in a volume (whose scale was defined by the popular arbitrary system at the time and the personal choice of atom) was chosen as the definition. Avogadro's number has a value equal to $6.02214179(30) \times 10^{23} \, mol^{-1}$. The mass of the electron m_- has a value of $9.10938215(45) \times 10^{-31} \, kg$, and the mass of the proton m_+ has a value of $1.67261637(83) \times 10^{-27} \, kg$.

[10] When I first discussed this with my friend and mentor Phil Emmi, he curiously said, "So what you're saying then, is that in Nature all is one."

[11] There are two constants referred to as "Planck's constant" among physicists. Both are named after Max Planck. The original Planck constant (denoted h) is commonly used in applied or optical physics. It describes quanta of light and reflects the discrete unit of energy emitted as photons, as in the equation $E = h\nu$, where E is energy, h is Planck's constant, and frequency ν is in units of Hz or $\frac{1}{s}$. Theoretical physicists use a version of this constant that is in the form of natural units. They refer to it as Planck's constant, but it is also known as the reduced Planck constant and Dirac's constant.

In natural units Planck's constant is denoted $\hbar$ (pronounced "h-bar") and is equivalent to $\frac{h}{2\pi}$. Unless otherwise signaled, by the symbol h, (the one without the bar) *Planck's constant* refers to $\hbar$ throughout this book.

If we translate Planck's equation $E = h\nu$ in terms of angular frequency (units of radians/second) of ω (omega), then $E = h\nu = \frac{h\omega}{2\pi} = \hbar\omega$. Because of this $\hbar$ has been called the "quantum of angular momentum." $\hbar$ is said to be in natural units because it is directly tied to the other universal constants c and G (without any messy multiplication amplifiers) by way of the minimum discrete values found in spacetime known as the Planck length, Planck time, and Planck mass. John D. Barrow. (2002). *The Constants of Nature; From Alpha to Omega—The Numbers that Encode the Deepest Secrets of the Universe* (in English). Pantheon Books. ISBN 0-375-42221-8; Wikipedia, Planck constant, http://en.wikipedia.org/w/index.php?title=Planck_constant&printable=yes).

[12] Einstein to Arnold Sommerfeld, January 14, 1908; Walter Issacson. *Einstein*, p. 155.

[13] Lee Smolin. (2006, September 23). Never Say Always. *New Scientist*.

[14] David J. Chalmers. *The Conscious Mind—In Search of a Fundamental Theory*, pp. 50–51, 76.

[15] Spoken by the character Dilios in the movie 300. Screenplay written by Zach Snyder and Kurt Johnstad.

Chapter 17 — Deterministic vs. Stochastic

"What sorts of laws shape the universe with all its contents? The answer provided by practically all successful physical theories, from the time of Galileo onwards, would be given in the form of dynamics—that is, a specification of how a physical system will develop with time given the physical state of the system at one particular time."

Roger Penrose

"No Victor believes in chance."

Friedrich Nietzsche

Winter, Federal Prison Camp, Florence, Colorado.

Things had gotten pretty bad. I had become convinced that I was remembered only as a great disappointment to everyone in the outside world, assuming any of them remembered me at all. Although it had been years since I'd heard a word from the shadowed personas of my past—including the woman that I gave the moon—every waking moment of my life continued to be powerfully defined by an aching desire to be forgiven, to be accepted by the people I missed so badly, to somehow be a part of their world. Heartbroken and depressed I walked among the living dead, blending in among the forgotten, being talked down to by uneducated, power hungry "correctional" officers who seemed to get a kick out of making up random rules and then changing them the next minute just to make a show of their power.

A constant drizzle of snow fell from the dark gray sky. The richest color on the compound, the grass on the baseball field, was buried beneath a blanket of three inches of wet cold white. Snow used to be a thing of beauty. Rosy cheeks, laughter, last minute drives up the canyon in search of a great hill to tube down, fireplaces, snowmen—they had all been good memories facilitated by this soft white powder. But here the snow was never played with. Not a single snowball had been created in the three winters I had seen here, and the sole impressions in the snow covered field were the footprints left by ravens—the only creatures that could find their laughter in this place.

Wearing prison issued greens, David Cantu and I circled the dirt and asphalt track getting colder with each lap.

"Have you ever read the prison handbook?" I asked.

"I've read parts of it. Why?"

"Do you remember reading anything about not being allowed to build a snowman?"

"No, not specifically, but that doesn't really mean anything because there's the rules, then there's *the rules*," Dave replied as he gestured with his hands.

I knew what he meant. If anything counted as a trustworthy rule in this puritanical place it was that if you're having fun, you're probably breaking a rule. Childhood prepared me well for that rule—obey, follow, don't question, and for God's sake don't think for a moment that anyone is important enough to deserve a dream. Walking the track I remembered back to the day that I first defied those rules, the day that I first felt like a real person. I missed feeling that way.

"Let's ask Cordova if he's heard of any rule," I said.

Cordova was referred to as *number one on the compound*, because no one had been here longer than him. It didn't take long to find him.

"Cordova, have you ever heard of someone making a snowman on the compound?"

"No, I'm pretty sure you'd go to the hole for that," Cordova said.

"But do you know of any rule against it?"

"I remember hearing something about a rule against throwing snowballs, but rule or no rule, you'd still go to the hole. Tarnaski is on tonight."

Out of all the correctional officers Tarnaski had the biggest reputation for being trigger-happy. Because of his seniority, this chain-smoking alcoholic got away with just about anything. Beyond being a scrawny cowboy with something to prove, he was a very mean drunk—the kind that regularly gave out shots for things like not having your shirt tucked in properly (whatever that means) and sending people to the hole for having the wrong look on their face.

"Thanks for your advice Cordova."

Dave and I walked slowly toward our cells. After a few minutes of silence I asked, "would you rather go through another winter without Christmas, or have Christmas in the hole?"

Dave knew what I meant, so instead of answering my question he replied, "let's go spread our stuff around."

Whenever anyone went to the hole, the cops thought of it as their job to ransack their locker and randomly throw away some of their books, letters, magazines, or anything else that might have sentimental value. The only action we could take to reasonably protect ourselves from this assault was to loan our sentimentals out to other inmates in the wing ahead of time.

After I relocated my handwritten chapters of this book, it was time for us to find out whether or not snowmen were really forbidden in a Federal prison. Without gloves, snow pants, or snow boots, we headed out to the baseball field. Kneeling down at the edge, we smashed clumps of snow into the shape of rolling pins and got to work. We slowly lost feeling in our fingers, but we didn't stop working. We were on a mission. After two hours our faces were bright red from the chilled wind, our hands could barely move, yet we had managed to put together a fairly impressive, six-foot snowman. Although this monument didn't have a scarf, hat, or even a carrot for a nose, its classic snowman shape was unmistakable.

It was strange to have done something public and meaningful without even being reprimanded. The only thing left to do was wait for our punishment.

At 3:50 all inmates were recalled to their cells for the four o'clock count. On a normal day count took about thirty minutes, at which point we were released, one building at a time, to the chow hall for dinner. Today things were different. Fifteen minutes into the lockdown period, those of us in Summit Unit (the unit furthest from the baseball field) began to hear a faint roar coming from the other building (Teller Unit). Creative speculations quickly spread around in an attempt to figure out what was going on. Around five o'clock Tarnaski finally walked down our wing with his sycophant in tow, mumbling something about how the snowman screwed up his count, and how inmates aren't allowed to have Christmas spirit.

When we were released to the chow hall, the story of what had happened spread like wildfire. Everyone from Teller Unit, whose cell window faced the baseball field, told the tale of how Tarnaski went out to the field with a baseball bat and approached the snowman with an angry look on his face. The inmates watching this spectacle began booing and hissing from inside their cells as Tarnaski attacked Frosty with an incredible fury. (By universal agreement the snowman had already acquired a name.) Swinging the bat again and again, he finally managed to topple it. Not content with bringing it down, he began to stomp on its remains with his military boots. When he finished, he admired his work for a moment, then strutted away, proud as a cowboy.

Hearing this retelling of events, Dave looked at me and sighed.

"Oh well, at least we didn't go to the hole," he said.

The next day was Wednesday, which meant two things: the big wigs would be on the compound for a few hours, and the quality of lunch would noticeably improve in order to give an impression that things were different from how they really were. This meant that I had access to the warden. At lunch I went up to the warden and asked, "Is there a rule against making snowmen?"

"As long as you don't take it off my compound, I don't care," he said.

Satisfied with this answer I sat down and ate my lunch with a smile. Immediately after lunch Dave and I went to the weight pile and recruited half a dozen inmates to help us resurrect Frosty. Most of these recruits had witnessed Tarnaski's murderous rage firsthand. Working together, we rolled large snowballs, broke them into pieces and carried them to our rising pillar of snow. We threw our frozen treasures up to Joey, who, standing on top of our project, caught the pieces and then thoroughly stomped them into place. We rolled every last inch of snow on the field. When we were done, Frosty was twelve feet tall and six feet wide.

As the four o'clock count approached we all began to be curious about what would happen. Dave and I were still half-expecting Tarnaski to discover our role in the Frosty project, handcuff us and drag us off to the hole. We didn't care. For some reason this was worth getting locked up in solitary for the rest of our bits.

Predictably, Tarnaski showed up for his shift, saw the massive snowman in the middle of the baseball field—and threw an absolute fit. Shouting drunken obscenities toward the prisoners that were watching him from their cell windows, he grabbed his baseball bat and began to wail on Frosty. After ten minutes of huffing and puffing Tarnaski looked at the minimal damage he had caused, dropped the baseball bat, and walked off. A few minutes later he came back with a shovel. Swinging the shovel as hard as he could, he attacked Frosty again. Eventually, he managed to topple Frosty over. Proudly, he glared at the prison windows, then sauntered off.

At dinner, eyewitnesses to Frosty's second murder recounted the story again and again. Somehow the chow hall began to feel different. The tension on the compound was transforming into something I'd never felt before. And something that had once been alive in me was resurrecting.

Then, in the middle of the night, a miracle happened. It snowed. The miracle wasn't that the heavens were providing us with more material to resurrect Frosty. The miracle was that for the first time in years I, and over five hundred others, cared about whether or not it snowed. When the snow started to fall people began to laugh and shout in glee. Throughout the night, random inmates would enthusiastically yell down the wing, "It's still snowing!" With a smile on my face I stared out my window tracing the path of fat snowflakes as they danced and flurried about. I was too restless to fall asleep. My heart was racing. I felt different. And for some reason the shackles that had so intimately defined my life began to feel like socially constructed illusions. We may have been forgotten, cast away in shame and swept under the rug, but today eight inches of glorious, beautiful snowflakes of freedom were falling straight from the heavens to remind us who we really are.

The next day Bob Gilstrap (an educated UFC fighter serving time for kayaking from Canada to the US in the middle of the night with several kilos of cocaine) joined us at the edge of the baseball field. Staring out into the field of snow, and speaking more to himself than anyone else, he said he was tired of seeing this dipsomaniac cop bully us around. Then Mike Ritter (an x-pirate from the Thailand marijuana business) joined us. Soon a dozen accomplices, from a wide variety of backgrounds, were all working together to accomplish one goal.

The next few hours transformed Frosty into more of a pillar than anything humanoid in shape. It must have looked like we were constructing a big middle finger in the center of the field. After our tower was over sixteen feet tall and ten feet wide, Joey crafted a small 3-foot snowman on top of the tower, where it would be safe from Tarnaski's reach.

Our next project was to carry buckets of water across the compound in order to slick down the sides of our pillar and smooth out the steep slope that surrounded it creating a hard-packed, icy arena. We slipped and fell several times during its construction. When we were done, it was nearly impossible to reach the central pillar on stable footing.

The next two nights were special for two reasons. They were Tarnaski's days off, and they were very very cold. As the temperature dropped, Frosty froze to the core—transforming into a solid block of smooth ice. It became an art piece, reflecting our newfound sense of unity, the purity of having purpose, and reminding us what it is like to be a part of something you are proud of.

On the third night Tarnaski arrived just in time for the four o'clock count. The wait from Summit Unit was unbearable. For nearly an hour we could hear a faint roar of emotion coming from the other building, but we couldn't make out what was happening. After two hours had passed we were finally released to the chow hall—before Teller Unit. When the Teller inmates were finally allowed to join us, word of the events filled the room.

According to eyewitnesses, Tarnaski approached the monstrosity with his trusty shovel, but the icy incline caused him to fall flat on his face. The Teller inmates burst out into laughter behind their windows, infuriating Tarnaski further. Swinging his shovel from a distance, he wasn't able to leave so much as a distinguishable mark on the pillar. For several minutes they watched him huff and puff in frustration, falling all over himself as he tried to get close to the pillar to destroy it. Finally he went to work on cutting some footholds in the ice beneath him. This took several minutes and tired him out substantially. Then, when he secured his footing, he swung his shovel with all his might. It simply bounced off.

For another ten minutes Tarnaski's rage fueled his wild fit. Then he apparently gave up. Twenty minutes later he returned with a bag of salt. He tore open the bag, dumped its contents around the base of the snowman, but it didn't make any difference. Salt was not going to bring down Frosty. Tarnaski kept at it—huffing and puffing, taking breaks, slipping around, and the occasional face plant. Eventually he became exhausted. Standing in the long shadow of this giant, icy middle finger, Tarnaski shook both of his fists in the air as he screamed at the top of his lungs. Then he returned to his counting task.

Although Tarnaski's next shift was during the day, it was a weekend, which meant that he could get away with even more than usual because there was no chance that any suits were going to be dropping in. At about two o'clock in the afternoon Tarnaski left the compound and then returned with the maintenance truck. To make a great show of his intentions, he revved his engine at the edge of the baseball field. The entire compound stirred as inmates rushed to the weight pile to watch this spectacle. Spurred by alcohol, frustration and vicious anger, Tarnaski continued to rev his engine. This was his prison. No inmate had the right to have Christmas spirit or build a snowman without his permission.

Standing shoulder to shoulder on the weight pile, nearly every inmate on the compound silently watched as Tarnaski put the truck into gear. Speeding up to about thirty miles per hour, Tarnaski ran straight into the solid block of ice, but it didn't budge an inch. The entire front hood of the prison truck crumpled in on itself, the bumper tore off, the windshield spider-webbed, and the horn went off, braying like a wounded moose through the moonlit night. Little Frosty watched the whole thing from his safe perch.

As Tarnaski staggered out of the mangled truck with a bewildered look on his face, a victorious roar erupted on the weight pile. Tarnaski stared us down, gesturing for us to move out, but we continued delighting in our symbol of freedom—tasting the little bit of humanity that had seeped into this place.

The truck had to be towed away, and Tarnaski was officially reprimanded for ruining the work truck. Frosty ended up being a part of the compound until late June. When baseball season started, the inmates added a new rule to the game. "Hit frosty with the ball and it counts as a home run." We had found a window of hope and possibility. We had begun to weave the fabric of our futures in a place that was designed to take our futures away.

Level of Description

> *"The apparent peculiarity of quantum theory arises from mistaking an incomplete description for a complete one."*
>
> *Detlef Dürr[1]*

As time passed, Frosty's stand stirred a poignant curiosity in me, which led me to engage in a process of personal reflection. Sitting quietly for hours each day, trying to pay attention to my mind, thoughts, impulses, emotions, and their patterns, I began to see the interconnectedness of the world. Gradually, the boundaries by which I had so faithfully used to define the world around me, and even myself, began to blur. Where once I had seen fixed entities, partitions, and borders I began to see assemblages, differentiation, processes, and emergent properties. This clarity helped me recognize that, despite the emotional involvement that I personally experienced, the events that led to Frosty's triumph were deterministically controlled in an elaborate process of bottom-up emergence. My responses may have naïvely appeared to come from some autonomous sense of "free will", disconnected from the rest of the realm of cause and effect, and Tarnaski's actions may seem to be isolatable and blameworthy, a loose wheel in the machinery; but the more I thought about it the more I began to understand that my brain was tricking me (like it constantly does with time, or as it does in the McGurk effect[2]), attempting to convince me that my own actions are only schizophrenically part of the endless chain of cause and effect. This opened the door to the discovery that *blame* is a misguided notion.

The universe, or any closed (isolated) subsystem within it, has a state that evolves either deterministically, or stochastically. These are the only choices. Moment to moment these states are either connected by the laws of cause and effect (a position elucidated by Bohm's interpretation of quantum mechanics), or they are not (a position backed by the Copenhagen interpretation and others). Interestingly, both of these options call into question the legitimacy and logical soundness of the notion of *blame*.

In a deterministic universe blame is an unsound concept because it is logically inconsistent to blame anything or anyone for actions that could not have been otherwise—occurrences that are deterministically controlled. People cannot legitimately be blamed for actions they ultimately had no control over. On the other hand, if the universe is stochastic, then the notion of blame is completely incongruous. We cannot logically hold people responsible for actions that came from nowhere. People cannot legitimately be blamed for actions that they had no control over. Blame requires us to assume that an agent's actions cause effects (that determinism holds) while simultaneously assuming that the agent's actions were not the effects of other causes (that determinism does not hold). It is incoherent. Nevertheless, the question remains—is Nature deterministic or stochastic?

To claim that a system is deterministic is to claim that it evolves in a manner that is entirely fixed once the dynamics of that system are spelled out and a complete state is specified for at least one moment. In other words, knowledge of a single fixed state enables us to compute the time evolution of a deterministic system in either temporal direction, because within the system "final conditions are just as good as initial ones for determining the evolution of a system."[3]

All systems characterized by classical physics are deterministic (e.g. Newton's laws of motion, special relativity, general relativity, electrodynamics, thermodynamics, chaos theory and nonlinear dynamics). Some of these systems are complex, and may be difficult to predict in practice, but if their state is specified with enough detail, the evolution of that state is predictable.

In chaotic systems small perturbations in the initial state can result in significant changes to the system's state at a later point in time. These perturbations lead to changes in the state that manifest exponentially with time rather than linearly. For this reason, it is difficult to predict the future state of a chaotic system without precisely identifying its initial state. Nevertheless, the evolution of chaotic systems is completely fixed (determined) by the initial state of the system.

When we label a system "chaotic", we mean that it would be very impractical for us to ascertain its state with enough precision to predict its evolution over the time span we are interested in. How soon the evolution of such a system will misalign with calculation or prediction depends upon how sensitive the system is to small perturbations, and how accurately its state was ascertained. Technically, because no system's state is ever measured with absolute precision, all systems that don't have significant dampening effects can be considered chaotic on some time scale. Nevertheless, we traditionally reserve the term *chaotic* for systems that misalign with calculation during short or familiar time scales.

Weather patterns are good examples of chaotic systems because they depend critically on the tiniest details of the system's state. In order to map the system's time evolution (how clouds and weather systems change over time) we would have to know the dynamics that control the system and be able to meticulously specify the smallest details of its state for at least one moment—e.g. the velocity and position of every molecule in the entire atmosphere. Without this precise information the state of the system will appear to evolve chaotically (hence the name). But there is no doubt that those changes actually have direct causes.

> *"We ought to regard the present state of the universe as the effect of its antecedent state and as the cause of the state that is to follow. An intelligence knowing all the forces acting in Nature at a given instant, as well as the momentary positions of all things in the universe, would be able to comprehend in one single formula the motions of the largest bodies as well as the lightest atoms in the world: to him nothing would be uncertain, the future as well as the past would be present to his eyes. The perfection that the human mind has been able to give to astronomy affords but a feeble outline of such an intelligence."*
>
> Pierre Simon de Laplace[4]

What about quantum mechanical systems? Are they deterministic or stochastic? Although some might be tempted to respond "stochastic" to this question, it turns out that the answer depends on how we treat the state vector. If we assume that the state vector $|\psi\rangle$ is a fundamental descriptor, that it completely and exhaustively describes the state of a quantum system, then we are forced to conclude that it is impossible to have an equation, or program, that deterministically encodes the time evolution of our universe. This is a theoretical assertion, not a practical one. If the state vector is fundamental, if there is nothing beneath it, then no such equation exists, and effects can, at best, be probabilistically tied to causes. Under this condition the evolution of the state vector is intrinsically unpredictable (stochastic). From this it follows that determinism does not hold in Nature.

The other option is to assume that the state vector is *not* a fundamental descriptor of the system's state. One way to translate this is to say that the state vector $|\psi\rangle$ is a statistical ensemble of possible states, and that beneath that ensemble a more exact specification of the system's state exists, which can be elucidated by additional variables. Under this assumption the system's precise state evolves deterministically according to the Schrödinger equation. This casts quantum mechanical systems as classically deterministic (independent of our inability to obtain a direct specification of this exact state). From this it follows that determinism holds in Nature.

Which of these options best describes Nature? Is the universe stochastic or deterministic? Addressing that question will allow us to whittle down the logical space of candidate frameworks for physical reality. The first step in that process is to become cognizant of what is at stake if we assume that the state vector is, or is not, a fundamental descriptor. We need to grasp the ontological outfall of both options.

If the state vector is taken as a fundamental descriptor of the system, then the fact that systems have objective properties (at least while we are examining them) makes state vector reduction a necessary add on. However, by adding state vector reduction to quantum theory we introduce a stochastic quality into the inner workings of Nature. This is a rather violent upset. Every other dynamic equation in physics is strictly deterministic. General relativity, special relativity, Fourier series and functions, Heisenberg's equations of motion, Cauchy equations, all holomorphic functions, and even the Schrödinger equation are entirely deterministic descriptors.

The assumption that the state vector is a complete and exhaustive descriptor of quantum mechanical systems interrupts the deterministic beauty of physics by requiring us to tack on state vector reduction to our description—thereby introducing "quantum uncertainty". The fact that quantum systems behave in non-deterministic ways is only a "fact" if we grant that the state vector is fundamental. The lack of determination in quantum mechanics comes about only in the application of state vector reduction. "It is not to be found in the time-evolution of the quantum state, as described by the Schrödinger equation."[5]

In short, stochastic state vector reduction is balanced against all of the other deterministic equations of physics. If it is necessary, if the state vector truly is fundamental, then Nature is not deterministic. And if this is the case, then there does not exist, even theoretically, a map of physical reality, cause and effect do not ultimately drive the evolving cosmos, and physicists are searching for a treasure that exists only as a dream.

There are serious reasons to doubt that state vector reduction occurs, thereby doubting that the state vector is the most accurate descriptor of quantum mechanical systems. In addition to how this assumption gives birth to the previously mentioned epistemic disaster, state vector reduction also turns out to be an ontologically empty notion. It purchases us no grip on reality.

The assumption that the state vector is the most accurate descriptor of quantum mechanical systems, forbids the existence of a specific state of space. But we always measure objective properties that conform to specific states of space, instead of smeared out probabilistic properties that stem from probabilistic assortments of states. State vector reduction was crafted to bridge this gap, but it fails to accomplish this task. Reduction is asserted, not explained. It is entirely "ad hoc". It wasn't naturally derived from any symmetry arguments. It doesn't elegantly fit and, as of yet, no one has been able to propose a possible mechanism that could explain it.

Furthermore, according to relativity, this "collapse can be instantaneous in at most one reference frame, leading to two possibilities: either some feature of the situation picks out a preferred reference frame, with respect to which the collapse is instantaneous, or the collapse is not instantaneous at all."[6] Because, nothing intrinsic to spacetime picks out a preferred reference frame, the assertion that state vector reduction is instantaneous is suspect.

For these reasons, many physicists "doubt that the evolution of the state vector can possibly be taken seriously as an adequate description.[7] Instead of thinking of state vector reduction as real, they generally believe that the time-evolution of the quantum state is to be taken as an "underlying truth" and that we must come to terms with state vector reduction, in one way or another, as being "some type of approximation, illusion, or convenience."[8]

This leads to the idea that the state vector is *not* the most accurate descriptor of quantum mechanical systems; that instead, it represents a statistical ensemble of the possible states of space. Beneath this state vector is a precise state (one of the possible states within the state vector). Under this assumption state vector reduction (wave collapse) never occurs because the system was never actually in a statistical ensemble of states. It possesses a single precise state that evolves according to the Schrödinger equation.

Once we assume that the state vector is a blurred representation, that it describes the system in terms of its possible states (from a perspective that knows nothing about its actual state), state vector reduction dissolves. We may continue to describe the state of the system in terms of the state vector because we don't know its precise state, but when we make a measurement, we no longer need to invoke state vector reduction to explain the fact that we measure objective properties.

When we take a measurement we catch a partial glimpse of the actual underlying state, taking a strobe light image of the state of our system and observing its configuration at a given moment. Because the system has an exact state, its properties are objective. Taking a measurement is like observing a system of particles and glimpsing their momentary arrangements, but picking up no data on how each particle is moving about—their individual velocities. This glimpse shows us an objective state, but it leaves us completely unable to predict how that state will evolve.

To unveil the underlying precise state, to get beyond the statistical ensemble that defines the next possible state (the state vector), additional variables are necessary. This is a direct consequence of the fact that the complex vectors referenced by the state vector live in the space of states—instead of phase space (Chapter 12: The State Vector).

> *"I am, in fact, rather firmly convinced that the essentially statistical character of contemporary quantum theory is solely to be ascribed to the fact that this (theory) operates with an incomplete description of physical systems… [In] a complete physical description the statistical quantum theory would… take an approximately analogous position to the statistical mechanics within the framework of classical mechanics…"*
>
> Albert Einstein[9]

Accessing the state of the system is a matter of ascertaining the positions and velocities of the base constituents of that system. In traditional classical systems these positions and velocity vectors are defined in x, y, z space. But the state of space—the positions and velocities of the vacuum quanta—are defined in the superspatial dimensions, or what has been traditionally called the space of states.

If we are restricted to x, y, z measurements, then we are also restricted from elucidating more than the system's configuration at a single moment. To complete our knowledge of the system's state we need to have access to the velocity vectors that belong to each quantum in the system. These vectors are complex because they are defined in additional dimensions. To unveil them we must construct an additional variable theory. Without the contribution of these additional variables we cannot specify the exact state of the system, nor can we predict its evolution.

Structure beneath the state vector necessitates a subquantum (additional variable) theory. In other words, the assumption that the state vector is *not* the most accurate descriptor of quantum mechanical systems persuades us to accept additional variables (e.g. additional dimensions) as physically real. When we include these additional variables and provide a description of that structure, quantum mechanical probabilities naturally become interpreted as epistemic probabilities of the sort that arise in ordinary statistical mechanics. Therefore, additional variables allow a natural explanation of the statistical character of quantum mechanics and provide an elegant resolution of the infamous measurement problem.

By contrast, the assumption that the state vector *is* the most accurate descriptor of quantum mechanical systems carries us to an epistemic disaster, requiring us to admit state vector reduction, which introduces a stochastic element to Nature, without any explanation for how this reduction occurs.

Why then have so many physicists accepted the assumption that the state vector is a fundamental descriptor? If the assumption that the state vector is not a fundamental descriptor clearly has more ontological value than the opposing assumption, then why has it been seemingly abandoned? The answer, outside of any pressure to side with what has become popular, is that for decades physicists thought that additional variable theories had been proven impossible. They thought their only choice was to entertain the less desirable option.

> *"If all this damned quantum jumping were really here to stay then I should be sorry I ever got involved with quantum theory."*
>
> Erwin Schrödinger

In 1966 John Bell produced a theorem[10] (independently and almost simultaneously introduced by Simon Kochen and Ernst Specker[11]), which was widely interpreted to forbid additional variables in quantum mechanics. The theorem noted that if one assumes *noncontextuality*, *value definiteness*, and *value realism*, then it can be shown that additional variables cannot be used to construct a theory that reproduces the predictions of quantum mechanics.

Noncontextuality holds that if a quantum mechanical system possesses a property (an observable with a set value), then it does so independently of any measurement context. In other words, its value is independent of all the other observables that the experimenter may decide to measure at the same time. The property pre-existed in the system before any measurement.[12] Value Definiteness holds that all observables defined for quantum mechanical systems have definite values at all times. Value realism holds that if there is an operationally defined real number a, associated with a self-adjoint operator $\boldsymbol{A}$, and if, for a given state, the statistical algorithm of quantum mechanics for $\boldsymbol{A}$ yields a real number β, with $\beta = prob(v(\boldsymbol{A}) = a)$, then there exists an observable $\boldsymbol{A}$ with value a.[13]

As it stands, the BKS theorem is accurate. Nevertheless, when it comes to ruling out additional variable theories it is empty and irrelevant. Bell,[14] Bohm,[15] and Mermin[16] have pointed out that this "impossibility proof" is logically unsatisfactory because it arbitrarily imposes conditions that are relevant to the standard interpretation of quantum mechanics, but are not relevant to the theories they aim to dismiss—any theory with additional variables.[17]

Additional variable theories are not restricted by these assumptions. For instance, instead of assuming value definiteness an additional variable theory can assume what is called partial value definiteness, which can be accomplished by choosing a set of observables that can be assigned definite values without contradicting the BKS theorem. The best-known example of this is found in Bohm's interpretation, which prescribes positions and functions of position with definite values. Another approach is to let the set of definite observables vary with the state of the system, an approach taken by various modal interpretations.[18]

Mistaking these *impossibility proofs* as relevant, the quantum physics community redirected nearly all of its attention to the Copenhagen interpretation (the favorite interpretation of the most influential physicists at the time). Today that camp has become so popular that we call it the "standard interpretation," or the "orthodox interpretation".

Driven by their great desire to convince, the authors of that interpretation went too far, making unsupportable authoritative statements, claiming that the Copenhagen interpretation gave the only possible ultimate description of physical reality and that no finer description would ever become possible. Overwhelmed by the political momentum of that interpretation, it took a long time for the physics community to realize that the *impossibility theorems* were irrelevant.[19]

> *"Quantum mechanics is certainly imposing. But an inner voice tells me that it is not yet the real thing. The theory says a lot, but it does not really bring us any closer to the secrets of the Old One. I, at any rate, am convinced that He does not play dice."*
>
> *Albert Einstein*[20]

John Bell himself, the original author of the *impossibility theorem*, recognized its irrelevance. Yet even he was systematically misquoted, misunderstood, or ignored as he tried to call attention to it. He wrote, "in 1952 I saw the impossible done. It was in papers by David Bohm. Bohm showed explicitly how parameters could indeed be introduced, into nonrelativistic wave mechanics, with the help of which the indeterministic description could be transformed into a deterministic one. ... [W]hy then had Bohm not told me of this 'pilot wave'?... Why did von Neumann not consider it? More extraordinarily, why did people go on producing "impossibility" proofs, after 1952, and as recently as 1978?... Why is the pilot wave picture ignored in textbooks? Should it not be taught, not as the only way, but as an antidote to the prevailing complacency? To show us that vagueness, subjectivity, and indeterminism, are not forced on us by experimental facts, but by deliberate theoretical choice?"[21]

Bohm's construction is a valid example of precisely what the impossibility proof was supposed to forbid, but physicsts are largely unaware of its existence. Quantum mechanics is still taught through the lens of the Copenhagen interpretation, and due to the historical trajectory of physics, many physicists mistakenly think that this lens captures the entire discussion.

Now that we know that additional variable theories are not ruled out, now that we don't have to treat the state vector as a fundamental descriptor, we can back out of this historically entrenched path and explore more ontologically satisfying solutions. More importantly, now that we have an additional variable model by which to elucidate the complete state of a system (via vacuum quantization), now that we have theoretical access to both the configuration of the quanta of space at a particular moment, and their individual velocity vectors (which point in the superspatial dimensions—the additional variables), we can describe the probabilities of quantum mechanics in a way that is epistemically identical to those that arise in ordinary statistical mechanics. This return to a deterministic picture of physical reality carries us to a simple and elegant resolution of the infamous measurement problem.

The Geometry of Determinism

> *"We look upon a thing as the effect of chance when we see nothing regular in it, nothing that manifests design, and when furthermore, we are ignorant of the causes that brought it about. Thus, chance has no reality in itself. It is nothing but a term for expressing our ignorance of the way in which the various aspects of a phenomenon are interconnected and related to the rest of nature."*
>
> *Pierre Simon de Laplace*

Stochastic interpretations of quantum mechanics emanate from the assumption that the state vector is a fundamental descriptor. That premise is ultimately motivated by the assumption that the vacuum is infinitely and smoothly connected. By contrast, the assumption that the vacuum is quantized carries us to a construction with additional variables, which makes room for a coherent description of a system's complete state.

The possibility of specifying the positions and velocities of the system's ultimate constituents (the quanta) naturally carries us to the claim that beneath the state vector an exact state exists. It is this state that evolves according to the Schrödinger equation. It is this state that we partially access when we take measurements.

Quantum mechanical systems appear stochastic because our four-dimensional investigations are not privy to all the dimensional information that allows resolution of the underlying deterministic dynamics. At best, they have access to what's going on in a strobe-light manner. This is why reduced dimensional perspectives weave stochastic threads into the fabric of reality.

To illustrate this point, let's imagine that we only have access to a thin stripe of some system—a flatlander's view. As we watch this stripe we notice different colors flashing at different locations on the stripe. Every time a flash occurs along our stripe we can record its color, where it is, and when it took place. As we watch these flashes appear and disappear, it won't take us very long to recognize our inability to precisely predict exactly when and where the next flash will occur, or what color it will be.

We may discover a way to make probabilistic claims about the next flash, but we could never deterministically predict a specific event (when, where, or what color). Noticing this, we might think that we are forced to conclude that the system is fundamentally dynamically stochastic (probabilistic). But logically there is room for another conclusion.

Deterministic systems are only guaranteed to appear dynamically deterministic from perspectives that capture their full state. Incomplete access to the state of the system can create the illusion that the system is dynamically stochastic. Therefore, the observation that we are limited to describing the dynamics of a system in a stochastic sense could mean that the system is fundamentally stochastic, or it could mean that there is more to the state of the system than we are accessing (that there are additional variables).[22]

In our example, the stochastically threaded dynamics reflect an incomplete picture. The complete system is a pocketless and frictionless pool table on which many colored balls are bouncing around (Figure 17-1).

"Everything is determined, the beginning as well as the end, by forces over which we have no control. It is determined for the insect as well as for the star. Human beings, vegetables, or cosmic dust, we all dance to a mysterious tune, intoned in the distance by an invisible player."

Albert Einstein[23]

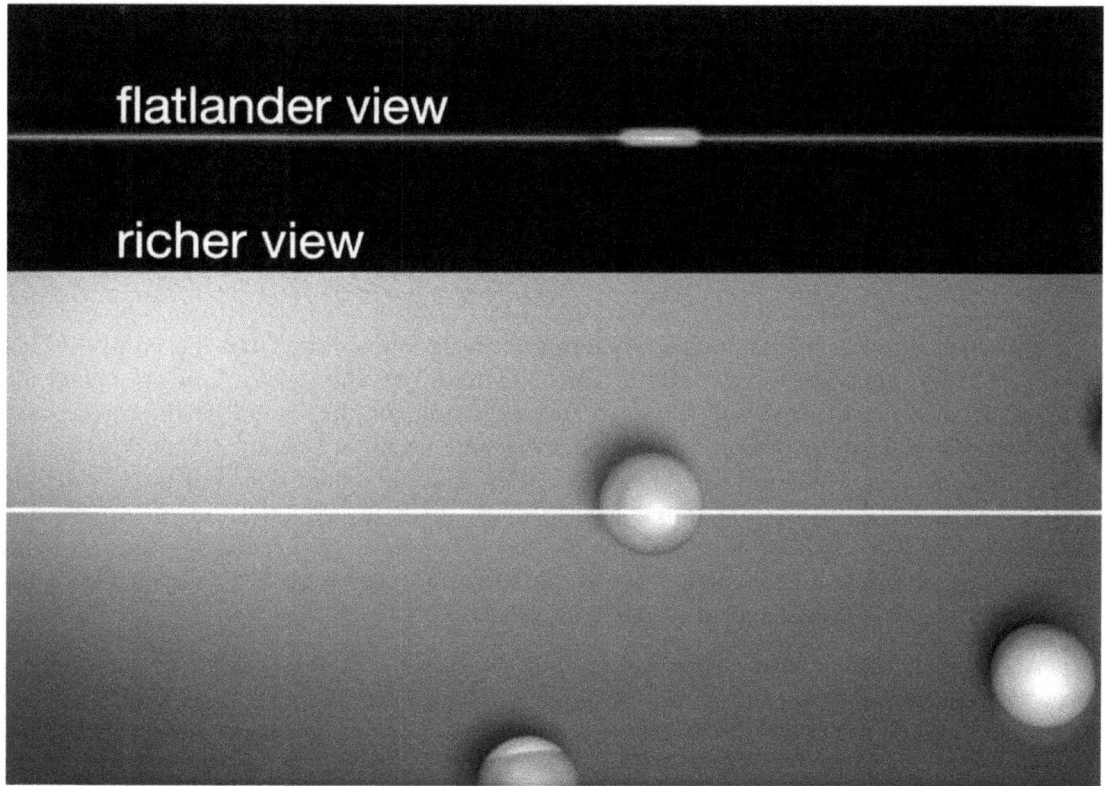

Figure 17-1 Accessing the full dimensions of the system allows us to deterministically model its dynamics

To reinforce this point, let's consider a particulate system—a container of gas. Macroscopically this system will always be found obeying simple deterministic laws that connect its pressure, volume, and temperature as long as it is in a state of equilibrium. However, if the system is not in a state of equilibrium then it can violate these laws. To describe the time evolution of a system that is in a state of nonequilibrium we must appeal to statistical mechanics—making our evolution description probabilistic.

The level of our description forces us into this conundrum. The ultimate explanation of the system's behavior exists on the molecular level, which is not resolved by the macroscopic properties of pressure, volume and temperature. An explanatory barrier is reached because the system's evolution depends upon details that are being ignored by the descriptors of that system.

Notice that even a system that is macroscopically in a state of equilibrium loses stability on microscopic scales. When we consider a system on a scale that allows us to resolve its individual molecules, it can no longer be described as being in *equilibrium*. As we approach this molecular scale, the notions of pressure, volume, and temperature dissolve. On these scales the system becomes dynamically probabilistic, unless we transcend those macroscopic descriptors and describe the system in reference to its underlying microscopic behavior.

The fact that the vacuum appears to lose its deterministic qualities as we approach the quantum scales suggests that on a deeper level there is more structure to the vacuum. To restore determinism, we have to access the dynamics of that underlying, richer geometric structure. This motivates us to admit an additional variable theory.

Interpretations

"The Copenhagen Interpretation is hopelessly incomplete... as well as a philosophic monstrosity."

Hugh Everett III[24]

The most popular interpretation of quantum mechanics sprang from the minds of Niels Bohr, Werner Heisenberg, and John von Neumann in the 1930s, and is called the Copenhagen interpretation, the standard interpretation, or the orthodox interpretation.[25] The Copenhagen interpretation assumes that the state vector is a fundamental descriptor of quantum mechanical systems and, therefore, holds that "a system does not possess definite properties until we, as it were, force it to declare them by carrying out an appropriate measurement."[26] To explain the fact that we observe definite properties, the interpretation admits state vector reduction, describing it "as something to be taken as merely describing the experimenter's 'knowledge' of a quantum system."[27]

Accordingly, this interpretation asserts that the observer's *consciousness* has something to do with the physical state of the universe. Such a conclusion is not only confusing (because it buttresses itself to the individual phenomenal experience), but it is dangerously suggestive of history's most stubborn conviction—the belief that we are the *center of the universe*. Under Nietzsche's warning that "convictions are more dangerous enemies of truth than lies,"[28] it would be an extremely poor practice for any scientist to adopt such a viewpoint. As Roger Penrose puts it, any theory that demands the presence of a conscious observer to explain state vector reduction "leads to a very lopsided (and, I would argue, highly implausible) picture of the universe."[29]

Technically, the Copenhagen interpretation doesn't even qualify as a lopsided, implausible picture of the universe, because it pushes the idea that peering "behind the scenes to see what is really happening"[30] is pointless so long as the statistical, mathematical structure of a formalism correlates with measurement. In other words, the Copenhagen interpretation doesn't just fail to provide a description of reality, it attempts to forbid one.[31] In Leggett's words, it would be more correct to refer to the Copenhagen interpretation as "the Copenhagen non-interpretation, because its whole point is that any attempt to interpret the formalism in intuitive terms is doomed to failure..."[32]

The many worlds interpretation was first put forth by Hugh Everett as his Princeton Ph.D. thesis, and is currently the favored choice of luminaries such as Stephen Hawking. Like the Copenhagen interpretation this interpretation assumes that the state vector is a fundamental descriptor of quantum mechanical systems, but it interprets that claim in a unique way. Instead of assuming that the system's state is a blurred combination of all of the possible states that live in the state vector, the many worlds interpretation assumes that each of those possible states lives in its own universe. Therefore, it takes the state vector to be fundamentally describing "a grand quantum linear superposition of alternative universes."[33] In other words, instead of interpreting the state vector as a generator of an endlessly proliferating number of possibilities, it takes it to be representing "an endlessly proliferating number of different branches of reality."[34]

According to this interpretation, when a measurement takes place, all the alternative outcomes actually coexist in reality, each in a unique universe. This set up transforms state vector reduction into an expression of how the experimenter's awareness state ends up in one of those universes moment to moment. It claims that state vector reduction stems from the experimenter's requirement to have a consistent *awareness state*, which forces the impression that there is just "one world" in which the reduction process appears to take place.[35] In this sense the observer's *consciousness* is taken to *control* the universe's state, but no explanation is given for how this control is acted out.

Other approaches (e.g. environmental decoherence, the consistent-histories approach, relational quantum mechanics, the transactional interpretation, stochastic mechanics, objective collapse theories, the von Neumann/Wigner interpretation, the many minds interpretation, quantum logic, quantum information theories, modal interpretations, time-symmetric theories, branching spacetime theories, the Calogero conjecture, the Semiotic interpretation, the Landé interpretation, quantum Bayesianism, etc.) attempt to regain a sane synchronization with the assumption that the state vector is a fundamental descriptor by crafting up different ways to account for state vector reduction. The problem is that these interpretations also explain state vector reduction in a way that critically relies, in one way or another, on the observer's consciousness. As a result, these "explanations" are epistemically empty.

It is worth noting that if the Universe is not deterministically controlled, if the state vector *is* a fundamental descriptor, then the notion of a *Universe* is incoherent. To assume that a single Universe exists with a specific state, and to assume that it is best represented by the state vector, is to embrace a logical contradiction. If the state vector is a fundamental descriptor, then physicists must swap out the notion of a *Universe* for some sort of *Probable-verse* or *Statistical-verse*. At minimum they become tied to the assertion that future events apply to some sort of Probable-verse or Statistical-verse, while past events belong to a Universe—an entity that has a single, smoothly connected history. This presents an array of logical problems.

Should the present be held responsible for transforming the entire Probable-verse or Statistical-verse into something completely different—a Universe? What mechanism could be responsible for this remarkable transition? Here modern physicists either say nothing or point to *consciousness*. This is a problem. The idea that consciousness is a trigger of state vector reduction has no assertoric force. It is not useful, or meaningful, to claim that an observer's consciousness is responsible for transforming the entire Probable-verse or Statistical-verse into a Universe unless a causal story is told. This sort of quasi-logic is not productive. It tends to confuse us with our own reflection by catering to the distracting goal of self-glorification.

In contrast to the aforementioned interpretations, which all assume that the state vector is a fundamental descriptor, two interpretations go the other route and assume that the state vector is *not* the most complete descriptor of quantum mechanical systems. These interpretations make clear and precise claims about the nature of the state vector, and refrain from relying on the observer's consciousness in any critical way. Therefore, they both warrant our attention.

The ensemble interpretation, also known as the statistical interpretation, assumes that the state vector is a statistical quantity that only applies to an ensemble of similarly prepared systems or particles. In other words, instead of assuming that the state vector is a complete and exhaustive description of the state of the system, it assumes that the state vector represents an *ensemble of systems*, a measurement ensemble, something that comes into existence as many (essentially) identical experiments are set up and performed.

The weakness of this interpretation is that it offers no explanation for why individual systems (like a single photon in a double-slit experiment) behave quantum mechanically instead of classically—why a single photon lands on the screen in one of the places that quantum mechanics says it should, instead of one of the places where classical physics says it should. It only allows the state vector to statistically apply to many similarly prepared experiments. This limitation is inscribed by the assumption that the state vector references a statistical collection of similarly prepared systems, instead of an ensemble that statistically represents the system's possible underlying states.

The other interpretation that departs from the assumption that the state vector is a complete and exhaustive descriptor of quantum mechanical systems is Bohm's interpretation (also called the de Broglie-Bohm theory, the pilot-wave theory, Bohmian mechanics, or the causal interpretation). This interpretation, which has been notoriously misunderstood and misrepresented in the physics literature, is entirely deterministic. Out of all of the previous interpretations of quantum mechanics it carries the most ontological value and assertoric force. It is significant enough to warrant its own chapter (Chapter 24).

Consequences of Determinism

"All things are hidden, obscure and debatable if the cause of the phenomena be unknown, but everything is clear if this cause be known."

Louis Pasteur

Determinism rewrites how we understand the role of conscious "intentions." It helps us get past the tendency to confuse correlation (of conscious self-awareness) with ultimate causation (of our actions). Humans possess a will that is richly determined (by many factors). Even when we recognize a will that sufficiently exists before an action, it is incoherent to assert the will as an ultimate cause for those subsequent actions. (Recall that this is the case if Nature is deterministic or stochastic.) As Spinoza puts it, "In the mind there is no absolute or free will, but the mind is determined by this or that volition, by a cause, which is also determined by another cause, and this again by another, and so on ad infinitum."[36]

Also, if Nature is deterministic, then consciousness logically and naturally supervenes on the physical, arising by virtue of the functional organization of the brain.[37] When all of the physical facts are taken into account, consciousness is entailed by those facts. In other words, it is logically inconsistent to posit a world physically identical to ours that is void of consciousness.[38] If two universes share the exact state at a given moment, then their states evolve identically throughout.

Determinism dictates that every event is intimately connected to the whole, that even chaotic occurrences, such as those that led me to believe that I chose the food I ate for dinner last night,[39] are strict obeyers of the deterministic system we call the universe. In a deterministic Universe physical laws also subsume psychological laws. Once we recognize that it is incoherent to claim that choices can cause subsequent events without being caused themselves, once we recognize that there is an exact state to the Universe, which evolves deterministically, we are motivated to update our perspective.

The insight that "A man can do as he wills, but not will as he wills;"[40] or that we are "psychologically incapable of knowing what is good and not doing it,"[41] can profoundly change how we interact with the world and how we see ourselves. A new perspective may not change our propensities for love, fear, altruism, greed, friendship, empathy, etc., but it can change how we intellectually interact with those propensities, or the feedback loops that stem from those propensities. A deterministic mindset helps us avoid the inflation of our fears, and it tears down our ability to use those fears as rational justifications for blame, resentment, bigotry, racism, jingoism, etc. The recognition that, "A man's life, in all its events great and small, is as necessarily predetermined as are the movements of a clock"[42] offers us a new way of being.

> *"The religious inclination lies in the dim consciousness that dwells in humans that all nature, including the humans in it, is in no way an accidental game, but a work of lawfulness. That there is a fundamental cause of all existence."*
>
> *Aaron Bernstein*

Our biological interface may continuously generate the illusion of free will, but we don't have to be intellectually confined by that illusion. A better understanding of that illusion can give us the ability to use its utilitarian function, while dismissing its negative consequences.

In order to make good decisions we have to balance the seemingly antithetical forces of emotion and rationality. In practice, predicting the future requires us to accurately perceive the present situation, have insight into the minds of others, and deal with uncertainty.[43] The illusion of free will is a consequence of the inherent errors in this process.

The ever-present impulses that convince us that we have free will are rooted in the brain's limbic system where they unify the brain's emotional center and allow decision-making. The interpretive emotions that stem from this process require and reinforce the sensation of autonomy. In other words, when we study the mental processes that lie behind our decision-making we find that our emotional impulses strengthen our convictions—allowing us to escape perpetual indecisiveness. Our emotions are not the enemy of decision-making; they are integral to it. "Our most basic emotions evolved to enable us to make rapid and unconscious choices in situations that threaten our survival."[44]

Decision-making primarily entails predicting the future—something that we are poorly equipped to do. This ineptitude stems from two sources. First, the processing power and storage capacity of our brains is far too weak to take in the complete state of the universe for any given moment. Second, our sensory organs do not access that complete state. They are theoretically limited to sensing (usually with poor resolution) four-dimensional reduced versions of that state. Therefore, we are biologically ill equipped to predict the future precisely, and can only roughly predict events that unfold non-chaotically.

Yet even without the necessary information to accurately predict, our brains are constantly predicting the future and making decisions. How do they do this? The answer is that our emotional impulses serve to heuristically circumvent our lack of information and push us to a decision. The price we pay for gaining the ability to make decisions is that we end up stuck with an emotionally reinforced illusion of free will.

Many modern studies have reinforced this claim. For example neurobiologist Antonio Damasio from the University of Southern California in Los Angeles "studied people with damage to only the emotional parts of their brains, and found that they were crippled by indecision, unable to make even the most basic choices, such as what to wear or eat."[45] There are also a plethora of studies that back up the original findings of Benjamin Libet, a physiologist at the University of California, San Francisco, who found that brain signals associated with motions chosen randomly by the volunteers "occurred half a second before the subject was conscious of deciding to make them."[46]

The order of brain activities turned out to be perception of motion, followed by decision, rather than the other way around. Libet had shown that "the conscious brain was only playing catch-up to what the unconscious brain was already doing. The decision to act was an illusion, the monkey making up a story about what the tiger had already done."[47] With the advantage of contemporary brain scanning technology, modern research has only increased our confidence in this conclusion. In fact, in 2008 scientists were able to predict whether subjects would press a button with their left or right hand up to 10 seconds before the subject became aware of having made the choice.[48]

"Not one man can, consciously, act against his own interest…"

Fyodor Dostoevsky

Today no one would treat their car as foolishly as they treat other human beings. We would never attribute a car's annoying behavior to "bad character" or imagine fixing the problem by punishing the car. Instead we would rationally attempt to find out what caused the car to behave undesirably and set it right. This sort of reaction automatically stems from our understanding that the car's actions are strictly a result of causes.[49]

In a deterministic map every action in the universe is intimately united through cause and effect—including human actions. Determinism encourages us to invest ourselves in reality, and to stop judging others as if any other action could have been the result of the causes involved. It shows us how shallow it is to say that people do things that we don't like because they are *sinners*.[50] In as much as the belief in free will affords us "an outlet for sadism by cloaking cruelty as justice,"[51] determinism deprives us of a way to coherently wash laziness and apathy under the rug with the blood of others. It beseeches us to expand our empathy.

Our moral landscape is significantly dehumanized by a belief in free will. This belief favors those who lack genuine concern for others and, in the long run, it supports the very behaviors it claims to abhor. The more we swing our sledgehammer into the engine of our car the more trouble it will give us. Determinism gives us the option of reacting to the world based on an understanding of interconnectedness. It encourages us to invest ourselves in the task of identifying the causes involved, to gain a rich understanding of Nature's causal structure, and to interact based on this understanding.

The fact that there are no uncaused causes does not mean that there is no human "freedom." Freedom means doing what one wills. It is a description of when one's actions align with one's innermost impulses, or what the philosopher Thomas Hobbes called "the last appetite."[52] Thus the free person is the one whose experiences do not oppose his or her natural volitions. This person lives in harmony with their physical, natural, and complete self.

> *"That which approaches nearest to its nature is nearest to supreme."*
>
> Socrates[53]

Determinism also doesn't mean that humans do not have "wills." It means that our wills are part of the deterministic state of the universe. The state of the universe may align with the projection of our will, or it may not. Either way, our will at present does not lie directly on the causal chain of our actions at present, nor does it play an explanatory role for the state of the universe at present. Nevertheless, because our wills are caused, the will of our past may carry influences that are integrated into the causal chain of the present.

Transcending the illusion of free will means recognizing that we are an intimate part of the universe's total state. It means discovering that we are directly connected to the infinite expanses of dimensional cascades and that we are a vital part of Nature's causal chain. When this perspective is intuitively integrated it becomes a significant contributing factor in shaping both our behavior—our will—and the likelihood of obtaining freedom in our life.

In a deterministic Universe wishing to change one fact about our past, or our present, means wishing to change everything through the infinite expanse of time. On the other hand, to accept a single moment in time is to accept all of time. As Nietzsche put it, "If we affirm one single moment, we thus affirm not only ourselves but all existence. For nothing is self-sufficient, neither in us ourselves nor in things; and if our soul has trembled with happiness and sounded like a harp string just once, all eternity was needed to produce this one event and in this single moment of affirmation all eternity was called good, redeemed, justified, and affirmed."[54] Affirmation of determinism allows us to "become one of those who make things beautiful."[55]

[1] Detlef, Dürr, Sheldon Goldstein, & Nino Zanghí, Quantum Physics Without Quantum Philosophy.

[2] The McGurk effect is a perceptual phenomenon that occurs when our visual inputs do not line up with our auditory inputs. For example, when a person watches a video of a person speaking, but listens to audio that doesn't match the movement of the lips, they often experience hearing something different than the sound they were exposed to. If this combination is recorded, and played over and over, the observer can have a completely different experience depending upon whether or not they just listen, just watch, or do both simultaneously. https://www.youtube.com/watch?v=jtsfidRq2tw

[3] Roger Penrose, *The Road to Reality*, p. 687.

[4] Pierre Laplace. (1820). Essa: Philosophique sur les Probabilités, forming the introduction to his Theorié Analytique des Probabilités. Paris: V Voureier, repr. Translated by F. W. Truscott and F. L. Emory. (1951). A Philosophical Essay on Probabilities. New York: Dover.

[5] Roger Penrose, p. 530.

[6] Tim Maudlin, *Quantum Non-Locality and Relativity*, Second Edition, Blackwell Publishing, p. 196.

[7] Ibid., p. 520.

[8] Ibid., p. 529.

[9] P. A. Schilpp, Ed., *Albert Einstein, Philosopher-Scientist.* Library of Living Philosophers, Evanston, III., 1949, pp. 666, 672; Detlef Dürr, Sheldon Goldstein, & Nino Zanghí, Abstract *Quantum Physics Without Quantum Philosophy.*

[10] J. S. Bell. (1966). On the problem of hidden variables in quantum mechanics. *Rev. Mod. Phys.* **28**, 447–452; reprinted in *Quantum Theory of Measurement,* J. A. Wheeler & W. H. Zurek editors, Princeton University Press (1983), 396–402; and in Chapter 1 of J. S. Bell, *Speakable and Unspeakable in Quantum Mechanics,* Cambridge University Press (1987); second augmented edition (2004), which contains the complete set of J. Bell's articles on quantum mechanics.

[11] S. Kochen & E. P. Specker. (1967). The problem of hidden variables in quantum mechanics. *J. Math. Mech.* **17**, 59–87.

[12] Franck Laloë. *Do We Really Understand Quantum Mechanics?*, p. 114–115.

[13] http://plato.stanford.edu/entries/kochen-specker/

[14] J. S. Bell. (1966). On the problem of hidden variables in quantum mechanics. *Rev. Mod. Phys.* **28**, 447–452; reprinted in *Quantum Theory of Measurement,* J. A. Wheeler & W. H. Zurek editors, Princeton University Press (1983), 396–402; J. S. Bell, *Speakable and Unspeakable in Quantum Mechanics,* Cambridge University Press (1987); second augmented edition (2004), which contains the complete set of J. Bell's articles on quantum mechanics.

[15] D. Bohm & J. Bub. (1966). A proposed solution of the measurement problem in quantum mechanics by a hidden variable theory. *Rev. Mod. Phys.* **38**, 453–469; D. Bohm & J. Bub. (1966). A refutation of the proof by Jauch and Piron that hidden variables can be excluded in quantum mechanics. *Rev. Mod. Phys.* **38**, 470–475.

[16] N. D. Mermin. (1993). Hidden variables and the two theorems of John Bell. Rev. Mod. Phys. 65, 803–815, in particular see § III.

[17] "[T]he assumptions of Kochen and Specker... appear, in fact, to be quite reasonable indeed. However, they are not. The impression that they are arises from a pervasive error, a naïve realism about operators..." Sheldon Goldstein. Bohmian Mechanics, published 10-26-2001; substantive revision 5-19-2006, Stanford Encyclopedia Of Philosophy.

[18] In a variant approach some observable R is chosen to be always-definite and the set of definite observables is expanded to the maximal set that avoids a Kochen-Specker obstruction; J. Bub. (1997). *Interpreting the Quantum World*. Cambridge University Press.

[19] Franck Laloë. *Do We Really Understand Quantum Mechanics?*, p. 37. A large proportion of the physics community still hasn't figured this out.

[20] Albert Einstein in a letter to Max Born, Dec 4, 1926; Walter Isaacson. (2007). *Einstein*, p. 335; AEA 8-180.

[21] J. S. Bell. (1987), p. 160.

[22] In some regards this is the exact opposite of Immanuel Kant's assertion that the *phenomenal world* is determined and the *noumenal world* is free. The world of our phenomenal experiences is codified by the four-dimensional world of our sense inputs. This dimensionally reduced realm is 'free' (at least by a self-referential definition) in the sense that it retains at least some irreducible stochastic character. The richer, eleven dimensional world, however, removes the tenuity that the four dimensional stochastic character relies upon and reveals an underlying deterministic structure.

[23] Viereck, 375; Walter Isaacson. (2007), p. 392.

[24] Peter Byrne. (2007, December). The Many Worlds of Hugh Everett. *Scientific American*, pp. 98–105; Hugh Everett in response to criticisms from Bryce S. DeWitt, editor of the journal: Reviews of Modern Physics, 1957.

[25] This name reflects the dominant influence that Niels Bohr, who was from Copenhagen, had on this school of thought.

[26] A. J. Leggett. (1987). *The Problems of Physics*, Oxford University Press.

[27] Roger Penrose. *The Road To Reality*, p. 806.

[28] Friedrich Nietzsche. Human, All-Too-Human. *The Portable Nietzsche*, trans. Walter Kaufmann. (1968). New York: Penguin, section 483.

[29] Roger Penrose. *The Road To Reality*, p. 806.

[30] Gary Zukav. *The Dancing Wu Li Masters*, p. 79.

[31] Ibid.

[32] A. J. Leggett. (2002). Testing the limits of quantum mechanics: motivation, state of play, prospects. J. *Phys. Condens. Matter* **14**, R415-R451.

[33] Roger Penrose, p. 783–784.

[34] Gary Zukav. *The Dancing Wu Li Masters*, p. 83.

[35] Roger Penrose, p. 784.

[36] Baruch Spinoza, *Ethics*, part 2, proposition 4B.

[37] In a deterministic map, even the phenomenal quality of consciousness is logically supervenient on the physical state of that map. As a consequence, interpretations that heavily rely on a presupposed ambiguity of consciousness are superseded.

[38] This ultimately disallows David Chalmers "conceivability" argument (The Conscious Mind: In Search of a Fundamental Theory). It is a contradiction to claim that something nonphysical can be conceived. To conceive is to conceive physical properties. It cannot mean anything to imagine what the ghost looks like, sounds like, how it interacts with other materials, how it smells, etc., if that ghost is defined as having no physical properties. The conceivability argument is fundamentally flawed.

[39] This example might not be the most appropriate for me at the time of writing given the fact that the circumstances I was under at the time afforded me only the slightest illusions that the food I ate was in any way up to me. However, I assume/hope that the majority of readers will be fortunate enough to relate to the comfortable illusion that they are 'choosing' what food they eat.

[40] Schopenhauer in *Einstein* by Walter Isaacson, p. 391; Einstein. The World As I See It. Einstein 1949 and Einstein 1954. Meno, 776–786, trans. Benjamin Jowett in Plato's Meno: Text and Criticism, ed. Alexander Sesonske and Noel Flemins (Belmont Calif: Wadsworth. 1965), pp. 12–13.

[41] Ibid.

[42] Schopenhauer. *On Ethics*. Parerga and Paralipomena: Short Philosophical Essays. New York: Oxford University Press. (2001), 2:227; Walter Isaacson. *Einstein*, p. 618.

[43] Kate Douglas & Dan Jones. (2007, May 5). How to make better choices. *New Scientist*, p. 35.

[44] Ibid.

[45] Ibid., p. 38.

[46] Dennis Overbye. (2007, January 2). Free Will: Now You Have It, Now You Don't. The New York Times, p. D4.

[47] Ibid. For more on this see: Sukhvinder S. Obhi and Patrick Haggard. (2004, July—August). Free Will and Free Won't, Motor activity in the brain precedes our awareness of the intention to move, so how is it that we perceive control? *American Scientist*, Volume 92.

[48] C. Soon, M. Brass, H. Heinze & J. Haynes. (2008). Unconscious determinants of free decisions in the human brain. *Nature neuroscience* **11** (5): 543–545. Doi:10.1038/nn.2112.

[49] Bertrand Russell. *Why I Am Not a Christian*, p. 40.

[50] The word "sinners" comes to us from the early Roman campaign to force its subjects to be united in one religious structure. They had particular trouble forcing their father-son Moon God (Elohim and Baal) as a replacement of the already in place Moon God "Sin". Rome chose this God out of the desire to attain the same sort of power that their neighboring Jewish tribes possessed. They reasoned that this power was primarily a function of their religious unity. So to promote this unity in their own lands they began to stamp out this rival moon god by calling anyone that believed in Sin a "sinner." This derogatory term labeled someone as punishable for not being part of the group.

[51] Bertrand Russell. *Why I Am Not a Christian*, p. 43.

⁵² See: Thomas Hobbes. (1971). *Leviathan*, Edited by C. B. Macpherson. Baltimore: Penguin Books.

⁵³ Socrates in Xenophon's Memorabilia; Douglas J. Soccio. Archetypes Of Wisdom, p. 94.

⁵⁴ Friedrich Nietzsche. *Will to Power*, no. 1032; Diane Barsam Raymond. *Existentialism And the Philosophical Tradition*, p. 186.

⁵⁵ Friedrich Nietzsche. *The Gay Science*, p. 160.

Chapter 18 — Emergent Reality

"As we look at Nature at levels of greater and greater complexity, we see phenomena emerging that have no counterpart at the simpler levels."

Steven Weinberg

"There is a fundamental disparity between the way we perceive the world, including our own existence in it, and the way things actually are."

The Dalai Lama[1]

In Chapter 1 we talked about the differences between the traditional Greek and Hebraic modes of thought. The Greek tradition assumes that physical reality is built from elemental *things*, or ultimate nouns. By contrast, the Hebraic tradition assumes that ultimate reality is a verb—that all the forms of Nature are products of interaction and are therefore void of any inherent, or permanent noun-like existence. The Greek tradition concludes that without space, physical reality does not exist, and the Hebraic tradition concludes that interaction (and therefore time) is required in order for physical reality to exist.[2] Although both views have merit, the question remains—which one more accurately portrays Nature?

When we process this question through the fractal pattern of vacuum quantization, we arrive at a unique solution. Nature and its map comprise both a verb *and* a noun. This conclusion is not merely a reference to Einstein's marriage of space and time. The issue we are considering is whether or not physical reality is composed of fundamental, elemental, and indivisible constituents. If it is, then those constituents make physical reality fundamentally a noun. But if those constituents are emergent phenomena whose properties, in turn, depend upon the interactions of even lower-level constituents, and those constituents emerge from the interactions of even lower-level constituents, and so on, then Nature is entirely a verb.

The emergence written into our fractal model of vacuum quantization claims that Nature is a verb. But when we attempt to map that verb it necessarily transforms into a noun. This is the case because in order to exhaustively frame a system we must represent it as being composed of a finite number of dimensions. Each order of perspective contains minimum constituents, elemental building blocks, or ultimate nouns, that embody the highest resolution within that map.

"In any system there is always some "protected" level which is unassailable by the rules on other levels..."

Douglas Hofstadter[3]

To increase our resolution we must include dimensions that are internal to the minimum dimensional constituents (the quanta) of our previous perspective. We do this by resolving the internal structure within those constituents—the interactive parts that give rise to the assemblage we thought of as a whole. This increase in resolution pushes the fundamental constituents of the map to an even finer level and increases the dimensions of the map.

With each increase in resolution, some of the system's properties will be reducible to the new lower-level description, while others will not. Higher-level properties that are not reducible—properties that exist on the higher level, but are nowhere to be found on the lower level—are called emergent. They arise not only as a collection of the underlying constituents, but also with a strict dependence on the interactions of those constituents.

As long as our dimensional perspective remains finite, fundamental nouns will construct the base of the model. Higher-level properties will either directly supervene on the properties of the system's foundational constituents, or they will be emergent—meaning they will also depend on the collective interactions of those constituents.

Emergent properties may gain their essence as a verb, but any map we use to conceptualize those properties will necessarily retain a base of elemental nouns (because we cannot exhaustively frame an infinite dimensional framework). This is why the Greek tradition, which is concerned with mapping physical reality, has been forced into the assumption that ultimate reality exists as a noun. This restriction does not apply to the Hebraic tradition because it does not specifically concern itself with map construction. To fully appreciate this interpretation, let's discuss the process of emergence and explore the meaning of supervenience.

Supervenience

There are two ways for a property to supervene on the system's lower-level properties—reducibly and irreducibly. Reducible supervenience is a trivial notion. Properties on one level are derived from a lower level, which already possess that property. For example, spectral absorption of a stellar cloud of hydrogen gas is explained in reference to the spectral absorption of its constituent molecules. By contrast, irreducible supervenience references assemblages—wholes with properties that are not present in their parts.

A classic example of an assemblage is found in a water molecule. Hydrogen and oxygen make up the parts of the water molecule whole. As John Stuart Mill pointed out, independently hydrogen and oxygen have the property of exciting fire, but when they are combined they gain the property of extinguishing fire. This is an example of an emergent property. Mill didn't use the word emergent, but he did start the emergentism movement.[4]

The concept of assemblages was introduced by Gilles Deleuze to counter the Hegelian idea that parts of the whole are fused into a seamless totality. Hagel's totalities were irreducible and nondecomposable. By contrast, Deleuze's assemblages are irreducible and decomposable.

The identity of an assemblage is retained in reference to the interactions of its parts. These interactions exercise capacities and give rise to actualized properties. If the interactions change, the manifest properties change. Without these interactions the whole would be just a collection of things—not more than the sum of its parts, and the properties of the whole would reducibly supervene on the properties of its parts.

When parts interact with one another in such a way that they yield a whole with properties of its own—properties not found in the individual parts—that whole is an assemblage. The parts of an assemblage can be detached and swapped, without changing the identity of the assemblage, so long as the interactions between parts is not altered.

The map we have been considering depicts the vacuum as composed of subsystems (the quanta), which by first approximation evolve independently, and interact elastically. These quanta all have identical capacities. Between interactions each subsystem "quivers unto itself" acting like a harmonic oscillator, which suggests that the evolution in each domain will be periodic.[5] This picture leads us to the expectation that assemblages in the vacuum are decomposable—quanta from one assemblage can be swapped out for another without changing the properties of the assemblage.

If we describe an assemblage from a perspective that does not resolve its parts or the interactions of its parts, that description will manifest information loss. Most notably, that description will suffer from an inability to describe how structures in that system *self-organize* or *come into existence*. To avoid this problem we have tasked ourselves with describing the system from a perspective that resolves the parts of the system and their interactions. The most ideal description of reality is one that portrays a wholly invertible map—a cascading composite of interacting subsystems that are always folding, unfolding and differentiating.

Complexity theory explains how properties and structures arise out of the spontaneous self-organization of smaller things. "This occurs, according to the theory, not only in spite of a chaotic universe and the lack of a central planner, but because of these conditions."[6] Applying the insights of complexity theory to vacuum quantization allows us to describe how a natural system, which is composed of interacting subsystems, self-organizes into structures and behaviors without a central planner. The structure and behavior enlaced throughout physical reality supervenes on the cascading dimensional foundations that underlie it.

To understand the richness of this insight, let's take a look at a familiar example of a complex adaptive system—an ant colony.[7] Ant colonies are elaborately structured societies that have dumps for refuse and even cemeteries for dead ants. Lines of workers busily carry food into the colony and trash/corpses out. They build structured habitats, organize highways, and even stage epic raids. Some of this organization is sophisticated enough to suggest "intentional order" or "intelligent planning". For example, the distances between anthill, cemetery, and refuse dump are always maximized—and this reasonable community desire is a feat that takes a complex mathematical calculation for a planner to achieve. So how is it that the ants achieve this without central organization?

"Ants aren't smart. Ant colonies are."

Deborah M. Gordon[8]

When we watch an ant marching along, we might think that it knows where it is going, or that it has a plan, but it doesn't. Individually, ants aren't clever little engineers or architects. In fact, most ants don't have a clue when it comes to deciding what to do next. Yet from countless interactions between these individual dummies, each of which is following simple rules of thumb, intelligence arises on the collective level.

The colony's structure has no top-down central planner; it self-organizes from the bottom up. The ants are members of a complex adaptive system. They self-organize into larger-scale emergent structures and behaviors, such as food lines and hill builders, which are ever shifting and adapting to changes in environmental conditions, allowing the colony to survive through many generations of ants. Complexity emerges from these simple rules. This is called swarm intelligence.

The absence of overt planning for this kind of organization has been confirmed by computer models of individual "virtual ants", which spontaneously self-organize, and create "virtual ant colonies" with all the same elaborate structures of natural ant colonies. The computer programmers do not program any organization of the virtual ant colony, only the behaviors of the individual virtual ants, which then self-organize just like real ants do.

Ants are only one example of this emergent self-organization. The same sort of phenomena can be found throughout the natural world. Flocks of starlings, swarms of honeybees, schools of Bigeye Jack, herds of wildebeests, clouds of locusts, and synchronized flashes of fireflies (which resemble the synchronized firing of the cells in our heart muscle), all exhibit patterns and structures that emerge in response to self-organization.

Whenever interacting individuals of any size (molecules, cells, individual animals and plants, social groups, cultures, or quanta) fulfill certain simple criteria, self-organizational emergent behavior arises. In his article From the Bottom Up, Neil Theise explains these criteria as:

1—There must be numerous individuals, and it matters how many there are. Colonies of different sizes show different organization, just as a village is not a city.

2—The individuals must interact with each other and with their environment by negative feedback loops, like a thermostat that turns off the heat when a room gets too warm. There can be positive feedback loops too—imagine that the warmer a room gets the more the heat gets turned on, but if these aren't balanced by sufficient negative feedback loops, then self-organization may start to occur, but it won't work: it will be energy-expending and nonadaptive, quickly burning out the system—think hurricanes, tornadoes, or cancer.

3—The individuals must directly respond to their local environments without monitoring the group as a whole. There is no single individual monitoring the whole group/colony. Rather, each individual is merely responding to what it senses in its immediate surrounding: food, water, dirt, other ants, and so on.

4—There must be a small degree of randomness in the system, often referred to as quenched disorder.

Within determinism, there are two ways to get quenched disorder in a system. The first is for the system to be affected by inputs that are *external* to the system, and the second is for the system to be affected by inputs that are *internal* to the system. Systems are usually defined as closed or isolated, but in practice this is not always the case. External inputs can affect the system and cause it to evolve unexpectedly. For example, an ant colony may experience environmental changes that cause its collective form to change. If we were modeling only the colony, if we were not privy to those external factors, these changes would appear to be random.

Quenched disorder can also come about in a more subtle way. As we noted earlier, modeling a system has the side effect of ascribing a fundamental noun at the base of that system, a lowest-level constituent. The problem with this is that it cuts off part of the internal structure of the system. Therefore, the modeled system will slightly differ from the actual system, introducing a small degree of randomness in its evolution.

If the vacuum is quantized, and its quanta are quantized, and so on, then quenched disorder will inevitably be a characteristic of any description of Nature that has a finite number of dimensions. Because all maps must frame a finite number of dimensions, it will always remain possible for inputs that stem from the ignored dimensions to affect the system. As dimensional resolution increases, fundamental nouns are revealed as emergent phenomena. These bottom-up constructs automatically arise from simple internal interactive parameters, which are, in turn, directly determined by the parameters of their internal dimensional constituents, and so on.

This bottom-up explanation of complexity theory applies to all the *things* in Nature—from quantum patterns, atoms, molecules, rocks, life forms, planets, solar systems, to the cosmos on a grand scale, and even to that which extends beyond the borders of our universe.

The Emergence of Structure

"If you want to understand function, study structure."

Francis Crick[9]

As an example of the explanatory power complexity theory contributes to the process of emergence, consider the formation process of snowflakes. The basic constituent of structure for this system is, of course, the H_2O molecule. Because we are interested in discovering how complexity arises in the natural realm, without intentional orchestration, we shall consider whether or not we can end up with the complex patterns and breathtaking beauty that we find in snowflakes, simply by defining a simple set of interactive parameters for the basic constituents of the system—the H_2O molecules. This is a perfect example of bottom-up construction because it helps us grasp the emergence of the enormous complexity we see in the real world from simple compressed properties responding to an assortment of interactions.

One of the first things we notice about snowflake formation is that, although it is commonly said that "no two snowflakes are exactly alike", there is a dramatic resemblance between all snowflakes that form under the same conditions. The most crucial parameter for snowflake formation is temperature. Temperature dictates the crystalline structure of forming snowflakes. Between -1 °C and -3 °C, snowflakes crystallize as plates and dendrites. At about -5 °C, needles and hollow columns appear. At even colder temperatures, the flake design returns to dendrites and plates. Flakes that start forming at around -5 °C begin to construct columns, but if they encounter cooler or warmer temperatures after this process is under way, then the crystals end up as capped columns—part column, part plate[10] (Figure 18-1).

To understand bottom-up explanation for the emergent snowflake complexity, imagine a hexagonal grid representing the possible geometric arrangements that the H_2O molecules can have relative to one another. Now imagine that a large number of H_2O molecules are randomly dispersed throughout this grid space—a condition that we can represent by coloring in a scattered assortment of locations on our grid. Next, we allow each molecule to either move to one of the adjacent grid spaces, or stay put, but in order to eliminate the possibility of a *designer* we require random motions from each molecule.

Figure 18-1 Different environmental conditions lead to different snowflake structures

This set up reproduces "chaotic" origins. From here, the emergence of snowflake structure, requires only that we assign some simple rules of interaction. With the right rules, the H_2O molecules will construct the complex symmetrical snowflakes we find in Nature.

Because we know these properties must be temperature dependent, our first rule will dictate that above 0 °C all randomly colliding molecules of H_2O do not stick together. However, when the temperature drops so will that restriction. Because temperature is equivalent to thermal agitation, it is really only the manifestation of an inherent property that changes with temperature, not the property itself. This distinction will not significantly impact our discussion.

Let's say that when the temperature drops just below 0 °C, the molecules of H_2O begin to stick, but only when they collide in a very specific way. When a molecule collides into another molecule it will stick only if there happens to be an odd number of molecules adjacent to it in the moment of its collision. (This rule can fill in for the motivator of covalent bonding.)

When we allow this system to evolve, we end up with the formation of dendritic shapes—six fold symmetric snowflakes growing all throughout the grid (Figure 18-2). Because the arrival of each molecule is random, each structure is unique in its form. However, the emergent structure of all the possible snowflakes shares the same overall symmetry and apparent "design".

Figure 18-2 Snowflakes randomly generated from a hexagonal cellular automaton: Silvia Hao

To utilize this process to explain all the various forms of snowflakes found in Nature, we need to extend our grid to a three dimensional one and allow a variance in the tendency for the molecules to stick as the temperature (thermal agitation) changes. What we end up with is an enormous array of complex forms emerging from nothing other than the parameters that control the interactions between the base constituents of the system.

Note that, although we set the constituents of the system to arrive randomly, the reaction to that arrival is not random. The inherent properties of the constituents are responsible for authoring the intricate and highly ordered forms that they construct. The claim here is that this condition is ubiquitous. It applies, to the constituents of every substrate that is composed of many repeating parts that interact. Most importantly, it applies to the vacuum quanta.

The compression of these properties can explode into the complexity of forms that decorate the canvas all around and within us, using entropy to drive the wheels of evolution, in an always-differentiating process, leading to an origami cosmos—folding, unfolding, and refolding.[11]

For another demonstration of how simple parameters can lead to the complex patterns found even in living things, we start with a particular collection of simple affine transformations,[12] iterate them randomly, and end up with a unique fractal figure. Michael Barnsley of the Georgia Institute of Technology and his colleagues used computers to follow this process, and were able to create remarkable 3D renderings of patterns that are biologically produced in Nature. One of these renderings created a graceful model of a black spleenwort fern (Figure 18-3). This pattern is the result of a simple collage application of four affine transformations—each a combination of a translation, a rotation, and a contraction.[13]

This demonstrates that it is possible to reproduce the intricate design of a spleenwort fern (and many other varieties of fern) simply by assigning a few basic interactive parameters to the building blocks of a particular substrate. (Bifurcation patterns in trees can be similarly described.) When it comes to the opulence of complexity that is supported by the bottom-up explanation of complexity theory, these elaborate patterns only scratch the surface of possibilities.

Figure 18-3 Reproducing a black spleenwort fern

Not all systems possess the same potential for complexity through self-assembly via the bottom-up process. In addition to the four qualifiers we listed earlier, a system's ability to increase in complexity depends upon its "energy rate density". When we look at a system's energy flow in relation to its mass, we find a real and impressive trend of increasing energy per time per mass for all ordered systems.[14] More complex structures have higher energy rate densities. Stable structures with higher complexity also result in higher contributions to the universe's entropy (per unit of mass). This suggests that entropy is the engine of complexity and, when it comes to life forms, the driving force of evolution.

> *"Life forms process more energy per unit mass than any star, and increasingly as they evolve... the greater the complexity of a system, the greater the flow of energy density through that system."*
>
> Eric Chaisson

In this sense, life forms represent the acme of efficiency of the entropy engine. As highly ordered systems they do more per unit mass toward increasing the entropy (disorder) of the universe than all other systems. This, of course, is how they were "selected". Their ability to command energy resources without burning out—"not so much energy as to be destructive and not so little as to be ineffective" drives their formation. The entropy engine selects for its own increase.

Interaction itself may find its best description as random (at least in four dimensions), but the result of that interaction is not random. When these processes favor an increase in the energy rate density of the entire system, structure and complexity will emerge from chaos. In Eric Chaisson's words, "in an expanding non-equilibrated universe, it is "free" energy that drives order to emerge from chaos."[15]

Bottom-up Structure

> *"We do not live in the universe, we are the universe, arising directly from its substance in an endless revivifying self-organization, from the smallest to the largest. Within unity, there is differentiation, the absolute and the relative."*
>
> Neil Theise

Until fairly recently, astronomers were working under the assumption that top-down scenarios dictated large-scale structure formation. They had assumed that large-scale structure somehow accounted for all other scales of structure. But recent observations of the universe's large-scale structure have conclusively ruled out top-down organization as the process responsible for the formations found in Nature. Scientists now believe that small structures self-assembled to form larger objects over time—slowly constructing the organization we see today.[16]

One of the major occupations of our scientific quest is to reductively explain the phenomena in Nature—to explain high-level phenomena in terms of underlying properties. If the bottom-up scenario applies to all the various-level forms in Nature, then the intricate complexity and astounding beauty in Nature supervenes on the more simplistic, foundational parameters of physical reality. Biological phenomena are explained in terms of cellular phenomena, which in turn can be explained in terms of biochemical phenomena, which are explainable in terms of chemical phenomena, which are explainable in terms of physical phenomena.[17]

To assume that these physical phenomena are brute facts is to assume that they cannot be reductively explained. If the geometric structure of the vacuum is self-referential, if it conforms to a fractal, then there is no bottom level to explanation, and symmetries define the infinities.

The reductive ladder of explanation is infinitely extended through these symmetries. Consequently, the supposed physically "brute" phenomena (e.g. the constants of Nature) can be explained by the more basic parameters of the vacuum quanta. These quanta become emergent phenomena of the collectively interacting subquanta that compose them, which in turn are emergent phenomena of their internal constituents. At every level we continue our reductive ladder of explanation, but as we transcend dimensional boundaries the explanations repeat symmetrically.

> *"The very essence of being a part requires a whole to be part of, and there can be no whole without parts."*
>
> Douglas J. Soccio

When we start with a framework composed of subsystems, or parts, that interact to self-organize into complex adaptive systems, or assemblages, we come upon the requirement that several of those complex adaptive systems have the potential to come together to form other emergent systems that gain definition on scales of different resolution. Without a mechanism to impose dimensional restriction, this process continues indefinitely—extending the depths of supervenience into the cascades of infinity and giving us a looking glass into the whole.

Within an eleven dimensional description, the vacuum quanta self-organize—forming things like localized plane waves, which represent photons, and superfluid vortices, which represent elementary particles. These photons and elementary particles then interact to organize into higher orders of emergent phenomena: atoms, molecules, the cells of an embryo, postnatal life, tissues, organs, bodies, social groups, societies, species, and on and on. At all levels (except the one selected as foundational by the specific dimensional truncation we select) inherent existence—the noun quality—does not belong to the object being described.

In general, complex systems depend on hierarchical levels. To parse this a little more let's return to our consideration of the ant colony. If we observe a line of ants from a distance, we might notice that it looks like a unified thing, an entity, a dark shape shifting in the sand. However when we look closer we discover that the line of ants is not a unified thing at all, but rather a multitude of individual ants organizing themselves in space and time. Zooming in further, we find that the individual ant now appears to be the concrete *thing*. Then, as we look even closer we find that the individual ant's body as a thing disappears. It becomes the product of the self-organization of its component cells. On this scale, cells appear to be the concrete things. The pattern continues.

In general, "What we take as the essence of an individual thing, be it an ant, person, or planet, is nothing more than the emergent self-organization of smaller things."[18] This means that "Inherent existence as a thing, rather than as an organizational phenomenon of smaller things, depends on the scale of observation."[19]

The cascading quantized structure of the vacuum enhances the reach of emergent self-organization. It asserts that at any given scale, higher-order phenomena can be said to supervene on lower-level properties. From this we gain a reductive explanation that is not truncated—which gives us deeper conceptual access to the complexities of physical reality.

Eliminating Illogical Infinities

"Heard Melodies are sweet, but those unheard Are sweeter."

John Keats[20]

"...we long to make music that will melt the stars."

Gustave Flaubert[21]

Vacuum quantization allows us to connect to the process of emergence and to poetically capture the complex structure that constantly emerges from the underlying substrate. It also naturally eliminates illogical infinities, while avoiding an overwhelming increase of functional freedom from additional dimensions.

It has been said that string theory gains its strongest motivation from the desire to eliminate the infinities of quantum field theory.[22] Nevertheless, this condition has not been met within string theory. Roger Penrose explains that this shortfall stems from the "enormous increase in functional freedom in higher-dimensional theories $\left\{ \infty^{P^{\infty^M}} \infty^{P^{\infty^M}} \right\}$, for a (1+M)-dimensional spacetime, where one has to envisage some means of freezing out this extra freedom."[23] In short, a functional freedom of $\left\{ \infty^{M^{\infty^3}} \infty^{M^{\infty^3}} \right\}$ would be acceptable because it would align with the functional freedom we observe in Nature. This value assumes that no object (emergent phenomenon) can move about in more than three spatial dimensions within a specified (fixed) perspective.

Dr. Penrose notes that we cannot assume such a restriction in a higher-dimensional theory simply because we desire our mechanics to match observation. In a quantized vacuum this problem is naturally overcome because a mechanism for "freezing out this extra freedom" is implicitly ingrained within its geometric structure. That mechanism is simply a function of dimensional resolution, which means that it has something in common with the process of emergent phenomena.

To develop this idea further, let's consider an object, which is given as an assumed rigid entity—like our previous line of ants. As we examine this line of ants we might notice that the object's dimensional motion becomes restricted because it retains its definition as a rigid object only on a very specific dimensional perspective. How many ways can a line of ants move? Not the individual ants, just the line.

Let's reduce this even further. Imagine that our object of interest is a single quantum. This object is incapable of moving through intraspatial dimensions—because those dimensions are defined on a scale that is enveloped by the single quantum. It is also impossible for a single quantum to move through spatial dimensions, because if it is by itself, then there are no spatial dimensions to move through.

Therefore, on the scale that we can most directly define a single quantum, we discover that it can only move about through the three superspatial dimensions. These dimensions are the very dimensions that allow us to define the quantum as an *object*. In this sense, functional freedom becomes restricted because of the inherent hierarchical structure of the vacuum's geometry. This separation of dimensional function is the mechanism that forces our functional freedom to remain of the order of $\left\{\infty^{M^{\infty^3}} \infty^{M^{\infty^3}}\right\}$ although our map is composed of M spatial dimensions.

Therefore, according to vacuum quantization, Nature and all its forms are endless cascades of emergent phenomena—self-organizing entities that build from the smallest to the largest—being selected upon by the engine of entropy. The finite and the infinite are harmoniously connected. When we map it with a finite dimensional perspective it becomes a noun. When we don't, it remains an endlessly cascading verb.

This insight unites the pillars of physics (general relativity, and quantum mechanics) to the third pillar of enlightened human thought—evolution via natural selection. As Darwin so famously wrote in the closing lines of *The Origin of Species*, "There is a grandeur in this view… from so simple a beginning endless forms most beautiful and most wonderful have been, and are being evolved."[24] The constant mixing of the cosmos (of the superfluid vacuum) spurs on that evolution and gives birth to the emergent beauty that we see all around us.

[1] *NewScientist*, (2006, January). The Om of Physics, pp. 46–47.

[2] For an in-depth analysis of this division of thought see: Thorlief Roman. (1970). *Hebrew Thought Compared with Greek*. New York: W. W. Norton & Company.

[3] Douglas R. Hofstadter. *Gödel, Escher, Bach: An Eternal Golden Braid*, p. 688.

[4] Manuel DeLanda. (2011). Assemblage Theory, Society, and Deleuze.

[5] Gerard 't Hooft. *The Mathematical Basis for Deterministic Quantum Mechanics*.

[6] Neil Theise M.D. (2006, Summer). From the Bottom Up. *Tricycle*, p. 24.

[7] This example is borrowed from Neil Theise's discussion in *From the Bottom Up*, and Peter Miller's *Swarm Theory*.

[8] Deborah M. Gordon is a biologist at Stanford University who studies red harvester ants (*Pogonomyrmex barbatus*). Reported by Peter Miller. (2007, July). Swarm Theory, *National Geographic*, p. 130.

[9] Francis Crick. *Discover, Science Almanac*, p. 346.

[10] Michael Klesius. The Mystery of Snowflakes. *National Geographic*. See also SnowCrystals.com, or http://www.its.caltech.edu/~atomic/snowcrystals/class/class.htm

[11] Gilles Deleuze. *A Thousand Plateaus*.

[12] An affine transformation behaves somewhat like a drafting machine that takes in a drawing—that is, takes in the coordinates for all the points making up the lines in the drawing—then shrinks, enlarges, shifts, rotates, or skews the picture and finally spews out a distorted version of the original. Ivars Peterson. *The Mathematical Tourist*, p. 128.

[13] Ivars Peterson. *The Mathematical Tourist*, pp. 130–131.

[14] Eric Chaisson. (2006, January 7). The Great Unifier. *NewScientist*.

[15] This quote and the one in the previous paragraph are also from Eric Chaisson.

[16] James Trefil. (2006, July). Where is the Universe Heading. *Astronomy*, p. 38. See also: G. Bertone, D. Hooper, J. Silk (2005). Particle dark matter: Evidence, candidates and constraints. Physics Reports **405** (5–6): 279–390. arXiv:hep-ph/0002126

[17] David J. Chalmers. *The Conscious Mind*, p. 51.

[18] Neil Theise M.D. (2006, Summer). From the Bottom Up, p. 26 Emphasis added.

[19] Ibid.

[20] John Keats. (1820). *Ode on a Grecian Urn*.

[21] Gustave Flaubert. (1857). *Madame Bovary*.

[22] Roger Penrose, p. 884.

[23] Roger Penrose, p. 923.

[24] Charles Darwin. *The Origin of Species*, closing remarks.

Chapter 19 — **The Hierarchy Problem**

"It could well be that there's some big piece of reality that we don't fully understand."

Christopher Stubbs[1]

The officer's keys rhythmically clanked as he casually made his way up to our level.

"Hey, CO... CO... What time is it?" my cellmate yelled through the bars as he passed.

The officer responded without even turning his head, "Why, you got somewhere to be?" He chuckled to himself as he continued his swagger.

My cellmate approached the bars and yelled down the walk, "Yeah I got a hot date, and I don't want to be late." The officer continued on his way and yelled back, "Four eleven."

Near the top of our tall, slender window a small circular hole let in a beam of light through the paint that covered the exterior side of the glass. With my stubby golf pencil I circled the spot where the beam ended on the floor, and recorded the time next to it.

"Weak man, that mutha-fucka is weak," my cellmate said.

"Yeah, but he holds the keys," I said. He continued to watch the officer make his rounds.

"You see how he walks around here, never looking anyone in the eye, pretending he don't know what this place is?"

I looked up and said, "If it weren't for him we wouldn't have this." I pointed to the sundial that was being pieced together on our floor. He shrugged me off. The rest of our cellmates were lying in their beds ignoring us both.

After a long pause I looked up at him and said, "Would you like to know about a mystery in science in which the weakest of them all holds the key?"

Probing the Hierarchy Problem

"Why is the strength of gravity acting on elementary particles so weak?"

Lisa Randall

The enormous difference in strength between gravity and, say electromagnetism, can be illustrated by the fact that it takes the entire weight of a star to overcome the electromagnetic repulsion between the atoms in the center of that star. How do we explain this vast division in strength? The answer to that question is expected to be both profound and insightful because it is the mismatch between gravity and the other forces that divides our description of physical reality into two incompatible parts.

Despite the fact that particle physicists have devoted decades of intense research to solving the hierarchy problem, the question of how the feebleness of gravity interlocks with the rest of the picture remains a mystery. The standard model of particle physics makes it easy to treat all forces as the result of an interchange of *force particles*. With regard to the electromagnetic, weak, and strong nuclear forces, all of our experiments have shown an absolutely stunning alignment with this theoretical depiction. This alignment becomes the supporting foundation for an underlying symmetry in Nature because it links the strengths of these forces into a relatively tight range and unifies the source of their origination and the proposed mechanics responsible for them.

All of this is aesthetically beautiful and pleasing, except for the fact that we have a rather serious upset when we attempt to compute the strength of gravity through the same model. Paradoxically, when we treat gravity like we treat the other forces—as a similar exchange of some kind of force particle—we find that the standard model clusters gravity's expected strength in range with the other known forces. It predicts that the symmetry underlying the other forces should also belong to gravity, and it spits out a value for the strength of gravity that is astronomically different from what we observe it to be.

> *"While the other forces of the universe (electromagnetism, the weak nuclear force, and the strong nuclear force) are roughly all the same strength, gravity is wildly different."*
>
> Michio Kaku

Comparing gravity's actual strength to the standard model's theoretical prediction of its strength, we end up with a discrepancy that spans sixteen orders of magnitude. This is a serious problem. Such an enormous misalignment suggests that the standard model of particle physics is still missing something big.

Over the years, two popular approaches have attempted to make sense of this enormous discrepancy. The first approach assumes that gravity does in fact belong clustered with the other forces in symmetry and strength—that the true strength of gravity is as the standard model predicts.[2] To account for the feebleness of gravity that is observed, this approach then makes the claim that gravity undergoes an enormous dilution by way of additional dimensions. In other words, gravity is attenuated, which means that its strength is primarily dispersed elsewhere.

In order to make this approach work, theorists have been forced to assume two critical conditions. First, in order to sufficiently dilute gravity, the extra dimensions have to be very large, or very many. Second, gravity must be the only thing that is capable of being diluted throughout these extra dimensions. This assumption ensures that "everything that doesn't involve gravity would look exactly the same as it would without extra dimensions, even if the extra dimensions were extremely large."[3]

The problem with this approach is that without a framework by which to uniquely select a specific number of extra dimensions, or to explain why gravity is the only thing that becomes diluted, these conditions introduce mysteries that are just as big as the one we set out to explain. Therefore, these assumptions merely reword the hierarchy problem.

> *"Their models don't actually solve the hierarchy problem because you still have to explain why the dimensions are so large."*
>
> Lisa Randall

Nevertheless, this idea posits an interesting prediction. It says that deviations from Newton's law of gravity should exist on distances that depend upon the size of those extra dimensions, which is correlated to the total number of extra dimensions that gravity is diluted through. "If there were only one large extra dimension, it would have to be ... as large as the distance from the Earth to the Sun in order to dilute gravity enough. That's not allowed... If there were just two additional dimensions, they could be as small as a millimeter and still adequately dilute gravity... With more additional dimensions, it can be sufficiently diluted even if those extra dimensions are relatively small. For example, with six extra dimensions the size need only be about *10^{-13}* centimeter, one ten thousandth of a billionth of a centimeter."[4]

To date, gravity's alignment with Newton's inverse square law has not been tested on a scale capable of ruling out, or supporting, this prediction.[5] Because of this, supporters of this approach for solving the hierarchy problem hope that more accurate measurements will one day discover deviations on scales smaller than a millimeter and vindicate the idea. Any such evidence would be interesting, but wouldn't bring us the full ontological clarity we are after.

We are trying to gain an explanation for why things are as they are. We are after an intuitively accessible picture of physical reality that enables us to causally understand all of the things we now consider mysterious. To understand those mysteries we need to explain the mechanism that gives rise to them and the geometry that supports them. Positing the existence of additional dimensions, without constructing an intuitively accessible model of those dimensions, leads us to a very shallow reward. It may be a good start, but it doesn't carry us far enough.

The second popular approach for solving the hierarchy problem also assumes that the standard model's treatment of forces (being created by the interchange of force particles) applies identically to gravity, but it attempts to account for the feebleness of gravity by suggesting that the force particles responsible for gravity somehow have unique properties that must effectively weaken its strength. Because the particles that are imagined responsible for this, called gravitons, have thus far escaped all attempts to measure them, there has not been much progress made on this front.

Both of these attempts are trying to treat gravity as though it were fundamentally the same as the other known forces, despite the fact that in the physical world gravity manifests itself as characteristically different. The motivation behind this comes from the desire to uncover deeper symmetries hidden in Nature and to use those symmetries to enhance our grasp of the natural realm. But what if there is a simpler way to unite the four forces? What if they are connected by a different kind of symmetry?

The assumption that the vacuum is a superfluid could be the key to unification. If every force corresponds to a way in which the natural geometry differs from Euclidean geometry, then gravity can be understood to be unique among those differences because it is the only one that comes into focus macroscopically. That is, gravity is specifically offset from the other three forces because it arises as a "small-amplitude collective excitation mode of the non-relativistic background condensate."[6] In other words, it represents how the density of the vacuum slowly changes from one region to another, which necessitates a smooth representation that is only accurate in the low-energy, low-momentum regime.

To understand why an accurate description of gravity is restricted to the low-energy, low-momentum regime, it is useful to be aware of the fact that fluid mechanics is an emergent consequent of molecular dynamics within its low-energy, low-momentum limit. In other words, fluid mechanics is not a fundamental descriptor of any of the systems we apply it to. Those systems are actually driven by an underlying microphysics. Fluid mechanics exists only as an emergent approximation of the low-energy and low-momentum regime of the molecular dynamics that drive the system's evolution.

Likewise, a velocity field (a vector field) and a derivative density field (a scalar field), which the Euler and continuity equations critically depend upon, do not exist on the microscopic level. They are emergent properties that are only resolved on scales larger than the mean free path and the mean free time.

If the vacuum is a superfluid, whose metric is macroscopically describable by a state vector (a velocity vector field), then the density gradient of that fluid is an emergent approximation of the system instead of a fundamental descriptor. The cohesion of that approximation requires macroscopic scales, and molecular dynamics that are defined within the low-energy, low-momentum regime. Gravity becomes an expectation because, if the vacuum is a superfluid, if it can be modeled as an acoustic metric, then small fluctuations in that superfluid will obey Lorentz symmetry even though the superfluid itself is non-relativistic.[7]

The assumption of vacuum superfluidity fully reproduces expectations of compressibility (the ability for the metric to curve or warp), while projecting an *internal* velocity restriction. It also sets up an expectation of acoustic horizons, which turn out to be analogous to event horizons with the notable difference that they allow for certain physical effects to propagate back across the horizon, which might be analogous to, or responsible for, Hawking radiation. Therefore, if the vacuum is a superfluid, then gravity can be viewed as a macroscopic emergent expression, a collective property of the vacuum that supports long-range deformations in the density field.[8] This small-amplitude characteristic is responsible for the feebleness of gravity.

The strength of a *force* reflects the degree to which the geometric properties that author it contrast with Euclidean projections. Gravity is the weakest force because it only comes into focus on macroscopic scales, and therefore only slightly deviates from Euclidean expectations. The strong nuclear force, electromagnetism, and the weak nuclear force, are much stronger because they are all authored by geometric characteristics that deviate from Euclidean projections on even microscopic scales.

Another way to put this is to say that metric distortions that qualify as gravity fields are inherently incapable of directly accessing the degrees of freedom that belong to the underlying molecular dynamics that drive the system. The metric distortion that leads to gravitational phenomena is capable of existing statically—the density gradient it represents is blind to the molecular dynamics that give rise to it—while the strong force, electromagnetism, and the weak force, are strictly sustained dynamically—they explicitly reference the underlying molecular dynamics. The magnitude of gravity (the degree to which this geometric distortion differs from the static Euclidean space) is, therefore, comparatively diluted. This is a consequence of the average-over process that gives rise to its geometry.

Therefore, in as much as we consider underlying molecular dynamics to be an explanation of fluid mechanics (on low-energy and low-momentum scales), the assumption that the vacuum is a superfluid comes with a natural explanation for why gravity is so feeble compared to the other forces.

Geometric Unification

"Not I but the world says it: all is one."

Heraclitus

Setting the strength of the strong force at unity, the respective interaction strengths of the four forces are:

Strong	$= 1$
Electromagnetic	$\sim \frac{1}{137}$
Weak	$\sim 10^{-6}$
Gravitational	$\sim 10^{-39}$

If the vacuum is a superfluid, then this succession can be explained as follows. The strong nuclear force references the propensity for connectivity distortions, lattice waves that ripple through the medium, to couple together—forming stable quantum vortices. In acoustic metric terminology, these metric waves are called phonons. Once quantum vortices are formed, this coupling mechanism also becomes responsible for binding them together (when they are in phase with each other), forming vortex composites—such as protons, neutrons, and nuclei. Quantum vortices represent the most severe departure from Euclidean geometric expectations because they entail a geometry that actively twists, turns and folds back on itself.

Electromagnetism expresses the presence of divergence or curl in the vacuum flux. More specifically, electric fields map to divergent flow in the vacuum, sources or sinks for fluid flow and magnetic fields map to the circulation density of the superfluid vacuum, otherwise known as the curl. These flow descriptors notably depart from Euclidean expectations, but not as severely as the presence of quantum vortices.

The weak nuclear force becomes a direct consequence of spacetime nonlocality (of how the vacuum quanta are spread throughout superspace). Quantization represents a departure from Euclidean geometry that is weak on all macroscopic scales, but plays a significant role as we zoom into microscopic scales.

All three of these forces make direct reference to, and can be defined on, a microscopic section of the metric. They are also dynamically sustained. Gravity, on the other hand, is an emergent geometric expression that is statically defined (changes in that distortion are dynamically communicated, but radial density gradients do not need to be changing to exist).

On the scale of a few quanta, the geometric distortion that gives birth to gravitational effects is nowhere to be found. Density gradients only come into focus as we zoom out to macroscopic scales.

In general, a force's magnitude reflects how sharply its geometric distortion contrasts with Euclidean expectations (within the realm of their influence). Because a change in density (curvature) is a macroscopic descriptor, gravity's strength is defined only on macroscopic scales and, therefore, comparatively represents a very minor upset to Euclidean expectations.

This explains the succession of force strengths. The greatest geometric departure from Euclidean expectations is found in quantum vortices (the strong nuclear force). Divergent flow and curl in the superfluid (electromagnetism) represent the next greatest departure from Euclidean geometry. Quantization follows, giving rise to the quantum tunneling effects of the weak force. And vacuum density gradients represent the smallest departure, which are responsible for the effects of gravity.

Unification is found in the fact that the strong, electromagnetic, weak and gravitational forces each represent a characteristic way in which the geometry of the superfluid vacuum differs from Euclidean projections. Gravity is different only in that it is defined as a macroscopically emergent distortion compared to the three microscopically defined metric distortions. In other words, because gravity references only small-amplitude collective excitations in the background condensate, radial density gradients, which are responsible for gravitational effects, only come into focus on macroscopic scales.

This model makes predictions of its own. Under the assumption that the vacuum is a superfluid, curved spacetime becomes a small-amplitude collective property of the quantum dynamics—a consequence of how the quanta of spacetime collectively combine to create macroscopic geometric distortions. Within the low-energy, low-momentum regime, these low-amplitude distortions reproduce Einstein's famous spacetime curvature (as shown by several analogue gravity models). Outside of that regime, however, they dictate some rather interesting, and testable, deviations from relativity.

At low energies the superfluid vacuum has a locally Minkowskian structure that is expected to guarantee the presence of special relativistic effects. General relativistic effects are negligible at short distances. Nevertheless, according to our model, the locally Euclidean geometry of that manifold is also expected to break down as we approach ultra-high energies and momenta. In a way, this result is analogous to how fluid mechanics exists only as an emergent approximation of the low-energy and low-momentum regime of the molecular dynamics, because the small fluctuations in the superfluid obey Lorentz symmetry only in the low-energy, low-momentum regime. At high enough energies and momenta the microscopic structure and dynamics of the superfluid vacuum takes on an arrangement that violates Lorentz invariance. This leads to a modified expectation for the dispersion relations of elementary particles. A complete formulation of the microscopic physics that define the metric, and its evolution constraints, will enable us to determine exactly what sort of modified dispersion relation to expect.

[1] Christopher Stubbs is an astronomer at Harvard University. He was quoted by David Appell. (2008, May). Dark Forces at Work. *Scientific American*, pp. 100–102.

[2] This approach was developed by the Stanford physicist Savas Dimopoulos in collaboration with Nima Arkani-Hamed and Gia Dvali, and then extended by Lisa Randall, Raman Sundrum and others.

[3] Lisa Randall, *Warped Passages*, p. 365.

[4] Ibid., p. 373.

[5] "Newton's law of gravity works fine over astronomical distances, but it has never been tested down to the size of a millimeter." Michio Kaku.

[6] G. E. Volovik. (2003). *The Universe in a helium droplet*, Int. Ser. Monogr. Phys. **117**, 1–507; K. G. Zloshchastiev, Spontaneous symmetry breaking and mass generation as built-in phenomena in logarithmic nonlinear quantum theory, Acta Phys. Polon. B **42** (2011) 261–292 ArXiv:0912.4139.

[7] K.P. Sinha, C. Sivaram, & E.C.G. Sudarshan. (1976). Found. Phys. 6, 65; (1976). Found. Phys. 6, 717; (1978). Found. Phys. 8, 823.

[8] M. Novello, M. Visser, & G. Volovik. (2002). Artificial Black Holes. World Scientific, River Edge, USA, p. 391; K. G. Zloshchastiev, Spontaneous symmetry breaking and mass generation as built-in phenomena in logarithmic nonlinear quantum theory, Acta Phys. Polon. B **42** (2011) 261–292 ArXiv:0912.4139; G. E. Volovik, The Universe in a helium droplet, Int. Ser. Monogr. Phys. **117** (2003) 1–507.

Chapter 20 — **Beyond Forces**

"Their 'scientists' would concoct a clever invention called a 'force' in order to hide their ignorance."

Michio Kaku[1]

When we model the motion of a butterfly's wing, trace the reverberating echoes of a crow's caw between canyon walls, predict the path that lava will flow after it oozes to the surface of the newest hot spot on earth, or gauge how the particulates of an erupting geyser on Enceladus will trickle toward Saturn and replenish its outermost rings, we are explaining those phenomena in terms of theoretical axioms. If theory and observation fully agree, and if it is possible to coherently grant the theory's axioms epistemological status, then we have reason to believe that the theory gives us ontological access to the causal story behind the phenomena.

If our observations contradict our theories, then belief in the projected causal stories of those theories is no longer justified. But a contradiction is not necessary to rob us of our explanatory power. Epistemologically empty axioms automatically cut us off from a causal story, even when the theory built from those axioms grants us accurate predictability.

The equations we use to model the phenomenal world contradict Euclidean expectations, because Euclidean geometry is unable to naturally (internally) account for the effects we blame on gravity, the weak nuclear force, electromagnetism, and the strong nuclear force. To make up for this, we might attempt to model Nature as though it is Euclidean *and* something else, but to do this is to invent additional (and unexplained) ingredients—to contradict the possibility of obtaining a coherent causal story.

Whenever we assume a spacetime metric that geometrically differs from the vacuum's actual geometry it becomes necessary to create suppositions of mystical *force fields* that exist on top of that background spacetime metric to explain our observations. The more our assumed geometry differs from Nature's true geometry the more we have to invent *forces* to mimic the phenomena in it. The inclusion of a force contradicts the premise that Nature conforms to that geometry. If forces are necessary, we have the wrong geometry.

When we switch from Euclidean to Minkowskian assumptions (the axiomatic foundation of general relativity), gravitational phenomena go from being the result of a mystical force (mapped by equations that exist independently from the system), to intrinsic aspects of spacetime's geometric character. Even though the three microscopic forces are not enveloped by Minkowskian geometry, the fact that this transition necessitates one less force suggests that it is closer to Nature's true geometry.

Superimposing force equations on top of our assumed geometry may enable accurate representations of *how* systems in the universe evolve, but this process cannot give us insight as to *why* they evolve that way. Forces do not offer an intuitive source for, or explanation of, the phenomena they model. They tell us nothing about *why* those effects exist because, as Michio Kaku points out, forces are merely concoctions designed to hide our ignorance.

Einstein's conception of spacetime reduces that ignorance. The geometry of general relativity has no need for a magical gravitational force pulling on bodies. Objects orbit because they are following straight paths in curved spacetime. Gravitational phenomena are intrinsically accounted for in this geometry. Can this insight be extended? Can we move beyond forces altogether? Can we geometrically explain all force phenomena?

A New Perspective

> *"A field of force represents the discrepancy between the natural geometry of a coordinate system and the abstract geometry arbitrarily ascribed to it."*
>
> Arthur Eddington[2]

Forces do not exist. Phenomena only appear mysterious, or require *forces* to account for them, when those phenomena are explored through a lens of axiomatic tenuity. As Arthur Eddington pointed out, there would be no need to conjure up forces if our expectations sprang from the natural geometry of spacetime. All effects would be internal to the system—natural and expected consequences of the vacuum's geometry.

The most useful way to understand forces, or force fields, is to recognize that they hint at how the natural geometry of spacetime differs from the geometry we've ascribed it. To explore this point, let's consider an example. Suppose Dave and Zia are sliding without friction near the equator of a perfectly spherical ice (frictionless) world. Dave is ten meters north of the equator, going twenty km/hr to the east, while Zia is ten meters south of the equator, also going twenty km/hr to the east. Because Dave and Zia both believe that they are traveling on parallel paths, they expect no chance of collision. To them it is clear that, so long as no forces act on them, causing them to deflect their trajectories, they should coast around the world, keeping a constant twenty meters apart. But after a while, they notice that even though they haven't turned at all they are slowly drifting toward each other. Because of this, they conclude that there must be a mysterious force that is attracting them toward each other.

If we were watching this spectacle from the International Space Station, where the curvature of Earth is visible, we would discover just what is going on. With no forces attracting them, Dave and Zia are each traveling along paths that are defined as perfectly straight by the geometry of the Earth (remaining perpendicular to the surface). These paths are called great circles or *geodesics*.[3] (Figure 20-1) Each great circle completes a loop around the world.

Latitude lines are not straight lines on Earth—they are not geodesics.[4] This fact can easily be seen near the poles where latitude lines become tight circles (Figure 20-2). Because Dave and Zia have misunderstood their local geometry, because they have approximated the shape of the Earth to be cylindrical near its equator, they believe that latitude lines are straight. In response to this oversight they might invent a magical force to explain why they have deviated from their expected latitudinal paths, but that 'explanation' would be entirely empty.

If Dave and Zia are inclined to investigate this new force, they can set up another experiment. This time they might ask ten people of varying masses to follow them. When they do, they will find that each of those individuals follows the same collision path regardless of their mass. Because Dave and Zia are attributing their attraction to a force, and because each person requires a different magnitude of force (to accelerate their individual mass), they conclude that this attractive force is proportional to the mass of the object being attracted.

Figure 20-1 Geodesics

In reality there is no force at play here. As always, this *force* is an illusion—a consequence of the observer's ignorance of the system's true geometry. Mistaking latitude lines for geodesics Dave and Zia have conjured up this force to align their predictions and observations. But when they realize that great circles are the geodesics (the straight paths) for the geometry of the Earth, the apparent attraction between them dissolves—becoming nothing more than the manifestation of a geometric condition that they were unaware of. Under the correct geometry, no force is necessary to explain their intersecting geodesic paths.

According to Einstein's general theory of relativity, Newton made this exact mistake when he formulated his law of gravity. Correcting for the geometric property of curvature, we find that gravity isn't a force at all. Rather, it is a manifestation of how the geometry of spacetime macroscopically differs from the geometry of Newtonian (or Euclidean) space.

In Einstein's picture, Earth's mass deforms the geometry of spacetime around it. This geometric deformation explains why it is the case that as we stand on the ground we do not feel gravity pulling us down. Rather, we feel the ground pushing up on us, deflecting us from traveling a straight path through spacetime. Similarly, Dave and Zia cannot feel any force between them. But if they were to fasten a stick between them, keeping them a fixed distance apart, then they would be able to feel the stick pushing on them, deflecting them from their natural great circle paths.

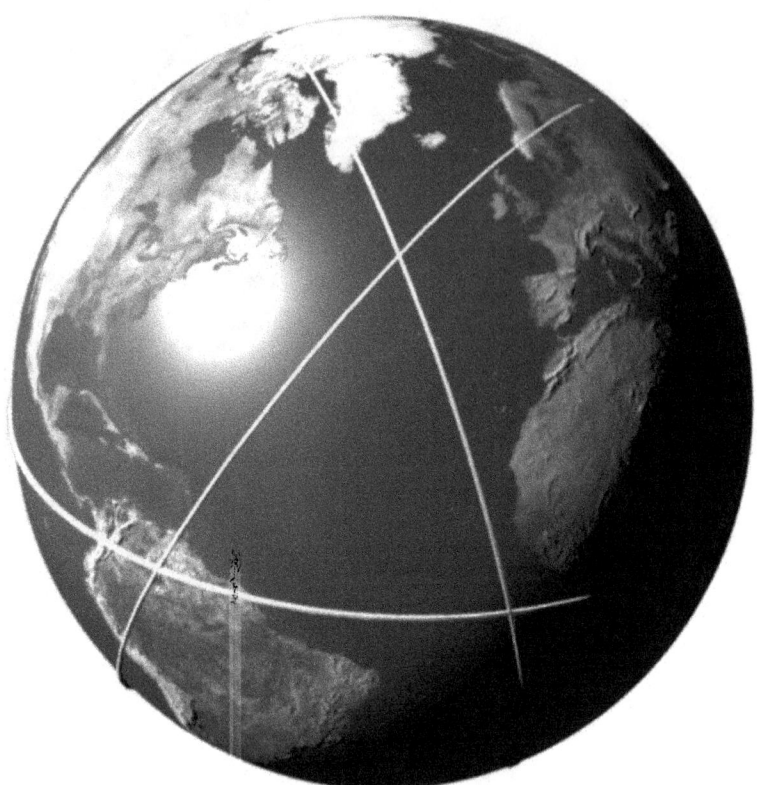

Figure 20-2 Latitude lines

Interpreting the world via forces means embracing a geometry that is different from the natural geometry. Objects appear to change their motions, or properties, due to *forces* only when we fail to accurately map the true geometric character of spacetime. Incorrect axiomatic assumptions lead to inaccurate expectations. Inventing magical *forces* is about covering up the errors in our foundational assumptions. This act may allow us to realign our predictions with our observations, but it also cuts us off from satisfying our curiosity. Phenomena cannot be "explained" via forces because forces are merely placeholders of our ignorance.

When we start from the right set of assumptions about Nature's geometry we no longer project inaccurate expectations about the world and, therefore, no longer need to postulate the existence of mystical forces. The fact that four forces are needed to cover up the errors that spring from the assumption that the vacuum is Euclidean, suggests that the vacuum's true geometry supports four kinds of distortions that Euclidean geometry has no room for.

A superfluid vacuum has an interactive geometry that is naturally characterized in terms of vector and scalar fields. A vector field representing the vacuum state details the physical positions and velocities of every quantum of space at a given moment. A scalar field that references the vacuum state assigns numbers to each region in the vacuum, representing either the magnitude of vacuum fluid flow in or out of that region, or the density of that region.

Variances in these fields characterize nonuniform arrangements of vacuum quanta, or nonuniform fluid flow in the vacuum. These variances give rise to force phenomena because they reference characteristics that have no counterpart in Euclidean geometry. This suggests that to fully encode force phenomena we need only to complete a description of the geometry of the vacuum—a description that is exhaustively defined by the relative positions, velocities, and torques of the vacuum quanta.

Gravity

> *"I seem to have been only like a boy playing on the seashore, and diverting myself in now and then finding a smoother pebble or prettier shell than ordinary, whilst the great ocean of truth lay all undiscovered before me."*
>
> Isaac Newton[5]

When aiming to unveil geometric origins for force phenomena we must remember that a force is a mathematical construct used to mimic/predict observed changes in an object's motion (for example, a change in its velocity). It is common to mistake a force as an explanation for these changes. Forces do not explain changes—they merely act as placeholders for the mystery presented by those changes. For example, if we ask the question, "why does the volcanic moon Io orbit Jupiter?" The response we might get is "because of the force of gravity." This is an empty answer. It simply renames the question. "Gravity" becomes a name for an unexplained causal connection. Using our equations for gravity we may successfully model *how* Io orbits Jupiter, *how* its path depends upon its velocity and distance from Jupiter, but we do not *explain* these occurrences by referencing a force.

The question is, "why does Io appear to follow a curved path instead of just going straight through space?" Anyone who responds to this question by assigning responsibility to a force is technically sidestepping or ignoring the question.

What's really going on is that, despite naïve Euclidean projections, Io isn't following a curved path. It is going straight through x, y, z space. Consequently, we do not need a magical force to explain its motion. What we need to explain is why it appears to us that Io is following a curved path. If Io is going straight, then why does it look like Io is following a curved path? The answer to this question is that we're still conceptualizing spacetime as if it were Euclidean.

In our eleven-dimensional superfluid picture, straight paths through x, y, z space don't always align with Euclidean expectations. Moving straight through space means having identical experiences of space. If one side of an object is consistently experiencing space identically to its other side, then the object is moving straight through space (and time). If it is sitting still, not moving through space at all from a given reference frame, then the whole object is still moving straight (all parts identically) through space and time. However, if one side is experiencing more quanta of space than another, then the object is moving on a curved path through x, y, z space. It is also moving on a curved path through time because the side experiencing more space is experiencing less time.

If the vacuum is quantized, then the density of space (defined in relation to the superspatial dimensions) is naturally characterized in terms of a scalar field. Gravitational phenomena arise when macroscopic density gradients are present in the vacuum. Curved spacetime, or gravitational fields, reference changes in vacuum density.

For an object to follow a straight path through space, all of its parts must interact with an equal amount of space—even in the presence of a spatial density gradient. Therefore, wherever a spatial density gradient exists, wherever the quanta of space are not lined up in a perfect periodic arrangement, a straight path through the familiar dimensions of space no longer conforms to the expectations of Euclidean geometry. This means that Io's orbit is a straight path through space because, for its particular velocity, its path precisely traces how space is warped in that region. In other words, Io is moving straight through curved space—it is following a spacetime geodesic (Figure 20-3). The ever-changing geometric connectivity of the superfluid vacuum dynamically determines the meaning of *straight* for each moment, and through each specific region, in the universe—making geodesics velocity dependent entities.

Getting beneath *forces* means explaining the ways in which the vacuum's true geometry categorically differs from the structure of Euclidean geometry. If the vacuum is a compressible inviscid fluid (a superfluid), then it can support different densities—marking our first categorical difference between Euclidean geometry and the geometry of a superfluid vacuum.

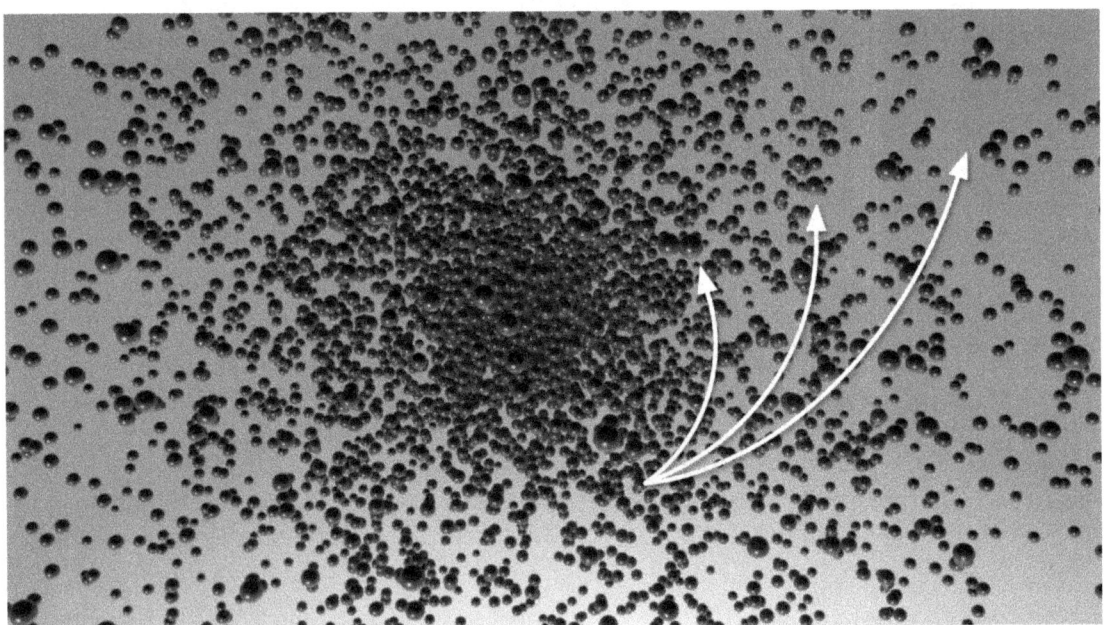

Figure 20-3 Straight paths through the vacuum (geodesics) depend on how the quanta are arranged in superspace and velocity.

Because density is a small-amplitude collective property of the system's compressibility, density gradients are emergent consequences of how the quanta of spacetime collectively interact to create and sustain macroscopic geometric distortions in the fluid density. These low-amplitude distortions accurately reproduce Einstein's expectations of spacetime curvature within the low-energy, low-momentum regime. Outside of that regime, however, they deviate from those expectations, offering a testable prediction of the theory.[6]

The Weak Nuclear Force

> *"The degrees of freedom... in the geometry of spacetime itself must also take part in the quantum physics."*
>
> Roger Penrose[7]

Can modeling the vacuum as a superfluid also give rise to the phenomena that we blame on the weak nuclear force? Does it introduce a geometric property that is responsible for those phenomena? As it turns out—it does.

When an electron disappears from the connectivity of space at one point, and then reconnects at a distant point, without having been located at any of the locations inbetween, it is said to have traveled through a wormhole. Also, when a portion of a nucleus escapes the atom (say an alpha particle made up of two protons and two neutrons) it is said to have penetrated the barrier of the strong nuclear force via a quantum tunnel. In both cases the weak nuclear force is deemed responsible. Any time something travels through a microscopic wormhole (a quantum tunnel), we attribute this occurrence to the weak nuclear force. Where do these wormholes come from? How do they form?

Quantum tunneling (travel via wormhole or radioactivity) occurs with statistical consistency in the microscopic realm. In 1896 Henri Becquerel first observed radioactivity in the element known as uranium (U^{238}). Subsequently, the half-life of uranium has been found to be about 5 billion years; which means that, on the average, half of the uranium atoms in any given mass will spontaneously disintegrate into thorium atoms and helium nuclei (alpha particles) in 5 billion years. There are several other examples of radioactive decay. For instance, a free or *unbound* neutron will, on the average, spontaneously decay into a proton, an electron, and an antineutrino in about 15 minutes. This occurrence is called beta decay.

Perhaps the most popular example of something that decays is C^{14}, an isotope of carbon that is created when cosmic rays strike Earth's atmosphere. This isotope decays to C^{12} with a half-life of 5,730 years. Because all life on Earth is constantly exchanging the carbon in its body during its life (through digestion and respiration), all living things retain the atmospheric concentration of C^{14}. After death, however, this radioactive isotope is no longer replenished and consequently the amount of C^{14} relative to C^{12} in an organic mass can reveal how long it has been dead—as long as the amounts are measurable. With modern techniques, radiocarbon dating can be used to accurately measure something that is between 500 and 50,000 years old. Potassium (K^{40}) decays to Argon (Ar^{40}) and Calcium (Ca^{40}) with a half-life of approximately 1.25 billion years, which makes it a prime tool for dating basalts (volcanic rock beds or ash deposits) that are more than 50,000 years old and so on.

A superfluid vacuum dynamically accounts for quantum tunneling—revealing wormholes to be a region of superspace with reduced quantum density—or even regions vacant of quanta. In a quantized fluid vacuum these tiny wormholes are constantly being created and destroyed as the quanta shift around—microscopically altering the state of the vacuum. This geometric mixing ensures that quantum tunneling is a short-ranged and microscopic effect. Occasionally vacant pathways open up that are large enough for conglomerate particles of matter, like alpha particles, to move through (Figure 20-4). The likelihood that a tunnel will form depends on its size. Larger (longer or wider) tunnels become exponentially less likely.

Figure 20-4 Quantization gives rise to quantum tunnels

Physicists conjured up a non-zero field of weak *charge* in order to reclaim predictability—superimposing this weak field on top of the Euclidean metric. But the assumption that a field of weak charge interweaves with, and somehow permeates, a Euclidean vacuum can be replaced by the assumption that the vacuum is a quantized medium—explaining the effects in terms of the geometry itself. This is a more elegant and ontologically satisfying solution. The microphysics of the vacuum—its internal degrees of freedom—are responsible for creating and destroying microscopic tunnels in the medium. Weak force phenomena are expectations of vacuum quantization.

Electromagnetism

> *"...the 4-year old Albert Einstein was amazed. When his father showed him a compass, it was young Albert's first clue, he later wrote, that there was "something behind things, something deeply hidden," and he spent his life trying to find it."*
>
> Bruno Maddox[8]

Assuming that the vacuum is Euclidean cuts us off from obtaining a genuine explanation of electromagnetic effects—requiring us to invent a force in order to align our predictions with our observations of those effects. By contrast, the assumption that the vacuum is a fluid allows us to internally generate the effects of electromagnetism—elegantly explaining those effects as hydrodynamic consequences of the vacuum's fluid properties. In fact, when we treat the vacuum as a superfluid, the electric field becomes an expression of divergent flow in the vacuum, and the magnetic field becomes an expression of curl, or circulating flow.[9]

Divergence is characterized by the extent to which each point in a fluid behaves as a source or a sink (Figure 20-5). Fluid flowing away from a source, like a spring of water, represents positive divergence, and fluid flowing towards a point, like a drain, represents negative divergence—or convergence.

Figure 20-5 A vector field with positive divergence

Contractions and expansions are also espressions of divergence. For example, when a gas is heated it expands in all directions such that the net velocity vector field that characterizes the motion of all its particles points outward from that region. If the divergence at a particular point of the velocity field has a positive value, then it is a source. If it has a negative value, like a gas that is cooling and contracting, then it is a sink.

If the vacuum is Euclidean, then vacuum divergence and curl is zero everywhere, and the electric field ***E*** and the magnetic field ***B*** present genuine mysteries. But if the vacuum is a fluid, then in physical terms, the electric field represents divergence in that fluid, and the magnetic field represents curl.

To develop this insight in more detail, we must model the vacuum as a fluid by specifying all of the positions of the quanta and attaching little velocity vectors to each of them—representing how they are moving about in the superspatial dimensions—producing a velocity vector field ***A***. The divergence of that three-dimensional vector field characterizes the extent to which the flow of that vector field behaves like a source or a sink at each point, in terms of a signed scalar.

To fully capture that divergence we also need to represent the fluid in a scalar sense. That is, instead of keeping track of all the velocities of all the quanta that make up the vacuum, we partition the fluid volume into many sub-volume regions, let those regions shrink to some minimum size, and assign a number to each region corresponding to the amount of flux through each boundary. If the amount of fluid flowing into a particular boundary is equal to the amount flowing out, then the region is assigned the number zero. If the outward flow across a boundary is greater than the inward flow, then the region gets a positive number with a magnitude that reflects the net outward flow. And if the inward flow is greater than the outward flow, then the region is assigned an appropriate negative number. Mapping out these numbers gives us the potential scalar field φ.[10]

Once we have done this, the divergence of $\boldsymbol{A}$, written as $\boldsymbol{\nabla} \cdot \boldsymbol{A}$, is equal to the negative gradient of this potential scalar field minus the partial derivative of $\boldsymbol{A}$ with respect to time.[11]

$$\boldsymbol{\nabla} \cdot \boldsymbol{A} = -\boldsymbol{\nabla}\varphi - \frac{\partial \boldsymbol{A}}{\partial t}$$

Checking this equation against Maxwell's equation for the electric field,

$$\boldsymbol{E} = -\boldsymbol{\nabla}\varphi - \frac{\partial \boldsymbol{A}}{\partial t}$$

we discover that the electric field is equal to the divergence of $\boldsymbol{A}$: $\boldsymbol{E} = \boldsymbol{\nabla} \cdot \boldsymbol{A}$

Likewise, the presence of a magnetic field tells us that there is curl or circulating flow in the vacuum, like an eddy or a whirlpool. In general, curl can be described as infinitesimal rotations in the three-dimensional vector field $\boldsymbol{A}$ (Figure 20-6). It is expressed as the vector product of the nabla operator and the vector field.[12]

$$\boldsymbol{\nabla} \times \boldsymbol{A} = \hat{x}\left(\frac{\partial A_z}{\partial y} - \frac{\partial A_y}{\partial z}\right) + \hat{y}\left(\frac{\partial A_x}{\partial z} - \frac{\partial A_z}{\partial x}\right) + \hat{z}\left(\frac{\partial A_y}{\partial x} - \frac{\partial A_x}{\partial y}\right)$$

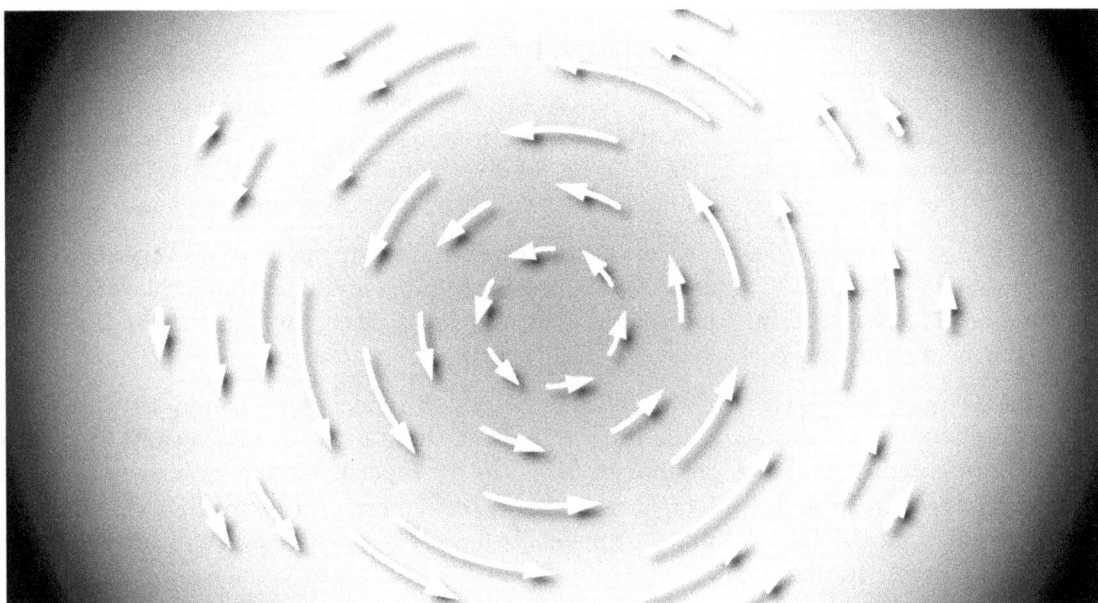

Figure 20-6 A vector field with curl

At every point in the velocity vector field, the curl of A, written as $\nabla \times A$, is represented by another vector, which characterizes the rotation at that point. This vector points in the direction of the axis of the rotation, as determined by the right-hand rule, and its length represents the magnitude of rotation. When we take a collection of these vectors we end up with a new vector field that represents the circulation density at each point in the vacuum—otherwise known as the circulating flow, or curl. This vector field is the magnetic field B.[13]

$$B = \nabla \times A$$

In short, the vector field A represents the flow velocity of the vacuum fluid, the divergence of A represents the electric field E, and the curl of A represents the magnetic field B. All three of these *fields* are physically real and free of mystical connotations.

This is straightforward condition of our model. If the vacuum is a fluid then, according to the Helmholtz theorem, it can be mapped by the rules of vector field theory—which means that it can be represented by a general vector field $F = \nabla \times A - \nabla \varphi$, where A is a velocity vector field of that fluid and φ is a scalar field representing the fluid flux. The observable properties of any such fluid (its divergence and curl) is governed by four field equations:

$$\nabla \cdot (\nabla \times A) = 0 \qquad \nabla \cdot (\nabla \varphi) = \frac{\rho}{\varepsilon}$$

$$\nabla \times (\nabla \varphi) = -\frac{\partial (\nabla \times A)}{\partial t} \qquad \nabla \times (\nabla \times A) = \mu j + \mu \varepsilon \frac{\partial (\nabla \varphi)}{\partial t}$$

Where ε and μ represent the permittivity and permeability of the specific fluid, ρ represents the charge density and j represents the current density.

Inserting the permittivity and permeability values that are specific to the vacuum (ε_0 and μ_0), and interpreting divergence in the vacuum as the electric field, $-\nabla \varphi = E$, and vacuum curl as the magnetic field, $\nabla \times A = B$, we end up with Maxwell's equations for electromagnetism, which are necessary and sufficient for a specification of electromagnetic theory.

$$\nabla \cdot B = 0 \qquad \nabla \cdot E = \frac{\rho}{\varepsilon_0}$$

$$\nabla \times E = -\frac{\partial B}{\partial t} \qquad \nabla \times B = \mu_0 j + \mu_0 \varepsilon_0 \frac{\partial E}{\partial t}$$

This model grants us immediate ontological purchase of electric and magnetic fields, anchoring them directly to real physical properties.[14] It also enables us to understand why electromagnetism is gauge invariant, why light is an electromagnetic wave, why photons remain localized instead of diluting as they travel through the vacuum, why like charges "repel" and opposite charges "attract", and why there are no magnetic monopoles in Nature.[15]

To unravel those additional insights, we note that some flow descriptors are explicitly reference frame dependent while others are what we call *gauge invariant*. For example, from

301

one reference frame we might characterize the global flux of a fluid—the general direction and speed of its flow—to be eleven kilometers per hour in the x direction, yet from another reference frame we could simultaneously describe it to be flowing zero kilometers per hour, having no direction. This is an example of a flow descriptor that is *not* gauge invariant, because it explicitly depends upon the observer's reference frame.

By contrast, springs and eddies are *gauge invariant* descriptors of fluid flux. Divergent flow from a spring still manifests as divergence no matter what reference frame we observe it from, and twisting whirlpools look like twisting whirlpools whether or not we are moving relative to them.

A change in reference frame may induce curl from a source of divergence, or induce divergence from a curl, but when there is a source of divergence or curl in one reference frame they cannot both be eliminated by a gauge transformation—a change in reference frame. [16] Therefore, the fact that electric and magnetic fields are gauge invariant automatically follows from the fact that they represent vacuum divergence and curl.

Understanding electric and magnetic fields in terms of vacuum divergence and curl also helps us understand what it means to say that light is an electromagnetic wave. Light travels through the vacuum in the form of localized plane waves—vacuum phonons that we call photons. The propagation of each of these distortions necessarily induces divergence and curl into the vacuum. Because of this, propagating vacuum phonons (photons) can be interchangeably called electromagnetic waves.

The fact that photons stay localized as they propagate through the vacuum is also a consequence of this model. In 1958, Philip Anderson showed that the propensity for lattice waves to diffuse depends on two factors, the number of dimensions occupied by the phononic waves, and the degree of randomness in the medium (which could be a function of defects or impurities, or just a lack of lattice arrangement in the medium).[17] Any two-dimensional phonon (plane wave) propagating through a medium, whose internal lattice structure is highly random, will remain localized.

Therefore, the fact that photons do not spread out (dilute or delocalize) as they travel through the vacuum, the fact that they are held together by *Anderson localization*, implies that the vacuum is a fluid whose internal degrees of freedom allow for lattice mixing—producing a high degree of randomness. Conversely, the assumption that the vacuum is a fluid leads to the expectation of localized photons.

In addition to this, vacuum superfluidity makes the fact that like charges repel each other, while opposite charges attract, a consequence of fluid dynamics. In a superfluid vacuum elementary particles are modeled as stable quantized vortices, or sonons—smoke rings with a twist (see Chapter 21). The simplest of these sonons represents the electron or, if we reverse the chirality, the positron. The magnitude of geometric distortion defines the mass of the particle, but the flow character of that distortion defines its electric and magnetic fields. In other words, quantized vortices are fluid-dynamically stabilized with a specific mixture of divergence and curl—a mixture that defines their electric field and magnetic moment.

Vortices in stable configurations that swirl in the same way hydrodynamically pack space up between them. In other words, when two vortices with identical chirality are side-by-side, the amount of space between the vortices, in terms of density, is dynamically increased. In the upper hemisphere of the region between them, the vortex on the left contributes a downward vacuum flow, while the vortex on the right contributes an upward flow. These flows interact with each other, causing the space between vortices to agglomerate. In the lower hemisphere of the region between them, the vortex on the left contributes an upward vacuum flow, while the vortex on the right contributes a downward flow (Figure 20-7).

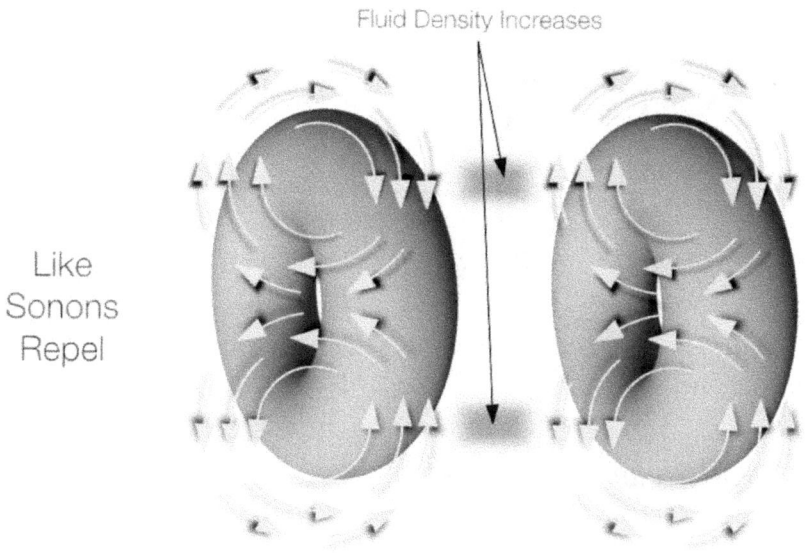

Figure 20-7 *Like particles (sonons) naturally 'repel'*

Four dimensionally we say that particles of like charge are pushed apart by some mysterious force. But if this picture is correct, then instead of saying that like charges repel each other, it may be more appropriate to say that like charges cluster space between them—actively increasing the amount of space between them—fluid-dynamically separating the particles.

Vortices with opposing chirality, representing opposing charges, have the opposite geometric tendency. In the upper hemisphere of the region between them, both sonons contribute downward vacuum flow, and in the lower hemisphere they both contribute upward flow. In terms of fluid dynamics (Bernoulli's principle) this constructive increase of flow creates a low pressure between the vortices, which introduces a metric tendency to draw them together—leaving less space between them (Figure 20-8). This allows us to hydrodynamically explain how and why like charges repel and opposite charges attract.

Fluid dynamics is also responsible for the fact that magnetic monopoles do not exist. When a curl is induced in a fluid, the opposing curl is adjacently induced. The circulating flow inside a quantized vortex is reversed outside it. That is, the fluid inside a vortex rotates one way while the fluid outside that vortex rotates the other way. This is why magnetic fields always have simultaneous positive and negative "poles". By contrast, when a point of positive divergence is induced, it is not automatically locally offset. Divergence is globally balanced, which is why the total electric charge in the universe is balanced, but that balance is not maintained in a localized way. Therefore, points of electric charge (divergence) exist in Nature without reference to a corresponding convergence, but magnetic charges (curl) always come in pairs.

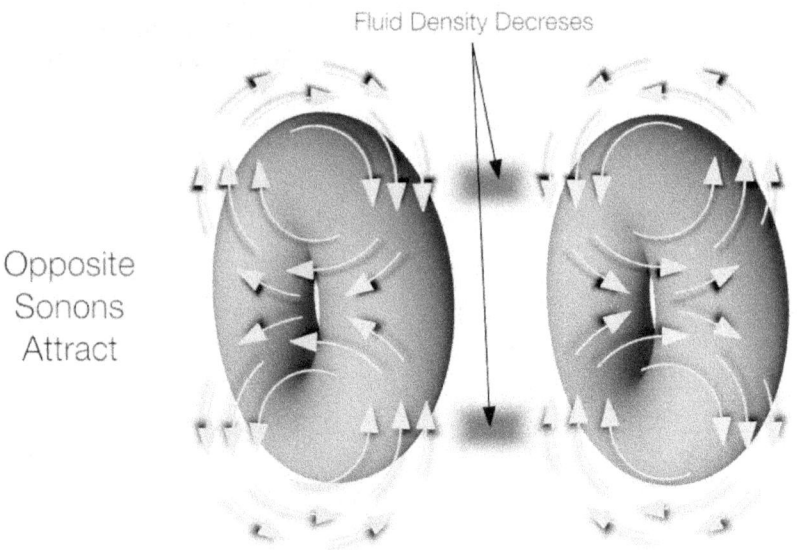

Figure 20-8 Opposite particles (sonons) naturally 'attract'

The Strong Nuclear Force

"The history of human discovery is characterized by the boundless desire to extend the senses beyond our inborn limits. It is through this desire that we open new windows to the universe."

Neil deGrasse Tyson[18]

The strong nuclear *force* is held responsible for holding together the quarks that make up protons and neutrons, and for holding proton-neutron pairs together within nuclei. The amount of energy it takes to separate a quark from an antiquark increases significantly with distance, yet we don't observe strong force effects macroscopically. In practice, the strong nuclear force is very short ranged. When a free proton gets closer than about 10^{-15} meters to the nucleus of an atom it gets sucked in, but at greater distances it is electromagnetically repelled. Once 10^{-15} meters is reached—which is about the size of the proton itself—the proton is pulled into the nucleus with a force one hundred times more powerful than the repulsive electromagnetic force. How do we explain these effects?

Hydrodynamically, strong force effects can be explained in terms of quantum vortex formation (which is also responsible for the emergence of the property of mass—see Chapter 21: The Higgs Mechanism) and vortex trapping. When three phonons intersect in the vacuum just right they can transform into a cohesive quantized vortex. This vortex becomes a localized distortion, which means that it is no longer traversing the vacuum at the speed of light, and that its rest energy is captured in the twists and turns of the metric.

Once these vortices form, energy considerations lead to a natural proposal for how they come together to form composite vortices. Each vortex has an extended energy wave and the energy waves of nearby vortices overlap. To minimize the total energy of these interacting waves the vortices must be aligned in a way that allows their waves to destructively interfere at large distance. The existence of these lower-energy arrangements introduces a coupling mechanism—explaining how and why fundamental particles combine into composite particles.

Under this paradigm, gluons become a reference to the mechanism of vortex trapping, a process that geometrically stabilizes its parts against background impulses that might otherwise disrupt them individually. Exploring that stability we note that when we attempt to separate a quark from an antiquark, when we try to smooth out the swirling vortex, we discover that it is much more energy-efficient to create additional quarks and antiquarks from the vacuum than it is to separate already existing pairs. (Just like it's easier to create a peak and a trough in a medium—a whole wave—than it is to just create a peak.) In short, the strong force encodes the ability for the metric to contain vortices—extremely stable distortions.

When it comes to grand unification the end game is to move beyond the illusion of *forces*—to obtain a fully accessible geometric explanation of force phenomena. Under the assumption that the vacuum is a superfluid, curved spacetime is a small-amplitude collective property of the quantum dynamics, the weak nuclear force is a direct consequence of spacetime nonlocality (of how the quanta of space are spread out through superspace), electromagnetism is an expression of divergence and curl in the medium, and the strong force captures the tendency for quantum vortices to form and combine in the superfluid. This proposal has the flavor of the unification we seek.

Bridging the gap between imagination and reality brings us closer to the possibility of having knowledge with deductive certainty—the clearest kind of certainty—whose inheritance comes from geometry. We may never be able to answer John Locke's question,[19] and know for sure that our map accurately portrays that which it is intended to map, but without constructing a portrait of the stage on which Nature's phenomena play out, without building a coherent theory based on clear assumptions, we sabotage our imaginative reach.

With vacuum superfluidity we no longer have to invent *forces*, empty mathematical incantations, or suppositions of mystical fields that exist on top of a flat background spacetime metric to explain our observations. Most importantly, the possibility that the vacuum is a superfluid enables us to pursue explanations instead of settling for mere descriptions.[20]

[1] Michio Kaku. *Hyperspace*.

[2] Arthur Eddington. (1923). *The Mathematical Theory of Relativity*. Cambridge, England, Cambridge University Press, pp. 37–38.

[3] Instead of following curved paths in the Cartesian coordinate system, orbiting objects are following a straight path (a geodesic) in a curved coordinate system. "Gravity, therefore, is not really a *force* but is a manifestation of curvature in the geometry of spacetime." Peter S. Shawhan. (2004, July-August). Gravitational Waves and the Effort to Detect Them. *American Scientist*. Volume 92, p. 351.

[4] Except for the latitude line that traces out the equator.

[5] Sir David Brewster. (1855). *Memoirs of the Life, Writings, and Discoveries of Sir Isaac Newton*. Volume II, Ch. 27.

[6] At low momenta vacuum superfluidity perfectly reproduces the curved spacetime of general relativity. At high momenta the eikonal approximation can be used to extract a well controlled Bogoliubov-like dispersion relation, recovering non-relativitic Newtonian physics. Carlos Barceló, S Liberati & Matt Visser. (2001). Analogue gravity from Bose-Einstein condensates. Institute of Physics Publishing, Class. Quantum Grav. **18**, 1137–1156, PII: S0264-9381(01)18993-8.

[7] Roger Penrose. *The Road to Reality*, p. 827.

[8] Bruno Maddox. (2008, May). Three words that could overthrow physics: "What is magnetism?" *Discover*, p. 70.

[9] Divergence and curl are orthogonal (perpendicular and independent) parameters of flux, or fluid flow. To map this flow we use a vector field. The curl-free component of the vector field (also referred to as the longitudinal component) is the component of divergence, and the divergence-free component (also called the transverse component) is the component of curl.

[10] We could also construct a scalar field that maps the density of the fluid throughout. In other words, instead of referencing the flux through each boundary we could label the density of the fluid in each region. The gradient of this scalar field maps vacuum curvature—reproducing general relativity.

[11] In three dimensions, if the temperature and pressure are held constant, then the initiator of a persistent source of positive divergence must also be a sink (a source of negative divergence)—like a sonon or a smoke ring.

[12] Technically $x, y,$ and z should be replaced with $\sigma, \mu,$ and δ—the superspatial dimensions.

[13] Divergence can be represented by a scalar field and have a pointlike source, but curl requires a vector field and cannot be traced to a pointlike source without vanishing. This is why there are pointlike sources of electric fields in Nature, like electrons, but there are no magnetic monopoles.

[14] It also reveals that a changing gravitational field is an electric field.

[15] To further explore this topic see: Thad Roberts. (2015). Fluidic Origins of the Magnetic and Electric Fields: A physical interpretation of B and E. Academia.edu. https://www.academia.edu/12637409/Fluidic_Origins_of_the_Magnetic_and_Electric_Fields_A_physical_interpretation_of_B_and_E

[16] This gauge relationship between divergence and curl—between the electric and magnetic fields—is characterized by the Helmholtz theorem, which holds that any sufficiently smooth, rapidly decaying vector field in three dimensions can be resolved into the sum of a solenoidal (divergent-free) vector field $\boldsymbol{A}$ and an irrotational (curl-free) vector field $\boldsymbol{\nabla}\varphi$. Therefore, if $\boldsymbol{F}$ is a general vector field characterizing the flow of a fluid bounded domain V in $\mathbb{R}^3$, which is twice differentiable, and if S is the surface that encloses the domain V, then

$$\boldsymbol{F} = \boldsymbol{\nabla}\times\boldsymbol{A} - \boldsymbol{\nabla}\varphi$$

[17] P. W. Anderson. (1958). Absence of Diffusion in Certain Random Lattices. *Phys Rev* **109** (5): 1492-1505.

[18] Neil DeGrasse Tyson. *Death By Black Hole*, p. 152.

[19] Douglas J. Soccio. *Archetypes of Wisdom*, p. 291.

[20] In the standard model of particle physics, the CPT theorem is based on the premise that the metric is flat Minkowskian spacetime and that the fields leading to the microscopic forces are just somehow superimposed on top of this flat background. Roger Penrose. *The Road to Reality*, p. 818.

Chapter 21 — **Quantized Vortices**

"I believe Einstein was on the right track. His idea was to generate subatomic physics via geometry."

Michio Kaku

"[E]lementary particles are not the fundamental building blocks of matter. Instead, they emerge from the deeper structure of the non-empty vacuum of space-time."

Zeeya Marali, paraphrasing the insights of
physicists Xiao-Ganag Wen and Michael Lavin[1]

Under the assumption that the vacuum is an inviscid fluid, *mass* and *energy* become distinct references to geometric distortions in that fluid. *Mass* specifically denotes the presence of a localized distortion (of increased density). Distortions that are not localized, ones that require transverse propagation in order to be sustained—like plane wave pulses of increased density that propagate through the medium—are referred to as *light*, or more generally as *energy*. Pulses with densities below the background density, or inverse pulses, represent *negative energy*. In short, if it is possible to be in the same reference frame as the geometric distortion in question, if it is possible to point to it and assign it a fixed position, then the distortion is to be labeled *mass*. If this is not possible then the distortion is to be labeled *energy*.

In a fluid metric the total magnitude of any geometric distortion is expected to vary depending upon its speed (defined relative to the observer). When a mass particle (e.g. a soliton in the inviscid fluid, like a stabilized smoke ring) is not moving relative to us, the magnitude of that distortion chacterizes the particle's rest mass, also known as its intrinsic mass. When the particle is moving relative to us, there is an additional vacuum wavefront built up in front of it, adding to its total distortion. The faster the soliton moves, the greater its total distortion. The additional portion of that distortion characterizes the particle's kinetic mass (Figure 21-1).

As the velocity of the soliton approaches the propagation speed of the medium, the total metric distortion associated with it approaches an infinite value. This is why it takes an infinite amount of energy to accelerate a particle with non-zero rest mass to the speed of light in the vacuum.[2]

Therefore, once we have particles with rest mass, it is trivial (given the vacuum's fluid nature) to explain how they acquire kinetic mass—also known as relativistic mass. But how do we explain the emergence of these particles in the first place? What are elementary particles? How do they form? Why do they only come in certain sizes? Do all of Nature's elementary particles represent solutions to the vacuum's characteristic wave equation? If so, are these solutions all Lorentz covariant?

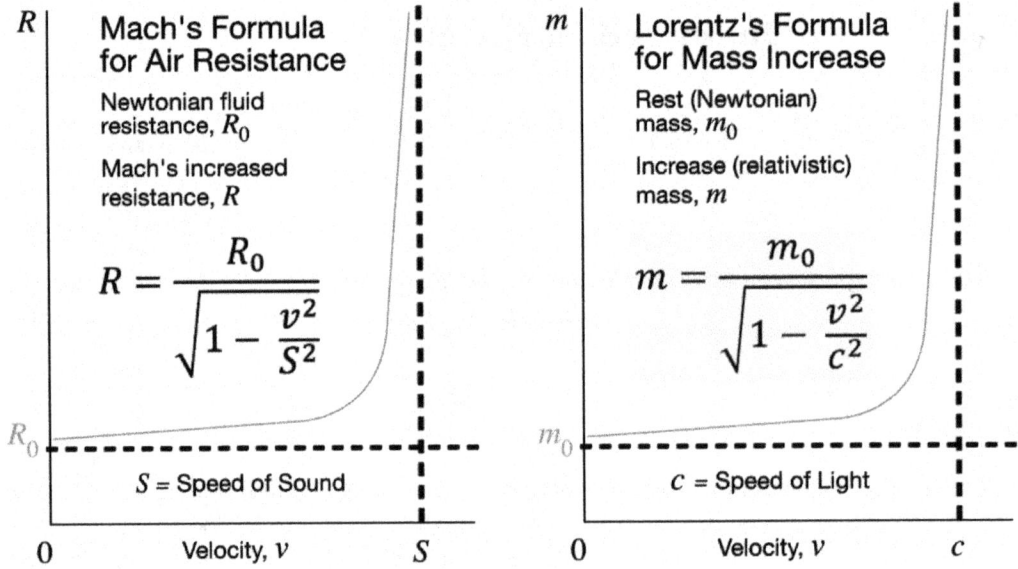

Figure 21-1 Air resistance vs. mass increase
If the vacuum is quantized (like air), then relativistic mass has a fluid dynamic explanation.

Lord Kelvin's Idea

> *"Are particles nothing more than tangled plaits in space-time?"*
>
> Lee Smolin[3]

In reference to these questions, Frank Wilczek, a physics Nobel Laureate, noted that William Thomson (also known as Lord Kelvin) postulated one of the most beautiful 'failed' ideas in the history of science when he suggested that atoms might be vortices in an aether that pervades space.[4] Believing in an aether, an invisible medium *in* spacetime that sustained electromagnetic waves, Thomson became intrigued by the work of Hermann Helmholtz, who demonstrated that "vortices exert forces on one another, and those forces take a form reminiscent of the magnetic forces between wires carrying electric currents."[5] As he explored this connection, he recognized that vorticity was the key to obtaining a model that could explain how a few types of atoms, each existing in very large numbers of identical copies, could arise in Nature.

To get his theory of vortex atoms off the ground, Thomson assumed that the aether was endowed with the ability to support stable vortices. Following Helmholtz' theorems, he then noted that distinct types, or species, of vortices would persist in the medium, and that these fundamental vortices could aggregate into a variety of quasi-stable molecules.

Thomson's idea—the idea that stable quantum vortices, whose topologically distinct forms and sizes are naturally and reproducibly authored by the properties of the medium itself, are the building blocks of the material world—is quite appealing. Sadly the idea has faded into obscurity, cloddishly dismissed and rejected, because the aether, the background fluid that these vorticities were thought to critically depend on, was abandoned. Scientists assumed that if the aether is out, then Kelvin's quantized vorticities are also out. But they may have mistakenly thrown the baby out with the bath water.

Providentially, the elegance of Thomson's quantized vorticities is resurrected when we trade the aether assumption, that there is a medium *in* the vacuum that supports electromagnetic waves, for the assumption that the vacuum itself *is* a superfluid medium with a metric that is macroscopically describable by the wave function.[6] The assumption that the vacuum is a superfluid, also called a quantum fluid, instinctively establishes vortex stability. It also leads to the expectation that the structure of the material world is written into the substrate of the vacuum itself—explaining that as quantized vortices form in the vacuum, supersymmetry is broken and subatomic particles emerge with very specific properties.

We are just beginning to explore some of the promising new possibilities offered by quantum fluids. Current research is focused on, among other things, theoretically understanding the formation of quantum vortices in Bose-Einstein condensates (and how they combine to form stable unions), linking those quantum vortices to a concept of matter origins,[7] and using BEC's to model black holes and their related phenomena in the lab.[8]

If vortices in the vacuum correspond to particles then "concentrated energy in empty space can transform virtual particles into real ones."[9] If this is what is going on, then the mechanism behind this transformation (the Higgs mechanism) needs to be explained. We need to explore how massless particles with two physical polarizations acquire a third stable polarization in the longitudinal direction. We need to figure out how the property of *mass* (locally maintained geometric distortions, or quantized vortices) springs into existence.

To push us towards a possible answer, we note that if we spin a beaker containing a superfluid, we end up with an array of vortices scattered about in that fluid. The number of vortices introduced is proportional to $\hbar/m$. Interestingly, superfluidity breaks down within each of these vortices, while the rest of the fluid retains its full superfluid characterization, remaining stationary (in the macroscopic sense). Therefore, the rotational energy becomes contained within these quantized vortices or solitons.[10] The difference between this kind of rotation and the rotation of a rigid body can be more precisely quantified by noting that the tangential velocity of the quantized vortices has a modulus that decreases with r:

$$v = \frac{\hbar}{m}|\nabla S| = \frac{\hbar}{m}\frac{1}{r}\frac{\partial}{\partial \varphi}S = \frac{\hbar}{m}\frac{s}{r}$$

whereas the tangential velocity of a rigid rotator has a modulus that increases with r: $\mathbf{v} = \mathbf{\Omega} \times \mathbf{r}$.

This is what allows us to claim that the vortices are localized. This, combined with the fact that vortices are defined as certain geometric distortions in the vacuum that spontaneously break or hide the underlying higher symmetric state, makes them perfect candidates for particles that inherit their rest mass via the Higgs field. Vacuum superfluidity, therefore, gives teeth to the hypothesized Higgs mechanism by offering a physical process that explains how energy is converted to mass.

The Higgs Mechanism

"One might as well speculate on the origins of matter."

Charles Darwin[11]

"...the precise model that gives particles their mass—is one of the biggest puzzles facing particle physicists today. One of the attractions of extra dimensions is that they might help solve this mystery."

Lisa Randall

The Higgs mechanism gives rise to the Higgs field (also called the Higgs boson, or the God particle),[12] which is used to codify the mysterious fact that particles possess rest mass. This mechanism is held responsible for causing certain geometric distortions in the vacuum[13] and thereby spontaneously breaking or hiding the underlying higher symmetric state of spacetime. How this field spontaneously breaks the symmetry associated with the weak force to give elementary particles their mass,[14] how it lowers the total energy state of the universe, or how viscosity is introduced into the system, has not been made clear.[15]

The Higgs boson was introduced into the electroweak theory as an ad hoc way of giving mass to the weak boson. Ironically, this insertion keeps the theory from solving the mass generation problem. Instead of explaining the origin of mass in the Higgs boson, the theory introduces this mass as a free parameter via the Higgs potential, ultimately making the value of the Higgs mass just another free parameter in quantum mechanics and leaving us completely in the dark about its origins.[16]

The value of this Higgs parameter has only been indirectly estimated, and many different estimates have been posited by the standard model (and its extensions), but even if theorists knew how to pick among these values—even if the mass of the Higgs boson were theoretically fixed—we would not have a satisfactory solution to the mass generation problem. The Higgs postulation only reformulates the problem of mass generation, pushing the question back to, "How does the Higgs boson get its mass?" The question we need to answer is—how do massless phonons in the vacuum transform into something that has mass? The process of that transformation is the heart of the Higgs mechanism.

This is where vacuum superfluidity comes to the rescue. Vacuum superfluidity naturally postulates a fundamental mechanism for mass generation. In short, elementary particles can be thought of as solitons—self-reinforcing solitary waves, or wave packets—in the inviscid fluid vacuum, or quasi-particle sonons.[17] In other words, they are persistent, localized solutions of the vacuum's wave equation. Relativity emerges in this picture because these solitons satisify the wave equation, and quantum mechanics emerges because of the quantized nature of the allowed solitons. This suggests that elementary particles acquire their mass directly from the fluid properties of the vacuum condensate—much like the gap generation mechanism in superconductors or superfluids.[18]

In other words, the assumption that the vacuum is a compressible fluid, subject to Euler's equation for compressible fluids, makes way for the possibility that the internal properties of that fluid give rise to stable geometric distortions that we call fundamental particles. "The key insight is that Euler's equation for a compressible fluid possesses quasiparticle solutions with chirality."[19] These chiral quasiparticles are persistent localized solutions of the wave equation for the vacuum that resemble twisting smoke rings.[20]

To derive the character of these solitons we consider a compressible inviscid fluid that obeys Euler's equation:[21]

$$\frac{\partial u}{\partial t} + (u \cdot \nabla)u = -\frac{1}{\rho}\nabla P$$

where P is pressure, ρ is density, u is velocity and $\frac{\partial \rho}{\partial t} = -\nabla(\rho u)$.

For low amplitudes this yields the wave equation

$$\frac{\partial^2 \rho}{\partial t^2} = c^2 \nabla^2 \rho$$

This wave equation has linear solutions, eddy-like solutions that resemble smoke rings (Figure 21-2a), and chiral solutions that resemble smoke rings with a twist (Figure 21-2b). The general solutions are called sonons (ξ_{mn}).

$$\xi_{mn} = \psi_0 R_{mn}$$

where

$$\psi_0 = A e^{i\omega_0 t},$$

$$R_{mn} = \int_0^{2\pi} e^{i(m\theta' - n\phi)} j_m(k_r \sigma) \, k_r R_0 \, d\phi$$

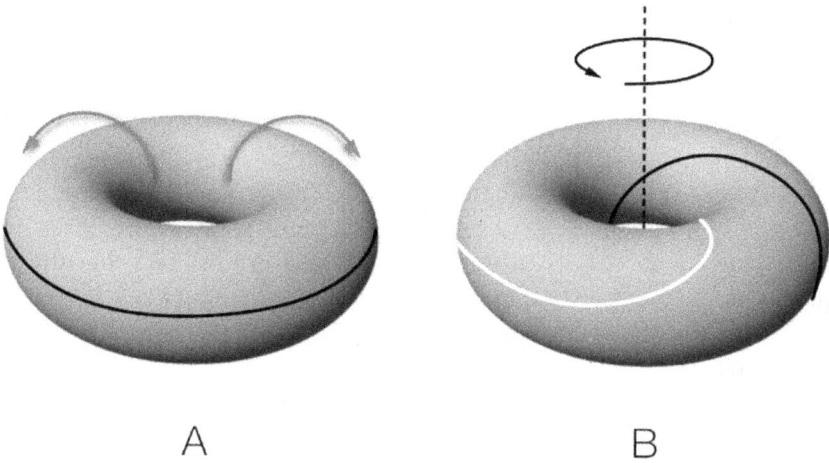

A B

Figure 21-2 Vortices quantized by chirality

Only certain sonons are stable, or capable of persisting—those in which chirality (m) and spin direction (n) are preserved by continuous transformations. This stability restriction explains why fundamental particles of mass are quantized, and gives us a way to model the fundamental particles. Beginning that process, Ross Anderson and Robert Brady have show that the most basic of these sonons, sonon R_{11} ($m = 1, n = 1$), represents the electron (Figure 21-2b).[22]

The possibility that the fundamental particles can all be explained in terms of the quantized sonons is majestically alluring. Vacuum sonons are the simplest and most elegant topological explanation for mass generation.[23] That is, once we take on the assumption that the vacuum is a superfluid, the symmetry breaking of the relativistic scalar field—the very thing that is responsible for mass generation—is explained by the fact that stable metric braids combine to create twisting vortices. Sonons cannot form without breaking the underlying symmetry of the invicid fluid. This symmetry is broken when vacuum phonons come together such that they form a braided twisting vortex—transforming into a geometric distortion that is localized and has a third polarization state. This suggests that matter generation is a consequence of quantum vortex formation in the superfluid vacuum (like the generation of dark solitons in one-dimensional BEC's).[24]

This solution to the mass generation problem opens up the possibility of topologically interpreting color charge as the twisting modes that are available to individual ribbon braids.[25] It also suggests that an incident photon will couple with an existing sonon (add or take away from its momentum), based on whether or not the two are in or out of phase with each other. From this it follows that it is the electron's charge (e) that sets its amplitude (likeliness) to emit or absorb (couple with) a photon.[26] This amplitude can be interpreted as the eleven-dimensional swirling momentum of the particle.[27]

As previously noted, this picture also provides us with a natural proposal for how fundamental particles come together to form composite particles—like protons and neutrons. When a single sonon is isolated, its surrounding energy wave is automatically minimized, but sonons in close proximity have energy waves that overlap. To minimize the total energy of these interacting waves the sonons must be aligned in a way that allows their waves to destructively interfere at large distance. This minimum energy arrangement introduces a coupling mechanism—explaining how and why fundamental particles combine into composite particles.

In other words, because similar sonons behave as oscillators, which are weakly coupled through the nonlinear term in Euler's equation, they have two modes of coupled oscillation—aligned phases or anti-aligned phases. The combined system is pushed to a lower energy state via coupling because energy is carried away by these waves as the alignment takes place. This process is entropically driven by the fact that the carrier waves have a frequency that significantly differs from the natural frequency of the nearby sonons, so they will have a long mean free path. For sonons of similar type, this mechanism locks their frequencies together, even if the sonons initially have slightly different frequencies and different sizes.[28]

Therefore, when it comes to the mass generation problem, vacuum superfluidity is a thriving contender among a swarm of competing theories. But the extraordinary ontological clarity of this interpretation—the way that it accomplishes the task of explaining mass and energy strictly in terms of hydrodynamic transformations of the vacuum's geometry—makes it stand out among those contenders. This advantage is more than enough reason to resurrect Lord Kelvin's beautiful idea.

[1] Zeeya Merali. (2007, March 17). The Universe is a String-Net Liquid. *New Scientist*, pp. 8–9.

[2] The natural phonon speed of a medium is determined by

$$v_s = \frac{1}{\sqrt{\beta \rho_f}}$$

where β represents the compressibility of the fluid and ρ_f represents the fluid density. For the vacuum, these parameters are equal to ε_0 (the vacuum permittivity) and μ_0 (the vacuum permeability) respectively.

[3] Davide Castelvecchi. (2006, August 12). Out of the Void. *New Scientist*, pp. 28–31. Paraphrasing Lee Smolin.

[4] Frank Wilczek. (2011, December 29). Beautiful Losers: Kelvin's Vortex Atoms. NOVA: http://www.pbs.org/wgbh/nova/physics/blog/2011/12/beautiful-losers-kelvins-vortex-atoms/ Much of this section follows this article.

[5] Ibid.

[6] K.P. Sinha, C. Sivaram, & E.C.G. Sudarshan. (1976). Found. Phys. 6, 65; (1976). Found. Phys. 6, 717; (1978). Found. Phys. 8, 823.

[7] K. G. Zloshchastiev. (2011). Spontaneous symmetry breaking and mass generation as built-in phenomena in logarithmic nonlinear quantum theory. Acta Phys. Polon. B **42**, 261–292 ArXiv:0912.4139; V. Dzhunushaliev & K.G. Zloshchastiev. (2012). Singularity-free model of electric charge in physical vacuum: Non-zero spatial extent and mass generation. ArXiv:1204.6380.

[8] M. Novello, M. Visser, & G. Volovik. (2002). Artificial Black Holes, World Scientific, River Edge, USA, p. 391.

[9] Michio Kaku. *Parallel Worlds*, p. 328.

[10] These quantized vortices are theoretically similar to flux lines in superconductors.

[11] Paul Davies. (2006, February 11). In Search of a Second Genesis. *New Scientist*, p. 48.

[12] The Higgs particle/field is named after Peter Higgs of the university of Edinburgh, who mathematically described the spontaneous symmetry breaking that such a field requires. Steven Weinberg. *Dreams Of A Final Theory*, p. 197.

[13] Steven Weinberg. *Dreams of a Final Theory*, p. 25.

[14] Lisa Randall, p. 212.

[15] In reference to this Volovik interestingly notes that, "particles and antiparticles which can be created from the quantum vacuum are similar to quasiparticles in quantum liquids." Volovik, p. 18.

[16] V. A. Bednyakov, N. D. Giokaris & A. V. Bednyakov. (2008). Phys. Part. Nucl. **39**, 13–36 ArXiv:hep-ph/0703280.

[17] This term was coined by Robert Brady. See Ross Anderson & Robert Brady. (2013). Why quantum computing is hard—and quantum cryptography is not provably secure. arXiv:1301.7351v1.

[18] K. G. Zloshchastiev. (2011). Spontaneous symmetry breaking and mass generation as built-in phenomena in logarithmic nonlinear quantum theory. Acta Phys. Polon. B 42, 261–292 ArXiv:0912.4139; A. V. Avdeenkov & K. G. Zloshchastiev. (2011). Quantum Bose liquids

with logarithmic nonlinearity: Self-sustainability and emergence of spatial extent, J. Phys. B: At. Mol. Opt. Phys. 44, 195303. ArXiv:1108.0847.

[19] Ross Anderson & Robert Brady. (2013). Why quantum computing is hard—and quantum cryptography is not provably secure. arXiv:1301.7351v1.

[20] Ibid.

[21] This derivation follows Robert Brady. The irrotational motion in a compressible inviscid fluid. arXiv:1301.7540 [physics.gen-ph].

[22] Ross Anderson & Robert Brady. Why quantum computing is hard—and quantum cryptography is not provably secure; Ibid.

[23] To examine a related approach see the relativistic Coleman-Weinberg approach; S.R. Coleman & E.J. Weinberg. (1973). Phys. Rev. D7, 1888.

[24] The non-linear term $|\psi(\vec{r})|^2$ in the Gross-Pitaevskii equation

$$i\hbar \frac{\partial \psi(\vec{r})}{\partial t} = \left(-\frac{\hbar^2 \nabla^2}{2m} + V(\vec{r}) + U_0|\psi(\vec{r})|^2\right)\psi(\vec{r})$$

allows for stable vortices. These plaits of quantized angular momentum are most naturally represented by a wavefunction of the form $\psi(\vec{r}) = \phi(\rho, z)e^{i\ell\theta}$, where ρ, z, and θ are representations of the cylindrical coordinate system and ℓ is the angular number. This is the form that is generally expected in an axially symmetric (harmonic) confining potential. To generalize this notion we determine $\phi(\rho, z)$ by minimizing the energy of $\psi(\vec{r})$ according to the constraint $\psi(\vec{r}) = \phi(\rho, z)e^{i\ell\theta}$. In a uniform medium this becomes $\phi = \frac{nx}{\sqrt{2+x^2}}$

where n^2 is density far from the vortex, $x = \frac{\rho}{\ell\xi}$, and ξ is healing length of the condensate.

A singly charged vortex ($\ell = 1$) in the ground state, has an energy ϵ_v given by $\epsilon_v = \pi n \frac{\hbar^2}{m} \ln\left(1.464 \frac{b}{\xi}\right)$, where b is the farthest distance from the vortex considered. A well-defined energy necessitates this boundary b. For multiply charged vortices ($\ell > 1$) the energy is approximated by $\epsilon_v \approx \ell^2 \pi n \frac{\hbar^2}{m} \ln\left(\frac{b}{\xi}\right)$.

Such vortices tend to be unstable because they have greater energy than that of singly charged vortices. There may, however, be metastable states with relatively long lifetimes, and it may be possible for unstable vortices to come together and create stabilized unions. See Ross Anderson & Robert Brady. (2013). Why quantum computing is hard—and quantum cryptography is not provably secure; Robert Brady. The irrotational motion in a compressible inviscid fluid. arXiv:1301.7540 [physics.gen-ph].

[25] Bilson-Thompson, Sundance O.; Markopoulou, Fotini; Smolin, Lee (2007). Quantum gravity and the standard model. *Class. Quantum Grav.* **24** (16): 3975–3993, arXiv:hep-th/0603022, Bibcode 2007CQGra..24.3975B, doi:10.1088/0264-9381/24/16/002.

[26] Richard Feynman. *QED*. pp. 91, 122, 125, 126, 129.

[27] Lisa Randall. *Warped Passages*. p. 319.

[28] Ross Anderson & Robert Brady. Why quantum computing is hard—and quantum cryptography is not provably secure; Robert Brady. (2013). The irrotational motion in a compressible inviscid fluid. arXiv:1301.7540 [physics.gen-ph].

Chapter 22 — **Superfluidity**

> *"That one body may act upon another at a distance through a vacuum without the mediation of anything else, by and through which their action and force may be conveyed from one to another, is to me so great an absurdity that I believe no man who has in philosophic matters a competent faculty of thinking could ever fall into it."*
>
> Isaac Newton

Before Einstein published his theory of special relativity scientists believed that electromagnetic waves were sustained by an invisible medium *in* spacetime called the aether. This medium was assumed to have a definite velocity at each point in spacetime and was, therefore, expected to exhibit a preferred direction. This expectation was contradicted by the Michelson-Morley experiment[1] and is incompatible with the relativistic requirement that all directions within the light cone are equivalent.

In 1951 Paul Dirac smoothed out this contradiction with the suggestion that the uncertainty principle should be applied to the aether medium. He argued that when we take into account quantum fluctuations in the flow of the aether, its velocity is no longer a well-defined quantity.[2] Instead, velocity becomes a diluted notion, distributed over various possible values. In Dirac's view, the aether is best elucidated by a wave function that represents the perfect vacuum state for which all aether velocities are equally probable. Dirac's arguments initiated a conceptual transformation in which the aether went from being understood as a medium *in* spacetime to the medium *of* spacetime itself.

In 1975 K. P. Sinha, C. Sivaram and E. C. G. Sudarshan extended Dirac's ideas by formally suggesting that spacetime (still loosely referred to as the aether) was a superfluid with a metric that is macroscopically describable by the wave function.[3] Sinha et al. recognized that small fluctuations in the superfluid background obey Lorentz symmetry even though the superfluid itself is non-relativistic. Subsequently, it has been shown that the assumption that the vacuum is a superfluid allows gravity to be described as a collective effect of a specific type of small fluctuations in the superfluid vacuum.[4] Since then, it has also been shown that the assumption of vacuum superfluidity allows us to derive Schrödinger's wave equation from first principles.

Superfluidity

Superfluidity is a consequence of having zero viscosity. Superfluids and Bose-Einstein condensates share this quality. A Bose–Einstein condensate (BEC) is a state, or phase, of a quantized medium normally obtained via confining bosons in an external potential and cooling them to temperatures very near absolute zero. Bosons are particles that are governed by Bose-Einstein statistics and are not restricted from occupying the same state. As bosons cool, more and more of them drop into the lowest quantum state of the external potential (or "condense"). As they do, they collectively begin to exhibit macroscopic quantum properties. Through this process the fluid transforms into something that is no longer viscous, which means that it has the ability to flow without dissipating energy. At this point the fluid also loses the ability to take on homogeneous rotations. Instead, when a beaker containing a BEC (or a superfluid) is rotated, quantized vortices form throughout the fluid, but the rest of the fluid's volume remains stationary.

The BEC phase is believed to be available to any medium so long as it is made up of identical and indistinguishable particles with integer spin (bosons). The statistical distributions of such media are governed by Bose-Einstein statistics. It has been observed to occur in gases, liquids, and solids made up of quasiparticles, and it also accounts for the cohesive streaming of laser light. In short, all fluids, whose constituents are subject to Bose-Einstein statistics, will undergo BEC condensation once their particle density and temperature are critically related.[5]

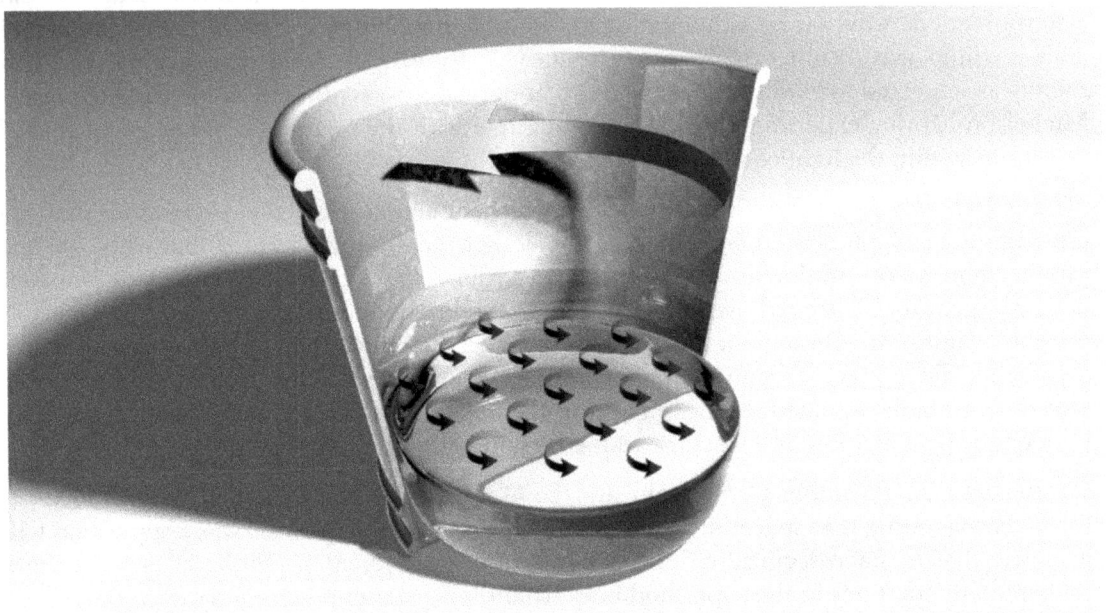

Figure 22-1 Superfluid response to rotation

Superfluids differ from regular fluids in some important ways. For example, superfluids have zero viscosity—zero internal friction—and therefore cannot dissipate energy. When we create a vortex in a regular fluid, by stirring it, it begins dissipating as soon as we stop stirring. By contrast, superfluids can sustain vortices indefinitely.

If we spin a beaker filled with a regular fluid the entire fluid ends up swirling about a single axis. A single vortex forms with a fluid velocity that is radially dependent. Near the center of the vortex the fluid moves slowly and it gradually speeds up towards the outer edge. This is not what occurs in a superfluid. Superfluids cannot take on bulk homogeneous rotations. Instead, when we spin a beaker filled with a superfluid we produce many minute quantized vortices (Figure 22-1).

These vortices appear throughout the medium while the rest of its volume remains stationary. The number of vortices that form is proportional to the rotational energy put into the system and $\frac{\hbar}{m}$, where $\hbar$ is Planck's constant and m is the mass of one quantum of the medium. Inside these quantized vortices the fluid velocity is greatest at the center and it dissipates radially (Figure 22-2).

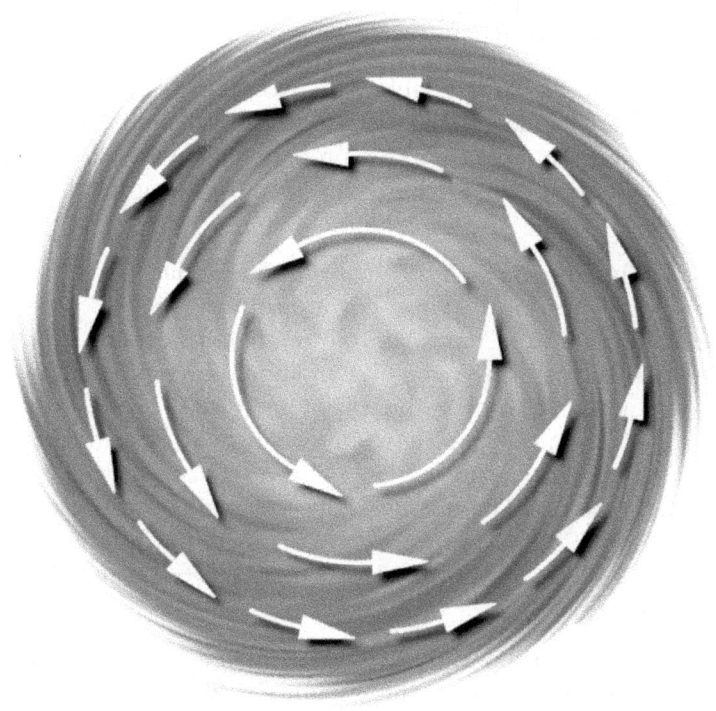

Figure 22-2 Inside a quantized vortex the fluid velocity is greatest at the center.

These vortices break the symmetry of the superfluid—they become distinguishable from the rest of the medium. It is worth mentioning that mass generation may be automatically built into a theory of vacuum superfluidity as a consequence of spontaneous symmetry breaking—as a consequence of the rules for how these quantized vortices form. In other words, the assumption that the vacuum is a superfluid leads to the expectation of metric distortions whose properties are suspiciously similar to the properties of fundamental mass particles.

Deriving Schrödinger's Wave Equation

Although the Schrödinger equation is one of the most important equations in physics, the essence of what it captures remains ontologically vague. We know that it mathematically details the behavior of the wave function ψ (or the evolution of the state vector $|\psi\rangle$), but this knowledge gives us no purchase on the question of what the state vector is. As previously mentioned, the metaphysical ambiguity beneath the wave function has given rise to dozens of interpretations of quantum mechanics, each starting from wildly different assumptions about what the wave equation tells us about the world.

Schrödinger originally introduced the wave function as de Broglie's pilot wave. He then removed de Broglie particles from it, leaving only a wave function field defined on configuration space, which significantly muddied the conceptual waters. Schrödinger tried (unsuccessfully) to make up for this lack of clarity by interpreting ψ as a charge density. Max Born subsequently interpreted it as a probability amplitude, whose absolute square is equal to a probability density.[6] To the dismay of every physics student, this definition stuck.

A "probability amplitude" is arguably just as ontologically inaccessible as the wave function it is supposed to elucidate. This has become the elephant in the physics classroom. We can lucidly talk about the amplitude of a probability wave, or the probability of finding an amplitude, but it is not so easy to claim that you are being lucid when talking about a probability amplitude.[7]

One of the beautiful consequences of the assumption that the vacuum is a superfluid is that it dissolves the vagueness of the wave equation. In a general sense, we note that if the vacuum is a superfluid, then it becomes natural to model its evolution under an expectation that it will behave as an acoustic metric. This conditional carries us to the logical expectation of an inherent wave equation that plays the role of being an intrinsic descriptor of any system that is internal to the vacuum. A metric composed of particles (molecules, atoms, quanta, etc.), which collectively contribute to the emergent notions of a density field and a velocity field, necessarily has waves. These waves are analogous to sound waves, and their propagation is a function of the compressibility of the medium. Therefore, a wave equation is a descriptor of the state of the vacuum and how that state—the connectivity of the vacuum itself—is allowed to evolve.

The hypothesis of vacuum superfludity is strengthened by the fact that we can indirectly and directly use the assumption that the vacuum is a superfluid (or a BEC) to derive Schrödinger's wave equation. Indirectly this can be achieved by taking the attributes that follow from vacuum superfluidity as fundamental guiding principles. In other words, because energy conservation, de Broglie relations, and linearity must hold within the superfluid vacuum, we can use those constraints to derive a wave equation.[8]

To develop this derivation, note that energy conservation can be expressed through the Hamiltonian H, which is classically used to represent the total energy E of a particle in terms of the sum of its kinetic energy T, and its potential energy V.

$$E = T + V = H$$

If a particle is restricted to one-dimensional motion its energy can be described as:

$$E = \frac{p^2}{2m} + V(x,t) = H$$

with position x, time t, mass m, and momentum p.

In three dimensions the equation becomes:

$$E = \frac{\boldsymbol{p} \cdot \boldsymbol{p}}{2m} + V(\boldsymbol{r},t) = H$$

where $\boldsymbol{r}$ is the position vector and $\boldsymbol{p}$ is the momentum vector.

When we extend this formalism to represent a large collection of particles, we have to keep track of the fact that potential energy is a function of all the relative positions of the particles (versus being a simple sum of the particle's separate potential energies). Consequently, the general equation becomes:

$$E = \sum_{n=1}^{N} \frac{\bm{p}_n \cdot \bm{p}_n}{2m_n} + V(\bm{r}_1, \bm{r}_2, \dots \bm{r}_N, t) = H$$

The assumption of vacuum superfluidity also naturally captures de Broglie relations,[9] because in any acoustic metric the energy of any phonon (a collective excitation in the medium's elastic arrangements) is proportional to the frequency v (or angular frequency $\omega = 2\pi v$) of its corresponding quantum wave packet. Therefore, in a superfluid vacuum we expect that

$$E = hv = \hbar k$$

Furthermore, any wave can be associated with a particle such that, in one dimension, the momentum p of the particle is related to the wavelength λ by

$$p = \frac{\hbar}{\lambda} = \hbar k$$

In three dimensions the relation becomes

$$\bm{p} = \frac{h}{\lambda} = \hbar \bm{k}$$

where $\bm{k}$ is the wave vector (the magnitude of $\bm{k}$ relates to the wavelength).

Finally, we can achieve generality in our derivation by noting that, in an acoustic metric, waves can be constructed via the superposition of sinusoidal plane waves. In other words, the superposition principle is expected to apply to the waves we mean to characterize, and more generally to any acoustic metric.

With these attributes we can structure an equation that characterizes the state of the vacuum over time, without abandoning ontological clarity as to what the wave function is. We simply need to construct an equation that is capable of translating the possible Hamiltonian of a system's particles (their possible kinetic and potential energies constrained by the aforementioned characteristics of the system) into a function of the state of the system ψ.

To trace out the steps of that derivation we would be well advised to express the phase of a plane wave in the superfluid vacuum as a complex phase factor via:

$$\psi = Ae^{i(\bm{k}\cdot\bm{r} - \omega t)} = Ae^{i(\bm{p}\cdot\bm{r} - Et)/\hbar}$$

Next we take the first order partial derivatives of this equation with respect to space and time and we get:

$$\nabla \psi = \frac{i}{\hbar} \boldsymbol{p} A e^{i(\boldsymbol{p}\cdot\boldsymbol{r} - Et)/\hbar} = \frac{i}{\hbar} \boldsymbol{p} \psi$$

$$\frac{\partial \psi}{\partial t} = -\frac{iE}{\hbar} A e^{i(\boldsymbol{p}\cdot\boldsymbol{r} - Et)/\hbar} = -\frac{iE}{\hbar} \psi$$

When we combine this with our earlier expression for energy, this leads to:

$$-i\hbar \nabla \psi = \boldsymbol{p}\psi \rightarrow \frac{\hbar^2}{2m} \nabla^2 \psi = \frac{1}{2m} \boldsymbol{p} \cdot \boldsymbol{p} \psi$$

$$\frac{\partial \psi}{\partial t} = -\frac{iE}{\hbar} \psi \rightarrow i\hbar \frac{\partial \psi}{\partial t} = E\psi$$

Now if we multiply our three-dimensional energy equation by ψ we get:

$$E = \frac{\boldsymbol{p} \cdot \boldsymbol{p}}{2m} + V \rightarrow E\psi = \frac{\boldsymbol{p} \cdot \boldsymbol{p}}{2m} \psi + V\psi$$

which naturally collapses into the following wave equation:

$$i\hbar \frac{\partial \psi}{\partial t} = -\frac{\hbar^2}{2m} \nabla^2 \psi + V\psi$$

This equation is exactly what Schrödinger came up with. Nevertheless, in this form it is slightly incomplete. To capture the full range of possible metric evolutions this equation must be augmented to allow for inter-particle interactions. To derive the wave equation in its complete form we can take a less scenic route by directly assuming that the vacuum is a BEC whose state can be described by the wave function of the condensate ψ. The particle density of this system would then be represented by $|\psi|^2$. If all of the particles in this fluid have condensed to the ground state, then the total number of atoms N in the system will be $N = \int |\psi|^2 \, d\vec{r}$. Treating these particles as bosons, and using mean field theory, the energy E associated with the state ψ takes on the following expression:

$$E = \int d\vec{r} \left[\frac{\hbar^2}{2m} |\nabla \psi|^2 + V|\psi|^2 + \frac{1}{2} U_0 |\psi|^4 \right]$$

By minimizing this expression with respect to infinitesimal variations in ψ, and fixing the number of particles in the system, we end up with the Gross-Pitaevskii equation (also called the nonlinear Schrödinger equation, or the time-dependent Landau-Ginzburg equation).

$$i\hbar \frac{\partial \psi}{\partial t} = \left(-\frac{\hbar^2 \nabla^2}{2m} + V + U_0 |\psi|^2 \right) \psi$$

where: m is the mass of the bosons, ψ is the external potential, and U_0 represents inter-particle interactions. Note that when inter-particle interactions go to zero this equation reduces to Schrödinger's original equation.

From this we see that it is possible to derive Schrödinger's wave equation from first principles. More significantly, if the assumption of vacuum superfluidity is correct, then this connection offers us the unprecedented ability to ontologically access what the wave equation means and where it comes from. The state of the vacuum is explicitly tied to this wave equation because it behaves as a superfluid. In short, the wave equation characterizes the state of the vacuum, excitations in the periodicity of the medium's particles, and all of its possible evolutions. It describes how distortions within the metric evolve by detailing how the vacuum itself evolves.

Analogue Gravity

Analogue gravity models arose out of the recognition that supersonic fluid flow can generate a *dumb hole*, which can be treated as the acoustic analogue of a *black hole*. These dumb holes are consequences of the fact that moving fluids drag sound waves along with them. If a fluid flow becomes supersonic within a certain region, the sound waves within that region will be dragged in such a way that it is impossible for them to propagate back upstream.[10] Such fluid flows form acoustic horizons (also known as sonic horizons) that are analogous to the event horizons found in general relativity. This analogy is so precise that it even mimics the presence of phononic Hawking radiation. But what if it is more than just an analogy?

> *The search for a quantum theory of gravity is fundamentally a search for an appropriate mathematical structure in which to simultaneously phrase both quantum questions and gravitational questions.*
>
> Matt Visser[11]

The assumption that the vacuum is an isotropic inviscid fluid (a superfluid) makes it possible to model the cosmos with equations that have ontologically accessible interpretations. As we saw in the previous section, treating the vacuum as a superfluid allows us to specify an effective metric and accommodate a wave equation as a descriptor of that metric. In general, the phenomena that we associate with superfluidity (propagations of collective excitations, interference effects that originate from the phase of the condensate wave function, etc.) are encoded, in simple and compact form, by the Gross-Pitaevskii equation.[12] We now turn our attention to the suggestion that an analogue of general relativity can be constructed from the geometrical acoustics of this non-relativistic theory.

The formal encasement of vacuum superfluidity makes it possible to capture the general relativistic notion of spacetime curvature, while converting that notion into a precise mathematical and physical statement. Analogue models, which make use of geometrical acoustics, turn out to naturally "capture and accurately reflect a sufficient number of important features of general relativity."[13] In their paper, *Analogue gravity from Bose-Einstein condensates*, Carlos Barceló, S Liberati, and Matt Visser mathematize the claim that superfluidity encodes a general relativistic system.[14]

According to their findings the generalized nonlinear Schrödinger equation that falls out of the assumption that the vacuum is a superfluid is what actually constrains the acoustic metric, as opposed to Einstein's equations of general relativity. The great insight here, according to Barceló et al., is that this generalized nonlinear Schrödinger equation "has a $(3 + 1)$-dimensional 'effective' Lorentzian spacetime metric hiding inside it."[15] And the cherry on the cake—in terms of testability—is that the effectiveness of this Lorentzian spacetime metric appears to be momentum dependent.

If the vacuum is a superfluid, then (according to Barceló et al.) at low momenta the generalized dispersion relation is expected to have the usual non-relativistic limit, at intermediate momenta the dispersion relation takes on a relativistic form, and at large momenta the dispersion relation returns to a non-relativistic form.[16]

From this we should expect a regime in which phase perturbations of the wave function behave as though they are coupled to an 'effective Lorentzian metric'. The existence of this regime is a generic consequence of vacuum superfluidity.[17] This means that it is not *a priori* implausible to mimic a gravitational field within an acoustic system. Nevertheless, to get that field to be completely accurate it may be necessary to incorporate a notion of anisotropy into the acoustics (e.g. by resolving vorticity, or curl, within the superfluid metric). In other words, spacetime curvature cannot be reproduced in its full glory unless we admit quantized vortices in the superfluid vacuum, i.e. mass particles.

This approach has the ability to solve a problem that has plagued attempts to formulate a theory of quantum gravity. If the vacuum is a superfluid, then its quantized nature naturally cuts off the ability to resolve an "effective metric" for gravity fields to be defined on. This both accounts for, and explains the failure to resolve such a metric. If the vacuum is quantized, then gravitational effects are emergent collective distortions in the metric, which do not have any direct underlying correspondence in terms of that metric. This is why wavelengths that are "long compared with the acoustic Compton wavelength see a Lorentzian "effective metric," while wavelengths [that are] short compared with the acoustic Compton wavelength probe the "high-energy" physics (which in this situation is the non-relativistic Schrödinger equation)."[18]

This perspective is in line with Sakharov's suggestion that gravity is no more fundamental than fluid dynamics[19]—a claim that our model extends to all of the forces. Fluid mechanics is not considered fundamental because there is a known underlying microphysics (molecular dynamics) that gives rise to it, and because it is just the low-energy low-momentum limit of that underlying microphysics.[20] Similarly, our quantized model provides us with a concrete explanation of the transition from the short-distance discrete realm to the long-distance 'continuum'. In this, it reveals how the spacetime manifold and metric come into focus, as a low-energy approximation (as we take an averaged-over collection of space quanta). The underlying microphysics allows for possibilities that do not conform to that approximation, but those deviations only show up at extremely high energies.

The prediction of energy dependent deviations from Lorentz invariance, which can be measured via precision tests of dispersion relations, adds to the experimental testability of this model. At low energies the spacetime manifold should have a locally Minkowskian structure. This means that it should exhibit special relativistic effects. As the ultra high energies are approached, the locally Euclidean geometry of the manifold should break down as the discreteness of spacetime begins to contribute to violations of Lorentz invariance. This will modify the dispersion relations for elementary particles.[21]

[1] Albert Abraham Michelson & Edward Williams Morley. (1887). On the Relative Motion of the Earth and the Luminiferous Ether. American Journal of Science 34: 333–345.

[2] P. A. M. Dirac. (1951). Nature 168, 906; (1952). Nature 169, 702.

[3] K.P. Sinha, C. Sivaram, & E.C.G. Sudarshan. (1976). Found. Phys. 6, 65; (1976). Found. Phys. 6, 717; (1978). Found. Phys. 8, 823.

[4] M. Novello, M. Visser, & G. Volovik. (2002). Artificial Black Holes. World Scientific, River Edge, USA, p. 391; K. G. Zloshchastiev, Spontaneous symmetry breaking and mass generation as built-in phenomena in logarithmic nonlinear quantum theory, Acta Phys. Polon. B **42** (2011) 261–292 ArXiv:0912.4139; G. E. Volovik. (2003). *The Universe in a helium droplet*. Int. Ser. Monogr. Phys. **117**, 1–507. Carlos Barceló, S Liberati, and Matt Visser, Analogue gravity from Bose-Einstein condensates. Institute of Physics Publishing, Class. Quantum Grav. **18** (2001) 1137–1156, PII: S0264-9381(01)18993-8.

[5] That relationships is encoded by the equation $T_c = \left(\frac{n}{\zeta(3/2)}\right)^{\frac{2}{3}} \frac{2\pi \hbar^2}{m k_B} \approx 3.3125 \frac{\hbar^2 n^{\frac{2}{3}}}{m k_B}$ where: T_c is the critical temperature of condensation, n is the particle density, m is the mass per boson, and ζ is the Riemann zeta function.

[6] W. J. Moore. (1992). *Schrödinger: Life and Thought*. Cambridge University Press, p. 219–220. ISBN 0-521-43767-9.

[7] Paraphrasing Wil Biddle: http://einsteinsintuition.com/2011/twisted-passages/

[8] The following two derivations are more richly elucidated at:
http://en.wikipedia.org/wiki/Bose–Einstein_condensate and
http://en.wikipedia.org/wiki/Gross–Pitaevskii_equation

[9] L. de Broglie. (1924). *Recherches sur la théorie des quanta* (Researches on the quantum theory), Thesis (Paris); (1925). *Ann. Phys.* (Paris) **3**, 22.

[10] Ibid., p. 7.

[11] Carlos Barceló, Stefano Liberati & Matt Visser. (2005). Analogue Gravity. Living Rev. Relativity, **8**, 12. [Online Article]: cited [<11-26-2012>], p. 77, http://www.livingreviews.org/lrr-2005-12

[12] Carlos Barceló, S Liberati and Matt Visser. (2001). Analogue gravity from Bose-Einstein condensates. Institute of Physics Publishing, Class. Quantum Grav. **18**, PII: S0264-9381(01)18993-8, p. 1140.

[13] Carlos Barceló, Stefano Liberati & Matt Visser. (2005). Analogue Gravity. Living Rev. Relativity, **8**, 12. [Online Article]: cited [<11-26-2012>], p. 6, http://www.livingreviews.org/lrr-2005-12

[14] Barceló et al. start with the Gross-Pitaevskii equation (writing it in its time dependent form),

$$i\hbar \frac{\partial}{\partial t}\psi(t,x) = \left(-\frac{\hbar^2}{2m}\nabla^2 + V_{ext}(x) + \lambda |\psi(t,x)|^2\right)\psi(t,x)$$

and its associated effective action:

$$S = \int dt\, d^3x \left\{ \psi^* \left(i\hbar \partial_t + \frac{\hbar^2}{2m} \nabla^2 - V_{ext}(x) \right) \psi - \frac{1}{2} \lambda |\psi(t,x)|^4 \right\}$$

Then they generalize the Gross-Pitaevskii equation by: (1) replacing the quartic $\frac{1}{2}\lambda|\psi(t,x)|^4$ by an arbitrary nonlinearity $\pi(\psi^*\psi) = \pi(|\psi|^2)$, (2) permitting the nonlinearity function to be explicitly space and time dependent: $\pi \rightarrow \pi(x, [\psi^*\psi])$ with $x \equiv (t,x)$, (3) permitting the mass to be a 3-tensor: $m \rightarrow m_{ij}$, (4) permitting that mass tensor to depend on space and time, and (5) allowing a time dependence for the confining potential $V_{ext} = V_{ext}(t,x)$.

In response to these generalizations the action becomes:

$$S = \int dt\, d^3x \sqrt{det[^{(3)}h]} \left\{ \psi^* \left(i\hbar \partial_t + \frac{\hbar^2}{2\mu} \Delta_h + \frac{\xi\hbar^2}{2\mu}{}^{(3)}R(h) - V_{ext}(t,x) \right) \psi = \pi(x, |\psi|^2) \right\}$$

where $[^{(3)}h^{-1}]^{ij}$ is the inverse of the 3-metric $^{(3)}h_{ij}$, $\frac{\xi\hbar^2}{2\mu}{}^{(3)}R(h)$ is the DeWitt term, and Δ_h is the three-dimensional Laplacian defined by
$\Delta_h \psi = \frac{1}{\sqrt{det[^{(3)}h]}} \nabla_i \left(\sqrt{det[^{(3)}h]} [^{(3)}h^{-1}]^{ij} \nabla_j \psi \right)$. (Carlos Barceló, S Liberati and Matt Visser, "Analogue gravity from Bose-Einstein condensates" Institute of Physics Publishing, Class. Quantum Grav. **18** (2001), PII: S0264-9381(01)18993-8, p. 1142.)

Varying this generalized action with respect to ψ^* gives us a generalized nonlinear Schrödinger equation:

$$i\hbar \frac{\partial}{\partial t} \psi(t,x) = -\frac{\hbar^2}{2\mu} \Delta_h \psi(t,x) - \frac{\xi\hbar^2}{2\mu}{}^{(3)}R(h)\psi(t,x) + V_{ext}(t,x)\psi(t,x) + \pi'(\psi^*\psi)\psi(t,x)$$

[15] Carlos Barceló, S Liberati & Matt Visser. (2001). Analogue gravity from Bose-Einstein condensates. Institute of Physics Publishing, Class. Quantum Grav. **18**, PII: S0264-9381(01)18993-8, p. 1142.

[16] More specifically, Barceló et al. show that at low momenta $(k \ll m_0)$ the generalized Bogoliubov dispersion relation $\omega(k) = \sqrt{m_0^2 + k^2 + \left(\frac{k^2}{2m_\infty}\right)^2}$ has the usual non-relativistic limit

$$\omega(k) = m_0 + \frac{k^2}{2m_0} + O(k^4)$$

at intermediate momenta $m_0 \ll k \ll m_\infty$ the dispersion relation takes on a relativistic form, and at large momenta $k \gg m_\infty$ the dispersion relation returns to a non-relavistic form

$$\omega(k) = \frac{k^2}{2m_\infty} + m_\infty + O(k^{-2})$$

[17] Carlos Barceló, S Liberati & Matt Visser. (2001). Analogue gravity from Bose-Einstein condensates. Institute of Physics Publishing, Class. Quantum Grav. **18**, PII: S0264-9381(01)18993-8, p. 1153.

[18] Ibid.

[19] Sakharov, A.D. (1968). Vacuum quantum fluctuations in curved space and the theory of gravitation. *Sov. Phys. Dokl.*, **12**. 1040–1041; M. Visser. (2002). Sakharov's induced gravity: A modern perspective. *Mod. Phys. Lett. A*, **17**, 977–992. Related online version (cited on 31 May 2005): http://arXiv.org/abs/gr-qc/0204062.

[20] Carlos Barceló, Stefano Liberati & Matt Visser. (2005). Analogue Gravity. **Living Rev. Relativity, 8**, 12. [Online Article]: cited [<11-26-2012>], p. 76, http://www.livingreviews.org/lrr-2005-12

[21] Ibid. p. 76–77.

Chapter 23 — **Illuminating Dark Matter**

Junior year, one week before high school prom.

In accordance with our pact, I looked through the yearbook and chose for him. "Heather Thompson," I said. Heather was in choir with us, but neither of us had ever talked to her. This was the perfect opportunity to get to know her better. Brian was already getting nervous. The fact that we didn't have time to do anything creative made him even more nervous.

"I don't know where she lives," Brian said. We turned to the phone book, expecting "Thompson" to be a very common last name. But to our surprise there were only a few Thompson's listed in the area. This encouraged us. Brian dialed the first number. "Hi, is Heather there?" he asked. Then he quickly hung up. "What did they say?" I asked. "They just said "who?" so I hung up." Brian dialed the second number. "Hi is Heather there?" He hung up again. "What happened?" "They said just a minute," he said with a hint of detective pride in his voice. We wrote down the address and then drove to the store to buy some balloons, roses, and a card.

We parked the wooden panel station wagon a block away from our destination, so it wouldn't be seen, and started walking. I looked at Brian. He was tall and looked older than he was. "What?" he asked defensively. He was holding an armful of long stem roses, walking around in the dark with a bunch of colorful helium balloons, and to top it all off he was trying to force himself to smile, but it just looked goofy. "Nothing," I said as I tried to erase the expression from my face.

The house was a cookie-cutter split level structure with four block windows on the front. As Brian went up the stairs leading to the front door, I hid in the bushes below, just ten feet to his right. He straightened his posture and rang the doorbell. The room above me lit up, and several seconds later the father of the house opened the door wearing a robe, leaving the screen door closed. "Yes?" he said in a rather ornery tone. Not wanting to appear nervous or scared, Brian kept forcing his big smile and spoke in his deep voice, "Can I talk to Heather please?" "What for?" the man asked. Brian looked at the obvious balloons and flowers that he was holding and said, "I want to ask her a question." The man glared at Brian in a strange way, slowly closed the door and then locked the deadbolt. Shortly after, the light in the room upstairs turned off.

Brian was still standing there wondering what to do. "Let's go," I whispered. "This doesn't feel right." Brian disagreed. He wanted this to be over. He certainly didn't want to have to come back and try again. So he rang the doorbell again. The light came back on and the man came back down to the door. As he opened the door it was obvious that the chain lock was still in place. "I've already called the cops," he said. "I suggest you leave without any trouble." Just then the window above me slightly slid open and the barrel of a shotgun protruded out, pointing at Brian. "I just want to ask Heather a question," Brian pleaded. "She's asleep," the man said in a very forceful tone.

Later Brian told me that at this point he started to think that somehow Heather knew that she was going to be asked out and that she had told her parents to play this gag on whoever asked her out so that she could tell everyone the story. "Can you please wake her up? This is kind of important," Brian insisted. The man was getting very upset. Preparing to slam the door again, he clenched his fists and yelled, "She's twelve years old!"

Brian walked away stunned and confused, and apparently completely oblivious to the fact that I was still stuck in the bushes—beneath a shotgun. As I watched Brian's aimless saunter, listening to the sound of my rapidly beating heart, I realized that sometimes the stories in our heads are what blind us most from seeing how things really are.

In the Dark

"...as long as the identity of the most dominant form of matter in the universe today remains unknown to us, we cannot claim that our scientific picture of the world even approaches completion."

Lawrence Krauss

Today "dark matter" is believed to make up 26.8 percent of the mass-energy content in the universe. If this is true, then dark matter is more prevalent in our universe than normal matter, otherwise known as baryonic matter, which makes up only 4.9 percent. The remaining 68.3 percent is said to exist in the form of "dark energy". If all matter-energy can be explained as a geometric distortion in higher-dimensions, then specific forms of matter-energy should correlate to unique kinds of geometric distortions. Therefore, to understand dark matter we must identify the geometric differences that separate it from baryonic matter.

The first difference between baryonic and dark matter, is that baryonic matter has the property of being microscopically identifiable—its particles have specific, microscopic locations. Dark matter, on the other hand, is something that has only shown itself on galactic or supergalactic scales. Although the standard cosmological model currently assumes that solar systems, dwarf galaxies, etc. are seeded by dark matter, making dark matter the metaphorical pit of the peach, actual observations of dark matter phenomena give us reason to believe that it only exists as some sort of halo whose effects trail off to zero at about five to ten times the visible size of the host galaxy.[1]

Our first hints of the existence of dark matter surfaced in 1933 when Fritz Zwicky, an astronomer at the California Institute of Technology working at Mount Wilson, noticed that there was something amiss with the motions of the galaxies in the Coma cluster, some 320 million light-years away. The speeds of these galaxies relative to one another "were so great that the gravitational attraction due to the visible material in the galaxies should not have been sufficient to hold them together. Yet held together they were!"[2] Somehow there was more gravity holding this cluster together than the observable mass of the cluster could account for.

Zwicky realized that this observation had two possible explanations. Either gravity was different (stronger) in the Coma cluster than in our neighborhood, or the Coma cluster contained an unseen form of matter, which provided the additional gravitational tug needed to explain those galactic orbits. Understanding that the strength of gravity was coded by a universal constant (G), Zwicky surmised that unseen swaths of gas between the galaxies (undetected mass) accounted for the additional gravity.

Zwicky's guess was a reasonable one, but in the 1970s it was no longer an allowed solution. Astronomers now had high-resolution measurements of the rotational velocities of individual galaxies and the orbital velocities of individual stars within those galaxies. As they examined these new measurements, they were astonished to discover that the outer rims of individual galaxies were spinning faster than thought allowed. The velocities of the stars on the outer rims of these galaxies were so great that they should be escaping the galaxy they were in—slingshotting into the vast depths of space—instead of remaining gravitationally attached to their home galaxies. Unless, that is, there was much more gravity acting upon those stars than the sum of all the galaxy's constituent stars were providing.

These observations presented an intriguing mystery to say the least. That mystery was complicated by the fact that the same data set showed that the stars near the center of these galaxies orbited normally. The stars orbiting in the outer regions of these galaxies required the presence of more mass, or additional gravity, to explain their motions. But the interior of these galaxies had gravitational effects that appeared to match up with the observable quantities of mass. This was a truly strange development.

Zwicky's rather obvious explanation was no longer consistent with the data. Permeating swaths of unseen gas could no longer account for the missing mass, unless they were somehow concentrated around the outer edges of the galaxies in a dense halo. But if they were concentrated in this manner, then they would give off a characteristic infrared signature, which is not observed.

Our best formation models tell us that large swaths of gas cannot be associated with mature galaxies. If the missing mass were really made up of gas, which is composed of normal matter (protons, neutrons, and electrons), then the pressure of that gas would have completely stifled gravitational collapse and stellar formation. Galaxies still containing large swaths of gas should consequently have little to no stars. They should have never matured. But galaxies with dark matter phenomena have clearly matured. They are chock full of stars. Furthermore, if all the gravitational effects of dark matter stemmed from ordinary atoms, "then the large numbers of neutrons and protons and electrons would affect calculations of the abundance of the light elements produced in the first few minutes of the expansion of the universe, so that the results of these calculations would no longer agree with observation."[3]

This means that it is not just formation models of galaxies and galactic clusters that require the effects of dark matter to come from something other than ordinary matter. The inflationary Big Bang theory requires it also. Therefore, dark matter simply cannot be composed of huge swaths of gas (baryonic matter). Something else has to make up the haloes of these galaxies—something that has gravity but doesn't absorb or emit electromagnetic radiation. What could that something be?

For decades now, astronomers and particle physicists have fretted over that question to no avail. The best that can be said at this point is that *something* is causing these effects. The effects are undeniably real. But because we have no idea what causes them, we just call the responsible party "dark matter".

Today the existence of dark matter is a commonly accepted part of cosmology even though no one has ever detected a single particle of it in the lab, or figured out what the 'particles' of dark matter are. We keep collecting more and more astronomical observations that support our conclusion that dark matter must exist—that out there in the haloed outer regions of galaxies and galactic clusters something is providing the extra gravitational glue to hold these things together—but the nature of that "something" has remained a complete mystery.

Science has encountered a mystery similar to this before. In the 1840s astronomer Urbain Le Verrier noticed that Mercury's perihelion advancement was greater than it should be. In order to explain this effect Le Verrier proposed that a previously undiscovered body of mass—a planet he named Vulcan after the Roman god of fire—was to blame for the extra gravitational effects Mercury was experiencing. He and other astronomers spent decades looking for the missing body of mass, but there wasn't one.[4]

In the beginning of the 20th century, the idea that there was some kind of matter between Mercury and the Sun, causing the additional gravitational effects that were needed to explain Mercury's excessive perihelion advance, was a reasonable guess.[5] But it was wrong. The solution to Mercury's perihelion advance came seventy years later, when Einstein's general theory of relativity superseded Newtonian mechanics. In this solution Vulcan did not pull on Mercury's orbit. There was no object possessing the "missing" mass. Instead, the fabric of spacetime was geometrically richer than previous expressions allowed for, and its additional geometric character accounted for Mercury's strange behavior.

It is possible that Urbain Le Verrier's search for Vulcan is being echoed in our several decade long search for the thing we call dark matter. If we are to use the pages of history to our advantage, then perhaps our focus should not be so heavily placed on what these dark matter objects are. Maybe, instead, we should focus our attention on the possibility that a never-before considered warp in the geometry of spacetime is responsible for the effects of dark matter.

Some cosmologists have tried to avoid the trenches of this mystery by trying to link dark matter to the formation of galaxies, claiming that dark matter is responsible for seeding galactic evolution. These formation models are tied to the assumption that the amount of dark matter in a galaxy should be roughly proportional to its size. This requirement has been contradicted by observation, and by the theoretical problem that if dark matter had actually seeded galaxy formation, then there should be 10 to 100 times more small galaxies than there are in the deep recesses of space.

Dwarf galaxies, like the ones that orbit the Milky Way, tend to rotate so quickly that there is no escaping the conclusion that they are richer in dark matter than galaxies many times their size. Some larger, hotter galaxies have little to no dark matter in them. NGC 3379 fits into this category. "Measurements of the orbital velocity of gas clouds in NGC 3379 suggest it contains no dark matter at all. Yet star clusters circling further out do seem to be expressing an extra gravitational pull."[6] It's as if the haloed region of dark matter, if there is one, lies completely outside the reach of the stars in NGC 3379.

Additional oddities offer further clues. Although the shapes of galaxies vary from spiraled planar pancakes to cigar-shaped blobs, dark matter always seems to exist in a near-spherical halo around their host galaxies. These haloes are not, as one might naïvely expect, necessarily the same shapes as their parent galaxies.

As Lawrence Krauss puts it, "dark matter is distributed uniformly in a spherical volume when the luminous matter is restricted primarily to a disk."[7] A special type of spiral galaxy has confirmed that dark matter haloes are spherical in shape. These galaxies are similar to the more common spiral disk shaped galaxies, but they have an additional "polar ring" of luminous material that is oriented perpendicular to the main disk. By comparing the velocity of the objects in the polar ring with those in the main disk at the same radial distance from the center of the galaxy, astronomers have determined that dark matter halos are spherically symmetric within an accuracy of ten percent.[8]

What are these halos? Why do they surround galaxies of visible matter?[9] How do they form? Why do they only exist macroscopically? Why are they always spherical, even when their host galaxies are not? Better yet, why is the motion of these dark matter haloes determined by the gravitational dynamics of the baryonic matter in their host galaxies, when dark matter accounts for far more mass than the baryonic matter in these galaxies? And why is it the case, if what Pierre Sikivie has argued proves correct, that dark matter is only just now falling into the galaxy? Why would these haloes be shrinking over time?[10]

A Natural Explanation

> *"...whatever impression we have, say, of living in a three-dimensional space... is somehow flatly illusory."*
>
> *David Albert*[11]

The assumption that the vacuum is quantized opens us up to a natural explanation for the effects of dark matter. If all forms of matter-energy are geometric distortions in the vacuum—ways in which its geometry differs from Euclidean expectations—then dark matter haloes can be understood as a boundary between different geometric phases of space.

What would support the boundary between these different vacuum phases? To answer that question we need only to note that in neighborhoods populated by stars, the presence of mass (vortices) breaks the symmetry of the vacuum. These vortices align the motions of the quanta, increasing the vacuum density, but the magnitude of that alignment diminishes with distance.

In the cold depths of space, randomness of motion restores symmetry to the vacuum. This is a different vacuum phase—a different geometric arrangement, and density. This geometric metamorphosis defines a phase change (Figure 23-1).

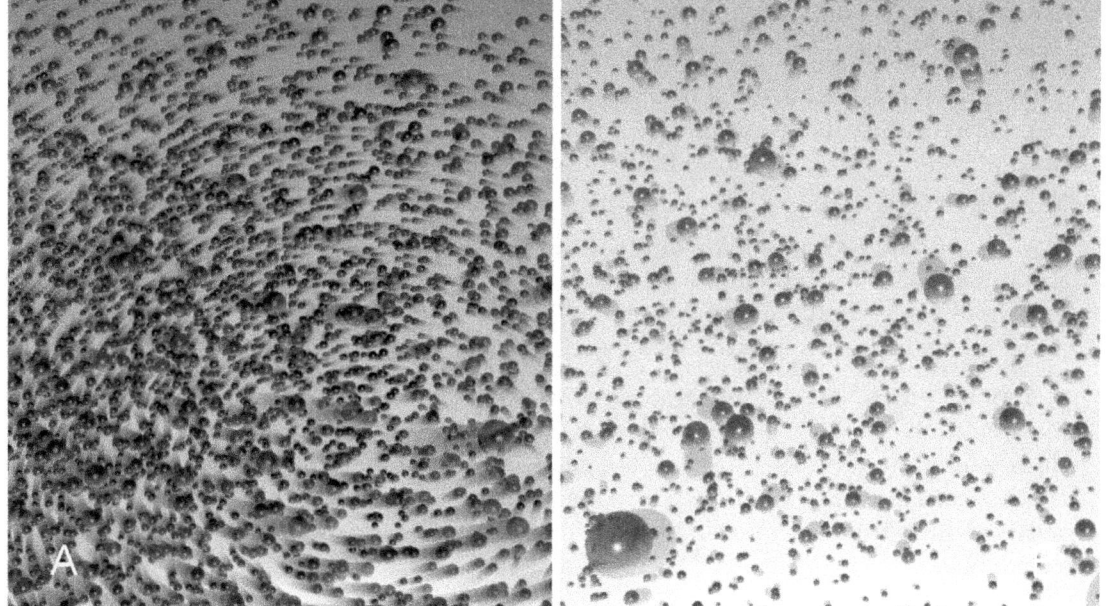

Figure 23-1 Vacuum phases near matter (A) and in the cold depths of space (B)

By understanding the arrangements of vacuum quanta in regions that support different degrees of vacuum symmetry we dissolve the mystery of dark matter. The conditions near galactic centers are different than those in the vast expanses that stretch far beyond galaxies, galactic clusters, and superclusters. This is because vacuum density varies in step with the degree to which symmetry is broken. Eventually, a phase change occurs in the vacuum. The transition zone between these two geometric vacuum phases is, by definition, a macroscopic spacetime warp. It exists in the form of a spherical halo that surrounds its host.

This simple explanation allows us to corral dark matter phenomena in a natural way. We don't find localized dark matter particles, because phases (and phase changes) are defined as averaged-over collections—they dissolve when we try to pinpoint them. The interactive mixing of the quanta that produce this averaged-over effect constructs haloed phase transition zones that are near spherical in shape, no matter what shape the interior galaxy is, because symmetry-breaking and temperature are radially communicated through space.

Dark matter swaths are centered on star filled regions because luminous matter plays an active role in authoring them. This is why luminous matter always exists within haloes,[12] or put the other way around, why dark matter is almost always "contained outside the region in which most of the light is emitted."[13] (Some time delay exceptions may exist in the aftermath of collisions between galaxies with these haloes.) These haloes provide extra gravity because a change in vacuum density is a warp in the vacuum—and gravity is nothing more than a warp in the geometry of the vacuum.

This gives rise to several testable predictions. For example, if a change of phase in the vacuum is responsible for dark matter effects, then the radius of each haloed transition zone should depend on the mass and temperature of its host galaxy. If a significant amount of stars in a particular galaxy are burned out, the radius of its phase transition zone—its dark matter halo—should be smaller than that of another galaxy of identical mass, whose stars are still shining. And when a supernova increases the output luminosity of its host galaxy, it should increase the radius of the galaxy's dark matter halo. Furthermore, dwarf galaxies that produce less heat should have smaller halo radii than warmer, larger galaxies—a condition that is supported by current data.

We should also find that the interior edges of dark matter haloes were further out in the distant past, because the background temperature of space was higher. As space cooled these haloed regions should have reduced their interior radii (at least within galaxies giving birth to little to no stars).

If we can find several similar Einstein rings in succession, or similar spiral galaxies with polar rings, dispersed throughout the vast regions of spacetime, then we should be able to check those observations against this prediction. On average, the inner radii of those haloed regions should have shrunk as the universe cooled.

Another way to test this model is to measure the internal temperatures of spiral galaxies compared to the temperatures inside bar-shaped galaxies. As P. J. E. Peebles and J. Ostriker pointed out in the early 1970s, spiral disk galaxies are unstable when left on their own. Over time they collapse into rotating bar-shaped galaxies unless they are stabilized by being embedded in a spherical distribution of matter (a warp in spacetime) of comparable or greater mass.[14]

If dark matter is really an expression of a phase change in the vacuum, then spiral galaxies that have collapsed, or are collapsing, into bar-shaped galaxies should be warmer in temperature than stable spiraled disk galaxies. Without dark matter haloes, spiral galaxies collapse relatively quickly (in astronomical terms).

An increase in temperature pushes the interior edge of the galaxy's dark matter halo outward—beyond the reach of the spiraled arms—and therefore, favors the collapse. Cooler galactic temperatures, on the other hand, produce dark matter haloes that intersect the spiraled arms and, consequently, stabilize the spiraled disk shape. By measuring how common spiral galaxies are, we can make statistical claims on how long the average spiral galaxy survives before collapsing into say a bar shaped galaxy.

Because dark matter haloes keep the outer stars in galaxies from going through with the collapse, they explain the abundant population of spiral galaxies that we observe. The stars that are stabilized by the halo gravitationally pull on the stars near them, and those stars pull on their neighbors, and so on, slowing the collapse and stabilizing the whole system. This means that if the radii of dark matter haloes are indeed temperature dependent, as our model predicts, spiral galaxies that are characterized by cooler temperatures should statistically remain stable longer than those characterized by warmer temperatures.

Haloes associated with individual galaxies are therefore predicted to have radii that correspond with the total mass and energy output of the stars in those galaxies. If the energy of a galaxy's output decreases, if the galaxy cools, the radius of its dark matter halo will decrease. If, on the other hand, the energy output of a galaxy increases, e.g. if one of its internal stars goes supernova, the radius of the halo surrounding that galaxy should increase. These haloes are simply interactive boundaries between vacuum phases.

Under this proposal the search for particles of dark matter becomes a bit like Urbain Le Verrier's search for the planet Vulcan. Dark matter is not composed of exotic elementary particles of mass. Dark matter does not exist in the form of particles at all. The galactic haloes that give rise to dark matter effects are born of non-localizable phase changes in the vacuum. And the radius of each dark matter halo depends on the mass and energy output of its host source.

[1] Lawrence Krauss, *quintessence—The Mystery Of Missing Mass In the Universe*, p. 219.

[2] Ibid., p. XXIII.

[3] Steven Weinberg. *Dreams Of A Final Theory*, p. 267.

[4] *The Economist*. (2009, March 8). Wanted: Einstein Jr., pp. 89–90.

[5] Steven Weinberg. *Dreams Of A Final Theory*, p. 94.

[6] Stuart Clark. (2008, March 8). Cosmic enlightenment. *New Scientist*, pp. 29–30.

[7] Lawrence Krauss. *quintessence—The Mystery Of Missing Mass In the Universe*, p. 74.

[8] Ibid., p. 75.

[9] See work presented at the American Astronomical Society in Austin, Texas on January 10, 2008 by Tommaso Treu of the University of California, Santa Barbra. (2008, January 12). Where the shadows lie. *The Economist*, p. 72.

[10] Lawrence Krauss. *quintessence—The Mystery Of Missing Mass In the Universe*, p. 289.

[11] David Albert. Elementary Quantum Metaphysics, p. 277; Craig Callender—Review of David Albert's work.

[12] Amanda Gefter. (2007, March 10). Don't mention the F word. *NewScientist*, p. 33.

[13] Lawrence Krauss. *quintessence—The Mystery Of Missing Mass In the Universe*, p. 74.

[14] Ibid., p. 75.

Chapter 24 — Bohmian Mechanics

"If the universe is not the way Bohm describes it, it ought to be."

John P. Briggs and F. David Peat[1]

Near the edge of the Upper Amazons, Yauliyacu, Peru.

"So what it is like in America?" my new acquaintances asked in Spanish. We were standing beneath the only streetlight in the town, which was tapping power from the line that fed the nearby mine. "Its different," I said. "Different how?" "Well, for one thing, when we go to a store to buy groceries, we don't see counters full of chickens, with blood dripping everywhere, and flies swarming about." They gave me a look of confusion. "You don't have chickens in America?" "We have chicken," I replied. "We just don't have the blood and flies." They looked at each other and began laughing. "Chickens with no blood, you're a funny man."

A light brown mangy dog, with extremely snarled-up hair, approached from the right, pranced down the dirt road, and passed us, as if it hadn't seen us at all. "What do you do for work," they asked. "I work for NASA," I said. They stared at me with expressionless faces. "Have you heard of NASA?" They looked at each other for a moment then returned their gaze to me, without saying a word. I looked up and waited to catch a satellite drifting across the heavens. "There, you see that," I asked, as I pointed to the satellite. "I work for the organization that put that up there." Their eyes widened. "You know," I added, "the people that put a man on the moon."

This led to convulsive laughter. "Man on the moon," one said to the other. Their whole bodies were participating in their response. "Funny man," one blurted out as he gasped for air. "Maybe that's where the chickens with no blood came from," the other replied. "Yes, yes, they traded one man for a hundred bloodless chickens."

Another light brown, mangy dog approached from the right, and pranced down the dirt road, just as the previous dog had. Still laughing, but trying to catch their breath, they asked, "And what... what kind of gold do you have in America?" "What do you mean?" I replied. "Our mine is where the gold comes from. We have twenty-four carat gold. What kind do you have—fifty carat?" They mockingly snickered as they waited for my response. "Usually twelve to eighteen carat I think." "Eighteen carat?" They sounded genuinely surprised. "You mean you don't have real gold in America?" They took on a slightly more proud posture.

Again the dog approached from the right, pranced down the dirt road, and passed us. "What is that dog doing?" I asked. "What do you mean?" they replied. "That's the third time I've seen that dog come from over there, pass us, and disappear into the darkness over there." They stared at me as if they were expecting a joke. "What is the dog doing out here?" I asked. In a matter-of-fact way they responded, "He's looking for bitches." I could tell my question was lost in translation.

Attempting to bridge the gap I said, "Where I come from, we don't see dogs walking down the street in the middle of the night." "Why not?" they asked. "Well in America people own dogs and they keep them on leashes when they are not in the house." "In the house?" they asked. "Yes, they are seen as pets." They didn't seem to understand.

337

"Let me put it this way," I said. "Our grocery stores have many aisles that stretch from one telephone pole to the next." I pointed to the poles around us for clarity. "One aisle is for spices, another for breakfast cereals, or canned vegetables, and another aisle is just for cat and dog food."

Once again, they burst into loud, uncontrollable laughter. "Why would anyone feed a dog?" one of them asked. "They feed themselves." The other laughing Peruvian reached out to balance himself against the wall and said, "Are all Americans as funny as you?"

"I'm serious," I asserted. "Sure, sure," he said with a wink. "You buy food for the cats and dogs, because you have so much money that you don't know what to do with it." "And what do you do with your eighteen carat jewelry?" the other one added. "Do you put it in the garbage?" They doubled over in laughter.

Coming out of the Dark

> *"Everyone takes the limits of his own vision for the limits of the world."*
>
> *Arthur Schopenhauer*

Physicists today are largely unaware of the fact that quantum mechanics is perfectly choreographed by the mathematics of the de Broglie-Bohm theory, otherwise known as Bohmian mechanics. Despite the fact that Bohm's formalism is entirely deterministic, and less vague than the standard interpretation of quantum mechanics, it has only been widely recognized and embraced among philosophers of physics.

This chapter will serve as an introduction to Bohmian mechanics and its ontological implications. Among other things, this introduction will include an expansion on the historical events, or "unfortunate accidents," that have led to the present ignorance of the fact that Bohm's formalism offers superior mathematical clarity. In part, this will serve as an explanation for why the orthodox or "standard" interpretation of quantum mechanics is still held by the majority of physicists today—something that is arguably one of the greatest intellectual tragedies of our time. It is hoped that, at minimum, this background knowledge will free the reader from any tendency to blindly accept an interpretation of quantum mechanics simply because it happens to represent the majority view. By reclaiming a sense of healthy skepticism, and a general awareness of how current views were formed, we stand to expedite future efforts to secure a theory of everything.

The mathematics supplied in this chapter should be useful to those sufficiently versed in the prerequisites. However, because my goal throughout this book has been to cover all significant concepts with as much intuitive accessibility as possible, an unfamiliarity of advanced mathematics should not inhibit the reader from following the value of the discussion in this chapter. I've made it my strategy to explain the concepts behind the math independent of the actual equations. Therefore, it is hoped that the equations in this chapter will satiate the hunger of those who have developed the taste for representing the world through symbols, while the explanations of those equations will allow everyone else to adequately follow the discussion.

In as much as the Bohmian formalism is aligned with the assumptions we have been exploring in this book, it offers us the chance to expedite the process of formally encoding our construction. This sort of alignment has played out several times in history. For example, when Einstein was developing his theory of general relativity he realized that he needed a mathematical model for curved spaces. Georg Freidrich Bernhard Riemann had developed such a model, known as Riemann geometry.

Similarly, when Werner Heisenberg was twenty-five years old he set out to develop a means of organizing experimental data into tabular form in order to formalize the infant theory of quantum mechanics. Soon after he began this task he discovered that an Irish mathematician named William Hamilton had already developed such a method. Hamilton's *matrices* were a method of organizing data into arrays, or mathematical tables—a method, as it turned out, that had been around approximately one year longer than Heisenberg himself.

Before Heisenberg applied Hamilton's matrices to the formalism of quantum mechanics, matrices were considered purely abstract—fringe mathematics that was inapplicable to the real world. But Heisenberg found this mathematical structure to be more like a "precut piece," a natural reflector of the microscopic realm.[2]

Had Einstein not learned of Riemann geometry, he may have had to invent it himself, and this task would have undoubtedly postponed the completion of his masterpiece. Had Heisenberg not discovered Hamilton's matrices he may also have had to invent what he subsequently called matrix mechanics for himself.[3] To say the least, this would have slowed him, and therefore quantum theory, down.

By contrast we note that when Newton was attempting to formulate his theory of gravity, the mathematics of his time was unable to weigh in on one of the central assumptions of his formalism—that the mass of a voluminous body can be treated as if it all belonged to a single point (called the center of mass). Being able to make this assumption would allow Newton to simply define the positions of voluminous bodies, and this in turn would reduce the equation for the gravitational *force* between two bodies to $f(g) = G \frac{m_1 m_2}{r^2}$, where r is the distance between the two centers of mass. But in order to mathematically justify this assumption Newton had to invent calculus from scratch.[4] This delayed the publication of his theory of gravity by twenty years (until 1687).

With our new vision for vacuum superfluidity, we find ourselves in a situation similar to what Einstein, Heisenberg and Newton faced. We need to find an existing mathematical construct that can relate to that picture, or we need to invent new mathematics capable of doing the job. This situation motivates us to look at David Bohm's mathematical approach, and to see if that approach can be logically extended to represent our complete model.

Finding Bohm

> *"Bohmian mechanics is empirically equivalent to orthodox quantum theory."*
>
> *Sheldon Goldstein*

In addition to the Schrödinger equation, which is shared among all quantum mechanical interpretations, Bohmian mechanics[5] is completed by the specification of actual particle positions, which evolve (in configuration space) according to the *guiding equation*. This combination elegantly restores determinism into the dynamics of physical reality; accounting for all the phenomena governed by nonrelativistic quantum mechanics—from spectral lines and scattering theory to superconductivity, the quantum hall effect, quantum tunneling, nonlocality, and quantum computing.

On top of this, Bohm's theory magnificently elucidates state evolution without elevating the role of the observer to something mystical.[6] This reveals that the stochastic property of the orthodox approach of quantum mechanics, which manifests in state vector reduction, is merely a reflection of the incompleteness of that approach.[7]

By declaring that a particle's wave function interacts with the particle and *guides* or *pushes* the particle around in a way that determines its subsequent motion, this approach explicitly captures nonlocality in a way that introduces a new level of clarity. For example, in the double-slit experiment Bohm's approach explains that each individual particle goes through one slit or the other, while its wave function goes through both and suffers interference. Because the wave function guides the particle's motion, the particle is likely to land where the wave function value is large and it is unlikely to land where it is small.[8]

State vector reduction never occurs in this model (the wave function never collapses) because the state vector exists as a separate element of reality. Orthodox quantum mechanical interpretations, which are plagued with state vector reduction, describe a system as having many possible outcomes prior to observation and only one outcome after observation. This introduces a definite temporal asymmetry. Bohm's model is not plagued with this problem. It portrays one single outcome as a possibility both before and after observation, because it sharply specifies an exact state of space. This restores time symmetry and allows a deterministic evolution.

Bohm's model has been praised as a cure to the conceptual difficulties that have plagued quantum mechanics because it elegantly does away with much of the subjectivity and vagueness found in the standard approach. Despite this, mainstream physicists haven't embraced this interpretation, or examined it in depth. In fact, the large majority of them haven't even heard of it. This is embarrassing, surprising and frustrating.[9] If Bohmian mechanics provides a cure to modern quantum mechanical philosophic complacency, then why have there been so few to study the richness of this elegant formalism?

James Cushing notes that, Bohm's formalism has been systematically ignored and misunderstood for "reasons having more to do with politics, positivism, and sloppy thought, than for reasons central to physics."[10] Several historically perpetuated fallacies have discouraged people from giving Bohm's formalism a real look. First off, the model suggests that there is something called *configuration space*, asserting additional variables and creating a dualism almost Platonic in scope.[11] This counts as a "strike against" Bohmian mechanics only in the sense that it conflicts with assumptions that have become popular among physicists. In addition to this, mainstream quantum physicists have been trying to map reality based on the assumption that wave functions somehow collapse upon measurement—contrary to the fact that Schrödinger's equation demands that they do not. Bohm's model denies wave function collapse. Therefore, although it is simple and in agreement with Schrödinger's equation, it has been overlooked because it has not been in accord with popularized mainstream efforts.

"New opinions are always suspected, and usually opposed, without any other reason but because they are not already common."

John Locke

Physicists also compulsorily reject Bohm's construction because it explicitly builds nonlocality into its framework—even though violations of Bell's inequality have conclusively shown that the vacuum of our universe is nonlocal.[12] This is perplexing. Nonlocality is unavoidable in any theory that recovers the predictions of quantum theory.[13] Therefore, any criticism of a theory that displays Nature's nonlocal feature in an obvious way is both unfounded and counterproductive. Despite this, Bohm's inherent explication of nonlocality continues to be obnoxiously mistaken as a strike against it instead of for it.

"That the guiding wave, in the general case, propagates not in ordinary three-space but in a multidimensional-configuration space is the origin of the notorious 'nonlocality' of quantum mechanics. It is a merit of the de Broglie-Bohm version to bring this out so explicitly that it cannot be ignored."

John Bell

Finally, and most significantly, Bohm's theory has been neglected by physicists who thought that additional variable theories had been proven impossible.[14] *Impossibility theorems*, like the one produced by John Bell,[15] or the one independently and almost simultaneously introduced by Simon Kochen and Ernst Specker,[16] or John von Neumann's original no-go theorem,[17] were widely interpreted to forbid additional variables in quantum mechanics (Chapter 13: Level of Description). What these theorems actually show is that additional variable formulation of quantum mechanics must be nonlocal, and that "quantum theory itself is irreducibly nonlocal."[18] To cite Bell's inequality as something that forbids additional variables is to show a gross misunderstanding of the theorem. When it comes to ruling out additional variable theories, the theorem is empty and irrelevant.

As Bell,[19] Bohm,[20] and Mermin[21] have pointed out, these impossibility proofs are logically unsatisfactory because they arbitrarily impose conditions that are relevant to the standard interpretation of quantum mechanics, but are not relevant to the theories they aim to dismiss—any theory with additional variables.[22] Nevertheless, it took a long time for the physics community to realize that the impossibility theorems were irrelevant.[23]

John Bell himself, the original author of one of the impossibility theorems, recognized its irrelevance, but he was systematically misquoted, misunderstood, or ignored as he tried to call attention to it. Ironically, he was then portrayed as being against Bohmian mechanics, despite the fact that he was its prime supporter during his lifetime.[24] He said:

"But in 1952 I saw the impossible done. It was in papers by David Bohm. Bohm showed explicitly how parameters could indeed be introduced, into nonrelativistic wave mechanics, with the help of which the indeterministic description could be transformed into a deterministic one. More importantly, in my opinion, the subjectivity of the orthodox version, the necessary reference to the 'observer,' could be eliminated...

But why then had Bohm not told me of this 'pilot wave'?... Why did von Neumann not consider it? More extraordinarily, why did people go on producing "impossibility" proofs, after 1952, and as recently as 1978?... Why is the pilot wave picture ignored in textbooks?

Should it not be taught, not as the only way, but as an antidote to the prevailing complacency? To show us that vagueness, subjectivity, and indeterminism, are not forced on us by experimental facts, but by deliberate theoretical choice?"[25]

The rest of the story as to why Bohmian mechanics is not currently favored as the mainstream interpretation of quantum mechanics can be traced back to orthodox philosophical intransigence. Those that fail to comprehend, or factor in, the ontological advantages that come from the determinism and mathematical clarity of Bohmian mechanics often attempt to downplay the formalism by pointing out that it "doesn't make any predictions that differ from those of ordinary quantum mechanics." Technically, that's not much of an objection because we could equally argue that empirically "the standard theory doesn't go beyond Bohm's theory."[26]

In light of this empirical equivalence, physicist Hrvoje Nikolic of the Rudjer Boskovic Institute in Zagreb, Croatia has said, "If some historical circumstances had been only slightly different then it would have been very likely that Bohm's deterministic interpretation would have been proposed and accepted first, and would be dominating today."[27] The standard interpretation has simply become *the standard* as a happenstance of history. The tragedy is that, because of the overwhelming political momentum of the standard interpretation, valid alternative interpretations (of which there are many) have largely been ignored.

The fact is that Bohmian mechanics completely accounts for nonrelativistic dynamics. It choreographs every dance in the quantum mechanical realm, and does so deterministically. For these reasons alone it is worthy of our attention. But we also might raise a brow in response to the way it frees us from the limiting assertion of the orthodox interpretation.

The most controversial aspect of orthodox quantum mechanics is not the formalism itself, but rather "a further assertion to the effect that we cannot get beneath this formalism, to account for it in microscopic terms."[28] The quantum formalism is touted as a "measurement" formalism. "Thus it is a phenomenological formalism describing certain macroscopic regularities."[29] In this, and in many other ways, it is analogous to thermodynamics.

The thermodynamic formalism details the dynamics and interrelated properties of the larger macroscopic system based on assumptions about the underlying behavior of a large number of microscopic constituents that it takes to be in equilibrium. For example, the ideal gas law $PV = nRT$ relates the macroscopic properties of an ideal gas (pressure, volume, and temperature), and ultimately explains that relationship based on an underlying assumption that the system (the ideal gas) is made up of microscopic constituents (molecules) that interact elastically and are in a state of equilibrium.[30]

Several averaged-over macroscopic mathematical relations automatically follow from such assumptions. Because these mathematical relations have reliably held up in our laboratory experiments our confidence in the substrate of elastically interactive constituents (molecules) is strengthened. We now believe that we can intuitively access what lies beneath the thermodynamic formalism by accounting for its microscopic substrate. Whether or not anyone ever directly sees a molecule, or an atom, having a picture of the underlying microscopic dynamics greatly improves the intuitive access we have of physical reality.

Clearly, as we derive a quantum formalism, it is in our best interest to retain the ability to "get beneath it," and explain it in microscopic terms. One way to do this is to start with the assumption that the system (the vacuum in this case) is composed of a large number of microscopic constituents that (at least to first approximation) interact elastically. Interestingly, when we assume that the vacuum can be represented this way—as a quantum field, or an infinite collection of coupled harmonic oscillators—a quantum formalism similar to Bohmian mechanics "emerges in such an inevitable manner that we are almost forced to conclude that philosophical prejudice must have played a crucial role in its nondiscovery."[31]

The assumption that the vacuum is quantized can be powerfully expressed by the derivation we are about to undertake. To explicate that connection we simply need to allow the word *particles* to reference the vacuum quanta in the derivation. Clearly, this derivation can be done in the absence of a correlate map, as it was originally. Nevertheless, having a map to reference can give us an intuitive advantage during that derivation. Once again, let me state that the reader will not lose any continuity by simply skimming the equations that follow. The explanations are intended to be sufficient for clarity.

Deriving Bohm's Formalism

Let's begin by addressing the objective state of the wave function on the fundamental level—the Planck scale. If our system (a chosen domain of the vacuum) is composed of N particles, then a complete description of that system will necessarily include a specification of the positions Q_i of each of those particles. On its own, the wave function ψ does not provide a complete description of the state of that system. Instead, the complete description of this quantum system must specify both the particle positions and the wave function (Q, ψ), where

$$Q = (Q_1, Q_2, Q_3, \ldots Q_N) \in \mathbb{R}^{3N}$$

is the configuration of the system and

$$\psi = \psi(q) = \psi(q_1, q_2, q_3, \ldots q_N),$$

a (normalized) function on configuration space—the superspatial dimensions—is its wave function.

At this point, all we have to do to obtain our theory is specify the law of motion for the state (Q, ψ). Of course, the simplest choice we can make here would be one that is causally connected. In other words, one whose future is determined by its present specification, and more specifically whose average total state remains fixed—at least in the macroscopic sense of the familiar four dimensions of spacetime. To obtain this we simply need to choreograph the particle motions by first-order equations that assume elastic interactions. The evolution equation is Schrödinger's equation, which, as we showed in chapter 22, can be derived from the assumption of vacuum superfluidity

$$i\hbar \frac{\partial \psi}{\partial t} = H\psi_t = -\frac{\hbar^2}{2m_k}\nabla^2 q_k \psi_t + V\psi_t$$

where ψ is the wave function and V is the potential energy of the system.

Therefore, in keeping with our previous considerations, the evolution equation for Q should be $\frac{dQ_t}{dt} = v^{\psi t}$ with $v^\psi = \left(v_1^\psi, v_2^\psi, v_3^\psi, \ldots v_N^\psi\right)$ where v^ψ takes the form of a (velocity) vector field on our chosen configuration space $\mathbb{R}^{3N}$. Thus the wave function ψ reflects the motion of the particles in our system, in a macroscopic averaged-over sense, based on the underlying assumption of elastic interaction. These motions are coordinated through a vector field that is defined on our specified configuration space.

$$\psi \to v^\psi$$

If we require time-reverse symmetry and simplicity to hold in our system (automatic necessities for a deterministic theory) then,

$$v_k^\psi = \frac{\hbar}{m_k} Im \frac{\nabla q_k \psi}{\psi}$$

Notice that there are no ambiguities here. The gradient ∇ on the right-hand side is suggested by rotation invariance, the ψ in the denominator is a consequence of homogeneity (a direct result of the fact that the wave function is to be understood projectively, which is in turn an understanding required for the Galilean invariance of Schrödinger's equation alone), the Im by time-reverse symmetry, which is implemented on ψ by complex conjugation in keeping with Schrödinger's equation, and the constant in front falls directly out of the requirements for covariance under Galilean boosts.[32]

Therefore, the evolution equation for Q is

$$\frac{dQ_k}{dt} = v_k^\psi (Q_1, Q_2, Q_3, \ldots Q_N) \equiv Im \frac{\nabla q_k \psi}{\psi}(Q_1, Q_2, Q_3, \ldots Q_N)$$

This completes the formalism of Bohmian mechanics that David Bohm constructed in 1952.[33] As an extension of de Broglie's pilot wave model[34] this formalism exhaustively depicts a nonrelativistic universe of N particles without spin.[35]

Spin must be included in order to account for Fermi and Bose-Einstein statistics. The full form of the guiding equation, which is found by retaining the complex conjugate of the wave function, accounts for all the apparently paradoxical quantum phenomena associated with spin. For considerations without spin the complex conjugate of the wave function cancels because it appears in the numerator and the denominator of the equation. The full form of the evolution equation is

$$\frac{dQ_k}{dt} = \frac{\hbar}{m_k} Im \left[\frac{\psi^* \partial_k \psi}{\psi^* \psi}\right](Q_1, Q_2, Q_3, \ldots Q_N)$$

Notice that the right-hand side of the guiding equation is the ratio for the quantum probability current to the quantum probability density.[36] Given that the classical formula for current is density times velocity, "it requires no imagination whatsoever to guess the guiding equation from Schrödinger's equation."[37] This simplicity speaks to why Bohmian mechanics has been said to be "the most naïvely obvious embedding imaginable of Schrödinger's equation into a completely coherent physical theory."[38]

> *"[T]he idea of an objective real world whose smallest parts exist objectively in the same sense as stones or trees exist, independently of whether or not we observe them… is impossible."*
>
> Werner Heisenberg[39]

Today's physicists have been brought up under the orthodox shadows of characters like Niels Bohr, Werner Heisenberg, and John von Neumann.[40] These figureheads loudly declared that a deterministic formalism of quantum mechanics is physically, philosophically, mathematically, and logically impossible.[41] They set in motion the assumptions that physicists would carry for decades after them. For some reason they were so intransigently stuck to the idea that quantum theory demands radical epistemological and metaphysical innovations that they appear to have never truly considered getting beneath the quantum formalism and accounting for it in microscopic terms.[42] These men possessed extraordinary intellects, and contributed powerfully to the development of quantum mechanics, but they missed out on Bohm's obvious, elegant, and quite frankly trivial formalism.

In my opinion, that stubbornness is the primary reason that Bohm's interpretation of quantum mechanics is not the formal interpretation taught today. This intransigence has been quite lopsided. Craig Callender notes that, "For some reason or other, people often object to Bohm for reasons that they would never hold against other interpretations of quantum mechanics."[43] I suspect that this has something to do with the fact that, without a map of the underlying molecular dynamics, people have a tremendously difficult time elevating their intuition to a higher dimensional realm where nonlocality is automatic.

It is worth emphasizing, once again, that the biggest problem with the orthodox interpretation of quantum mechanics is that it maliciously severs the reach of our intuition. Its presumptions tautologically inhibit us from ever figuring out what is really going on by indefensibily asserting that Nature is not, and cannot be, described in a mathematically precise way. Many physicists and philosophers have felt the poignant sting of this truncation. Schrödinger himself never quite accepted the validity or completion of the wave function based on the intuitive damage it seemed to do. In reference to the wave function he said, "That it is an abstract, unintuitive mathematical construct is a scruple that almost always surfaces against new aids to thought and that carries no great message."[44]

A model's value is to be measured by its ability to provide us with salient ontological and mathematical clarity of the domain it portrays. Unlike the orthodox interpretation of quantum mechanics, which restricts our intuitive reach by importing vaguely defined beables (additional variables called classical terms), Bohm's formalism choreographs quantum mechanics in a way that is clear and mathematically precise. In short, instead of relegating Bohr's classical terms (the additional variables from the Copenhagen interpretation) to the surrounding talk,[45] Bohm makes them mathematically precise.

This leads to an interesting contrast. For example, despite the empirical equivalence between Bohmian mechanics and orthodox quantum theory, "there are a variety of experiments and experimental issues that don't fit comfortably within the standard quantum formalism but are easily handled by Bohmian mechanics. Among these are swell and tunneling times, escape times and escape positions, scattering theory, and quantum chaos."[46]

The more striking contrast comes from the fact that Bohm's model offers us a classical analogue by which to understand the quantum realm, while the orthodox interpretation attempts to forbid one. Let's explore this point. In the orthodox interpretation we are asked to believe that, for example, photons form an interference pattern on the back wall because they all *magically*, in a way we cannot comprehend, manage to go through both slits. For systems with more than two slits every photon magically manages to go through every slit.

In order to accept this interpretation we have to do more than abandon our notion of a particle—we have to accept that this *magic* is truly just that—*magic*. We have to accept that it really is impossible for us to ever have intuitive access to the process that causes photons, electrons, etc., to form interference patterns in the double-slit experiment—that it is impossible for us to comprehend, understand, or ever know what's really going on during these experiments.

The prevailing orthodox interpretation pushes this worldview upon us. Richard Feynman explains this by saying that the interference pattern made during the double-slit experiment is "a phenomenon which is impossible, *absolutely* impossible, to explain in any classical way, and which has in it the heart of quantum mechanics. In reality it contains the *only* mystery."[47] Feynman later said, "Nobody can give you a deeper explanation of this phenomenon than I have given; that is, a description of it."[48]

If this were true it would be a pretty large pill to swallow. But it is not true. The truth is that Einstein understood the Copenhagen interpretation of quantum mechanics perfectly—he just wasn't happy with its vagueness.[49] His intuition, that a deeper, more precise explanation is possible, has been fully justified. As we have seen, "Bohmian mechanics is just such a deeper explanation."[50]

From this precipice there is an apparent parallel between the advocates of the orthodox interpretation of quantum mechanics and the robed puppet masters of orthodox religions. Both preach that we are incapable of getting to know or discovering the truth for ourselves—that we should just give up and embrace unquestioning faith.

That attitude is detrimental to our personal journeys and catastrophic to the overall scientific quest. Bohm's interpretation frees us from the sins of orthodox unquestioning faith. It shows us that the path of the photon in our double-slit experiment reflects an interference pattern because the motion of that photon is governed by the wave function. Parts of the wave function pass through both slits while the particle passes through one slit. The parts of the wave function that pass through separate slits interfere with each other, developing an interference profile that guides the particle on its way.

The interference pattern we see is, therefore, an unavoidable consequence of nonlocality—of the fact that the vacuum is quantized. It is not a magical, unexplainable effect. If the particles are emitted one by one, then this interference pattern still builds up over time—provided that the trajectories of the ensemble have a random distribution, or equilibrium distribution, which we can denote as $\rho = \langle \psi \rangle^2$.

If every particle were to follow completely identical trajectories, then they would all end up at one spot, creating a single bright spot on our photographic plate (or the wall). In Nature, this is not a real possibility for photons because the substrate of the vacuum is composed of interactive quanta. For two particles to follow identical trajectories, identical trajectories must exist. They don't because the superfluid vacuum is not static. The quanta are constantly mixing about in configuration space. In quantum mechanics, the best information about available four-dimensional trajectories is given by $\rho = \langle \psi \rangle^2$ because in that formalism the vacuum is assumed to be in a state of equilibrium. The inherent mixing of the vacuum explains why extremely precise information about a trajectory in the familiar four dimensions can at best be described statistically, or probabilistically.

For two photons to follow identical paths through space (identical trajectories) the positions and velocities of all the intermittent space quanta along that path (the additional variables) would have to be identically configured. On macroscopic scales this is extremely improbable. So from the ontological vantage of Bohmian mechanics the interference pattern we see in the double-slit experiment is exactly what we should expect. That's a rather significant improvement over the orthodox assertion that we should just accept the double-slit experiment as something that we will never make sense of.

This beautiful formalism completely accounts for four-dimensional randomness, absolute uncertainty, familiar (macroscopic) reality, and the presence of the wave function within the vacuum, while explaining away the mysteries of wave-particle duality, wave function collapse, and providing a precise, sharp foundation for the uncertainty principle.

In this formalism, the transformation $\rho^\psi \rightarrow \rho^{\psi t}$ arises directly from Schrödinger's equation. If these evolutions are indeed compactable, then

$$\left(\rho^\psi\right)_t = \rho^{\psi t}$$

is equivariant. Therefore, under the time evolution ρ^Ψ retains its form as a function of Ψ.

The conclusion that $\rho\psi = |\psi|^2$ is equivariant follows immediately from the observation that the quantum probability current $J^\psi = |\psi|^2 v^\psi$, so that the continuity equation

$$\frac{\partial \rho}{\partial t} + div(\rho v^\psi) = 0$$

is satisfied by the density $\rho_t = |\psi|^2$. As a consequence, we find that if $\rho(q, t_0) = |\psi(q, t_0)|^2$ at some time t_0, then $\rho(q, t) = |\psi(q, t)|^2$ for all t.[51]

Note, that three assumptions were fixed in our derivation (1) that particles interact with perfect elasticity, (2) that they are in a state of equilibrium, and (3) that there are zero interparticle interactions. (The assumption of zero interparticle interactions was encoded in the non-linear Schrödinger equation and the subsequent guiding equation.)

We can interpret these assumptions as encoding a vacuum in an equilibrium state that contains zero quantized vorticies. These assumptions are valid as a first-order approximation of the system (because they completely map the quantum mechanical realm), but we are not logically bound to these assumptions. If we allow for second-order inelasticity,[52] or vorticity, we may be able to make room for relativistic effects. And if we are going to possess a complete formalism, then we need to be able to trace out the dynamic evolutions of all vacuum states—not just equilibrium states.

Going Beyond Bohm's Formalism

"If Bohm's physics, or one similar to it, should become the main thrust of physics in the future, the dances of the East and West could blend in exquisite harmony."

Gary Zukav[53]

In the same way that thermodynamics is not empirically equivalent to Newtonian mechanics, Bohmian mechanics is not empirically equivalent to a general physical model of a superfluid vacuum. Thermodynamics and Bohmian mechanics are both special cases, which emerge when we look at equilibrium situations. This implies that instead of being exact, quantum mechanics is a very accurate approximation of a deeper-level deterministic theory. To formalize that deeper-level theory—under the assumption that it models the microphysics of the superfluid vacuum—there are at least two adjustments that need to be made to Bohmian mechanics. If these adjustments lead to the correct theory, then quantum mechanics and general relativity should emerge as subsets of that deeper-level description.

To understand the first adjustment, note that if we take a box with N elastically interacting particles, and let them evolve, they will tend toward a uniform distribution (an equilibrium distribution), as long as they are under a uniform potential (e.g. not under the influence of gravity). If the system is a Newtonian one, then once this equilibrium is reached, we can sufficiently model its dynamics using thermodynamics. The story is identical for quantum mechanics.

Bohm's formalism explicitly assumes an equilibrium distribution of particles $\rho = \langle\psi\rangle^2$, because when it makes that assumption it recovers quantum mechanics. But the truth is, there is no reason to logically restrict ourselves to this assumption. The ultimate goal is not to recover quantum mechanics; it is to obtain a general theory of everything—a formalism of the underlying mechanics that allows us to completely model quantum mechanics, general relativity, and anything left out by those two. Therefore, as long as non-equilibrium configurations are possible, we should be interested in a model that is general enough to include them.

This more general, more fundamental, theory is expected to make more predictions than quantum mechanics, because it will make different predictions when it is modeling non-equilibrium states. Nevertheless, non-equilibrium states will be rare, because complex systems rapidly evolve toward their equilibrium state—relaxtion rates exponentially depend on complexity. Therefore, we should expect quantum mechanics to commonly apply, just as thermodynamics usually applies to Newtonian systems.[54]

The problem with restricting ourselves to the equilibrium states isn't that it doesn't usually work. It does. The problem is, that this restriction divorces us from the possibility of getting beneath certain interactions. For example, in ordinary thermodynamics, two boxes with different temperatures can be in their respective equilibrium states, but when we put them together the combined system must evolve, or relax, towards a new equilibrium state. This relaxation takes some time.

Instead of revealing the underlying dynamics that define the system's energy in terms of its kinetics, Bohmian mechanics portrays that energy in a smeared out way—as a constant surface on phase space. When two systems are combined with different energies, it projects a new constant surface, but is incapable of resolving any relaxation towards that new equilibrium state. As a result, Bohm's theory portrays two subsystems that are individually in quantum equilibrium as instantaneously reaching their new equilibrium state when they are put together.

This oversight is a consequence of the fact that there is no equivalent to temperature in Bohmian mechanics, no way to dynamically model how the systems relax to their new combined state of equilibrium. The whole system is just automatically depicted as being in equilibrium. In other words, in Bohm's theory, the distribution of each subsystem is just $\langle\psi\rangle^2$, an equilibrium distribution. There isn't anything like a temperature that is parameterizing the system. It is just a wave function modulus squared.

In order to distinguish equilibrium distributions we need to get beneath that formalism and reveal the underlying dynamics. We need to go from a high-level averaged-over thermodynamic description (Bohmian mechanics) to a lower-level molecular dynamic description. Only then can we explain how two isolated systems can be independently in equilibrium, but not the same. Only then can we resolve a relaxation time needed to reach a new equilibrium when we put these systems together.

To move towards that more fundamental description we must relax our equilibrium constraint, of $\rho = \langle\psi\rangle^2$, allowing for non-equilibrium states.[55] Doing this allows us to recover quantum mechanics when equilibrium states are realized, while retaining the ability to represent ensembles that are different from $\langle\psi\rangle^2$.

To generalize Bohm's formalism further, we note that it carries the assumption that the particle interactions are perfectly elastic. In other words, it assumes that there are no interparticle interactions. This assumption preserves symmetry, but it also unnecessarily narrows the scope of the theory. To rework the formalism to include possible interparticle interactions we simply need to trade the wave equation Bohm uses, the linear Schrödinger equation,

$$i\hbar \frac{\partial \psi}{\partial t} = -\frac{\hbar^2}{2m}\nabla^2\psi + V\psi$$

for the non-linear Schrödinger equation, also known as the Gross-Pitavaskii equation, which naturally follows from the assumption of vacuum superfluidity (see Chapter 22),

$$i\hbar \frac{\partial \psi}{\partial t} = \left(-\frac{\hbar^2}{2m}\nabla^2 + V + U_0|\psi|^2\right)\psi$$

where: m is the mass of the bosons, V is the external potential, *and* U_0 is representative of the inter-particle interactions. Using this non-linear equation, we then rederive the guiding equation.

This directs us toward a general formalism that has an exact classical analogy. In that analogy, equilibrium distributions reproduce the special case of thermodynamics, or in this case quantum mechanics. Non-equilibrium distributions make predictions outside of quantum mechanics, but rarely apply. And nonlinear solutions to the wave equation, most elaborately represented by quantum vorticity,[56] introduce small-amplitude deformations that collectively give rise to the curvature of general relativity on macroscopic scales.

The claim that non-linearity leads to a representation of induced gravity, or emergent gravity, means that "spacetime curvature and its dynamics emerge as a mean field approximation of underlying microscopic degrees of freedom, similar to the fluid mechanics approximation of Bose-Einstein condensates."[57] This intimately resembles Andrei Sakharov's 1967 proposal, supporting the claim that "gravity is not fundamental in the sense of particle physics. Instead... it emerges from quantum field theory in roughly the same sense that hydrodynamics or continuum elasticity theory emerges from molecular physics."[58]

This suggests that we can arrive at a more fundamental description than quantum mechanics or general relativity by modeling the underlying classical dynamics of the vacuum quanta. With that description in hand, quantum mechanics emerges as a special equilibrium case, just as thermodynamics does, and gravity (general relativity) emerges from quantum field theory in roughly the same sense that hydrodynamics emerges from molecular physics.

[1] John P. Briggs & F. David Peat. *Looking Glass Universe*.

[2] Gary Zukav. *The Dancing Wu Li Masters*, p. 109.

[3] "Within a year after Heisenberg developed his matrix mechanics, Schrödinger discovered that it was mathematically equivalent to his own wave mechanics." Today they are both used as independent modes of expression for quantum mechanics. Zukav, p. 110.

[4] Although Newton had constructed his theory of gravitation, he delayed its publication for twenty years (until 1687) "while he tried to justify a critical assumption of his theory: that the Earth's gravitational pull was exerted as if its mass were all concentrated at the center" (Randall, p. 87). Calculus is the mathematical tool that enabled this proof. During this period of delay, the German philosopher Gottfried Leibniz independently developed calculus. Because of this, Newton and Leibniz have traditionally both been given credit for independently discovering calculus. Newton was said to have invented calculus before Leibniz, but Leibniz's "notation was clearer and is the one used today". E. O. Wilson, pp. 31–32.

According to George Gheverghese Joseph, a historian of mathematics with the University of Exeter, a core concept of calculus (the infinite series) was first developed in the 14th century by Indian mathematicians. It was then transported, along with other scientific knowledge, from southern India to Western Europe by Jesuit priests. George Gheverghese Joseph & Dennis Almeida. (2007, June); Stephen Ornes. Calculus Was Developed In Medieval India. *Discover*, 100 top discoveries of 2007, p. 52; Lisa Randall. *Warped Passages*, p. 87.

[5] Bohmian mechanics is also called the de Broglie-Bohm theory, the pilot-wave model, and the causal interpretation of quantum mechanics. Louis de Broglie originally discovered this approach in 1927 and David Bohm rediscovered it 1952.

[6] S. Goldstein. Bohmian Mechanics. Stanford Encyclopedia of Philosophy. For more information on the identical success of Bohmian mechanics with the traditional quantum formalism see: Detlef Dürr, Sheldon Goldstein and Nino Zanghí. Quantum Physics Without Quantum Philosophy. Physical Review Letters, vol. 93, p 090402; Ward Struyve & Hans Westman, Proceedings of the Royal Society A, vol. 463, p. 3115; D. Bohm. (1953). Proof that probability density approaches $\langle\Psi\rangle^2$ in causal interpretation of quantum theory. Physical Review 89, 458–466; D. Bohm. (1952). A suggested interpretation of the quantum theory in terms of "hidden" variables', Physical Rev. 85, p 166–193; M. Daumer, D. Dürr, S. Goldstein, & N. Zanghí. A survey of Bohmian mechanics, Il Nuovo Cimento.

[7] Much of this chapter follows Sheldon Goldstein's publication, Bohmian Mechanics, found in the online Stanford Encyclopedia Of Philosophy.

[8] Brian Greene, (2004). *The Fabric of the Cosmos*, p. 206.

[9] I have had enlightening discussions with Sheldon Goldstein and Daniel Victor Tausk about this very matter. Both of them have devoted considerable energy toward correcting this historical problem. But they have run into a lot of resistance. They have noted to me that many people are just too intransigent to consider a solution to a problem they have been working on their whole life, even if it is placed right in front of them. After spending a career on the problems of quantum mechanics to no avail many of them would prefer that the problem remain unsolved.

[10] Cushing. (1994).

11 Vacuum quantization leads to a model that rides between Aristotelian naturalism and Platonic idealism. Aristotelian naturalism holds that reality consists only of the natural world. It is completely monistic and therefore denies the existence of a separate non-material order of reality. It also holds strongly to the belief that Nature follows orderly, discoverable laws. Platonic idealism, on the other hand, asserts that there is a non-material second transcendental realm. It is therefore dualistic. This non-spatiotemporal realm is believed to be accessible to the mind, but only to the mind. Vacuum quantization adjoins these two perspectives and ends up with a hierarchical monism. It proclaims that there is nothing outside natural order; there is no supernatural. It carries the explicit requirement for non-spatiotemporal realms that are directly accessible only to the mind (via vacuum quantization), but these realms are still part of the natural world—they follow orderly, discoverable laws.

12 Technically these violations show that the vacuum does not conform to local realism. To assume that realism is out is to assume that the entire scientific endeavor makes no connection to the real world (or that there isn't a real world to begin with). If we don't go that route, then we must assume that the vacuum is nonlocal.

13 For a presentation of the argument and the experimental results that secure this point, see: *Quantum Non-Locality and Relativity*, Second Edition, by Tim Maudlin.

14 Tim Maudlin. (2002). *Quantum Non-Locality and Relativity*, second Edition, Blackwell Publishing, MA, p.124. For more on this see Joy Christian, Disproof of Bell's Theorem by Clifford Algebra Valued Local Variables: www.arxiv.org/abs/quawnt-ph/0703179

15 J. S. Bell. (1966). On the problem of hidden variables in quantum mechanics. *Rev. Mod. Phys.* **28**, 447–452; reprinted in *Quantum Theory of Measurement*, J. A. Wheeler & W. H. Zurek editors, Princeton University Press (1983), 396–402; and in Chapter 1 of J. S. Bell, *Speakable and Unspeakable in Quantum Mechanics*, Cambridge University Press (1987); second augmented edition (2004), which contains the complete set of J. Bell's articles on quantum mechanics.

16 S. Kochen & E. P. Specker. (1967). The problem of hidden variables in quantum mechanics. *J. Math. Mech.* **17**, 59–87.

17 John von Neumann. (1932). *Mathematische Grundlagen der Quantenmechanik*. Springer, Berlin.

18 Quoted from personal discussions with Sheldon Goldstein.

19 J. S. Bell. (1966). On the problem of hidden variables in quantum mechanics. *Rev. Mod. Phys.* **28**, 447–452; reprinted in *Quantum Theory of Measurement*, J. A. Wheeler and W. H. Zurek editors, Princeton University Press (1983), 396–402; J. S. Bell, *Speakable and Unspeakable in Quantum Mechanics*, Cambridge University Press (1987); second augmented edition (2004), which contains the complete set of J. Bell's articles on quantum mechanics.

20 D. Bohm & J. Bub. (1966). A proposed solution of the measurement problem in quantum mechanics by a hidden variable theory. *Rev. Mod. Phys.* **38**, 453–469; D. Bohm & J. Bub. (1966). A refutation of the proof by Jauch and Piron that hidden variables can be excluded in quantum mechanics. *Rev. Mod. Phys.* **38**, 470–475.

21 N. D. Mermin. (1993). Hidden variables and the two theorems of John Bell. Rev. Mod. Phys. 65, 803–815; in particular see § III.

22 "[T]he assumption of Kochen and Specker… appear, in fact, to be quite reasonable indeed. However, they are not. The impression that they are arises from a pervasive

error, a naïve realism about operators..." Sheldon Goldstein, Bohmian Mechanics, published 10-26-2001; substantive revision 5-19-2006, Stanford Encyclopedia Of Philosophy.

In other words, supporters of the standard interpretation of quantum mechanics often fail to recognize that Bohr's classical variables are additional variables in their theory. Bohm takes these variables and makes them mathematically precise. Technically, attacking additional variable theories also attacks the standard model.

[23] Franck Laloë. *Do We Really Understand Quantum Mechanics?*, p. 37. There are still members in the physics community that are held back by this misunderstanding. I have interacted with many physicists that are deeply convinced that these no-go theorems forbid additional variable theories.

[24] See Wigner. (1976).

[25] J. S. Bell. (1987), p. 160.

[26] Mark Buchanan. (2008, March 22). No dice. *New Scientist*, pp. 28–31.

[27] Ibid.

[28] Detlef Dürr, Sheldon Goldstein, & Nino Zanghí. *Quantum Physics Without Quantum Philosophy*, p. 4. The mathematics to follow, as well as much of the remaining discussion, follows this work.

[29] Ibid., p. 4.

[30] Boyle's law is another example of this. Technically, pressure and temperature are macroscopic properties that also rely on this underlying assumption. These properties result from the group behavior of a large number of elastically interactive molecules in motion.

[31] Detlef Dürr, Sheldon Goldstein, & Nino Zanghí. *Quantum Physics Without Quantum Philosophy*, p. 4.

[32] Ibid., pp. 5–6.

[33] David Bohm. (1952).

[34] L. de Broglie. (1927).

[35] Of course in the limit $\frac{\hbar}{m} \to 0$, the Bohm motion Q_t approaches the classical motion. See: D. Bohm & B. Hiley. (1993). *The Undivided Universe: an Ontological Interpretation of Quantum Theory*. Routledge & Kegan Paul, London; Detlef Dürr, Sheldon Goldstein, & Nino Zanghí. *Quantum Physics Without Quantum Philosophy*, p. 7.

[36] Sheldon Goldstein. *Bohmian Mechanics*. For further examples of how easily spin can be dealt with in the Bohmian formalism see: J. S. Bell, 1966, 447–452; D. Bohm, 1952, 166–193; D. Dürr et all, A survey of Bohmian mechanics, Il Nuovo Vimento, and Bohmian mechanics, identical particles, parastatistics, and anyons, In preparation.

[37] http://plato.stanford.edu/entries/qm-bohm/

[38] D. Dürr, et al.

[39] Werner Heisenberg. (1958), p. 129; ibid.

[40] J. von Neumann. (1932); R. T. Beyer. (1955), pp. 324–325; J. S. Bell. (1982). 989–999; J.S. Bell. (1987), pp. 159–168.

[41] J. von Neumann. (1932); J. S. Bell. (1982), (1987).

[42] D. Dürr et al., pp. 4–6.

[43] Craig Callender. (1998). Review, Brit. J. Phil. Sci. 49, 332–337.

[44] Schrödinger, E. (1935). 23: 807–812, 923–828, 844–849.

[45] John S. Bell. (1976). The theory of local beables. Epistemological Lett. 9, 11-24; Reprinted in John S. Bell. (2004). *Speakable and Unspeakable in Quantum Mechanics*, 2nd ed. Cambridge U.P., Cambridge, pp. 52–62.

[46] Sheldon Goldstein. *Bohmian Mechanics*. For more on how the formalism of Bohmian mechanics naturally points to a formalism richer than the standard orthodox theory see: Anthony Valentini of the Perimeter Institute in Waterloo, Ontario, Journal of Physics A: Mathematical and theoretical, vol. 40, p. 3285; For discussion on escape times and escape positions see Daumer et al., 1997a, for scattering theory see Dürr et al., 2000, and for quantum chaos see Cuching, 1994; Dürr et al., 1992a.

[47] R. P. Feynman, R. B. Leighton, & M. Sands. (1963). The Feynman Lectures on Physics, I, New York: Addison-Wesley; Sheldon Goldstein. Bohmian Mechanics.

[48] Richard Feynman. (1967). The Character of Physical Law. Cambridge MA: MIT Press; Sheldon Goldstein. Bohmian Mechanics.

[49] It could be argued that Einstein initiated the pilot-wave approach with the concept of the Führungsfeld or guiding field. Wigner. (1976), 262; Goldstein. Bohmian Mechanics. Stanford Encyclopedia of Philosophy.

Einstein failed to complete the formalism of such an approach, but he independently encouraged both de Broglie and Bohm to press on with their efforts.

[50] Sheldon Goldstein. Bohmian Mechanics.

[51] D. Dürr et al., *Quantum Physics Without Quantum Philosophy*, p. 8.

[52] If a system only possesses a trace of inelasticity, then to a first order approximation it is elastic. In other words, if a system's behavior is reasonably captured by the dominant term of its evolution equation, and that term assumes perfect elasticity, then its inelastic evolution is overlooked by a first order approximation. In such systems inelastic properties are modeled only when we include the higher-order terms of its evolution equation. This kind of phrasing is common in math, as there are many systems that are represented by a power series, where the first term is the largest contributor to the whole and the successive terms become vanishingly less significant (but may still have non-zero values).

[53] Gary Zukav. *The Dancing Wu Li Masters*, p. 310.

[54] I thank Hans Westman for his in depth input on this section.

[55] For existing work in this area see Antony Valentini & Hans Westman. (2005). Dynamical origin of quantum probabilities. Proc. R. Soc. A, 461, 253-272 doi:10.1098/rspa.2004.1394; Ward Struyve & Antony Valentini. (2008, December 27). De Droglie-Bohm Guidance Equations for Arbitrary Hamiltonians. arXiv:0808.0290v3 [quant-ph].

[56] The electron is a spinning twisted torus in an inviscid fluid. It generates compression waves χ, which are in turn modulated by guiding waves ψ. And at large distance from

the sonon, χ may be approximated up to a phase factor as $\chi = \frac{1}{r} \sin k_r r$. Given that χ behaves like a carrier wave and ψ as its modulation, which is a complex function whose phase is important, this provides a physical model of the de Broglie-Bohm view that a particle moves through space surrounded by waves that obey the usual quantum equations. Ross Anderson and Robert Brady. (2013, Jan 30). Why quantum computing is hard—and quantum cryptography is not provably secure. arXiv:1301.7351v1 [quant-ph].

[57] A.D. Sakharov. (1967). Vacuum Quantum Fluctuations in Curved Space and the Theory of Gravitation.

[58] Matt Visser. (2002). Sakharov's induced gravity: a modern perspective. arXiv:gr-qc/0204062; Sakharov's idea was to start with an arbitrary background pseudo-Riemannian manifold (in modern treatments, possible with torsion) and introduce quantum fields (matter) on it but not introduce any gravitational dynamics explicitly. This gives rise to an effective action, which to one-loop order contains the Einstein-Hilbert action with a consmological constant. In other words, general relativity arises as an emergent property of matter fields, and is not put in by hand. A.D. Sakharov. (1967). Vacuum Quantum Fluctuations in Curved Space and the Theory of Gravitation.

Chapter 25 — **Symmetry and Symmetry Breaking**

"...the copious symmetries underlying natural law present a close runner-up to the atomic hypothesis as a summary of our deepest scientific insights."

Brian Greene

Thomas Range, Utah.

With each strike, white dust swirled away from my chisel, dancing gently through the air. This once majestic plume of volcanic ash was now a mountain of rhyolite. Time, pressure, and the slow percolation of dissolved rare minerals had crafted this range into a crystal nursery. I was searching for a hidden chamber within that nursery, a cavity within the matrix where rare minerals had collected and gave birth to crystals and gemstones. Though I'd been at it for hours, I felt like I was only moments away from success. This mountain was home to some of the world's best topaz and red beryl specimens, and it was the world's first known macroscopic cache of bixbyite.

Under the desert sun I spent the day combing through the eroded surface, finding small tablets of red beryl, broken cubes of bixbyite, and dozens of clear and rose topaz crystals, which had been bleached from their original champagne color by the sun's rays. Using the concentration of those minerals to point me towards their source, I was now digging into the most promising wall of rhyolite.

Although my shoulders and arms were tired, and sweat was stinging my eyes, I continued my systematic search. Over and over I pounded the chisel four inches deep into the rock, then wiggled it back and forth to loosen it for retrieval. I then placed it six inches away and started all over again. Finally, with one lucky hit, my chisel sank completely to its hilt. Letting out a yelp of excitement I pulled it back out and immediately began working to widen the hole. Forgetting all about my sore arms I kept at it until the hole was big enough for my hand. When I was done, I put down my tools and slowly reached in.

The cavity was roughly spherical and about eight inches in diameter. The bottom half was filled with clay that was cold and damp to the touch. After gently scooping out a handful, I used my Camelbak to drizzle some water on the contents in my hand. Beautifully truncated clusters of champagne colored topaz, a single red beryl, and a couple of small but pristine bixbyite crystals emerged. My heart raced as I saw the natural beauty of those crystals—how mesmerizingly symmetric they were. The red beryl was a perfectly hexagonal slab with a deep translucent red color. The topaz crystals were naturally terminated (some of them were doubly terminated) orthorhombic translucent champagne wonders. And the bixbyite crystals were shiny black isometric diploidal structures. As I examined my new treasures, rotating them in the sunlight, I found myself in awe of Nature's endless propensity for symmetry.

Symmetry

"If a theory is beautiful, this means it has a powerful symmetry that can explain a large body of data in the most compact, economical manner."

Michio Kaku

The fabric of physical reality is opulently enlaced with symmetries. These symmetries help us untangle the intrinsic laws of the universe and help us decode its ultimate structure. The common thread between all of Nature's symmetries is that they refer to manipulations that an object, or the metric, can undergo without being noticed.

For example, if we are moving with constant velocity, regardless of the magnitude or direction, our motion has absolutely no effect on the laws that explain our observations. Seeing through the cracks of Nature's Newtonian façade, Einstein extended this symmetry in a thoroughly unanticipated way, by elevating "light's speed to an inviolable law of Nature, declaring it to be as unaffected by motion as the cue ball is unaffected by rotations."[1] Einstein also discovered a symmetry hidden in the fact that "the force you feel from acceleration is indistinguishable from the force you feel in a gravitational field of suitable strength."[2]

Beneath those symmetries is something called translational symmetry (otherwise known as translational invariance), which means that there is no uniqueness derived from location in space. This symmetry reflects the fact that the laws of physics in one place are identical to the laws of physics in another place. As long as we stick to regions with zero curvature, all geometric relations and descriptions of interaction remain identical no matter where they occur.

It is not directly obvious that things had to be this way. One could imagine a universe in which the laws of physics change as you migrate from one region to another, but this is not how our universe behaves. Why? How is it that our universe has come to possess identical geometric and physical parameters from one side to the other (at least within regions of equal curvature)?[3] Why can any two points of space be swapped without changing the state of space, or the laws of physics?

Rotational symmetry, or rotational invariance, is another fundamental spatial symmetry. This symmetry encodes the fact that, in regions with no curvature, every spatial direction is on equal footing. In other words, the equations that describe motion in one direction are equally equipped to describe motion in the other spatial directions. What is the origin of this symmetry?

Another way to explain these symmetries is to say that, macroscopically, Nature is spatially homogeneous and isotropic. "Roughly speaking, *isotropic* means that the universe looks the same in all directions, so it has an $O(3)$ rotational symmetry group. Also, *spatially homogeneous* means that the universe looks the same at each point of space."[4] This means that if we somehow excised two patches of space of equal volume and switched them, or if we rotated the cosmos on the large scale, the universe would not behave any differently. From a four-dimensional perspective we wouldn't be able to tell the difference. It also means that every minimum location in space is identical to every other minimum location—even on the other side of the universe. How do we explain this?[5]

While we are trying to explain the universe's symmetries, let's not forget to explore the fact that, according to our best models, the laws of physics are not static in time. If we run the clocks backwards, toward the Big Bang, the symmetries that describe physical reality change. When certain thresholds are crossed, the vacuum becomes even more symmetric. In other words, the universe has less symmetry today than it originally had. Some of its initial symmetry has been broken or *frozen* out. How did this happen? Why were certain symmetries broken while other symmetries managed to remain intact?

Symmetry Breaking

> *"The secret of Nature is symmetry, but much of the texture of the world is due to mechanisms of symmetry breaking."*
>
> David Gross

When we push the clock back to the first moment (the Big Bang), the arrow of time—whose length represents the rate that time flows—shrinks toward zero. Each location in space experiences less and less time until the passage of time, in each location, completely stops. At this precise moment, the moment just before the first tick of time subsequent to the Big Bang, the universe possessed complete temporal symmetry. Every location in space was on equal footing (there were no spatial density gradients), and every location was experiencing time in an identical manner—not at all.

Thankfully, this perfect temporal symmetry no longer holds. Today, on microscopic scales each point in space experiences time differently. And on macroscopic scales spatial density gradients distort the experience of time from region to region.

Spontaneous symmetry breaking carried the universe from perfect temporal symmetry to its current asymmetric state and allowed evolution to proceed. In the distant future, as the universe cools to its lowest possible temperature, all matter will decay, all motion will cease, and complete temporal symmetry will once again reclaim the universe. Along with it, spatial symmetry will be restored. The entire universe will be in a state of maximum entropy. Each location in space will still be experiencing time independently, but at this point those experiences will be identical everywhere because there will, once again, be no warps in spacetime.

To familiarize ourselves with the concept of spontaneous symmetry breaking, let's examine the phenomenon of ferromagnetism. The constituent atoms in a spherical ball of solid iron are like little magnets that have a tendency to line up parallel with their neighbors, with the same north-south orientation. When the temperature is above 770 °C, the energetic thermal agitation of these atoms overwhelms their tendency to magnetically align. Above this temperature the material stops behaving as a large-scale magnet, because the little atomic magnets take on effectively random orientations—a highly symmetric configuration.

Below 770° C (the so-called 'Curie point'), it becomes energetically favorable for the atoms to line up, so, if it is cooled slowly enough, the iron becomes magnetized.[6] The system goes from being spherically symmetrical to possessing rotational symmetry about a single axis. Or in erudite-speak, "An $SO(3)$-symmetrical state (namely the original hot unmagnetized ball) evolves to an $SO(2)$-symmetrical one (the cold magnetized ball.)"[7] Spontaneous symmetry breaking occurs whenever a highly symmetrical state settles down into a state with less symmetry. Whenever a reduction in the ambient temperature induces an abrupt change in the nature of the stable equilibrium state of the material, reducing its symmetry, we call it a phase transition.

In general, a phase transition is a transformation in a system's organization—a reorganization of its parts. Vacuum quantization allows us to apply the concept of a phase transition to the vacuum itself. When the vacuum undergoes a phase transition it takes on a new vacuum state.[8] This allows us to understand the spontaneous symmetry breaking of the universe[9] in exactly the same way we understand other phase transitions.

Eras of Symmetry

> *"Symmetry plays a central role in phase transitions. In almost all cases, if we compare a suitable measure of something's symmetry before and after it goes through a phase transition, we find a significant change."*
>
> Brian Greene

The first era of symmetry represents a unique vacuum phase that lasted for 10^{-44} seconds after the Big Bang. It is called the Planck era, or the era of supersymmetry. During this phase the universe was in its highest symmetrical state because the vacuum quanta were compressed together so tightly that there was no independence of position—creating a vacuum state with just one unique position. During this brief 'era' the texture of spacetime was identical everywhere. There were no spatial gradients, no divergence or curl in the vacuum, no plane waves, no quantum tunnels, no sonons, and no time. Therefore, the four kinds of vacuum distortions (the four forces) were indistinguishable.

It has been said that during the Planck era the universe existed in a perfect phase of *nothingness*, but a better description would be a supersymmetric state of *potential*.[10] Although there were no geometric distortions in the vacuum, every quantum of space was maximally charged with energy. The quanta did not suffer this enormous compression for long. Instead, they did what all elastic conglomerations do after collision—they rebounded.

As this rebound ensued, a critical moment occurred when some individual quanta separated enough to be able to experience time—to resonate freely (completing at least one whole resonation before colliding with other quanta). The universe at this "first tick of time" was extremely different from how it is today. The vacuum density was enormously high, the temperature was approximately 10^{32} Kelvin, and the universe's temporal signature was just emerging. The introduction of a temporal signature carried the universe into what we call the *GUT* era (the Grand Unified Theory era), which began at $\sim 10^{-44}$ seconds. It also marked the beginning of a period of rapid expansion known as inflation.

To understand inflation, note that at the moment of the Big Bang the size of the universe was *effectively* one Planck length—because spatial size is a measure of the number of individually acting quanta of space (the number of unique locations in the system) and, in that instant, all of the quanta were maximally compacted and, therefore, coincident. As the individual quanta separated, independence of position emerged, astronomically ballooning the number of uniquely evolving spatial positions in the universe and triggering the flow of time. This emerging independence meant that the effective size of the universe "increased by at least a factor of 10^{30} during inflation, which means the effective volume of the universe increased by a factor of at least $(10^{30})^3 = 10^{90}$."[11]

As the number of unique locations grew, gravity became a distinguishable force, but the other three forces remained overwhelmed by the enormous kinetic energy of the quanta. When this happened the symmetry that allowed us to interchange gravity with the other forces broke. Temporal symmetry also broke at this point, because the existence of macroscopic density gradients meant that experience of time was no longer statistically identical everywhere. In other words, the spontaneous symmetry breaking that initiated the arrow of time, and split gravity off from the other three forces, occurred because the vacuum underwent a phase change—transitioning from a state of maximum compression, where the quanta were unable to individually operate, to a state with some separation.

The next era began around 10^{-34} seconds, which means that every quantum in the universe had, on average, experienced ten billion independent resonances—ticks of time. This was effectively the end of inflation. At this point, the quanta had completed their initial separation due to rebound, the pressure had dramatically dropped, and the temperature was ~10^{27} Kelvin.

From this point on cooling—which decreases the average number of collisions and, therefore, increases quantum independence—became the dominant cause of expansion. Cooling is a consequence of second order quantum inelasticity—a consequence of the fact that each quantum has internal structure that gets reordered during collision. That reordering absorbs a slight amount of the collision energy. While the quanta are sufficiently energetic, such that the energy absorbed is miniscule compared to the collision energy, we call the expansion "Friedman expansion".

During this era, the weak force became distinguishable from the strong force and the vacuum dropped closer to its superfluid state. At this point in history, quantum thermal agitation no longer completely overwhelmed vortex formation, but those vortices (mass particles) were created and destroyed with equal regularity.

At approximately 10^{-12} seconds another symmetry was frozen out. Before this time "an exact $U(2)$ symmetry held in which leptons and quarks were all massless, where *zig* electrons and neutrinos were on an equal footing with each other, and where the W and Z bosons and the photon could be appropriately *rotated* into combinations of each other according to a $U(2)$ symmetry."[12] Then, at about 10^{-12} seconds, the temperature dropped to just below the critical value throughout the universe.

Later, when the universe was about 3 minutes old, and thermal agitation was no longer sufficient to overcome mass formation, fundamental particles and atomic nuclei stably formed (think of the ferromagnetization example). At about 380,000 years the temperature had dropped to about 3,000 Kelvin, allowing atoms to be born. "Atoms formed as electrons settled around nuclei without being ripped apart by heat. Photons could now travel freely without being absorbed."[13] This is the radiation that the COBE satellite and WMAP spacecraft have measured. "The universe, once opaque and filled with plasma, now became transparent. The sky, instead of being white, now became black."[14] At this point that the electromagnetic force became distinguishable from the weak nuclear force—plane wave distortions became distinguishable from random thermal agitation. Consequently, another symmetry was broken.

After 1 billion years, stars condensed and the temperature of space dropped to 18 Kelvin. Then, around 6.5 billion years the quanta lost enough energy to begin gradually transitioning from Friedman expansion to an accelerated expansion that physicists call de Sitter expansion. By the time the universe was 6.5 billion years old the accelerated expansion was noteworthy. Today, roughly 13.8 billion years have passed (on average) since the first tick of time following the Big Bang and the temperature of space is now roughly 2.728 Kelvin.

The expectation is that the universe will continue to cool, accelerated expansion will continue, and the experience of time at every individual location in space will asymptotically grow more and more identical. Therefore, temporal symmetry will eventually be restored in the universe. This is the ultimate doomsday for the interests of intelligent species.[15]

As the universe approaches this "heat death" its broken symmetries will be mended. Eventually there will be no density gradients, no vortices, and no ripples in the vacuum. The four forces will be indistinguishable, but they will also no longer be necessary to characterize anything in the universe. The universe will have lost all of its energy. All motion will have ceased. Having reached complete independence, the vacuum quanta will be moving only through time.

The bad news for us, as beings that extract energy from spatial gradients and propagating distortion ripples, is that our existence becomes impossible in this final era. Nevertheless, as somewhat of a consolation prize for our inevitable absence, the universe will reclaim its intrinsic symmetries. At this point, trillions of years from now, the universe will find itself in a true state of *nothingness*—the kind without energetic *potential*. The quanta will have effectively reached their minimum interaction limit and will exist unchanged until something from outside of the universe comes along to change that state.[16]

Lorentz, Galilean and Other Symmetries

If the vacuum is a superfluid, if it can be modeled as an acoustic metric, then small fluctuations in that superfluid will obey Lorentz symmetry even though the superfluid itself is non-relativistic.[17] Below the excitation threshold the superfluid vacuum retains its identity as an ideal fluid. Under these conditions experiments like the Michelson-Morley experiment are expected to observe no drag force from the medium.[18] Nevertheless, with sufficient energy and momentum, Lorentz-breaking corrections are expected.[19] This implies that relativistic effects will be present so long as their corresponding fluctuations are small (compared to the phononic limit).

In reference to general relativity, Galilean symmetry (commonly used to describe our macroscopic, non-relativistic world) is understood as a valid approximation of the deeper underlying symmetry when velocities are small compared to the speed of light in a vacuum. Under the projection of a superfluid vacuum, Galilean symmetry is automatically inscribed as an approximation because the dispersion relations of superfluids exhibit relativistic properties at low momenta, which captures Galilean expectations at low speeds.[20]

More generally, vacuum superfluidity leads to the expectation that fluctuations in spacetime will behave like relativistic objects at small momenta, and like non-relativistic ones at large momenta. Here, the word *small* makes reference to the linearized or phononic limit.

A phonon[21] is a metric distortion, or a collective excitation in a periodic, elastic arrangement of atoms, molecules, or quanta. So long as an excitation is controlled by the elasticity of those arrangements it falls within the phononic limit.

To map an excitation (metric distortion) that exceeds the elastic capabilities of the metric's interacting constituents, Lorentz-breaking corrections must be introduced. Nevertheless, at low energies and momenta Lorentz and Galilean symmetries remain approximate. The point here is that a framework built from the premise that the vacuum is a superfluid allows us to ontologically penetrate the reason these symmetries are approximations of Nature's exact symmetry.[22]

> *"Since the symmetries are among the most beautiful and powerful tools at our disposal, one might expect that the theory of the universe must possess the most elegant and powerful symmetry known to science."*
>
> Michio Kaku

Exploring what is perhaps the most powerful symmetry in Nature, quantum physicists now suspect that the smallest meaningful distance in x, y, z space is the Planck length ($\sim 10^{-35}$ meters). M-theory gives us insight into what this *smallest limit* really is. Instead of being a cut-off point of confusion, it is an inversion point.

This implies that our mathematical description of the universe reflects at the boundary of the Planck length. In other words, if we could zoom into a single quantum, we would be moving beyond meaningful descriptions within the familiar dimensions of space. Nevertheless, a description would apply. From scales just below the Planck length a hypothetical observer would see what appeared to be an entire universe viewed from an apparently very large scale (observing practically a whole universe). From scales just above the Planck length the observer would see a very microscopic scale of just a small region of the universe he or she was viewing.

The transition between the two descriptions occurs at the scale of one Planck length. The further away our observers move from the Planck scale barrier, the more similar their descriptions of the universe become. As Brian Greene states, "This means that the physics within the Planck length is identical to the physics outside the Planck length. At the Planck length, spacetime may become lumpy and foamy, but the physics inside the Planck length and the physics at very large distances can be smooth and in fact are identical."[23] This means that within the *smallest distance* (a single quantum) there is an entire universe.

This repeating self-similarity is the most powerful kind of symmetry—fractal symmetry. Fractals have repeating geometric structures that cascade through an infinite number of scales in a self-similar way. If the fabric of the cosmos conforms to a fractal, then every complete universe can be described from another scale as being a single quantum. A collection of quanta come together to form a universe on another scale, which, in turn, represent a single quantum from another scale. The reflections go on and on.

According to the dimensional hierarchy equation an identical description of physical reality can be attained from any rung on the perspective ladder. This means that the laws of physics must also remain the same from every perspective of higher dimensional resolution, from all hierarchical scales. This depicts the underlying structure of physical reality (the vacuum itself) as possessing the highest possible symmetry—portraying it as a perfect fractal (Figure 25-1).[24]

Figure 25-1 Fractals replete with symmetry

This fractal geometry gives us a connecting fiber between the dimensions that echo far beyond the eons of time. It allows us to peer into the cyclic process of genesis and to understand the telescoping origins of emergent properties. It also allows us to see that the copious symmetries underlying natural law are directly tied to the atomic hypothesis (quantization). In this it exposes us to the echoes of infinity, where the songs of symmetry are composed.

[1] Brian Greene. *The Fabric of the Cosmos*. Random House, p. 224.

[2] Ibid.

[3] Hint: this symmetry also no longer strictly applies as we approach the Planck scale or when spatial density gradients are included in our system.

[4] Roger Penrose. *The Road to Reality*, p. 718.

[5] Until 1956 it was believed that all of the laws of physics obeyed three separate symmetries called charge *(C), parity (P),* and *time-reversal symmetry (T)*. Having a charge symmetry means that the laws are the same for particles and antiparticles. Having a parity symmetry means that the laws are the same for any situation and its mirror image (the mirror image of a particle spinning in a right-handed direction is one spinning in a left-handed direction). Time-reversal symmetry means that if you were to reverse the direction of motion of all particles and anti-particles in a system that entire system should go back to what it was at earlier times. This means that the laws of physics are the same in the forward and backward directions of time.

In 1956 two American physicists, Tsung-Dao Lee and Chen Ning Yang, suggested that the weak force does not strictly obey the parity symmetry. More specifically, they suggested that the weak force was asymmetric in such a way that would make the universe develop in a different way from which the mirror image of the universe would develop. This contradicted the symmetrical claims of parity. Later that same year Chien-Shiung Wu, proved that this prediction was correct. She did this by lining up the nuclei of radioactive atoms in a magnetic field, so that they were all spinning in the same direction, and then showed that the electrons were given off more in one direction than another. Lee and Yang both received the Nobel Prize for their idea one year later.

After this it was also found that the weak force did not obey charge symmetry. That meant that a universe composed of antiparticles would behave differently from our universe. But when these two symmetries were combined and considered together a symmetry remained that was not contradicted in Nature. This led physicists to believe that the weak force obeys the combined symmetry *CP*. That is, "the universe would develop in the same way as its mirror image if, in addition, every particle was swapped with its antiparticle!" Stephen Hawking. *A Brief History of Time*, p. 79–80.

This belief was torn down in 1964 when J.W. Cronin and Val Fitch discovered that the combined *CP* symmetry did not strictly hold in the decay of K-meson particles. Cronin and Fitch eventually received the Nobel Prize for their work in 1980. "A lot of prizes have been awarded for showing that the universe is not as simple as we might have thought!"

All of this led to the modern mathematical theorem that states that any theory that obeys quantum mechanics and relativity must always obey the combined *CPT* symmetry. The expectation is that Nature itself obeys the combined *CPT* symmetry. If this were the case then "the universe would have to behave the same if one replaced particles by antiparticles, took the mirror image, and also reversed the direction of time." This also means that, as Cronin and Fitch showed, that "if one replaces particles by antiparticles and takes the mirror image, but does not reverse the direction of time, then the universe does not behave the same." Somehow charge, parity, and time exist in some sort of entwined way. (Quotes, Ibid.)

[6] Roger Penrose. *The Road To Reality*, pp. 736–738.

7 Michio Kaku. *Parallel Worlds*, p. 194. As another example of $O(2)$ symmetry "think of a rocket designer who must create a sleek, streamlined vehicle to slice through the atmosphere. The rocket must possess great symmetry in order to reduce air friction and drag (in this case, cylindrical symmetry, so the rocket remains the same when we rotate it around its axis). This symmetry is called $O(2)$."

8 Roger Penrose. *The Road To Reality*, p. 736, 738.

9 Physicists have long known that "the universe went through a phase transition with an accompanying reduction in symmetry." Michio Kaku, *Parallel Worlds*, p. 267.

10 By observing the uniformity of the radiation dispersed throughout the universe we now have "observational evidence that in its earliest stages the universe was not populated by large, clumpy, high-entropy agglomerations of matter... [This] attests to the young universe being homogeneous...and when gravity matters...homogeneity implies low entropy." Brian Green. The *Fabric of the Cosmos*, p. 227.

11 Brian Greene, p. 312. Other estimates of this inflation give values of 10^{50} or even 10^{60} for the effective radius of the universe. See Michio Kaku. *Parallel Worlds*, p. 105.

12 Roger Penrose, p. 743.

13 Michio Kaku, p. 106.

14 Ibid.

15 Things change, however, when we take into account the inevitable reoccurrence of thirty-dimensional collisions resetting everything all over again.

16 When symmetry breaking is associated with a phase transition, several isolated regions are expected to independently seed that process—especially when the temperature is dropping fast. The faster the temperature drops, the greater the number of independent seeds.

As water is cooled past its freezing point several seeds of ice can start to form. These randomly dispersed seeds become platforms from which crystalline growths begins. These small crystals grow into larger and larger crystals of ice until the whole medium forms a solid lattice with several disjointed sections jammed together, in a manner that depends on the position and orientation of those original seeds. The same goes for our ferromagnetism example. As the hot ball of iron cools through the Curie point the symmetry is broken in several different locations that serve as seeds for the forming magnets. The final result is a patchwork of magnets aligned in an assorted manner.

Therefore, galactic seeds of mass scattered throughout the universe are to be expected if the universe is a quantized medium that cooled through its own respective "Curie point" in the past. This reflects what we see in the heavens. "Indeed, one of the most plausible models of galaxy formation—which has some significant observational support—proposes that they are largely 'seeded' by the supermassive black holes that now reside at their centers. This symmetry breaking is unlikely to take place 'all at once,' and domains in which symmetry is broken in different 'directions' may well occur." Roger Penrose. *The Road To Reality*, p. 741.

17 K.P. Sinha, C. Sivaram & E.C.G. Sudarshan. (1976). Found. Phys. 6, 65; (1976). Found. Phys. 6, 717; (1978). Found. Phys. 8, 823 (1978).

[18] Albert Abraham Michelson & Edward Williams Morley. (1887). On the Relative Motion of the Earth and the Luminiferous Ether. American Journal of Science 34: 333–345; P. A. M. Dirac. (1951). Nature 168, 906; (1952). Nature 169, 702.

[19] G. E. Volovik. (2003). *The Universe in a helium droplet*, Int. Ser. Monogr. Phys. **117**, 1–507.

[20] N.N. Bogoliubov, Izv. Acad. (1947). Nauk USSR 11, 77; N.N. Bogoliubov, (1947). J. Phys. 11, 23; V.L. Ginzburg, L.D. Landau. (1950). Zh. Eksp. Teor. Fiz. 20, 1064.

[21] Phonons are also referred to as quasiparticles. F. Schwabel, *Advanced Quantum Mechanics*, 4th Ed., Springer (2008), p. 253.

[22] In contrast to the traditional view, in which the low-energy symmetry of our world is traditionally described as being a remnant of a larger symmetry, which exists at high energy, and is broken when the energy is reduced, it has also been argued that starting from some energy scale (probably the Planck energy scale) higher energy leads to poorer symmetries, until even Lorentz invariance and gauge invariance are smoothly violated. Froggatt & Nielsen. (1991); Chadha & Nielsen. (1983).

From this point of view, relativistic quantum field theory is an effective theory. Polyakov. (1987); Weinberg. (1999); Jegerlehner (1998). In other words, it is an emergent phenomenon arising as a fixed point in the low-energy corner of the physical vacuum whose nature is inaccessible from the effective theory.

At lower energies the system acquires new symmetries, which it did not have at higher energy. From this point of view is possible that even Lorentz symmetry and gauge invariance are not fundamental, but gradually appear as we approach the low-energy, low momentum regime. From this viewpoint, Grand Unification schemes become incoherent if the unification occurs at energies where the effective theories are no longer valid.

In short, this view focuses on the idea that when the temperature of a superfluid approaches zero, the superfluid gradually acquires from nothing almost all the symmetries which we know today in high-energy physics: (an analog of) Lorentz invariance, local gauge invariance, elements of general covariance, and even the left-handedness and right-handedness are the emergent low-energy properties of quasiparticles. Grigory E. Volovik. (2003). *The Universe in a Helium Droplet*, Clarendon Press, Oxford, p. 2.

I think a bridge between these two views best captures the real process. That is, as laid out in this chapter, the high-energy early universe was supersymmetric, and evolved via broken symmetries to a less symmetric state. As cooling continues, the vacuum becomes more and more like a superfluid—leading to the gradual process of emergence that gradually returns the system to a state of higher symmetry.

[23] Brian Greene, p. 237.

[24] Laurent Nottale, a French astrophysicist and researcher at the Meudon Observatory in Paris has developed a theory that focuses on the fractal nature of spacetime. His theory is called "scale relativity" and it has some promising applications because of the powerful symmetries inherent in fractals.

Chapter 26 — Entropy

"Everything flows; nothing remains."
Heraclitus

"The geometry of space mixes the cosmos."
Amanda Gefter

Can the fact that the universe tends towards disorder be geometrically explained? Euclidean assumptions force us to take the second law of thermodynamics on as a brute fact. Does that change when we assume that the vacuum is a superfluid? In other words, does vacuum superfluidity naturally weave a tendency towards disorder into the fabric of our universe? Can it explain the law of entropy?

Every time we pour milk into our coffee we witness entropy at work. At first the concoction has relatively low entropy (high order) because all of the milk is initially separated from the coffee. Over time, however, the milk and coffee lose their independence and become more and more thoroughly mixed.[1] The rate that entropy increases depends on the dynamics of that mixing.

The second law of thermodynamics states that in our universe high entropy (disorder) is the natural state of being. The opposite of entropy is called "negentropy" (order). Continuous mixing may explain why the coffee and milk tend to remain thoroughly mixed once they have reached this state. But how do we explain the fact that the law of increasing entropy applies to every system in the universe?

The answer to that question is hidden in the fact that constant molecular rearrangement is responsible for the fact that the coffee and milk end up thoroughly mixed. This perpetual mixing is also responsible for the very small probability that the substances will coalesce. The key insight is that these conditions are completely general. Any system that persistently undergoes a process of internal mixing is endowed with the propensity to evolve from a state of low entropy to a state of higher entropy. But in order for mixing to occur, the system must be composed of quantized interactive parts. The very fact, then, that the law of entropy holds true everywhere in our universe, suggests that the medium of our universe (the vacuum) is quantized.

Quanta allow the vacuum to continuously mix and rearrange, and this quantum mixing encodes a time-symmetric tendency for increasing entropy as a law of Nature. But does quantization explain where the enormous amounts of order (low entropy) in our universe came from?

If Nature is fundamentally time-reverse symmetric, then how do we explain the millions of processes that we routinely witness unfolding in only one direction? For example, why do we always see coffee and milk mixing but we never see a mixture of coffee and milk fully separate? Shouldn't the very existence of unmixed coffee and milk be extremely unlikely in the first place? How can the second law of thermodynamics be a law of Nature when there are so many things that unfold asymmetrically in time?

Any explanation we might offer for the existence of low entropy (high order) becomes complicated by the fact that the second law of thermodynamics states that whenever a closed physical system happens to possess less than its maximum possible entropy it is extremely likely to possess more entropy subsequently and previously.[2] For example, once our coffee and milk have been thoroughly mixed, there exists a chance that at any specific moment some of the milk will regroup (separate from the coffee) due to random molecular interaction. The greater the separation the more unlikely it is to occur. But when it does occur, it does it in a way that is symmetric in time. How do we go from that explanation to an explanation of processes that are asymmetric in time?

Mixing Things Up

To work towards that answer, let's imagine a physical system that begins in a state of maximum entropy. To keep things simple let's say that our system consists of a pool table with no pockets and no friction. Including the cue ball, there are sixteen billiard balls in motion about the table. The balls continue to elastically collide and interact with each other and the four rails (because we have removed friction).

If I took a photograph of this system at a random moment (Figure 26-1) and asked you to describe the positions of the balls, you would likely concur that their arrangements appear random. This would not be surprising because random arrangements correspond with lowest order and therefore highest entropy. Suppose I continue photographing this system for a very long time until all the balls, for a brief moment, just happen to be tightly packed together—giving the system order and, therefore, decreasing its entropy (Figure 26-2).

Knowing that all these balls are in motion, what do you expect the system looked like just thirty seconds before this photo was taken? Also, what do you expect it looked like thirty seconds after? How about two seconds before, or two seconds after?

Figure 26-1 A random distribution has maximum entropy

Figure 26-2 An ordered distribution has less than maximum entropy

Because the system is in motion, it is extremely likely that the system will possess less order both before and after the photograph was taken—in a completely symmetric way. In fact, if I handed you two photos of the system, taken eleven and twelve seconds away from the moment of high order, you would have no means of determining the order of these photos. That is, you wouldn't be able to tell if they were taken twelve and eleven seconds before the clustering of the balls, or eleven and twelve seconds afterward (Figure 26-3). This is what we previously called time-reverse symmetry. It means that the probability of a sequence of events unfolding in one temporal order is equal to the probability of those events unfolding in the reverse order.

All of the laws articulated in classical, relativistic, and quantum physics are time-reverse symmetric.[3] For example, in Newtonian mechanics, forces that depend on particle positions that depend on time $x(t)$ can also be solved using negative values of time $x(-t)$.[4] Yet our experiences tell us that there are countless sequences of events that happen in one temporal order and never in reverse. Firewood burns, but never coalesces from ashes, gasses, and heat. Mirrors break, but never spontaneously form from scattered shards. Gasses thoroughly mix, but do not spontaneously thoroughly separate. Two bodies of unequal temperature equalize when brought into contact, but two bodies in contact never spontaneously or fully evolve into bodies of unequal temperature. (If they did it would be impossible to predict which body would become warmer and which would become cooler.)[5]

The laws of physics all declare that the probability of a set of events occurring in one order is equal to the probability of that sequence unfolding in the reverse order. So, if time-reverse symmetry truly exists in Nature, then why don't we see firewood coalescing from ashes, gases, and energy as often as we see firewood burning? This question is known as Loschmidt's paradox.

Once a system gains order, interactive mixing of its substrate can easily explain how that order decays, but the probability of a highly ordered object, such as ourselves, or firewood, spontaneously occurring within the system from random events is astronomically small.[6] So how is it that we have so much order in our universe? If the second law of thermodynamics truly holds, and time-reverse symmetry really is a quality of Nature, then why is it that our experience of time is so asymmetric?

Figure 26-3 Which frame comes first?

The solution to this question is that our universe is different from the given example in a significant way—it hasn't yet reached its maximum state of entropy. A more subtle difference is introduced by the notion that quantum collisions are not completely elastic—that each quantum has internal structure that gets reordered during collision, which absorbs a slight amount of the collision energy, introducing a smidgen of internal friction.

In our pool table example, we assumed that the system was frictionless and that it had already reached a state of maximum entropy (maximum disorder). For such a system, spontaneous occurrence of order would happen by chance every now and then, but it would be very rare. When it did arise, its reversal would be extremely likely—symmetrically returning the system to a state of maximum entropy.

Under these assumptions it automatically follows that if we cued the film of our billiard balls to a moment in which the system acquired a lower state of entropy (when they are all clustered together), we wouldn't be able to determine which way to run the film because we wouldn't be able to dynamically distinguish between the two directions in time.

By contrast, if the system started out with low entropy, if we were to rack the balls and then break them, it would be straightforward to determine the arrow of time for that sequence of events. This is the case because a continued decrease in entropy becomes astronomically more improbable with each new step (unless it has an exterior cause). Once maximum entropy is reached (the state referred to as equilibrium) we become incapable of distinguishing a preferred order of events.

We would also be able to determine an arrow of time in a system with a measureable amount of friction. A movie reel in which the balls slow down and eventually stop has a clear and preferred direction. According to our model, the internal friction in the vacuum is extremely small, but not zero. This suggests that time's arrow does ultimately have a magnitude, but that magnitude is so miniscule that we can safely say that Nature is roughly time-reverse symmetric. Therefore, for the most part, events occur asymmetrically in our universe because it has not yet reached its state of maximum entropy. The order of our universe is still decaying from the event we call the Big Bang.

Today that statement may not be very surprising, depending upon your familiarity with Edwin Hubble's discovery of other galaxies, and Vesto Slipher's observation that the spectral lines of those galaxies are routinely red-shifted.[7] Combined, these observations led to the inflationary Big Bang theory and the branch of physics called cosmology. The inflationary Big Bang theory is impressively supported by the observation of the cosmic microwave background [8] —a universal pressure of radiation permeating space and giving it a temperature of approximately 2.728 Kelvin. This radiation is a remnant of the original "flash" of the Big Bang that has been enormously attenuated (red-shifted).[9]

There was a time when claiming that the universe began with a Big Bang was considered blasphemous. In light of this, it is interesting to note that we can come to this conclusion—that the universe had a highly ordered beginning—simply by noticing that the laws of physics are time-reverse symmetric yet events occur asymmetrically (firewood always burns but never coalesces, and eggs splatter when dropped but never assemble from little broken bits and pieces). In short, the everyday sequences of events that we constantly witness demand that the universe had a beginning with low entropy. The order around us is a cosmologic relic—an entropic remnant of the beginning.

"Every time you break an egg you are doing observational cosmology."

Sean Carroll[10]

To absorb the full weight of this point, let's examine where the astonishingly high order in our bodies comes from. During our lives we intake raw materials and extract energy from them. This energy has low entropy and is the source of the order in our bodies. The energy balance for a person who maintains a constant mass is equal. This means that the amount of energy going in is matched to the amount of energy going out. However, the entropy of the energy consumed is very different from the entropy of the energy expelled.

The food, water, and air we take into our bodies have within them a significant degree of order and structure. But the products of the chemical reactions that go on inside us (leading to the expelled waste and, most significantly, the energy we give off to the environment in the form of heat) yield respectively highly disordered products. In this way, all life forms act as entropy converters. They take in low entropy ingredients and expel high entropy ones.

We derive our low entropy from the highly ordered food, air, and water we metabolize, but how do those ingredients acquire their order? The variety of foods that humans consume is quite large, but contemplating their individual sources of low entropy is unnecessary because we can trace the origin of low entropy for all carnivores, herbivores, and omnivores back to plants.

Where did the plants get their low entropy? The answer is that through photosynthesis plants acquired their low entropy from the energy of the Sun.[11] Therefore, we can trace the source of low entropy energy for life on Earth back to the Sun.

How did the Sun obtain its low entropy? The Sun (and the solar system) formed almost five billion years ago from a diffuse cloud of gas and dust that collapsed under gravity's influence.[12] In order to initiate this collapse the low-density cloud had to reach a fairly low temperature—less than about 100 degrees above absolute zero. At this temperature the constituent atoms of the cloud tend to stick together when they bump into each other, rather than careening off one another as they do at higher temperatures. Consequently, molecules began to form and grow into larger conglomerates. This, in turn, increased the density of the cloud and gave gravitational collapse the edge over counteracting pressures.[13]

For the term of our Sun's gestation, this collapse continued until the gas became so dense that, under its own weight, it ignited as a nuclear furnace, becoming a star by way of nuclear fusion. The radiation generated from this nuclear core created a pressure high enough to resist further gravitational collapse, stabilizing the new star. In short, the Sun derived its low entropy from a diffuse cloud of uniform gas having even lower entropy.

So where did this diffuse cloud of uniform gas come from? As a second-generation star, our Sun formed from the remnants of older, extinct stars whose lives ended with a powerful explosion called novae and supernovae.[14] These explosions spewed their contents out into space where they became precursors to many diffuse clouds including the one that formed our solar system.

So our food, air, water, and the very substances of Earth can trace their low entropy origins to elements that formed inside stars that have long since died, and the elements that initially made up those stars. Where did the diffuse clouds come from that proved adequate to form these earlier generation stars, or the first generation stars? According to our most advanced cosmological theories, the gas was formed during the aftermath of the Big Bang. When the universe was just a couple of minutes old it was filled with a highly ordered, uniform hot gas which was composed of mostly hydrogen, some helium, and trace amounts of deuterium and lithium. This initial state represents a condition of extremely low entropy—when, by analogy, all the coffee was separated from the milk.

According to observations made by the Cosmic Background Explorer (CBE) in the 1990s, and more recently, the Wilkinson Microwave Anisotropy Probe (WMAP), small irregularities in the density of the early universe under the influence of gravity, began a process of gentle accretion. As the dense gas cooled it fragmented, forming clouds that ultimately collapsed into stars. Therefore, the first generation stars formed from the dense gas that was created by the Big Bang.[15]

As we try to trace the entropic lineage further and further back, we discover that this is as far as cosmology goes. According to cosmology, the origin of low entropy in our universe is the Big Bang. Did the Big Bang have a cause? If it did, how did that event create such an ordered state?

From the second law of thermodynamics we have learned that within a closed or isolated system it is extremely unlikely to evolve toward a state of lower entropy. The lower the entropy the more unlikely it is that it will ever be reached. So how do we explain the fact that approximately 13.8 billion years ago our entire universe possessed an *extremely* low state of entropy? All physical processes that have unfolded since the Big Bang owe their existence and structure to the decay of that order—to the ever-increasing entropy of the universe. But how did our universe acquire this phenomenal amount of order?

In 1927, Arthur Eddington coined the term "time's arrow" to highlight this critical mystery and to add fuel to the fiery debate over what *time* really is. He also wanted to highlight the importance of figuring out whether or not Nature is really time-reverse symmetric, and how the first law of thermodynamics (the law of conservation of energy) and the second law of thermodynamics (the law of increasing entropy) can be reconciled with the asymmetry we see in the world. Where did all of that energy and order come from?

In response to Eddington's questions physicists have only been able to speculate with running imaginations. A natural solution, one that wouldn't violate any laws of thermodynamics, would be that the universe obtained its energy and low entropy from outside the system. The problem with suggesting this kind of solution is that traditional models run into a conceptual barrier when it comes to modeling something "outside of the universe". They are not equipped to explore this option and see only a stubborn paradox.

As Gary Zukav has said, "Paradoxes are the places where our rational mind bumps into its own limitations."[16] Consequently any attempt to attain "the understanding that we desire must involve passing the barrier of paradox."[17] Therefore, as long as the question of ultimate origins remains mysterious, the traditional way of framing the question will remain suspect.

In our model, the universe is an eleven-dimensional entity that, from a different level of perspective, is just a single quantum in another universe. This description leads to an expectation that the low entropy Big Bang had a cause that was entirely *outside* of our universe—as that quantum collided with another quantum (Chapter 27). This allows us to model the Big Bang, and the low entropy the universe acquired during that event, without contradicting the first or second law of thermodynamics.

Holograms

The holographic principle put forth by the Dutch Nobel laureate Gerard 't Hooft and Leonard Susskind in the early 1990s also relates to the mysteries of entropy. This principle suggests that the universe might actually operate like a hologram, which is a two-dimensional surface that can be illuminated to project a three-dimensional image. More specifically, the holographic principle claims that the three spatial dimensions of our experience might actually be holographic projections of physical processes taking place on a two-dimensional surface. The motivations for this suggestion are well grounded. For, as we have learned from black holes, the maximum entropy that a region of three-dimensional space can contain depends upon (scales by) its surface area, (a two-dimensional quantity) instead of its volume.

This observation suggests that basic degrees of freedom belonging to the universe's most fundamental ingredients (the entities that carry the universe's entropy) reside on a bounding surface. It also suggests that the information associated with the phenomena of a three-dimensional world can depend on a two-dimensional boundary.

What is this bounding surface? Where is it? Unable to discount the holographic principle, many have suggested that whatever this two-dimensional bounding surface is—the area of interaction that leads to all the patterns around us—it must be far away, perhaps on the surface of the universe itself. Under that assumption another question arises. How can such a distant boundary control the bulk of the cosmos?

Vacuum quantization introduces a new take on this matter. It suggests that the bounding surface described in the holographic principle is not located at some very distant location. Instead, the bounding surface, the two-dimensional interactive surface that controls the entropy of our universe, is precisely matched to the surface area of the universe's most fundamental ingredients—the quanta. This surface area controls the evolution of the vacuum because it determines the dynamics of the quantum collisions. The insights of the holographic principle are therefore directly tied to the entropy of our universe. This is important because it is by considering entropy through more expansive dimensional perspectives that we stand to gain insight into the cause of the Big Bang.

Furthermore, in order for entropy (S) to adequately represent something physically precise, it is necessary to have a fixed size for the minimum constituents within any system. A bounding surface, or a fixed minimum surface area of interaction, enables both an intuitive solution for why Nature manifests degrees of freedom based on surface area instead of volume, and a definitive interpretation for a measure of entropy. According to Boltzmann's formula entropy is explicitly equated to the logarithm of a phase-space volume $S = k \log V$ (where $k = 1.38 \times 10^{-23} \frac{J}{K}$).[18] Without a definitive minimum volume (individual quanta) the concept of entropy would remain entirely subjective—depending on some arbitrarily chosen scale—because the total combined surface areas of segmented volumes making up the entire phase-space volume are entirely dependent upon the size we choose for those fundamental base volumes.

For an infinitely smooth spacetime it is difficult to avoid this confusion and ambiguity. However, when we quantize the vacuum, entropy becomes a simple measure of the degrees of freedom within a region. This allows us to clearly define entropy.

Asymmetric States

By recognizing our universe as a system that has not yet reached a state of maximum entropy, we inadvertently solve some notorious asymmetric puzzles, known as the thermodynamic, electrodynamic, cosmological, and psychological temporal asymmetries.

The thermodynamic temporal asymmetry refers to the observation that heat flows from hot objects to cooler ones. It is an example of a common series of events that always appear to unfold in one order and never in reverse. It seems to claim that time is asymmetrical; due to the fact that time-reversal symmetry suggests that an equal probability of events unfolding in forward or reverse order exists. But, as we have found, such a description only applies to a system once it has reached a state of maximum entropy.

Such reasoning also applies to the electrodynamic temporal asymmetry, which is the observation that electromagnetic waves emanate outward from sources, but rarely converge inward on such points. For a system already in a state of maximum entropy electromagnetic waves would converge as often as they diverge. This means there would no longer be a thermodynamic or electrodynamic asymmetry between the past and the future.

The cosmological temporal asymmetry notes that the universe appears to be uniformly expanding and not contracting. An expanding universe—independent of its cause—is an indication that the universe has not yet reached its maximum state of entropy. To illustrate this, let's return to our pool table analogy. Instead of observing the system when it is already in a maximum state of entropy, let's imagine that the fifteen balls are racked near one end of the table, and the cue ball on the other side of the table.

After the cue ball strikes the cluster of balls, the entropy of the system increases because the balls go from having ordered to random arrangements. However, the system does not reach its state of maximum entropy until some time after the balls strike the rails. This is to say that if we observed the arrangements of the balls after they were set into motion, but prior to any interactions with the rails, the entropy of the system would still be less than the maximum. This fact enables us to distinguish a temporal direction.

If we filmed this process, cut that film into individual frames, and randomly shuffled those frames, it would be a straightforward task for anyone to rearrange that stack of frames into a correctly ordered sequence. This means that time's apparent "asymmetry is actually a property of states of the world, not a property of time."[19] It only manifests as an asymmetry until the "world" (the system) reaches a state of maximum entropy. When in a state of maximum entropy, time's "arrow" no longer measurably points in any direction.

If we had started with an even larger pool table, then the original entropy of this system would have been even lower—because the cluster of balls would have occupied an even smaller portion of the system. And if the rails of the table were to expand outward, starting from when we struck the cue ball, two things would occur. First, the rate at which the entropy of the system increases slows down. Second, the system will not reach its maximum entropy until the system's borders stop expanding.

For another example, imagine a closed canister of air in equilibrium (maximum entropy). If we suddenly expand the walls of the container, then the pressure and temperature will decrease as the molecules rearrange themselves to evenly occupy the larger volume. As this thermalization process takes place, equalizing the pressures in different regions, entropy increases. Therefore, as long as the canister is expanding, its interior state cannot be in complete equilibrium. Internal friction would play the same role as an expanding boundary. Therefore, instead of reflecting a quality of time, the cosmological temporal asymmetry reflects the universe's current state—the fact that it is either expanding or losing energy to internal friction. Once it reaches a state of maximum entropy and minimum energy perfect, time-reverse symmetry will be restored.

The psychological temporal asymmetry refers to the fact that we remember the past and not the future. Some have suggested that because our memories seem to form in only one direction, time itself must be asymmetric. This is an inaccurate conclusion. Our directional memory formation does not actually imply that temporal asymmetry is a property of time. Time-reverse symmetry would be broken only if memory formation retained a direction once the universe reached a state of maximum entropy.

Formation of memories is a unidirectional process because the universe itself is still increasing toward its maximum state of entropy. New memories add information and raise the entropy of the universe.[20] Therefore, if the universe, and everything in it, were already in a state of maximum entropy then there would be no mechanism by which new memories could form. Therefore, psychological asymmetric imprints reflect a physical state of the universe—not a quality of time.

In summation, there is overwhelming evidence, reflected in all the laws of physics, that Nature is time-reverse symmetric, but this quality only completely manifests itself when the system is in a state of maximum entropy. When we combine this with the fact that our universe has not yet reached its maximum state of entropy we are led to the inescapable conclusion that the dynamic trajectory of our current universe had a beginning defined by a state of extremely high order.

[1] Amanda Gefter. (2005, October 15). The riddle of time. *NewScientist*, p. 30.

[2] The first law of thermodynamics asserts that energy is always conserved within an isolated system. It is a law of equality. The second law of thermodynamics is a law of equality only if it describes a system that has already reached its state of maximum entropy. Otherwise, the second law manifests as a law of inequality. This speaks directly to time's apparent arrow. The third law of thermodynamics states that absolute zero can be asymptotically approached but never reached.

[3] Except for the ad hoc description of wave collapse which, as we discussed in chapter 17, is an unwarranted interpretation of what's really going on in Nature due to dimensional tenuity. Bohm's interpretation restores determinism time reverse symmetry to quantum mechanics.

[4] For example, if $x(t)$ solves $\frac{d^2x(t)}{dt^2} = F(x(t))$, then $x(-t)$ solves $\frac{d^2x(-t)}{dt^2} = F(x(-t))$. $x(-t)$ represents particle motion that passes through the same positions, but has reverse order and reverse velocity in reference to $x(t)$. Equivalently, the time derivative operator of Schrödinger's equation $i\hbar \frac{\partial}{\partial t}$ can be equally solved using $-t$ instead of t, or $-i\hbar \frac{\partial}{\partial t}$.

[5] It's not that such occurrences are impossible, it's that they are incredibly unlikely.

[6] This statement is a bit misleading because it refers to a system that has already reached a state of maximum entropy. In such a system the occurrence of highly ordered structures assembling from random events would be extremely small. In a system where entropy is still increasing, order is taken from objects of higher total order to construct resultant structures.

[7] Vesto Slipher is also the discoverer of Pluto.

[8] "This radiation was first theoretically predicted by George Gamow in 1946, on the basis of the Big Bang picture, and more explicitly by Alpher, Bethe, and Gamow in 1948; then again independently by Robert Dicke, in 1964. It was discovered observationally (accidentally) by Arno Penzias and Robert Wilson, in 1965, and immediately interpreted by Dicke and his colleagues." Roger Penrose. *The Road to Reality*, p. 733.

[9] Roger Penrose, p. 704.

[10] Sean M. Carroll. (2008, June). The Cosmic Origins of Time's Arrow. *Scientific America*, pp. 48–57.

[11] There are some ecosystems that are supported by non-photosynthetic processes. For example, near thermal vents in deep oceanic trenches, where no light from above penetrates, there are ecosystems supported by bacteria that metabolize pyrite and other minerals. Even these low-entropic sources can be traced back to the formation of the solar system and, therefore, their existence does not affect the result of our discussion.

[12] By radiometrically dating the retrieved Apollo Moon Rocks, we attain an age of 4.58 billion years for the Moon, and by extension the solar system. The tectonic activity on Earth is constantly erasing old rock records and replacing them with new formations. Consequently, the oldest rocks remaining on Earth, called cratons, are roughly 3.5 billion years old. However, within some of those cratons, sedimentary layers have been found to contain zircon crystals that eroded out of previous formations. These zircon

crystals are age dated to 4.1 billion years. Jack Hills in Western Australia is home to the oldest known craton.

[13] Neil DeGrasse Tyson. *Death By Black Hole*, pp. 187–188.

[14] Astronomers can easily distinguish a second-generation star from first-generation star—one that formed from the early remnants of the Big Bang. Elements heavier than helium are almost exclusively created by the fusion processes taking place in the interior of stars, or the violent supernovae explosions at the end of their lives. Therefore, any star, or solar system, that forms in the presence of heavier elements (which includes any solar system with planets, comets, asteroids, etc.) must have taken their ingredients from the remnants of dead stars. Today the remaining first-generation stars are all small stars. This is because smaller stars have longer life spans. The minimum mass limit that still enables nuclear fusion occurs in a star approximately one-tenth the mass of our Sun. Any star with a mass comparable to our Sun also has a lifespan comparable to our Sun's lifespan—approximately ten billion years. Therefore, any sun-sized star born around 13 billion years ago would have died around 3 billion years ago.

[15] For a more thorough discussion on this evolution see Brian Greene. *The Elegant Universe*.

[16] Gary Zukav. *The Dancing Wu Li Masters*, p. 205.

[17] Ibid.

[18] The Boltzmann constant appears in many equations employed to describe systems composed of many fine parts. For example, the evolution of a closed system of gas or a liquid, which are both composed of many identical molecules, is accurately mapped with equations that depend upon the Boltzmann constant (k), whose value is determined by the quantization of spacetime (Chapter 16).

[19] Paul Davies. (2002, September). That Mysterious Flow. *Scientific American*—Special Edition, A Matter of Time.

[20] Ibid.

Chapter 27 — **Genesis**

"A common misconception is that the big bang provides a theory of cosmic origins. It doesn't."
Brian Greene[1]

"In an isolated system there is always the possibility of an input from the outside world that can serve to reduce entropy from time to time."
Roger Penrose[2]

Gene Shoemaker stood in silence on a rocky ledge and scanned the out-of-place structure before him. His guide, a native Australian Aborigine, watched him as he tracked the curving cliff edge and then slowly focused in on the island of rock defining the depression's center. To Shoemaker this fossilized scar looked suspiciously like many other structures he had scouted out around the world. To his trained eyes those structures also mimicked the marks that covered the face of the moon.

Shoemaker was convinced that the surface of the moon was textured by impact craters, but the prevailing wisdom of the day held that the moon was pocked by calderas—characteristic depressions that form inside the rims of volcanoes. To convince others that the marks on the moon were impact craters, Shoemaker had to battle against the belief that the heavens were static. This belief portrayed the planets and moons as safe from bombardment because nothing in the heavens had changed since the creation. Impacts were thought impossible because the heavens were declared permanent and non-interactive.

"What do you think made this?" Shoemaker asked.

"Oh, I know what made this," his Aboriginal guide said with assurance.

"Oh really?"

"Yup... many, many years ago a giant caterpillar, with rows of huge teeth, came up from underground to see Her world. When She did she saw what mankind had been doing while She slept in the deep and it infuriated Her. She began spinning." He animated his words with his hands above his head. "Voom, voom, voom... faster and faster, until She violently pulled back into the earth—leaving this hole to remind us that She will return."

They both continued to stare off the cliff's edge. Then, after a few moments, the Aborigine asked, "What do you think made it?"

After a pause, Shoemaker lifted his hands above his head and began to motion. "I think a huge rock flew around the heavens for many, many years from many times beyond the Sun, until one day it came crashing down through the air as a big ball of fire. It struck the ground so powerfully that it obliterated the earth beneath it—leaving this hole.

Silence again, as they both stared. Then the Aborigine guide said, "I like mine better."

The Universe of Imagination

"The universe is just one of those things that happens from time to time."
Edward Tryon

Gene Shoemaker's concept of the heavens was eventually, and undeniably, verified when the Apollo Astronauts visited the moon, but it took a long time for people to fully absorb how interactive the cosmos are. Shoemaker himself was selected and trained as an Apollo Astronaut, but just before he quarantined for flight, NASA's doctors diagnosed him with a virus that grounded him. As his understudy took his place, Shoemaker watched his boyhood dream of touching the moon be destroyed by a virus that never developed into a sickness.

After that, Shoemaker and his wife began observing uncharted asteroids and comets. (Carolyn Shoemaker still has the distinguished honor of naming the most asteroids.) They spent years laboriously and meticulously photographing and charting faint objects in the heavens—hoping to change the world with an idea. Then one night, while riding the doldrums of monotonous routine, a regular examination of their photographic plates produced a miracle. They, along with David Levy, discovered what came to be known as Comet Shoemaker-Levy 9. As they tracked this comet, and entered its changing coordinates into the computer, the computer predicted that the comet would swing close by Jupiter, go around the Sun, and then, on its way back out of the solar system, slam into Jupiter.

As time progressed the calculations were fine-tuned, but the prediction remained the same. Then, on July 16, 1994, the very man who had boldly claimed that the heavens were not static, that comets and asteroids collide with the planets and moons, watched as a comet that he, his wife, and his friend discovered crashed into Jupiter.[3] It was a spectacular show that was easily observable from backyard telescopes (Figure 27-1). This was the exclamation point on Shoemaker's work. It unequivocally put to rest any doubt as to whether or not celestial impacts have happened because it showed that they still happen![4]

Shoemaker helped us go from seeing the universe as a static realm to an interactive, and sometimes violent place. We now recognize the significant role that the extraterrestrial realm has played in the evolution of life on Earth. The geologic record reveals that the end of the Ordovician, Devonian, Permian, Triassic, and, most famously but not most significantly, the Cretaceous periods were all marked by global mass extinctions. Except for the mass extinction at the end of the Ordovician, each of these transitions is correlated with a violent meteor impact that left its mark in the rock record.[5] The Permian mass extinction, also known as "the Great Dying", has also been correlated with a volcanic eruption that covered huge areas in Siberia and roasted chlorine-rich salts like two-kilometer thick layers of gypsum and halite. Whether or not the eruption was triggered by the impact is not yet known. What is known is that the universe is not the static, halcyon place it was once believed to be.

If the universe is not static, if it has not always been as it is now, then what was it like before, and how do we explain its original state? As we trace our lineage back through the generations, and our origins back through time, we can trace our roots through our early hominid ancestors, through the first mammals, the first vertebrates, protochordates, bacteria, and the first forms of protolife. We can even imagine following this cause and effect lineage back through the formation of the Earth, through the formation of the elements in the primordial universe, and even to the Big Bang. But is this the end? Does the origin lineage come to finality in this moment? Are the cosmos as impermanent as we are? Or, does the Big Bang itself ultimately have a cause?

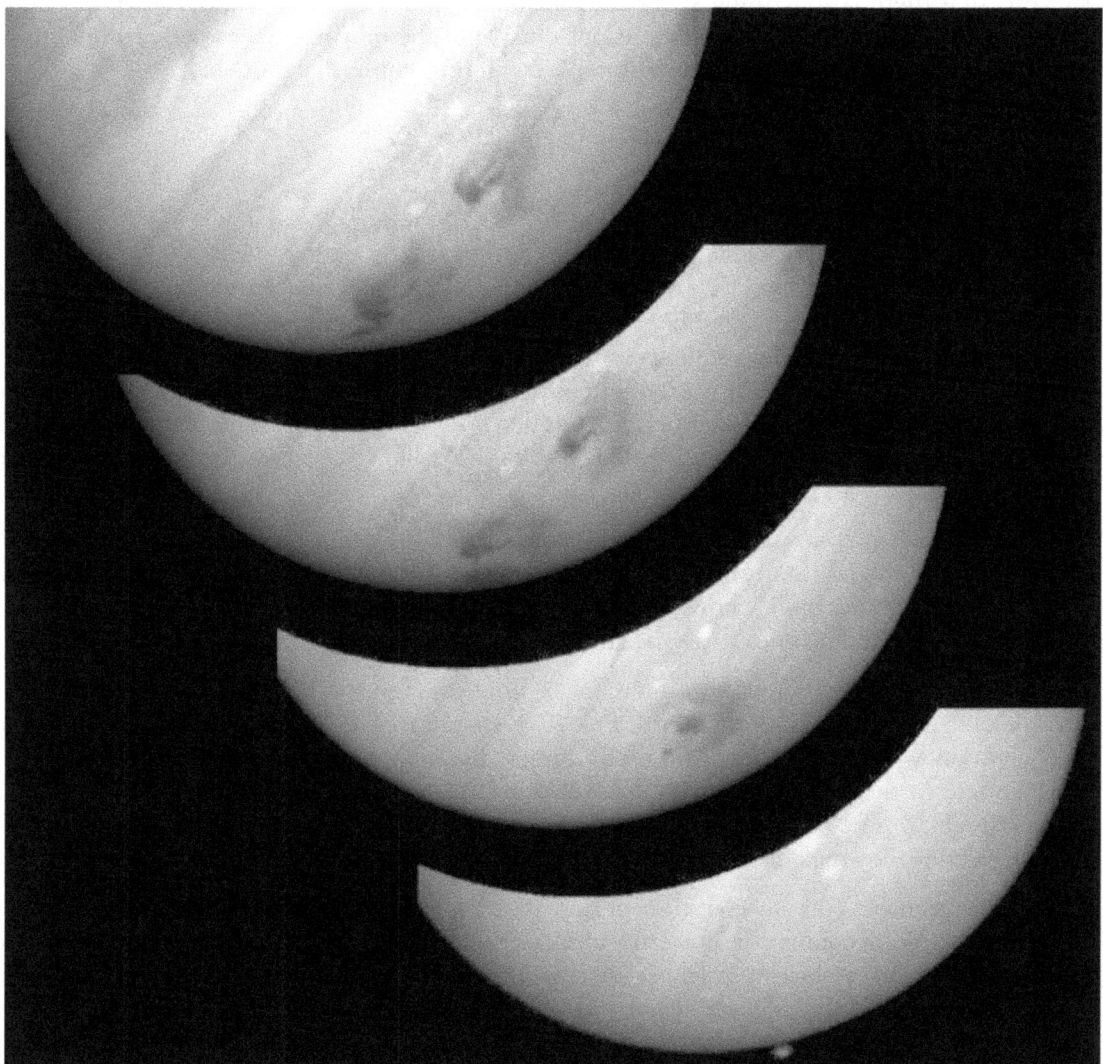

Figure 27-1 Impact of comet Shoemaker-Levy 9 (H. Hammel, WFPC2, HST, NASA)

Here our personal concerns connect to our cosmic ones. Up until recently science has been unable to offer us any substantial insight into the issue of ultimate origins. Standard cosmology has given us the very reputable inflationary Big Bang theory, but the Big Bang theory does not tell us how the universe was created. Instead, it tells us about the evolution of the universe since the bang. It describes our universe as a closed system that has evolved in terms of broken symmetries and increasing entropy for the last 13.8 billion years.

In short, the Big Bang theory leaves us wondering how the universe could acquire this enormous amount of energy and low entropy without violating the first or second law of thermodynamics. It also leaves us confused over the notion of cause (and time) in reference to the Big Bang. How could there be a cause for the Big Bang if in the moment of the Big Bang there was no time? How can there be cause and effect without time?

When a system acquires a significantly lower state of entropy it usually implies that it is a part of a larger system. The entropy of that larger system increases even when its constituents acquire lower entropy. Therefore, the low entropy associated with the Big Bang strongly suggests that there is something separate from our universe—something beyond it. This, in turn, submits that there are *other* dimensions outside of the universe. These other dimensions may be outside our purview,[6] but that doesn't mean they cannot be modeled.

Before the Bang

Gabriele Veneziano (and others), have introduced a model that attempts to speak to the process of genesis called the *pre-Big Bang scenario*. This model imagines that prior to the Big Bang our universe was collapsing into something like a supermassive black hole. Because the collapse of a black hole cannot be infinitely extended, Veneziano suggested that when the curvature (and therefore the temperature and density) of the whole universe reached its maximum allowed value (like inside a black hole) the collapse reversed, or bounced, leading to the Big Bang.

Inside black holes, space and time swap roles. The center of a black hole remains suspended in a single instant in time. This makes it difficult to trace the chain of cause and effect through a black hole. Similarly, as we run cosmologic models in reverse, toward the initial conditions, time comes to a screeching halt.

This suggests that imagining a moment of time prior to the Big Bang is like "asking for directions to a place north of the North Pole."[4] To avoid this mess, the pre-Big Bang scenario asks us to allow for some sort of time of another direction—a time dimension that remains independent of the geometric parameters of our four-dimensional universe. As the collapse ensues, regular time is squeezed out of the universe, but the other dimension of time continues on—allowing a causal dynamic evolution of the collapse.

In this scenario, the acceleration period that leads to the violent bounce accounts for inflation. It also accounts for how the universe acquired its extremely homogeneous and isotropic state. Nevertheless, it leaves the mechanism that ultimately initiated this process vague. What started the collapse, what started that, and so on. Therefore, the pre-Big Bang scenario falls short of being a true genesis solution, but it does push our question of ultimate origins back one step.

Another popular model that attempts to describe the state of the universe, before the Big Bang, is called the ekpyrotic scenario. The name comes from the ancient Stoic notion of ekpyrosis, a fire out of which the universe continuously gets reborn.[7] This scenario asks us to imagine the three familiar spatial dimensions of our universe as being represented by a flat brane that is suspended within a higher-dimensional spacetime, and which occasionally collides with some other hidden brane (Figure 27-2). According to this model, the most recent collision corresponds to the instant of the Big Bang.

A variant of the ekpyrotic model, called the cyclic model (developed by Paul J. Steinhardt and Neil Turok in 2002), describes the Big Bang as the result of the collision between two D-branes which may have occurred many times before.

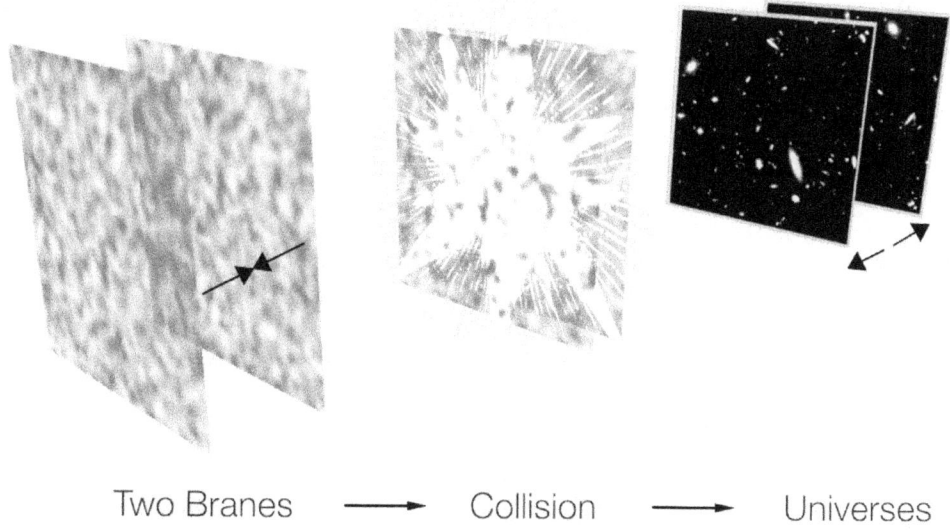

Figure 27-2 Ekpyrotic Big Bang

This model imagines our universe as a brane that is suspended in higher-dimensional space, in close proximity to some other brane. The space between the two branes is dimensionally separate from both those on our brane and those on the other brane (Figure 27-3).

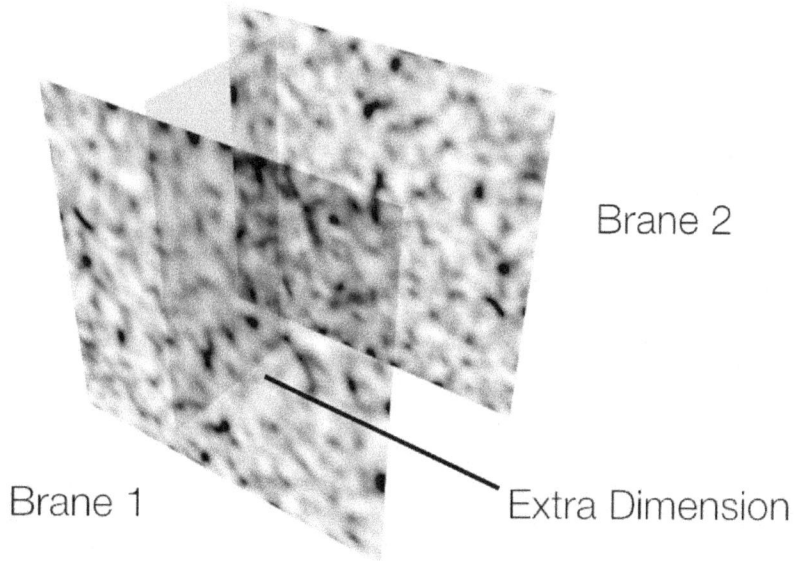

Figure 27-3 Universes separated by another dimension

Because the dimension of time is uniquely tied to the dimensions of space within our universe, this also suggests that the spatial dimensions between the two branes are not related to time. An additional temporal dimension is required by the claim that motion takes place through those additional spatial dimensions (as the branes come together to collide or separate). This independent dimension of time tracks motion in those outside dimensions, and therefore, facilitates the mechanics that lead to the Big Bang. Unfortunately, there is no extrapolation on this point in Steinhardt and Turok's model.

One advantage that the cyclic model has over its precursor ekpyrotic model is that it allows a mechanism to describe how the bang was so homogeneous—giving it an extremely low state of entropy. In the ekpyrotic model the colliding branes have to be almost exactly parallel to one another prior to collision in order to produce a sufficiently homogeneous condition for our early universe. The cyclic model has the advantage of being able to straighten out successively as the collisions continue. One problem, however, is that it remains vague whether the branes in this model extend to infinity (which introduces the question of how anything could move an infinite object so uniformly, or even at all), or whether the branes are finite (leaving the colliding ends waving back and forth).

Although the cyclic model doesn't depict an inflationary burst, it does have a longer, milder, period of accelerated expansion after the branes collide. This acceleration period solves the horizon, smoothness, and flatness problems that originally encouraged cosmologists to evoke an inflationary period into their cosmologic models.

Richard Tolman, of the California Institute of Technology, developed an older cyclic cosmological model in the 1930s. His model suggests that "the observed expansion of the universe might slow down, someday stop, and then be followed by a period of contraction in which the universe got ever smaller."[8] Instead of concluding that the universe would be doomed to reach a fiery finale in which it implodes on itself and comes to an end, Tolman proposed that the universe might undergo a bounce; that space might shrink down to some small size and then rebound, initiating a new cycle of expansion followed once again by contraction. A universe eternally repeating this cycle—expansion, contraction, bounce, expansion again—could elegantly avoid the thorny issues of origin. In such a scenario, the very concept of an origin would be inapplicable since the universe always was and would always be.[9] But a universe that has been around forever would look very different than ours.

According to Tolman's model, this mismatch means that our universe cannot be infinitely old. As we run the clock backward through these progressions, we find that the cyclic pattern of this model could have repeated for a while, but not indefinitely. During each cycle, the second law of thermodynamics dictates that entropy would, on average, rise. Therefore, if this cycle had been repeated an infinite number of times, then the whole universe should be in a state of maximum entropy. The contraction phase of this model is not the same as the expansion phase run in reverse. Physical processes such as falling leaves and melting snowmen would happen in the usual *forward* time direction during the expansion phases and would continue to do so during the subsequent contraction phase. Entropy would increase during both phases.[10]

There is another reason that this cycle could not have carried on indefinitely. General relativity states that the amount of entropy at the beginning of each new cycle determines how long that cycle will last. "More entropy means a longer period of expansion before the outward motion grinds to a halt and the inward motion takes over. Each successive cycle would therefore last much longer than its predecessor; equivalently, earlier cycles would be shorter and shorter. When analyzed mathematically, the constant shortening of the cycles implies that they cannot stretch infinitely far into the past."[11] Therefore, even in Tolman's cyclic framework, the universe would have a beginning—an unexplained beginning.

Tolman's cyclic model and the more recent pre-Big Bang and ekpyrotic scenarios (including the cyclic incarnation) subtly require that we include additional dimensions into our model of the universe. They depict a realm that possesses spatial dimensions outside our universe and a temporal dimension related to those external dimensions. But these scenarios leave the mechanics of these higher realms entirely subjective.

Our quantized model has an advantage here. It holds that, in a self-similar way, the individual quanta are composed of even finer subquanta, which are composed of sub-subquanta, and so on. Similarly, it suggests that there is a vantage from which our whole universe can be portrayed as a single brane (a superquanta), which combined with many other identical branes forms a megaverse. It also gives us dynamic access to the higher-dimensional framework that these branes are suspended in (Figure 27-4).

Figure 27-4 Multiple universes

According to this model, the collision that charged our universe with extremely low entropy is dimensionally separate from, but identical to (in the mechanical causal sense) the collisions that occur between two quanta. As universes collide their interior constituents are pressed together—creating an extremely homogeneous and isotropic spatial state—charging the internal system with high energy and low entropy without violating any laws of thermodynamics.

The collision of two spherical quanta will always internally appear to have been nearly head-on—naturally producing a sufficiently homogeneous condition that nonuniquely dictates an increase in entropy for the larger system, while injecting energy and low entropy into the internal system. In a fractal universe this process cascades through the levels of reality.

Therefore, the low entropy in our universe (the Big Bang) originated from a multi-dimensional collision that was caused by a previous multi-dimensional collision, which was eventually caused by an even higher-dimensional collision, and so on.[12] This paints the cosmos as eternal and beautifully symmetric. No matter how much we increase our dimensional perspective, the fundamental origin of low entropy is a multi-dimensional collision.

> *"This world... has always been, it is, and always shall be: an everlasting fire rhythmically dying and flaring up again."*
>
> Heraclitus[13]

According to this, there is no ultimate barrier to the chain of cause and effect. The state of the universe at the Big Bang represents an internal temporal limit (which is equivalent to saying a spatial boundary), but this temporal limit is not ultimately the first moment of time. By making use of the dimension of supertime (and higher dimensions of space) we can dynamically trace causes that go right through the moment when time is squeezed out of the universe.

After collision, the universe recoils and independently evolves, according to the laws of thermodynamics, until an inevitable multi-dimensional collision strikes again. This infinite cycle of the multi-dimensional collisions explains how the universe came to be imprinted with fantastically low entropy and enormous potential energy. It describes the mechanics of the Big Bang in a way that powerfully transforms the meaning of genesis.

Eternal Recurrence

> *"When on some gilded cloud or flower, my gazing soul would dwell an houre, and in those weaker glories spy some shadows of eternity."*
>
> Henry Vaughn[14]

If the assumptions of our model are correct, then *eternal recurrence* is captured by the omniverse. We are brought to this condition in the following way. The number of quanta that compose our particular universe is large but finite. According to the principle of dimensional hierarchy, our universe is but one of many that come together like water droplets in a sea—composing another universe (a megaverse) that gains resolution from an even more expansive dimensional perspective. This megaverse floats among another sea of similar megaverses and so on. The infinite collection of these perspectives is called the omniverse. On each level the number of constituents that make up the level of the universe in question is large but finite. However, because the hierarchies extend to infinity in both directions (greater expanse and greater resolution) the number of quanta that make up the entire omniverse is infinite.

Because each quantum is also a universe from another dimensional perspective, the number of universes in the omniverse is infinite. On every level, the evolution of each of these quantum universes is entirely determined by the geometric arrangements that its internal quanta acquired during its most recent Big Bang. Because there is a finite number of possible arrangements, and an infinite number of universes (each composed of a finite number of quanta), the set of possible evolutions is large but finite. This means that each possible evolution plays out identically an infinite number of times.

The consequences of this requirement can pierce into our very center. It can be either the most beautiful thought we have ever entertained, or the most frightening. Every single event that has unfolded in our universe since the Big Bang has been, is being, and will be played out exactly the same way an infinite number of times throughout the omniverse. Every moment replays perfectly as the "eternal hourglass of existence is turned upside down again and again..."[15] To fully embrace this condition is to come face to face with ultimate interconnectivity. Suddenly "every pain and every joy and every thought and sigh and everything unutterably small or great in your life"[16] is filled with the breath of infinity.[17]

The moment you first fell in love eternally echoes throughout the cosmos. All of the details that came together to create that moment, the feel of the desert air, the brilliant Milky Way above, the anticipation, the way her lips moved and her eyes danced, the secrets shared, even the incredible timing of that falling star as you first kissed, all will be repeated eternally. With each "echo" everything will be in its place and every action will be on cue. What could be more divine? What could resonate with more delight? Then again, eternal recurrence is nonselective. Every undesirable moment that has ever played out will also find itself eternally echoed. Each requires the other.

Friedrich Nietzsche reflected upon this idea of eternal recurrence in his book *The Gay Science* as a thought exercise to be used as a measure of the good life. He had the reader ponder the following question. If you had the choice to have your life relived eternally, each time identically, would you choose it? "Do you desire this once more and innumerable times more?" He then asked, "how well disposed would you have to become to yourself and to life to crave nothing more fervently than this ultimate confirmation and seal?"[18]

Our new picture transforms the fantasy behind Nietzsche's thought exercise into an inextricable physical part of Nature. It does more than ask us to consider whether we would find it divine or terrifying to learn that all of our experiences will be relived and experienced an infinite amount of times throughout the omniverse. It tells us that they will. Considering the depths of this connection can force us into some sublime quandaries.

"Eternity is not the hereafter... this is it. If you don't get it here, you won't get it anywhere."

Joseph Campbell

In 1584 an Italian monk and philosopher named Giordano Bruno was burned at the stake by the Catholic Church for supporting an idea that had been held by the Greek philosopher Democritus (5th century BCE), and the fifteenth-century cardinal Nicholas of Cusa. Later, the eighteenth-century philosopher Immanuel Kant, and the nineteenth century novelist Honoré de Balzac, would bolster the idea again. Bruno's claim was that there are innumerable suns and that "innumerable earths revolve about these suns."[19] This condition led him to conclude that life itself is innumerable throughout the vast cosmic ocean.[20]

The eternal recurrence of our model's dimensional cascades carries Bruno's idea one step further by proposing a physical mechanism in Nature that requires the literal manifestation of his claim. So perhaps we should dedicate this insight to Bruno's intuition and sacrifice. His vital impact will echo through the eternities, and his unwavering struggle against tyranny will have an infinite reach. His experiences have been, are being, and will again be integral to the very moment we now enjoy. They unfold within us, completely playing out as the infinite universes within us replay the song of the cosmos with rhythmic harmony.

Heraclitus of Ephesus (c. 500 BCE) once said, "To God all is beautiful, good and as it should be. Man must see things as either good or bad."[21] The dimensional cascades of our new map have given us access to the insight that every quantum is a universe and every universe is a quantum. If it is right, then right now, somewhere along that dimensional structure, an entire universe remains suspended in the moment of your first kiss. Every moment, each of our most personal experiences, is carried in the shadowed echo of an entire universe, playing out again and again.

The principles of eternal recurrence and ultimate interconnectedness compose the melodies of the cosmos and soundly connect us with the infinite. With a sense of an eerie yet somehow enlightening echo we find that vacuum quantization has become our intuitive and conceptual portal to undifferentiated reality.

This model reinforces the idea that our natural end is to be fully alive, to experience great exhilaration, great joy, great suffering, and great passions in every small moment, to be fully aware, vital and alert, to live a full life, not a pinched, restricted one.[22] It also brings new meaning to the phrase "living in the moment". For as T. S. Eliot said in *East Coker*, there is "a lifetime burning in every moment." And the flames of that fire reflect our "magnificent insignificance."[23]

> *"Did you ever say Yes to joy? O my friends, then you said Yes to all woe as well. All things are chained and entwined together, all things are in love..."*
>
> *Friedrich Nietzsche*

> *"So decide now that you are worthy of living..."*
>
> *Epictetus*

[1] Brian Greene. *The Fabric of the Cosmos.*

[2] Roger Penrose. *The Road To Reality.*

[3] Comet Shoemaker-Levy is also nicknamed 'the String of Pearls' because of the 21 pieces it broke up into as it passed close to Jupiter before rounding the sun. They first discovered it in March of 1993. After impact, the resulting scars lasted for months. Mario Livio. (2006, July). Hubble's Top 10. *Scientific American*, p. 44.

[4] After Gene (Eugene) Shoemaker died his remains were cremated and then shot on a rocket to the moon during the NASA Lunar Prospector mission. His remains are currently the only human remains on the moon. For $44,995.00 this service may soon be available through Space Services of Houston.

My version of the Shoemaker story is paraphrased. None of the quotes should be taken as exact.

[5] These impact sites are usually in carbon rich sediments.

[6] Lisa Randall. *Warped Passages*, p. 439.

[4] Gabriele Venziano. (2006, June). The Myth of the Beginning of Time. *Scientific American*, Special Edition—A Matter of Time, p. 73.

[7] The ekpyrotic scenario has been developed by Justin Khoury, Burt A. Ovrut, Nathan Seiberg, Paul Steinhardt, and Neil Turok.

[8] Brian Greene, *The Fabric of the Cosmos*, p. 405–406.

[9] Ibid.

[10] Ibid.

[11] Ibid.

[12] Evidence is already surfacing in support of a collision being the trigger for inflation. "Not all the properties of the microwave sky seem consistent with the [standard model of cosmology]. For example, the pattern of hot and cold spots should be structureless, having no preferred direction or arrangements, mirroring the random fluctuations of the primordial density perturbations. In reality, certain components of the pattern are inexplicably aligned as in the so-called "axis of evil" discovered in 2005 by Kate Land and João Magueijo of Imperial College London." Peter Coles. (2007, March 3). Boomtime. *NewScientist*, p. 37.

[13] Heraclitus, quoted by Douglas J. Soccio. *Archetypes of Wisdom*, p. 58.

[14] Henry Vaughn. *The Retreate.*

[15] Friedrich Nietzsche. *The Gay Science*, pp. 273–274.

[16] Ibid.

[17] This puts an interesting twist on Mallarme's words, "As eternity at last gives him back to himself." Or in his native tongue, "Tel qu'en lui--même enfin l'éternité le change." Quoted in *The Existential Background of Human Dignity*, by Gabriel Marcel, translated in *Existentialism And The Philosophical Tradition* by Diane Barsium Raymond, p. 337.

[18] Ibid. Nietzsche may have gotten the idea of eternal recurrence from the pre-Socratic philosopher Heraclitus or even from a contemporary of his, the German writer Heinrich Heine.

[19] Giordano Bruno. (1584). On the Infinite Universe and Worlds; Neil DeGrasse Tyson. *Death By Black Hole*, p. 83.

[20] Bruno also wrote, "Thus is the excellence of God magnified and the greatness of his kingdom made manifest; he is glorified not in one, but in countless suns; not in a single world, but in a thousand thousand, I say in an infinity of worlds." Michio Kaku. *Parallel Worlds*, p. 345.

[21] Douglas J. Soccio, *Archetypes of Wisdom*, p. 68.

[22] Ibid., p. 174.

[23] To further explore how intuitively accessing higher dimensions really unveils our magnificent insignificance consider how dimensional reduction results in the opposite effect. In Edwin Abbot's book the Square of Flatland discovered these effects when the Sphere from Spaceland showed him the creature of Pointland:

"Look yonder," said my Guide, "in Flatland thou hast lived; of Lineland thou hast received a vision; thou hast soared with me to the heights of Spaceland; now, in order to complete the range of thy experience, I conduct thee downward to the lowest depth of existence, even to the realm of Pointland, the Abyss of No Dimensions."

"Behold you miserable creature. That Point is a Being like ourselves, but confined to the non-dimensional Gulf. He is himself his own World, his own Universe; of any other than himself he can form no conception; he knows not Length, nor Breadth, nor Height, for he has had no experience of them; he has no cognizance even of the number Two; nor has he a thought of Plurality; for he is himself his One and All, being really Nothing. Yet mark his perfect self-contentment, and hence learn this lesson, that to be self-contented is to be vile and ignorant, and that to aspire is better than to be blindly and impotently happy." Abbot. (1884), p. 91.

Chapter 28 — **Dark Energy**

"We have no idea why [dark energy] exists or why it has the value it does."

Lawrence Krauss

"We may find out that dark energy vanishes in a puff of logic when we rid our mathematics of certain assumptions…"

Michael Brooks

Attempting to explain the mystery of dark energy, Angela takes Marie to the entrance of a long train tunnel, shows her a tuning fork that produces a middle C note, and then walks into the tunnel. Marie quietly listens for the sound of the tuning fork, but when the sound comes out of the tunnel she hears a B note instead of a C note. Somehow the C note has been mysteriously flattened. If Marie is confident that Angela didn't switch her C tuning fork for a hidden B tuning fork, then how could she explain the fact that she is hearing a B note from a tuning fork that can only produce a C note? What could possibly be changing (attenuating) the C note?

The textbook answer to this question is that Angela must have been moving away from Marie while the tuning fork was sounding. Angela agrees that an emitted frequency will appear attenuated to an observer if the distance between observer and source is increasing during emission, an effect known as the Doppler effect, but she claims that she was standing still while the tuning fork was sounding. Is it possible that Angela is telling the truth? Could she have been standing still?

It turns out that this question is vital to our dark energy investigation. A relevant comparison can be made between our tuning fork and distant stars (emitting specific colors of light). Although light travels for millions of years through space, instead of traveling a short distance through a train tunnel, the same general mystery applies. Does redshift automatically imply that the source and observer are getting more and more distant from each other? Or, is there another way to produce this effect?

The assumption that recessional velocities are uniquely causally related to redshifting is too narrow. Once we assume that redshift entails an increasing separation between source and observer, we are forced to conclude that the cosmos are expanding in size. Being completely in the dark about what could be responsible for this expansion, we are then forced to conjure up a fancy name for the responsible party like *dark energy*. But this responsible party might be illusory. It may be nothing more than an artifact of our assertions and assumptions.

In this chapter we will discover that there is another way for waves to become attenuated as they travel—a way that doesn't involve an increasing separation between source and observer. This discovery places new boundaries on the conclusions that we can draw from our observations of redshift. And those new boundaries are capable of making *dark energy* vanish in a puff of logic.

The History of Dark Energy

> *"The simplest (though, in some ways, most mysterious) possibility is that [dark energy] is a form of energy embodied in space itself, even when otherwise empty."*
>
> Mario Livio

The story of dark energy begins in 1917 when Einstein recognized that his field equation predicted a nonstatic universe.

$$R_{ab} - \frac{1}{2}Rg_{ab} = -8\pi G T_{ab}$$

Because Einstein believed that spacetime was unchanging he adjusted his field equation by introducing an additional term that was designed to keep the universe static. This extra term, represented by the Greek letter lambda (Λ), became the first incarnation of a modern mysterious *force* believed to control the fate of the heavens.

$$R_{ab} - \frac{1}{2}Rg_{ab} + \Lambda g_{ab} = -8\pi G T_{ab}$$

When scientists observed that the light from distant stars and galaxies was redshifted, they interpreted it to mean that the universe is expanding and subsequently threw out this cosmological constant. Then, when they discovered that expansion was accelerating, that redshift was increasing over time, the term had to come back. This time, instead of being held responsible for holding the universe still, it was held responsible for causing the accelerated expansion. This mysterious entity has gone by the names of cosmological constant, vacuum energy, quintessence, and dark energy.

The basic questions surrounding dark energy are still unanswered. Where does dark energy come from? In what way is it a manifestation of the properties of space? How does it manage to create more space? And why does it only seem to manifest its effects on cosmological scales?

The modern reincarnation of Einstein's cosmological constant, in the form of dark energy, is written into the field equation in the following way:

$$R_{ab} - \Lambda g_{ab} = -8\pi G \left(T_{ab} - \frac{1}{2}Tg_{ab}\right)$$

It is imagined to be a pervasive essence, suffusing all of space, and is held responsible for "expanding" space, or "creating" space. In other words, it magically increases the amount of space in our universe over time. But what does that mean?

Our inquiry, once again, begins with Einstein—an enigmatic free thinker who entered a world in which "the prevailing scientific and philosophical prejudice held that the universe was, on its largest scales, static, eternal, fixed, and unchanging."[1] This belief had its roots in

Euclidean intuitions. When we look to the heavens we don't see any changes. Consequently, it is natural to assume that the cosmos are fixed.

Many people feel that a static universe needs less of a philosophical explanation than a nonstatic one. If the large-scale universe does not change, then it has always been—end of story. But if it is expanding, or contracting, then this condition forces us to face the overwhelming mystery of its origin, and its possible demise. These implications may have been a cause for concern, but for Einstein the possibility of an expanding or contracting universe posed a threat more worrisome than cracking open the lid of Pandora's origin box. It carried the potential to violate the very foundation of general relativity—the equivalence principle.

According to the equivalence principle, all frames, all points of space, are on average on equal footing. An expanding universe, however, seems to suggest that a unique point exists—the point from which the expansion is directed away. Likewise, a contracting universe seemingly suggests a unique point that everything is moving toward. In either case, the existence of a *special* point threatens to violate the equivalence principle and reduce the symmetry of the universe. All frames, or locations, in space cannot be considered equal if there exists even one consistently unique point in space. And a theory cannot be considered self-consistent if it violates the very principle that it is based on.

Einstein introduced the cosmological constant to balance against the universe's tendency to expand or contract—a tendency that was embedded into his equations of spacetime. But this "easy fix" didn't really get rid of the problem. A few years after Einstein introduced the cosmological constant the Russian mathematician and meteorologist Alexander Friedmann, and the Belgian priest and astronomer Georges Lemaitre, independently analyzed Einstein's equations as they apply to the entire universe and found something striking. "They realized that the gravitational pull of the matter and radiation spread throughout the entire cosmos implies that the fabric of space must either be stretching or contracting, but that it could not be staying fixed in size."[2] In short, they found that even with a cosmological constant the universe couldn't be static. Geometric evolution was a condition that breathed throughout the very core of general relativity and could not be excised.

These insights made expansion a serious possibility. Then, in 1929, the question of whether or not the universe was eternally fixed was forever altered when Edwin Hubble turned his attention to the heavens. Using the 100-inch telescope at the Mount Wilson Observatory in Pasadena California, Hubble made discoveries that led scientists to recognize that there are galaxies completely separate from our own. When these distant galaxies were analyzed their electromagnetic signatures were found to be redshifted. Furthermore, the degree of redshift depended upon distance—further away stars were redshifted to a greater degree (Figure 28-1).

Today, more accurate measurements of these redshifts, combined with the implicit assumption that they are due to the Doppler effect, have allowed us to conclude that "galaxies that are 100 million light-years from us are moving away at about 5.5 million miles per hour, those at 200 million light-years are moving away twice as fast, at about 11 million miles per hour, those at 300 million light-years distance are moving away three times as fast, at about 16.5 million miles per hour, and so on."[3] Hubble's observations convinced scientists that, instead of being static, the universe is expanding. Einstein reacted to this information by withdrawing his support for the cosmological constant, calling it his "greatest mistake".

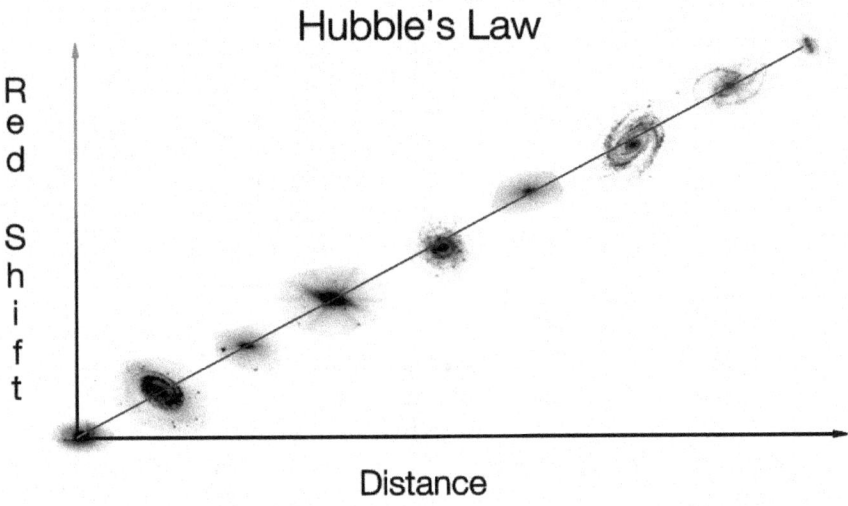

Figure 28-1 Hubble's Law: $redshift = H_0 D$, *where H_0 is Hubble's constant*

Expanding Space

What does all of this say about Einstein's equivalence principle? Doesn't an expanding universe imply a special point, from which everything is expanding? Doesn't expansion violate the equivalence principle, and consequently undermine general relativity? Before we answer this question we should also note that according to Hubble's observations everything in the cosmos is receding from *us*. Does this mean that Earth somehow represents a unique location in all of space? The answer to these questions is no.

If light stretches as it travels through vast distances of expanding space, then the further it travels the longer its wavelength becomes. We can represent the magnitude of this stretching by assigning a *recession velocity* to the light's source. This recession velocity encodes how the amount of space between the source and observer is increasing over time. Notice, however, that the claim that the amount of space between here and there is increasing is not necessarily the same as the claim that the source is getting further away because it is moving *through* space.

Today scientists commonly use a balloon anology to explain what is meant by *expanding* space. The analogy asks us to imagne that the surface of a balloon represents the familiar dimensions of space. It then asserts that we can simulate the condition of an expanding universe by blowing up the balloon.

This balloon analogy became popularized as a way to explain expansion in 1930, when it appeared in a Dutch newspaper following an interview with Willem de Sitter (of de Sitter and anti de Sitter space). The virtue of this model is that it helps us connect to the idea that in an expanding universe all points will appear to be the point from which everything recedes (from their own perspective). If we choose any point on the balloon as our reference, and watch all the other 'galaxies' as the balloon expands, we will notice that they all appear to recede from our position—no matter which position we start from (Figure 28-2). This satisfies Einstein's concerns because it means that all positions are equivalent (non-unique).

Figure 28-2 Expanding balloon model

This model also helps us see how the recession velocities we assign to each distant object are additive. If "the balloon swells during some time interval, doubling in size for example, all spatial separations will double in size as well... Thus, in any given time interval, the increase in separation... is proportional to the initial distance between them."[4]

Although this balloon model is commonly employed to explain the *expansion* of space, its explanation is empty. Technically, the explanation it offers is incapable of accounting for redshift. Here's why. If all of space is swelling, as the balloon model suggests, then everything must be growing in size, including our meter sticks (Figure 28-3). This makes it impossible to internally discover that expansion. That is, an observer trapped inside the universe would never see any effects of this expansion. The idea that dark energy is an essence that suffuses all of space cannot account for our observations.

The only way to make expansion account for redshift is to assume that space expands in the vast catacombs between galaxies, but for some reason doesn't expand where our rulers are. But why would the space in our rulers be different from the space outside of the Milky Way Galaxy?

One view suggests that matter held together by electromagnetic forces and nuclear forces does not swell because those forces overcome any possible swelling. Asserting this is a far cry from explaining it. Even if some regions of space somehow avoid swelling, we still have to explain what *swelling* space is.

When we say that the fabric of space is swelling, we mean that new space is coming into existence—that the number of unique locations in the universe is increasing over time. The balloon model doesn't reference this effect because the balloon material *thins* out as it stretches.

Figure 28-3 As space expands our meter sticks expand with it

If a galaxy's spatial distance is found as a measure of the spatial fabric between here and there, and the total amount of fabric between galaxies does not change when the balloon expands, then this balloon analogy completely fails to represent expansion. In other words, if the fabric of the balloon represents the fabric of space, then it isn't logically coherent to say that there is more space between objects after expansion occurs.

What about time dilation? Won't that be detected? In other words, won't the motion of galaxies that result from expansion cause the clocks to fall out of synchronization, as Einstein taught us with special relativity? The answer is no. Scientists are not claiming that the galaxies are moving *through* space due to expansion. Instead they are saying that the amount of space between the galaxies is increasing. Alternatively, we could say that the amplitude of energy (light) traveling through space is diminishing in excess of our expectations, which are based on $\frac{1}{r^2}$.

Since special relativity's time dilation is tied exclusively to motion *through* space, clocks that are situated in expanding space will stay synchronized even though the distance between them appears to increase. This assumes, of course, that our clocks aren't moving through space—that all of their redshifting is due to expansion. But how do we ensure this condition? A star's light will be redshifted if the star was moving away from us as it emitted that light. But the light from a stationary star will also be redshifted if the space it traverses swells during its journey. So how do we distinguish redshift caused by motion through space, from redshift caused by expansion?

One way to make this distinction is to use the uniformity of the microwave background radiation to test our motion compared to the cosmic flow of the vacuum. The microwave radiation is observed as very homogeneous throughout space as long as you are moving with the cosmic flow. If you are moving differently than that flow then you will observe the background microwave radiation as inhomogeneous. Just as the horn on a train has a higher pitch when it is approaching and a lower pitch when it is receding, a spaceship moving compared to the cosmic flow will have microwave crests and troughs hitting one end of the ship at higher-frequency than the ones hitting the opposite side of the ship. This sets up a temperature gradient across the ship because higher-frequency microwaves translate into higher temperatures.[5]

Because "spaceship Earth" is traveling around the Sun, and the Sun is moving around the galactic center, and the Milky Way is moving toward the constellation Hydra, the microwave background does appear to be a little warmer in one direction and a little colder in the opposite direction. When we correct for the effects of these motions, the microwave radiation exhibits exquisite uniformity. The temperature contributions of these microwaves end up being roughly equal in all regions of the sky (Figure 28-4).

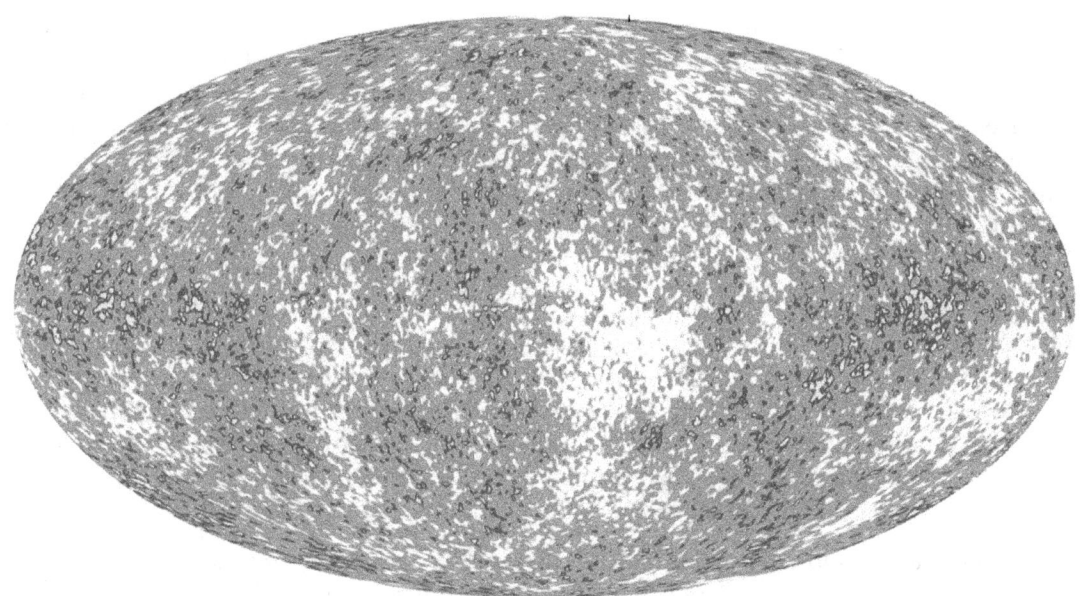

Figure 28-4 A map of the cosmic microwave background radiation (NASA)

If distant galaxies recede faster and faster (in additive fashion) as space swells or expands, does that "mean that galaxies that are sufficiently distant will rush away from us at a speed greater than the speed of light? The answer is a resounding, definite yes. Yet there is no conflict with special relativity."[6] Why? Because relativity only restricts how fast things can move through space. It says nothing about how fast new space can be created between them.

But is this really what is happening? Is space being uniformly whisked into existence throughout the universe—except for where our rulers are? Is dark energy, which comes with an energy bill 13.9 times larger than all of the ordinary matter in the universe, really the most elegant way to explain our observations? Can redshift—the kind we blame on the swelling of space—be explained another way?

Measuring Redshift

When we look at distant galaxies we are seeing them as they were when the light reaching our eyes was emitted (except for the change in color due to redshift). For example, the light reaching us from a galaxy 5 billion light-years away tells us where all the stars in that galaxy were positioned relative to each other 5 billion years ago.

If redshift implies expansion, then its magnitude tells us how much the space between the observer and the source has "expanded" since the light was emitted. Measuring the redshift from many different distances allows us to puzzle together how the universe's expansion rate has changed over time. The trickiest part of all of this is finding a way to determine the actual distance to a particular star or galaxy. How do we know that a galaxy is really 5 billion light-years away?

Looking at a distant galaxy is a little like standing on a dark highway and watching a car's taillights recede into the distance. We can find out how fast the car is going by measuring the redshift of the light, but in the darkness, there is no easy way to determine the distance.[7] The only real way to deal with this problem is to use a standard candle—an object whose total intrinsic output of light is known. Then, by measuring how bright this standard candle appears, we can figure out how far away it is. For example, if we knew the car's taillights were rated at 100 watts, we could measure the amount of light we see and deduce how far away the car is. Are there any standard candles in space? As it turns out, there are.

Type Ia supernovae occur in binary star systems in which one star, not too different from the Sun, has gone through its life cycle and shrunk down to a cinder about the size of Earth. When a star of this mass range has reached this stage in its life we call it a white dwarf. Some white dwarfs pull material (mainly hydrogen) away from their normal size companion star until a layer several feet thick builds up on its surface. When this layer grows to its critical limit "the whole thing explodes like a giant hydrogen bomb. For a period of weeks, this supernova can outshine the entire galaxy where it resides."[8]

By recording explosions of this type in nearby galaxies, whose distances are known from other techniques, we have learned that the longer the light from the supernova lasts, the brighter it intrinsically was. This allows us to use type Ia supernovae as standard candles. All we have to do is watch a distant supernova, record how long it shines, and then use this information to calculate how much light it emitted intrinsically. We can then calculate its distance by comparing its intrinsic brightness to the amount of light we actually receive (its apparent brightness).[9]

In 1998 two independent teams (the Supernova Cosmology Project team and the High-Z Supernova Search team) used this method to begin mapping the furthest reaches of the visible universe. Up until that point the physics community expected that the expansion of the universe was slowing down over time, so when these two teams published results that contradicted this expectation "there was a fair amount of alarm in the cosmological community."[10] Measurements from both groups found that the galaxies in the furthest reaches of the visible universe (billions of light-years away) are receding to a lesser degree (magnified by their distance) than the closer galaxies are. Somehow the expansion is accelerating.

Scientists now have data from over 300 distant supernovae that support this conclusion.[11] Faced with this evidence, which starkly contradicts the old view, cosmologists have concluded that there is something in the universe they hadn't accounted for—a mysterious missing ingredient that makes up approximately two-thirds of the total energy density of the universe.

To quickly review, we determine how far away a galaxy is by observing how long a type Ia supernova lasts in that galaxy, deducing from this how bright it was intrinsically. Then we compare this intrinsic brightness to its apparent brightness and, presto we have a distance value from our inverse square law.[12] To measure how fast a galaxy is receding, we measure its degree of redshift—how much the signature emissions from the gasses in that galaxy have shifted. The further this light is shifted toward the low energy side of the spectrum, the more redshifted it is, and the faster we claim its source is receding (Figure 28-5). So in the end, explaining expansion comes down to explaining redshift. The twist here is that if space is hierarchically quantized, then redshift can be explained via a mechanism that causes the vacuum pressure (the average superspatial velocities of the quanta) to decrease over time.

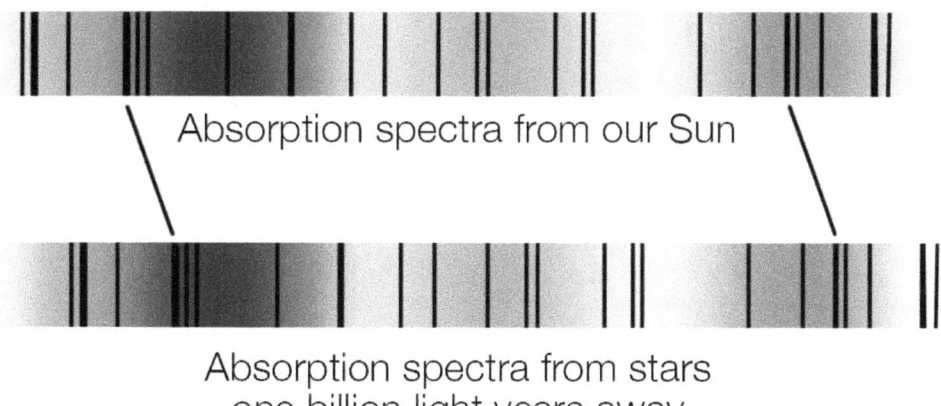

Figure 28-5 Shifted absorption lines

Another Way to Explain Redshift

> *"The inflaton field, by virtue of its negative pressure, 'mines' energy from gravity… It is impossible to excise the field, much as it is impossible to excise space itself."*
>
> *Brian Greene*

Energy is a quantifier of geometric distortions in the vacuum. In a closed system energy is always conserved. But this does not mean that the total energy of every level within that system will remain constant. In a hierachically structured system energy will escape from higher to lower levels. This means that if the vacuum is made up of quanta, which are made up of sub-quanta, and so on, then the energy of the vacuum, on any specific level, must decrease over time.

In other words, given that energy is a quantifier of geometric distortions, the rearrangement of subquanta that takes place during every quantum collision transfers energy from the quantum scale to the subquantum scale.[13] This energy transfer contributes an effective negative pressure to the vacuum, dampening the system and causing its temperature and pressure to drop. This dampening causes phonons in the vacuum (light) to lose energy—to redshift—as they travel. Redshift, therefore, directly advertises Nature's fractal structure.

"Anything that contributes to the energy density of the vacuum acts just like a cosmological constant."

Steven Weinberg[14]

As the pressure and temperature of the vacuum decrease, the number of quanta concurrently undergoing collisions also decreases. This leads to increased quantum independence, which infuses the vacuum with more unique locations—a trace of expansion left over from the inflationary period. But this expansion is only observable in as much as it occurs nonuniformly. That is, if warmer regions of space maintain their temperature while the catacombs of interstellar space continue to cool, redshift from expansion will be measured. Otherwise, if expansion occurs uniformly, it won't make an internally measurable contribution to redshift.

If a phonon travels through a medium while the pressure of that medium is decreasing, its wavelength will increase, its frequency will decrease, and its amplitude will decrease in excess of the expectations of the inverse square law.[15] If we knew the original wavelength, and amplitude of a wave as it left its source, but didn't know that the pressure of the medium had decreased as it traveled, observations of redshift could easily lead us to the conclusion that the source is further away than it actually is and that it is receding from us. But this would be an incorrect conclusion.

If the medium of space is hierarchically quantized, then this analogy directly applies to our observations of redshift in the universe. Redshift does not mean that distant galaxies are speeding away from us, or that the universe is expanding. Redshift—the kind we have blamed on dark energy—is a measure of how much phonons (light) lose energy during their trip via dampening. It reflects the degree to which the vacuum decreases in pressure during that propagation (due to quantum inelasticity).[16]

Redshift is increasing over time (accelerating) because the degree of relative dampening depends on the average energy of collision. A small, but steady amount of energy is transferred to the internal level with each collision—reducing the average thermodynamic velocity of quanta. This reduces the temperature of the system on the higher level and causes redshift. Because this bit of transferred energy represents a more significant fraction of the total energy of collision at cooler temperatures, dampening (redshift) manifests more readily at lower temperatures.[17]

If this picture is correct, if the vacuum is hierarchically quantized, then light from galaxies sufficiently distant from us will never reach us. If spacetime were an infinitely smooth continuum, then light's intensity would diminish in a way that was strictly proportional to $\frac{1}{r^2}$. Under this condition the energy of light wouldn't become zero until r was infinite. But when quantization plays a role, we end up with a minimum phonon threshold. The radius of our visible universe, how far out we can see, depends on the redshift signature—on the magnitude of dampening—because that signature determines how far light can travel before being completely diluted. Given that dampening is increasing, our model predicts that the radius of the *visible* universe is shrinking. Shorter emission wavelengths can travel further than longer wavelengths, but the maximum distance for each wavelength is decreasing.

In conclusion, redshift does not automatically imply that the source and observer are getting more and more distant from each other. There is another way for waves to become attenuated as they travel—a way that doesn't involve an increasing separation between source and observer. If the vacuum is hierarchically quantized, then its internal structure is directly responsible for redshift. This insight reanimates the idea that the fate of the cosmos is determined by its geometry, and from this perspective *dark energy* vanishes in a puff of logic.[18]

[1] Brian Greene. *The Fabric of the Cosmos*, p. 230.

[2] Ibid.

[3] Technically, light from a star that is 100 light-years away is slightly more red-shifted than half the value of redshift from a star that is 200 light-years away. The same applies for 100 million light-years vs. 200 million light-years. We discuss why shortly.

[4] Brian Greene. *The Fabric of the Cosmos*, p. 232.

[5] Ibid., pp. 235–236.

[6] Ibid., p. 237.

[7] James Trefil. (2006, July). Where is the Universe Heading. *Astronomy*, p. 39–43.

[8] Ibid.

[9] "A galaxy like our Milky Way exhibits about one type Ia supernova every few hundred years and its brightness fades in weeks making the search for them quite a challenge." David Appell. (2008, May). Dark Forces at Work. *Scientific American*, pp. 100–102.

[10] James Trefil. (2006, July). Where is the Universe Heading. *Astronomy*, p. 39–43.

[11] Ibid.

[12] There are a total of about 600 supernovae reported by astronomers every year. These supernovae are scattered across the universe. Some are near and others are far away (on astronomical scales) and a fraction of these are type Ia supernovae.

[13] The degree of inelasticity compared to the degree of elasticity is set by the difference in scale between the subquanta and the quanta—which are separated by some $60 - 120$ orders of magnitude.

[14] Morris Loeb. (1998). The Cosmological Constant Problem. Lectures in Physics. Cambridge, Mass: Harvard University Press; Steven Weinberg. (2005, November). Einstein's Mistakes. *Physics Today*; Aczel. (1999), 167; Krauss, 117; Green 2004, 275–278; Dennis Overbye. (1998, May 26). A Famous Einstein 'Fudge' Returns to Haunt Cosmology. New York Times; Jeremy Bernstein. (2001). Einstein's Blunder, 86–89; Walter Isaacson. *Einstein*, 356.

[15] "A typical scream on Earth can be heard up to a kilometer away, but on Mars, where the atmospheric pressure is lower, the same scream would only be audible for sixteen meters." Amanda Hanford & Hazel Muir. (2007, August 11). Noisy neighbors. *NewScientist*, p. 30.

[16] The Doppler effect also leads to redshift, and can be seen in orbiting systems. The point here is that the kind of redshift that has been blamed on dark energy is not due to the Doppler effect.

[17] Temperature and heat can easily be confused. Heat is the total energy of all the motions of all the quantized parts that make up the substance being described. The range of energies between the quantized parts is often large. Some are moving quickly and others slowly. Temperature is a measure of the average energy each part has. Temperature goes down because the average energy/velocity of the individual pieces of the medium goes down. Technically a single atom, molecule, or quanta cannot have "temperature", but it still has energy of motion.

[18] This insight may even offer us a way to measure the difference in scale between Nature's substrate levels—ultimately giving us a measure of the size of the universe.

Chapter 29 — **Intellectual Astronauts**

"We are explorers, and I cannot imagine how exciting it will be when we get even one step closer to true nature!! This belief and this desire is so huge that we will never give up."

Young-Kee Kim[1]

"Exploring the mysteries of our ever-fantastical universe—whether it's at the level of a cell, a molecule, or the whole amazing shebang—is not simply a sideline or vocation; it's at the soul of human experience... [it's] as central to life as breathing."

Sally Ride[2]

Science is an imaginative and investigative journey—a bold attempt to reach beyond the veil of our senses by creatively exploring new possibilities and checking them against our observations. So far, our inspection of Nature's topography has unveiled dark caves where time stands still (black holes), and wonders that challenge everything we believe (quantum tunneling, Heisenberg uncertainty, nonlocality, etc.). Some might see these upsets as nuisances, but these discoveries are responsible for informing our course corrections, filling our sails with curiosity, and ultimately carrying us to new perspectives.

Every mystery has the propensity to propel us forward, to reach new insights, and to help us escape our misconceptions. Every time we explore a new isle of thought, fight the dragon of scale, or summon the courage to risk broadening our horizons, we participate in the central theme of the human story.

The trek we have been taking in this book has exposed us to a new idea about how the world might work. The axioms beneath this idea have left us standing at the edge of powerful hierarchical cascades. Will the intrepid among us press on? Will we sail out against the prevailing sociopolitical current to test these ideas? Will we risk being wrong for a chance of obtaining clarity? Or will we play it safe and seek comfort in the familiar fog?[3]

I cannot say whether the axioms of qst perfectly match up with Nature's true form. I cannot say whether the ideas presented in this book will stand the test of time. What I can say is that scientific progress flourishes best in a world infused with imagination—a world in which we risk being wrong by throwing our ideas out there to be tested. Science cannot thrive behind a veil of fundamentalism, nor can it flourish when logical fallacies (like an appeal to authority) are used to justify ignoring new ideas.

By opening ourselves up to the higher-dimensional picture entailed by vacuum quantization we stand to amplify our grasp of Nature to an unparalleled extreme. But this extension comes with a price—it requires us to let go of what many consider humanity's favorite illusion. In short, gaining a direct link to the infinite, piercing the depths of interconnectedness, means losing our historic theistic or deistic truncation of infinite regress and dissolving the illusion of autonomous agency. As we ascend the turbulent cascades of this insight, as our goal comes within sight high above, each of us must choose whether or not we are going to pull up the anchor of our past. Risking the high adventure means following truth wherever it leads.

Einstein understood that a deterministic map would undermine the illusion of autonomous agency, but this did not deter him. Truth was far more beautiful to him than any projected mirage, and to him, seeing through that mirage was an insured eventuality. In 1929 he predicted that quantum physicists would one day reach "the limit of their mania for the statistical fad, [and] return full of repentance to the spacetime picture."[4]

To trigger that eventuality, we must escape the restrictive cocoon by lifting the anchor. We must sail into freedom and raise our consciousness to a new level—where we no longer need to grope through a mathematical jungle inadequately lit by physical intuition.[5]

Breaking the Conceptual Barrier

Nature is not Euclidean. Pretending that it is—grounding human intuition in the Euclidean paradigm—dampens our explorer's spirit. It does this, most notably, by insisting that the transcendent essence of physical reality—any part that cannot be squeezed into the Euclidean frame—is necessarily outside of Nature and, therefore, ultimately beyond exploration and comprehension.

Supporters of this view argue that causes will never be completely pinned down or understood, because they must all eventually be traced to a realm that is off limits to our exploration or comprehension—something outside of the four dimensions we are familiar with. But this view, which is bolstered by the standard interpretation of quantum mechanics, is merely an ad hoc assertion.

> *"So long as science appeals to something outside the universe, we must abandon any hope of ultimately understanding why the universe is as it is."*
>
> *Paul Davies*[6]

The most interesting aspect of this *limitation* argument is that it vitally depends on our inability to conceptually transcend the four-dimensional barrier. If humans really were incapable of comprehending higher-dimensional realms, then the argument might stand. But higher-dimensional realms are conceptually accessible to us (as we have seen throughout this book). Realms beyond the traditional four dimensions of spacetime are not logically out of bounds. The dimensions outside of familiar space and time are mappable parameters of Nature. The seen and unseen can both lie within our conceptual scope.

New axioms for space and time, like the ones explored in this book, can take us beyond this ad hoc truncation of infinite regress. Thomas Aquinas famously placed this truncation on what we can explore and comprehend, and he called this truncation "God". By doing this, he effectively defined God as something that is forever beyond our grasp—something that we can never come to know—instead of something that we increasingly come to understand as we continue our quest.

The model we have been exploring challenges the view that anything lies beyond the scope of investigation. The hierarchical structure of this map gives us a way to peer into any level of Nature's infinite regress, allowing us to trace causes to prior causes ad infinitum.

It is intellectually folly to select a point along an infinite regress and abruptly introduce an obtrusive truncation. All causes have causes. Nothing warrants the claim that any particular event, including the Big Bang, did not have causes. Truncations of infinite regresses are by definition paradoxes. Pretending to have an accurate representation in the face of these paradoxes leaves us in the darkness of ignorance.

> *"The recipe for perpetual ignorance is: be satisfied with your opinions and content with your knowledge."*
>
> *Elbert Hubbard*

Through the eyes of qst, the Big Bang is not conceptually troubling because it is not associated with an artificial truncation in the infinite regress of causation. In reflection of this, consider the following: imagine that we were deep underground, say more than a mile below the surface in the mines of Doe Run, near La Oroya, Peru. Now imagine that we pulled a large pyrite crystal from the hot blasted walls.[7] Holding this in our hands let's ask an interesting question. How many times can we cut it in half?

Let's say that the crystal was originally two hundred carats.[8] If we cut it in half we end up with two, one hundred carat crystals. Then, if we take one of those pieces and cut it in half again we end up with two smaller pieces weighing fifty carats each. Each time we take one of the halves and cut it in half again. Could we theoretically continue this process ad infinitum? Could we keep cutting indefinitely?

We must be careful when answering this question. If I had asked, can we cut pyrite indefinitely such that we always end up with smaller and smaller smidgens of pyrite, then the answer would have been a resounding no. This is because the smallest possible piece of pyrite is found on the molecular scale by definition. Once we reach one molecule of pyrite (FeS_2) we cannot cut any further if we wish to retain the chemical properties of *pyrite*. Therefore, based on our definition of pyrite there is a natural termination of this division process.

But what would happen if we didn't care about the label *pyrite*? Is this scale really an inherent end to our cutting process? Of course not. We can cut a molecule of pyrite into smaller pieces; we just end up with something that is no longer pyrite. But we still have something, and we can still continue our cutting process. Once we do, we find that our regress of division continues right through that transparent barrier.

In a very similar way qst helps us transcend the truncation of the infinite regress in reference to the origin of the universe. Our dimensional hierarchical structure helps us see that our *boundaries* are merely places where the properties of what we are describing, or dividing, need a more complete description. They don't inherently limit further division. Space is a perfect example of this. We can divide a region of space over and over until we reach one quantum. The next division is only special in that it changes our dimensional description of what is being divided. One quantum is simultaneously the smallest piece of our four-dimensional stage and an entire universe from another dimensional perspective. It lies at the boundary of dimensional perspectives. It is a simple marker along the infinite regress, a place of transitioning perspective, within the cascading levels of reality. It only appears to be a division barrier if we have restricted our vision of the vacuum's structure.

Gaining our Wings

"What science does for us is widen the window. It opens up so wide that the imprisoning black garment drops away almost completely, exposing our senses to airy and exhilarating freedom."

Richard Dawkins

According to this picture, the Big Bang was a physical phenomenon with prior causes, which in turn had prior causes, ad infinitum. The symmetries enlaced in our fractal geometry put us in touch with this "ad infinitum" because they imprint the dynamics of the Big Bang within every quantum collision. This cascading structure blows the need for a *first cause* out of the water. It helps us see that the illusion of a first cause is escapable.

As we expand our perspective (based on our new axioms) arguments for a truncated regression dissolve. We gain logical cohesion with the claim that there is no such thing as a first move, a first cause, or a time when physical reality came into existence from something non-physical. Consequently, we no longer need to think of any part of Nature as being beyond exploration.

The objective of our scientific quest is to free our imagination and intuition from the shallow limitations and imprisoning chains that have held us back. We sail to discover a deeper truth and, in doing so, discover our true selves.

The history of curiosity has opened this route to us. The next step is up to us. Will we brave the perilous journey? Do we have it in us to thoroughly challenge our old assumptions? Will we answer the call, step out on our own journey, and risk rejection, or failure, for the possible reward of gaining a new understanding of Nature?

"If we're ever going to reach that next great era of awakening, we'll need to be rescued from the devil of dimensionality."

Steven Strogatz[9]

"He who sees the abyss, but with eagle's eyes—he who grasps the abyss with eagle's talons: he has courage."

Friedrich Nietzsche[10]

In this book, we have explored a possible new set of axioms. Do not just adopt this way of seeing things. Challenge it, question it, and *if* you come to agree with it, make sure you own it for yourself. The process is the goal. By sailing into the unknown, and following our curiosity, we all have a chance to gain a richer comprehension of Nature—one filled with inspiration and insight.

Be forewarned. The tumbling cascades of a new idea may convey their power with a deep rumble. Do not fear this novelty. Before you turn around and dart back into the fog of ignorance simply to hold onto the illusion of autonomous consciousness, or the comfortable theistic, or deistic solution, allow yourself to deeply grasp the full beauty of that idea.

Mankind has passionately held onto the dream of intuitively comprehending the Cosmos. Countless people have felt their way through the world, following their curiosity to new discoveries, and slowly reshaping the worldview we have inherited. Every step along that path has been part of our great journey and every sacrifice offered by those before us has led to this precipice.

All of the significant changes in our course thus far have delivered severe blows to our self-image. But, when the storm clears, as it always has, a rainbow emerges in our new path and a deeper beauty surfaces. This is the process of loosening the grip of our convictions. Fear may be primally ingrained in us, but our intrepid spirit of curiosity is capable of helping us overcome those fears. By reclaiming the courage to explore new possibilities, together we can help science expand to the point that it redefines humanity's relationship to the universe and to the divine.[11]

[1] Young-Kee Kim. University of Chicago.

[2] Declarations of Sally Ride. Reported by K. C. Cole.

[3] William Laurence. (1939, March 14). Einstein Sees Key to Universe Near. *New York Times*, Quoted in *Einstein*, Walter Isaacson, p. 467.

[4] Walter Isaacson. *Einstein: His Life and Universe*.

[5] Banesh Hoffman. (1972), p. 227; Walter Isaacson. *Einstein*, p. 468.

[6] David van Biema. (2006, November 3). God vs. Science, *Time Magazine*; Paul Davies. (2007, June 30). Laying Down the Laws. *New Scientist*, p. 30–31.

[7] Pyrite means *fire rock*. It gets that name from its ability to produce sparks when it is struck with metal or another rock. It is golden in color and metallic in luster, has a hardness of 6.3 and is an iron sulfide (FeS_2). It is used as a marker for iron ores but is not mined as an ore itself. Instead it is considered semiprecious when in crystal form.

[8] A "carat" is a unit of weight. It originally comes from the Mediterranean Carob tree, which has very uniform seeds. Therefore, a one-carat diamond is equal in weight to one carob seed.

[9] Steven Strogatz. Fermi's Little Discovery and the Future of Chaos and Complexity Theory. *The Next Fifty Years*, p. 124.

[10] Friedrich Nietzsche. *Thus Spoke Zarathustra*, p. 247.

[11] Corey S. Powell. (2006). My Three Einsteins, an excerpt from his book as an article in *Discover*, p. 47.

Chapter 30 — **The Wilderness of Intuition**

"Preferring a search for objective reality over revelation is another way of satisfying religious hunger. It is an endeavor almost as old as civilization and intertwined with traditional religion, but it follows a very different course—a stoic's creed, an acquired taste, a guidebook to adventure plotted across rough terrain. It aims to save the spirit, not by surrender but by liberation of the human mind."

Edward O. Wilson[1]

"Even if the open windows of science at first make us shiver after the cozy indoor warmth of traditional humanizing myths, in the end the fresh air brings vigor, and the great spaces have a splendor of their own."

Bertrand Russell

The possibility that there is sacred divinity in the order of the universe propels us to seek a connection with it—to strive to grasp that which lies beyond the limits of our senses.[2] Looking outward from our blue-green sphere, we are drawn to fulfill the promise of the night sky—to reach into the depths of the black void and finally understand. The scientific method directly facilitates this aspiration, rigorously, and methodologically exposing us to new wonders, empowering the search for ontological clarity, and expanding the human mind.

Because it promotes the process of personal discovery, and cautions us to be wary of any archaic straightjackets of authority that claim to have a lock on divine experience,[3] science has become the enterprise that best satisfies our metaphysical hunger. It teaches us to think for ourselves and offers us protection from endlessly repeating the darkest chapters of history.[4]

Divinity is a matter of ascending the footholds of personal discovery.[5] It is a process that requires constant effort and substantial character, and it follows the path of freedom.[6] A free mind unwaveringly soars towards understanding, is a stranger to apathy, and continuously fights for independence of thought. By contrast, intellectual slavery comes with leisure time—requiring only that we settle for whatever we are told is true.[7]

Divinity awaits our naked curiosity. It charges us to take the reins ourselves, admit to that which we do not know, and find the vigor to press on. It inspires us to continue to strive to access a way of knowing that transcends our biological interfaces with the environment. Expanding beyond the breadth and depth of our senses is at the very heart of scientific exploration, while embracing ignorance underwrites fundamentalism.[8]

To reach that which is sacred we must follow our questions to wherever they lead. Curiosity is not to be feared, contained, or controlled. It is to be danced with. Ignorance is not a virtue, nor is it anything to be ashamed of. Ignorance gives our curiosity a quest and when we brave the terrain of that quest we are transformed.

"What can be more soul shaking than peering through a 100-inch telescope at a distant galaxy, holding a 100-million-year old fossil, or a 500,000-year old stone tool in one hand, standing before

the immense chasm of space and time that is the Grand Canyon, or listening to a scientist who gazed upon the face of the universe's creation and did not blink? That is deep and sacred science."

Michael Shermer

"When I peer into the depths of space and time, I am overwhelmed with a sense of reverence that I believe exceeds what others describe feeling when they enter a historical cathedral or mosque. I believe that searching to reveal the mysteries of the cosmos and unlocking the secrets of its internal physical laws is the highest spiritual quest."

Mike Ritter

Our explorations, the questions we ask, and the rigor to which we put our answers to the test, define us. Under this light it is clear that the scientific quest is far more than a detailed study of the natural realm. It is our opportunity to connect to the majesty and elegance that authors all of Nature; our chance to pull away from the security blanket of ignorance, and to sail through the treacherous doldrums of confusion. Our curiosity naturally compels us to journey to a higher precipice, to take the opportunity to discover our wings. It propels us to explore, to imagine, and to solve Nature's greatest mysteries.

Avoiding Moral Shipwreck

"Everything you believe is questionable. How deeply have you questioned it? The uncritical acceptance of beliefs handed down by parents, teachers, politicians, and religious leaders is dangerous. Many of these beliefs are simply false. Some of them are lies designed to control you. Even when what has been handed down is true, it is not your truth. To merely accept anything without questioning it is to be somebody else's puppet, a second-hand person."

Daniel Kolak and Raymond Martin[9]

As we learn to live life out of our own volitions, and embrace our natural curiosity, we discover that the question of what the meaning/purpose of life is drifts away. Once we take on the great human adventure, once we face the wind, life is validated in each and every act and a discovery is hidden around every turn. This process initiates a union with the infinite, lifting its participants beyond illusions of *meaning* in a way that echoes how questions like, "What creatures inhabit the underbelly of the Earth?" dissolve when we discover that the Earth is round.

"The question of the meaning of life is, as the Buddha taught, not edifying. One must immerse oneself in the river of life and let the question drift away."

Irvin D. Yalem

The assumptions explored in this book lead to a perspective that can help us overcome the dark shadows of our past. This perspective fights for a new way of being that rejects indoctrination, tyranny, and fear-based manipulations. From the heights of this new vista, moral progress is no longer stifled by the glorification of tradition, duty no longer outweighs love, relationships are no longer about control, and Nature is to be embraced instead of conquered. In this intellectual landscape people do not have to suspend their natural curiosity, or submit to mob mentality, to be socially accepted. They no longer have to give away their spark of potential to fit in.

Under the insight that the omniverse is deterministic, fixations on blame are misguided. Blame contradicts acceptance and exacerbates the harm it feigns objection to. Negative situations from the past, or the present, are best met with understanding. To move beyond our problems we must learn to understand the assumptions and misconceptions that gave rise to them.

In ancient times religion was considered responsible for moral instruction, consolation, inspiration, and explanation. As the scientific method matured, it became clear that science is much more equipped to meaningfully fill these roles. In as much as this changing of the guard has taken place, it has dissolved false claims, undercut fear-based manipulations, and exposed the negative impact of blame.

Although many people still use religion as a moral guide, it is clear that moral instruction does not require religion. In the broad sense, all of humanity has a shared impression of what is right and wrong to a surprising degree. As David Hume put it, our moral judgments ultimately rest on "some internal sense or feeling, which Nature has made universal in the whole species."[10] (We may need to make some exceptions here for psychopathy or other mental disorders.) It doesn't matter what culture a person grows up in, or if they are religious, the basic broad morals of human interaction remain the same. Nevertheless, our more narrow sets of moral principles change from place to place, and from time to time, in synchronization with changing worldviews. In this, moral progress is directly encouraged by the self-correcting process of science while, by contrast, the rigidity of scripture tends to stagnate it.[11]

"Religion is an insult to human dignity. With or without it you would have good people doing good things and evil people doing evil things. But for good people to do evil things, that takes religion."

Steven Weinberg

Our broad morals may be genetically ingrained in us, selected by eons and eons of social interaction, but our understanding of reality is malleable. New insights lead to greater understanding and our level of understanding underwrites our moral code.[12] Because the paradigms by which we understand the world evolve as science moves forward, the responsibility of moral instruction falls squarely on the shoulders of scientific exploration. Even in ethics, it is always "reprehensible to believe on insufficient grounds."[13]

Scientific exploration serves to enlighten and enhance our base set of morals, while fundamentalism in practice does just the opposite—suppressing our natural tendencies for empathy and kindness. How do they do this, you ask? Don't all religions teach love and kindness? Yes they do, but many of them also undermine those precepts by teaching that blind obedience is a virtue—that unquestioning faith (belief without, or even in opposition to evidence) is a good thing.

Archaic texts serve to keep their followers within a set of moral boundaries that reflect an outdated, narrow-minded worldview of the past—the worldview that was held (or was being promoted) when those texts were being written. Enlightened progress always leads to greater acceptance, understanding and kindness. On the other hand, anything that systematically rejects humanity's advancements and enlightened progress, discouraging people from exercising their innate curiosity, is, at best, only superficially concerned with love and kindness.

Richard Dawkins notes that, "Faith is an evil precisely because it requires no justification and brooks no argument." So why do so many people still consider unquestioning faith to be a good thing? The answer, it seems, at least in part, is based on humanity's deep and powerful intuition that answers exist. Without a deterministic map, science cannot claim to have access to answers to our ultimate questions. This situation presents a business opportunity for power-seeking individuals—a chance to sell an arbitrary set of 'answers' for the price of unquestioning obedience and blind faith.

"Insanity in individuals is something rare—but in groups, parties, nations, and epochs it is the rule."
Friedrich Nietzsche[14]

People that are unaware of the rigorous scrutiny that answers must pass through, people that blindly follow the command of others, are not necessarily bad, or stupid. They simply haven't had the fortune of discovering that an answer means nothing unless you intimately understand the logic behind that conclusion. They have not been afforded sufficient opportunity to recognize that the journey is more valuable than the destination, or that truth requires a personal journey.

"Just doing my job. Just doing what I'm told."
County Jail Guards

If we are to make moral progress we must keep in mind that humanity's natural moral inclinations are most readily overridden when authority figures step into the room. People tend to *follow orders*—even when those orders violate their deepest moral beliefs.[15] Stanley Milgram studied this tendency and vividly described his findings in his 1983 book *Obedience to Authority*. "When an authority figure, such as an experimenter in a white lab coat, ordered a subject to pull a lever to shock another person, the subject repeatedly did so, even though the other person—an actor in no actual danger—reacted dramatically to the 'shock'."[16]

Milgram found that many people respond to symbols of power by suspending their independent rationale and their sense of empathy.[17] He showed that our willingness to just do as we are told without question has the ability to transform us into monsters.[18] In response to the authority figure's commands, the people in Milgram's experiment continued to increase the voltage delivered to their victim despite their screams. Two out of three of them carried on until their victim appeared unconscious.[19] Similar experiments have produced consistent results across the globe.[20]

Humans have a wonderful capacity for empathy, but unquestioning faith and blind obedience often override that capacity. Relying on these tenets can make us less human, less noble, because when we fail to encourage curiosity, freethinking, and independence we lose out on ethical progress. This is why the scientific enterprise has usurped all others in its ability to adequately provide moral instruction. It, above all else, encourages us to think things through for ourselves.

> "Unthinking respect for authority is the greatest enemy of truth."
>
> Albert Einstein

> "Religion, as organized in its churches has been and still is the principal enemy of moral progress in the world."
>
> Bertrand Russell[21]

On March 16, 1966 Neil Armstrong and David Scott rode Gemini VIII into space and achieved the world's first docking. Approximately 20 minutes later the two spacecraft rapidly began spinning. Armstrong used his craft's thrusters to temporarily stop the spinning, but it soon resumed. The accelerated motion quickly became critical—endangering both the spacecraft and the Gemini VIII astronauts. Hoping that the problem was in the target vehicle (the Agena), mission control instructed them to separate from it. But after undocking, Gemini VIII began to spin even faster—about one revolution per second.

The astronauts were approaching unconsciousness and facing death if the spinning continued. So, in order to save their lives, Armstrong broke a strict mission rule: he turned off the thruster system and activated the reentry control system. As he had suspected, one of the maneuvering thrusters on Gemini VIII had become stuck in the open position, causing the spacecraft to spin.

By thinking independently, and appropriately acting outside of the rules, Armstrong was able to gain control of the spacecraft and stop the spinning. Despite the fact that his entire career had blossomed because of his ability to accurately and meticulously follow procedure, Armstrong was able to abandon the restrictive obedience mindset when it no longer benefitted the situation. I like to think that this quality helped NASA's selection committee decide who would best represent humanity on that historic first lunar landing.[22]

What about consolation? Can the scientific investigation assuage our fears? Can it be a source of emotional comfort? Of course it can. There is great consolation that comes with discovery. Every time someone personally puts together another piece of Nature's grand puzzle, or experiences reality in a new insightful way, they experience joy and a deep sense of consolation. To anyone that has joined the scientific quest this is self-evident. The more appropriate question is whether or not dogmatic fundamentalism is truly capable of providing any meaningful measure of consolation.

Can all that focus on hell, sinning, punishment, shame, etc., really foster a feeling of consolation or comfort? Can delusion ever be considered more comforting than the truth? Can separating yourself from your natural impulses and inclinations ever be truly consoling? Those who think that such acts are necessary to be part of a group might say yes, but for the intellectual who has chosen the path of transcendence, the ability to admit to that which one doesn't know (especially to yourself), and the freedom to actively search for the answers is quite consoling and comforting in and of itself. It can also lead to a much more fulfilling social life.

In order to argue that dogmatic faith, or as W. K. Clifford put it "the fatal superstitions that clog [every] race,"[23] is more consoling than the honest quest of science, we would have to claim that humans are inherently more comfortable in a state of delusion. In the long run, I do not think that this claim holds up. I don't think humanity, as a whole, prefers delusion on the grounds of consolation—especially those who have experienced both.

Personal Inspiration

"After a certain high level of technical skill is achieved, science and art tend to coalesce in aesthetics, plasticity, and form. The greatest scientists are always artists as well."

Albert Einstein

"Reason is immortal, all else is mortal."

Pythagoras

For quite some time now, science has been publicly portrayed as drab, boring, and pointless. It has become popular to think of scientists as socially awkward bean counters with a terrible sense of fashion and a complete inability to interact with members of the opposite sex. Even Einstein, the quintessential scientist, has been regularly misrepresented on this front. The truth is that Einstein had a deep fervor for the scientific quest, loved very passionately (he had many girlfriends and romantic affairs throughout his life), and had many close, lifelong friends. He was thoughtful, gentle, a pantheist, a musician, and so extremely socially and diplomatically equipped that he was offered the Presidency of Israel. He was also insightful enough to turn it down. [24]

Einstein balanced the eloquence required to reduce complex concepts into intuitive pictures with a creative way of looking at the world. Throughout his life he retained the "ability to write beautifully about simple things, like why the tea leaves collect in the middle of the cup when you stir it, and why there is a place on the sand between the water and the beach where it is easiest to walk."[25] Einstein was an icon of human life. He, like all other great scientists, artistically enriched humanity's vision.

The public misunderstanding of the scientist can be traced all the way back to Thales (585 BCE), who was the first to attempt to establish a rational cosmology and was therefore the first Western philosopher (today we would call him a scientist). Plato records that Thales fell into a well "when he was looking up to study the stars… being so eager to know what was happening in the sky that he could not see what lay at his feet."[26] Evidently a witty and attractive Thracian girl witnessed his fall and proceeded to publicly indignify Thales. Whether or not this actually happened, this story has continued to represent how those who remain primarily concerned with convention see those who actively pursue discoverable divinity.

"I want to know how God created this world. I am not interested in this or that phenomenon, in the structure of this or that element. I want to know His thoughts; the rest is detail."

Albert Einstein

It isn't difficult to find well-rounded, interesting, and passionate scientists. Scott Sampson is a perfect example. When I met Scott he was the lead paleontologist for the Utah Museum of Natural History. His office was filled with dinosaur bones—including a few species that he helped discover and name. He was well spoken, enthusiastic, and was in love with a beautiful doctor from South Africa. He spent his time organizing inspirational lectures, teaching classes, holding high society fund raisers, and going out in the field—prospecting and excavating dinosaurs in the most isolated corners of Earth (Madagascar, Mexico, southern Utah, Canada, Argentina, and more).

Along with Stony Brook University paleontologist David Krause and their team, Scott discovered the now famous German-shepherd-sized carnivore with forward protruding, conical teeth that curl out of its mouth in a way that is completely unique among therapods. After excavating it from Madagascar's late Cretaceous formations, the team named this amiable grotesquerie "Masiakasaurus *knopfleri*" after Mark Knopfler, the lead singer of the group Dire Straits—the music playing on their boom box when they first unearthed it. Scott was approximately six feet tall, had a muscular build, and wore attire that ranged in style from *GQ* to Indiana Jones.

Another example is Lindsay Zanno. When I met Lindsay she was twenty-four years old studying for her Ph.D. in dinosaur paleontology. Lindsay was slim, athletic, and had a tough-girl attitude. She dressed in popular fashion with a slight Greek influence and spoke with a New York accent whenever she said *coffee* or was in a hurry. Lindsay was approximately 5'4" tall, had long dark brown hair, and a thirty-third dynasty Egyptian hieroglyphic passage from the Book of the Dead (the Theban Recension) tattooed down her spine. Her physical feminine beauty was amplified by the way she would participate in and lead intelligent conversations—expressing her enthusiasm for the scientific exploration that she was part of. People like Lindsay make me wonder how anyone ever formed the opinion that scientists are boring or plain. In my experience scientists set the standard for the complete opposite side of the scale.

You might be tempted to say that these are atypical examples. After all we're talking about dinosaur paleontology here, right? My point is that every scientific field comes with its own thrill of discovery and life-enhancing passion. They are all facets of the quest for a deeper truth. People who know very little about science may be able to relate to some fields better than others at first, but when they dive into the scientific quest they quickly find that there are amazing discoveries to be made, and important questions to be explored, in every scientific field.

As an example, consider Frank Brown—an anthropologist/geologist and the Dean of the School of Mines at the University of Utah when I attended. Frank spent his summers in Kenya digging up bones with the Leaky family and exploring the structure of humanity's family tree. The last time I talked with him he was working on learning his sixteenth language—a little known bush language spoken only in Africa. In 2006 Frank and his team age dated the oldest fossil of Homo *sapiens* yet, pushing the date of earliest appearance back to 195,000 years ago.

Frank spoke with powerful authority, yet his voice was always inspiring, informed, and inviting. He conveyed the thrill of the quest extremely well. He was over six feet tall, always smiling, and thought on his feet as well as any person I have ever met. Everyone always seemed to learn something around Frank, but most of all they learned that the road of discovery is delightful—not boring.

Explorational geologist Erich Peterson is another example of an inspirational scientist. Erich was the man you would go to if you wanted to know where to dig for gold or other precious metals. I had the honor of going to Peru with Erich and exploring some of the world's deepest and most productive mines—where thick veins of silver striped the walls, thousands of huge pyrite crystals decorated the ceiling, and where I burned a hole in my jeans simply by sitting on the rocks at that depth. Erich was always preparing the next trip, trying to make sense of the last discovery, and his smiling eyes were constantly full of life.

Then there is George Cassiday, a master storyteller and avid cyclist who made a living as an astrophysicist, collecting data from the Flyseye II project located within the Dugway proving grounds, and teaching. His classes made me feel like I was sitting around a campfire, on the edge of my seat, listening to ghost stories. He had a gift for pulling students into the lives of historical figures, helping us see them as real people confronted with real dilemmas, and leading us down their paths of discovery. George made me giddy to join the quest. He completely redirected the path of my life.

And I cannot forget Tiffany Fowler. When I met Tiffany she was a biologist working as a stem cell researcher as an intern for NASA. Everything I learned about her made me aspire to be more. She was a track star, a skilled volleyball player, and an extremely enthusiastic, kind, attractive, and sylphlike young woman. When we went places she would show me a plant (or a lichen) that I had never noticed before and tell me its common name, its Latin name, what practical application it could be used for, (like using the soapy roots of the Mole-Yucca as hair shampoo and detergent for linen and wool), what other plants it was related to, if it was indigenous or not, and how it was tied to the rest of the ecosystem. When she got excited she spoke in French, and when she dreamed, she dreamed of exploring Africa and Australia where she would make new discoveries and bring medical advancements to the world. Her blonde hair fell just past her shoulders, and in the small of her back was a Chinese kanji tattoo for *truth*.

I could give dozens of more examples from my personal experiences of extremely interesting scientists who are passionate for life.[27] Not only do they set the example for grabbing life by the horns, many of them could give lessons on social interactions and fashion. These are genuine people who live life for the adventure that it is—free from the crutch of unquestioning faith.

Scientists may be popularly portrayed as socially awkward, sexually frustrated, boring, or mad individuals—something nobody would want to be, but the truth is that real scientists tend to have amazing lives. By rising above the restraints of command ethics (like blind obedience) scientists gain a greater understanding of their humanity. Because science encourages us to exercise our curiosity, and to actively continue the process of making personal discoveries, it fills our lives with inspiration and enhances our morality.

Seeking Explanation

"Every little increase in human freedom has been fought over ferociously between those who want us to know more and be wiser and stronger, and those who want us to obey and be humble and submit."

Phillip Pullman[28]

As long as people are drawn to strong leadership and shallow answers, rulers will continue to illegitimately claim to have unique access to the gods, or the secrets of Nature, to justify their authority. The only way to turn this around, the only way to escape being lied to by our leaders, is for us to change the game—to grow comfortable with uncertainty. We need to stop seeing intellectual cowardice and intransigent stubbornness as leadership traits, and start demanding that our leaders have the ability to logically consider all options and grasp their moral implications.

As humans we seek to know the universe. We want to have access to answers, but frequently misunderstand what it means to have an answer. Knowing a destination, without knowing the logical journey that takes you there, does not count as having an answer. Knowing that a specific event will occur, or that the world *is* a certain way, without knowing *why* is of little value. Similarly, labeling the cause of an event without clearly defining that label does not give us causal access to the event. Answers must give us causal access.

The need for explanation has historically been used as the trump card for religious apologists. In 1687 Isaac Newton pulled together the works of Copernicus, Galileo, Kepler, Descartes, and others,[29] and developed a powerful insight—the universe is governed by a few physical, mechanical, and mathematical laws. Newton's map of the universe was so simple that everyone could incorporate it into his/her intuition and begin to understand why Nature is the way that we see it. This "instilled tremendous confidence that everything made sense, everything fitted together, and everything could be improved by science."[30] It suggested that Nature can be unveiled by rational inquiry, which opened our eyes to the fact that the personal journey is intimately entangled with the systematic endeavor of science.

When phenomena were discovered that are incompatible with Newton's map (the perihelion advancement of Mercury, the consistency of the speed of light in all reference frames, etc.), the recognition that Newton's map was incomplete was used to persuade humanity to descend back into superstition. Under this shadowed promulgation humanity's spirit of exploration began to wane. In the darkness they began to believe that their lack of a candidate map meant that Nature was altogether unmappable.

This story has echoed through the ages. When Einstein unveiled a deeper map of the universe, humanity once again faced the possibility of gaining a deep understanding of Nature—of bringing the secrets of Nature within reach of intuition. But Einstein's map started out with a problem—it wasn't entirely intuitive. This problem was exacerbated by the fact that within a few years, discoveries of quantum mechanical effects popped up that could not be explained by Einstein's new map. In response to this, fundamentalists began to initiate another descent into superstition, rising to power along the way.

Einstein expressed a deep sadness over the overwhelming tendency of political forces to abuse the noble achievements attained by the foraging scientific endeavor. After Fritz Haber pioneered the use of chlorine gas as a means of inflicting human suffering, Einstein lamented, "our whole, highly praised technological progress, and civilization in general, can be likened to an ax in the hand of a pathological criminal."[31]

We are all responsible for safeguarding against this tendency. Science will continue to be abused and misused by those in power so long as the majority of people fail to think for themselves. To set the world free from the bonds of superstition, to protect it against such abuses, we need to promote a world of independent thinkers, and we need to contribute to that pool ourselves.

"Being superstitious is bad luck."
David Cantu

People are less likely to rely on superstitions when scientific explanations exist. But when full explanations don't exist, people can get impatient. Many of today's seasoned explorers have begun to claim that our goal is unreachable. More and more evidence suggests that a complete map of physical reality can only exist in higher dimensions. Those that believe that the human mind is physiologically incapable of visualizing a higher-dimensional map of physical reality tend to see this as an inherently dooming problem. It is not.

Our brains are "powerful enough to accommodate a much richer world model than the mediocre utilitarian one that our ancestors needed in order to survive."[32] Our minds are quite plastic and, as we have seen throughout this book, they are capable of intuitively accessing infinite complexity. Even if the geometry of our new map turns out to only approximate Nature, the construction itself shows that it is possible to intuitively access higher-dimensional frameworks.

The ability to comprehend higher-dimensional geometries resurrects the idea that answers to the big questions can be systematically unveiled by rational inquiry. Intuition can once again play its vital role in scientific exploration. A higher-dimensional perspective gives us a way to transcend the morass of superstition while exploring our deepest intuitions. It restores our hope, and propels us to continue seeking explanations.

The assumption that the vacuum is a superfluid with a hierarchical fractal structure poetically amplifies many insights that have been passed down through the mythic medium via music, art, and many myths from around the world. According to Joseph Campbell the "fundamental theme of all mythology is that there is an invisible plane supporting the visible one." As humans have attempted to obtain a *new way of knowing*, they have learned to recognize that there is more to the world than meets the eye. This is a visceral, magnetic and universal theme. Priests captivated by elaborate descriptions of heaven, and astronomers peering into the vastness of the heavens, recognize that there is more to the picture—that there is a realm extending beyond the reach of our senses, a realm of harmony and elegance.

"Modern physics impresses us particularly with the truth of an old doctrine which teaches that there are realities existing apart from our sense-perceptions, and that there are problems and conflicts where these realities are of greater value for us than the richest treasures of the world of experience."
Max Planck[33]

Given this theme, it might be interesting to see how some of history's philosophical, theological, and ideological passages relate to the superfluid vacuum framework. Please keep in mind that similarities between ancient passages and the superfluid model may offer different angles from which to interpret the model, but they should not be taken as support for it.

In our first comparison we note that, "cultures from ancient Babylon and Egypt to India and China all refer to an underlying, featureless, eternal substance from which all structure later emerged."[34] The Kybalion imagines that Nature, in its complete form, is deterministic. It also connects determinism to the existence of *many planes of causation* and imagines a linking symmetry between those *planes*. It says, "Every Cause has its Effect; every Effect has its Cause; everything happens according to Law; Chance is but a name for Law not recognized; there are many planes of causation, but nothing escapes the Law." It follows this up with the claim that "There are planes beyond our knowing, but when we apply the Principle of Correspondence to them we are able to understand much that would otherwise be unknowable to us."[35]

Other interesting similarities are found in ancient Asian cosmologies. Early Chinese metaphysical systems (e.g. Taoist texts) stress the intimate interconnectedness of all events, and describe Nature as an interwoven unity of two realms—heaven and earth, yin and yang—that constitute a single reality. The Tao, or heaven, is part of that single reality. It remains hidden from sensory experience, but it influences Nature and actively takes part in all that unfolds.[36]

Buddhists describe Nature's complete map as being composed of two parts. The first part consists of everything that we observe around us, and the second part is called nirvana. This other realm is "there, but hidden..."[37] Nirvana is said to be a realm that co-exists with the world of our experience, but lies beyond the reach of our physical senses. Eastern philosophies maintain that this realm is capable of affecting the realm of our experiences. They believe that physical reality is composed of at least two separate planes that intermingle and exist together at all times.

Buddhist tradition also holds that while sitting under a Bodhi tree, Siddhartha Gautama saw something that surpassed all previous forms of human knowing—the unison of the familiar realm and the *untouchable* great beyond. According to legend it was at this point that the new Buddha intimately realized that he and all of life was part of an unending process of change. He saw that everything was one, and that every part reflected the whole. Physical reality itself was revealed to him as a system of interconnected, inseparable parts, rich and complex—a place where previous expressions of contradiction translated into descriptions of harmony and equilibrium.

To relate this new understanding to us Buddha said, "In the heaven of Indra, there is said to be a network of pearls, so arranged that if you look at one you see all the others reflected in it. In the same way each object in the world is not merely itself but involves each other object and in fact *is* everything else."[38] It is said that this great awakening transformed Siddhartha into the Buddha (which literally means "the one who had awakened"). He had seen beyond a boundary that humanity had only previously known as *an impossible barrier to transcend*. By doing this he tapped into the full essence of Nature and being. This richer point of view triggered a change of heart and a change of habit—forever altering its discoverer's course with the wisdom that "As above, so below, as without, so within."[39]

Finally, we note that in the East, enlightenment is often considered a matter of overcoming dualism. In fact, within the Zen tradition the most concise summary of enlightenment translates as: transcending dualism.[40] With this in mind, it is interesting that the fractal conception of reality developed in this book automatically transforms dualism to a hierarchical monism. By granting us access to determinism it triggers a change of heart by helping us transcend the dualism of right and wrong, praise and blame, etc., and helps us accept reality as a whole. Having intuitive access to the realm that lies beyond our senses can give us a window into the labyrinthine byways of various mythical and poetic attempts to understand Nature. It puts us back on the path of explanation and allows us to intellectually and emotionally evolve.

Parsing Consciousness

Determinism refutes the idea that one can be free "from the constraints imposed by the physiochemical states of one's own body and mind."[41] What then does it mean to be *conscious*? What is *consciousness*?

Beyond the fact that *consciousness* is a high-level description of a system whose underlying structure and patterns are extremely hierarchical, the title *conscious* belongs to objects if they act in a manner that is externally unpredictable. More specifically, if an action is caused by effects that are not externally discernible, or theoretically understandable (from within the dimensional perspective available), then we say *consciousness* is the cause of the observed effects.

To fully develop this point, consider the following: When I was young my family had an old farm truck named Betsy that seemed to have a personality of her own. Sometimes she worked, and then just randomly quit. Sometimes she'd start right back up again and other times she wouldn't. Before turning the key we used to ask her, "What kind of mood are you in today Betsy?" She'd answer with a working radio and a bad turn signal.

Now that I'm older, I realize that there were causes for how Betsy acted from day to day—I just didn't know what those causes were. Does this mean that Betsy was *conscious*? Somehow we all know that she wasn't, but why? I suggest that the reason has two parts. First, we all believe that if knowledgeable experts were to examine Betsy's quirks, they would, at least in principle, be capable of determining the causes of her behavior. Second, (and this is a less important factor) despite the complexity required to have a functioning truck, as of yet no truck takes in raw data from the external world and filters that data through an identity program (a level of complexity highly removed from the level in which the physical inputs themselves are defined on). Trucks are not controlled by computers that internally reinforce a self-referencing notion of *experience*, which then, as part of a feedback loop, forwardly participate in the causal chain. In short, we believe that Betsy's actions do not stem from *consciousness* because we believe that ultimately they were derived from discernible physical inputs.

Although every effect has a cause, if those causes are internal to a system, or beyond detection, then we can end up unable to predict the behavior of that system. When this occurs it becomes useful for us to introduce the idea of *consciousness*, whether or not we assume that an *awareness* filter is internally translating those events as *experiences* and actively remaining in a self-referencing mode.[42]

Actions cannot escape being deterministically caused. Nevertheless, if physical reality is composed of more than four dimensions, then a four-dimensional map automatically places restrictions on what causes can be traced. Within a four-dimensional metric, we should not expect to be able to determine the causes of every action. The fact that we only directly perceive the world as four-dimensional may make the concept of *consciousness* practical, but an infusion of that practicality ultimately places limits on our understanding.

"Creativity intrinsically depends upon certain kinds of "uninterpretable" lower-level events."

Douglas Hofstadter

To further explore this point, imagine something that looks like a rock resting on the side of a sandy slope. It does not move, it is not animate in any way, and therefore we safely and naturally conclude that it is not conscious. Some might note that it has an energy associated with it (the energy of the atoms that make it up, its temperature, etc.), but we can probably all agree that the rock itself is not a conscious entity capable of escaping the laws of cause and effect. If a rainstorm begins, and a stream starts to rush past the rock with enough force to push it down the hill, will we now say that the moving rock is conscious? Of course not.

If we can easily observe and identify the cause (the flowing water) for the effect (the movement of the rock), then consciousness plays no necessary role. In other words, if causes are discernible then there is no utility to invoking the label *consciousness*. Now imagine that, suddenly, our rock starts jumping. No rain, no wind, no observable causes at all—it just starts wiggling and jumping up the hill. Would you change your mind about it being conscious? What if you picked up the rock, examined it, and found no way to make it repeat its behavior? Having no explanation for what caused the rock to move, the default claim is that consciousness is at work. As a rule, consciousness is invoked whenever the causes of actions are indiscernible. It inscribes our ignorance, encoding the fact that what caused the rock to wiggle up the hill—or, as it is commonly worded, why the rock *chose* to wiggle up the hill—is unknown.[43]

Whether or not the mechanical substrate of creativity is hidden from view, it exists.[44] Events that have a measurable dependence on superspatial inputs (events that make use of the uncertainty principle) are open to an interpretation of consciousness when observed four-dimensionally. From a four-dimensional perspective these causes cannot be dynamically fully mapped—they escape theoretical determiniation—and this gives rise to the impression of consciousness.

By contrast, when we use a fractal structure to frame reality in its entirety, we gain theoretical access to the deterministic origins of all actions, and therefore reduce consciousness to Betsy's quirks. Knowledge of what's going on eleven-dimensionally can, at least theoretically, enable someone to determine what's on my mind now—so long as they had access to a previous state of my mind and all of the relevant physical inputs that have occurred between those two states. My set of thoughts, feelings, and actions follow an entirely predictable and deterministic course within eleven dimensions. The illusion of autonomous consciousness, or the illusion of independent choice, dissolves in the full picture.

> *"The mind is a mirage that perceives itself as not being a mirage."*
> *Douglas Hofstadter*[45]

The central topic for an entirely separate book could be developed from how our new map opens the door for a unified theory of mind—and may even resolve the question of why, in evolutionary terms, there should be conscious experience at all. But that is another book. Whether or not the map we have been exploring in this book turns out to accurately represent Nature, whether or not we have found the true heading for our destination, humanity's quest for wisdom—our scientific journey—is the greatest story.

> *"Science isn't about being right, it's about being less wrong."*
> *Marcus Tofanelli*

The task now is to continue to sail through time as explorers, discovering continents that have never before been imagined, and escaping the bonds of supernatural explanation. As scientists, as curious humans, we must seek to meld objective, rational knowledge with transcendent, intuitive understanding. Each of us must personally reach out to touch the nerve endings of transcendent wonder in a way that far surpasses the entire panoply of superstition and continue to explore the wilderness of intuition.

From Columbus to Einstein, from those who have come before to those who have yet to come—the highest treasure is, and will always be, the journey of intuition. Let us continue that journey by opening our eyes to a view no longer tainted with the fog of self-delusion and wishful thinking. Let us examine the full texture of physical reality and discover all of its dimensions.

> *"Only the curious have something to find."*
> *Unknown*

The song of *Atlantis* is calling out to us, asking us to pull up the anchor of unquestioning faith and to find the strength we need to fill our sails with curiosity and imagination. Let each of us sail toward the distant horizon, in search of understanding. Let each of us discover our magnificent insignificance—one fathom at a time.

[1] Edward O. Wilson on the scientific rapture known as the Ionian Enchantment. *Consilience*, p. 7.

[2] F. H. Bradley, referenced in *Why I Am Not a Christian* by Bertrand Russell, p. 101.

[3] Mark Morford. (2008, March 5). Survey says we're losing our religion. Let's hope so. *San Francisco Chronicle*, p. E5.

[4] Or as Mark Morford puts it, learning to think for ourselves protects us against being broken up into "angry tribes who stomp our feet and wave our little gilded books and launch wars over promised lands and chosen peoples and crucifixes and crusades and witches and pagans and gays. Ibid.

[5] For those that read footnotes, I would add that attempts to distract us from this ascension are a disservice, and those that busy themselves with such intentional misdirections, are complicit with the moral corruption of mankind. It is contradictory to speak of the *divine*, and then "parse it and restrict it and beat it into submission and claim it for one people, one history, one country or church or authoritarian body." This is the "highest form of divine insult." Ibid.

[6] As Miles Davis once put it, "Knowledge is freedom and ignorance is slavery". Miles Davis, Miles: The Autobiography, 1990.

[7] Those that wish to enslave us insult our divinity. Case in point, consider the words of St. Augustine, "There is another form of temptation, even more fraught with danger. This is the disease of curiosity. It is this which drives us to try and discover the secrets of nature, those secrets which are beyond our understanding, which can avail us nothing and which man should not wish to learn." And in the words of Martin Luther, "Reason should be destroyed in all Christians." Richard Dawkins, 'The God Delusion.'

[8] Neil DeGrasse Tyson. *Death By Black Hole*, p. 26.

[9] Daniel Kolak & Raymond Martin. (1990). The Experience of Philosophy. Belmont, Calif,: Wadsworth, p. 2.

[10] David Hume. (1947). Dialogues Concerning Human Religion, ed. Norman Kemp-Smith. Edinburgh: Nelson, Bk II § 1.; Douglas J. Soccio. *Archetypes Of Wisdom*, p. 308.

[11] Apparently some people believe that without scriptures they would be incapable of being moral. In response to this claim we might note that modern laws forbid the killing of any person—instead of just a person of the same religion. Compared to the noxious motif of the Bible, this moral precept is advanced—not to mention our modern notion of women's rights, which clash with the misogynistic threads that hold the Bible together. Moral improvements result from changes in perspective initiated by scientific progress. For example, our internal prohibition against the killing of persons was universalized by the scientific claim that personhood is not a question of religious belief, sex, skin color, or cultural upbringing.

[12] See also Frans de Wall's recent book, *Primates and Philosophers*, (2006), and the recent neurological research of Mirror neurons, which completely undermines moral relativism—the view that there are no biologically set universal truths about what is right and wrong.

[13] W. K. Clifford. (1999). *The Ethics of Belief and other Essays*, Amherst, NY: Prometheus Books.

[14] Friedrich Nietzsche. *Beyond Good and Evil*, p. 156.

¹⁵ It has been argued that this tendency leaves us, in practice, less moral than other social species. As a rather telling example of this point consider the following: "In the mid-1960s, two groups began experiments on rhesus monkeys designed to explore how they would respond to seeing another monkey receive an electric shock. At about the same time, the social psychologist Stanley Milgram began testing people to see how they responded to authority—in particular, whether they would obey an authority figure who instructed them to administer an electric shock to another human being. In one of the rhesus experiments, an individual was trained to pull levers to obtain his daily ration of food. When the monkey learned this task, a second monkey was introduced into an adjacent cage. Now when the first monkey pulled the levers, he delivered a severe shock to the other monkey. Surprisingly, not only did the first monkey stop pulling, but he did so for several days, even though by not operating the levers he forfeited his daily meal. He was starving, but the guy next door benefited by avoiding shock. Monkeys in control of the levers were more likely to abstain from pulling if the other monkey was a familiar cage mate, than if he or she was an unfamiliar individual or a member of another species, such as a rabbit. Lastly, individuals who had been in the hot seat themselves and experienced a shock abstained longer than monkeys who had not had that experience." Marc D. Hauser, *Swappable Minds, The Next Fifty Years*, edited by John Brockman, pp. 53–54. See also Frans de Wall's recent book, *Primates and Philosophers*, (2006).

The rhesus monkey experiments are particularly striking in light of Milgram's diametrically opposed results with humans, vividly described in his 1983 book *Obedience to Authority*. "When an authority figure, such as an experimenter in a white lab coat, ordered a subject to pull a lever to shock another person, the subject repeatedly did so, even though the other person—an actor in no actual danger—reacted dramatically to the 'shock'. A Martian that had descended to Earth to watch these two experiments would be forced to conclude that rhesus monkeys empathize while humans do not. Rhesus monkeys appear to know what it's like to be another in pain, while humans either don't or simply don't care." (Ibid.)

¹⁶ Marc D. Hauser. *Swappable Minds, The Next Fifty Years*, edited by John Brockman, pp. 53–54. See also Frans de Wall's recent book, *Primates and Philosophers*, (2006).

¹⁷ By all accounts they were completely ordinary people.

¹⁸ Those that are well versed in history may not find this to be very shocking—given that the darkest pages in human history seem to all be related to that simple act.

¹⁹ Michael Bond. (2007, April 14). They Made Me Do It. *New Scientist*, p.44.

²⁰ Thomas Blass. (1991). Understanding behavior in the Milgram obedience experiment: The role of personality, situations, and their interactions, pp. 398–413. doi: 10.1037/0022-3514.60.3.398.

²¹ Bertrand Russell said this during a lecture he delivered on March 6, 1927 at Battersea Town Hall under the auspices of the South London Branch of the National Secular Society. The full quote is as follows: "You find as you look around the world that every single bit of progress in humane feeling, every improvement in the criminal law, every step toward the diminution of war, every step toward better treatment of the colored races, or every mitigation of slavery, every moral progress that there has been in the world, has been consistently opposed by the organized churches of the world. I say quite deliberately that the Christian religion, as organized in its churches, has been and still is the principal enemy of moral progress in the world." *Why I Am Not a Christian*, pp. 20–21.

22 Valerie Neal, Cathleen S. Lewis, & Frank H. Winter. *Spaceflight—A Smithsonian Guide*, pp. 98–99.

23 Clifford, 1999.

24 We could divide people that deal with the sciences into two categories. The majority of people fall into a category we might call technicians, the information gatherers and processors, the bean counters, or what my friend Matt Emmi would call the stamp collectors. These people may process and gather a lot of important scientific information but they are not necessarily overwhelmed with a divine curiosity and passionately driven to discover how physical reality is composed and how everything works. Those who are filled with this divine curiosity and the adventurous explorer's spirit define what I generally mean by the word scientist.

25 N. David Mermin. (2004, September). *Discover*, p. 80.

26 Plato. *Theatetus*, 174 A, translated by F. M. Cornford. (1961), in *The Collected Dialogues of Plato: Including the Letters*. New York: Pantheon, p. 879.

27 Jeff Chapple, Matt Emmi, David Cantu, Michael Ritter, Shae Lynn Saur, Katie Mary Peek, Zigmond Peacock, Ian Clark, Harrison Schmitt, Michelle DiBenitto, Edwin (Buzz) Aldrin Jr., Nathan Miller, Max Siker, Patrick Jones, Troy Koch, Richard Price, Brenda Dingus, Patrick Wiggins, Mike Zolenski, Wendel Mendel, Ron Carey, John Holfort, Matt Dalton, and Dillon Lee to name a few.

28 Voiced by John Parry in *The Subtle Knife*, by Phillip Pullman.

29 Peter Dew and M. Feingold point out that there were many more "giants" that Newton was standing on the shoulders of.

30 Richard Koch & Chris Smith. (2006, June 26). The Fall of Reason. *New Scientist*, p. 25.

31 Thomas Levenson. (2008, March). Albert the Icon. *Discover*, pp. 44–49.

32 Richard Dawkins. *The God Delusion*, p. 361.

33 Max Planck. (1931), p. 107; Neil DeGrasse Tyson. *Death By Black Hole*, p. 30.

34 Lawrence Krauss. *quintessence—The Mystery Of Missing Mass In The Universe*, p. 4.

35 *The Kybalion*, by Three Initiates, p. 24.

36 Warren Matthews. (1999). World Religions, 3rd ed. Belmont, Calif,: Wadsworth, p. 209; Michael C. Brannigan. (2000). The Pulse of Wisdom, 2nd ed. Belmont, Calif.: Wadsworth, pp. 23–27; Douglas J. Soccio. (2004). Archetypes Of Wisdom, 5th ed. Belmont, Calif.: Wasdworth, Chapter 2.

37 Thanissaro Bhikku. (2006, Summer). Faith in Awakening. *Tricycle*, p. 75.

38 This comes from the English translation of *The Flower Garland Sutra*, A "sutra" is a written account of the Buddha's teachings. Sir Charles Eliot. (1969). *Japanese Buddhism*, New York, Barnes and Noble, pp. 109–110; Gary Zukav. *The Dancing Wu Li Masters*, pp. 238–239.

39 Douglas J. Soccio, *Archetypes of Wisdom* pp. 33–50.

40 Ibid.

41 Edward O. Wilson. *consilience*, p. 130.

[42] This "awareness" filter might simply mean that the system is constantly being evaluated on a level that is far removed from the level in which the contributing inputs are noticeable or directly accessible.

[43] The word "chose" in this case is telling because it is a recognition of the entity's internal experiential filter.

[44] Douglas Hofstadter. *Gödel, Escher, Bach: An Eternal Golden Braid*, p. 673.

[45] Douglas Hofstadter. (2007). *I Am A Strange Loop*. Basic Books.

Afterword

> "Two roads diverged in a yellow wood,
> And sorry I could not travel both
> And be one traveler, long I stood
> And looked down one as far as I could
> To where it bent in the undergrowth;
>
> Then took the other, as just as fair,
> And having perhaps the better claim,
> Because it was grassy and wanted wear;
> Though as for that the passing there
> Had worn them really about the same,
>
> And both that morning equally lay
> In leaves no step had trodden black.
> Oh, I kept the first for another day!
> Yet knowing how way leads on to way,
> I doubted if I should ever come back.
>
> I shall be telling this with a sigh
> Somewhere ages and ages hence:
> Two roads diverged in a wood, and I—
> I took the one less traveled by,
> And that has made all the difference."
>
> Robert Frost, *The Road Not Taken*

When Einstein said, "science can be created only by those who are thoroughly imbued with the aspiration toward truth and understanding," he gave voice to the idea that real science comes from a bold assertion—an intrinsic volition. To be a scientist you must first authorize yourself to partake in the quest to unveil the mysterious, you must find the courage to explore the most terrifying regions of chaos and confusion, and you must commit to ascending the footholds of rationality in search of a perspective that makes room for those mysteries.

Armed with rational inquiry, systematic observation, rebellious creativity, and a healthy dose of the explorer's fire, the scientist climbs to higher precipice and with each step is transformed. This never-ending process of discovery can leave us trembling in fear, doubting the heights that challenge our most central beliefs and upset the status quo, but this is the process of enhancing our humanity, it is the route of moral progress.

If you've been on this journey you might have wondered, "Where do I find other rebellious and creative thinkers?" Are there any such explorers out there? Given the sociopolitical structure of modern science, is there really anyone out there that is crazy enough to plot a course into the unknown, to sail past the paradigm patrol ships that guard our present ways of thinking, and to take on the risk of (at least academically) sharing Bruno's flamed fate? Is there anyone out there that is enough of a scientist to risk expanding our humanity?

Yes, there are others out there—people like the late Benoit Mandelbrot. When Mandelbrot started his Ph.D. in mathematics he imagined breaking free from the traditional method of modeling the world as a composite of simple geometric shapes like: lines, triangles, rectangles, and ellipses. He suspected that Nature's complexities could not be fully captured by those oversimplified boundaries. In an effort to better come to grips with Nature's complexities, he toyed with the idea that the forms of Nature could be better represented by self-similar geometric structures called *fractals*.

When Mandelbrot proposed doing his Ph.D. thesis on fractals, his advisors threatened to disavow him. Such an idea, they said, was not worthy of a thesis, and if he wanted his degree he would never speak of it again. Fractals were just too much of a threat to mathematical tradition.

In 1952 Mandelbrot obtained his Ph.D. in Mathematical Sciences at the University of Paris. Then, to the annoyance of the mathematics community, he returned to the imaginative world of fractals. Prestigious journals quickly made it clear that if anyone published Mandelbrot's crackpot ideas their reputation would be forever tainted.

Mandelbrot's insight, like most new ideas, was too radical for the practitioners that were deeply committed to the assumptions it challenged—but those that were less committed to those assumptions found it universally intriguing. Artists were mesmerized by the fact that Mandelbrot's geometric structures captured a finite volume with an infinite perimeter, poets pontificated about how his patterns contain themselves in a never-ending fashion, and musicians connected fractals to their strange loops.

As a new generation emerged, people had become familiar with Mandelbrot's fascinating shapes, but many of them were completely unaware that these forms had "no real application"; they didn't know that fractals were a "waste of time", or that they were "impossible to apply to the real world", so they began to apply them.

One day, Nathan Cohen, a budding engineer and ham radio enthusiast, was told that he wasn't allowed to hang an antenna outside his apartment, so he bent a wire into the shape of a fractal and made an antenna that was so small that it wouldn't be noticed by his landlord. When he connected it, he discovered that it worked better than the larger antenna, and that it was sensitive to a wider range of frequencies. Now we all have fractal circuits in our electronics.

Loren Carpenter discovered that fractals were perfect for generating natural landscapes and then used fractals to generate an entire planet for *Star Trek II: The Wrath of Khan*, which became the first completely computer-generated sequence used in a feature film. Biologists also became amazed by how easily fractals captured the complexities of Nature's living forms. Despite the 'fact' that fractals had "no application to the real world", they kept popping up with real world applications.

Mandelbrot shared his vision by giving lectures aimed at the general public and by writing several books designed to vividly introduce the reader to the world of fractals. He was determined to share his insight because he knew that it possessed the potential to break us free from linear explanations. It was an idea that threatened to expand our comprehension of the world and, therefore, our humanity.

In 1982, Mandelbrot published *The Fractal Geometry of Nature*, which directly challenged those that dismissed fractals as "program artifacts." After a thirty-year struggle, mainstream mathematics was finally ready to accept that fractals were genuinely intriguing. But exactly how much of a role do they play?

A week before he died, Mandelbrot and I discussed that question in a radical and interesting way. What if the geometric structure of space, the very stratum of place, has a fractal form? What if the difference between this fractal structure and the smooth, continuous structure we currently assume space has is responsible for the unsolved mysteries in physics? What if we could gain intuitive access to those mysteries by mapping this fractal structure of space? Could it really be that simple? Could the intuitive understanding of Nature that we have been searching for be simply a matter of discovering the fractal geometry of space? This book has begun our exploration of those questions.

Like Mandelbrot's concept of fractals, quantum space theory needs to be allowed a chance to fill the hearts and minds of humanity's creative thinkers. If you are passionate about glimpsing the nexus where Nature's most hidden secrets are woven together, and if reading this book has given you insight that you find valuable, please share it with those that are ready to explore a new perspective. Together we can change the course of human momentum by offering a glimpse beyond the brim of the traditional worldview.

If you want to become a part of the effort to further develop this construction and to test its predictions, find me so that we might plot the course from here together. Likewise, if you'd like to use your creative talents to help share this perspective with the world—creating science tutorials, animations, videos, or public outreach presentations—or if you would like to support this research in other ways, I welcome you to the journey and look forward to hearing from you.

Sincerely,

Thad Roberts

To review the predictions of quantum space theory see:

    Chapter 8: Lorentz Contraction
    Chapter 11: Infinite Dimensional Cascades
    Chapter 14: Quantum Tunnels
    Chapter 15: Entropy and Radiation
    Chapter 16: Geometric Origins of the Constants of Nature
    Chapter 18: Eliminating Illogical Infinities
    Chapter 19: Geometric Unification
    Chapter 20: Electromagnetism
    Chapter 21: The Higgs Mechanism
    Chapter 22: Superfluidity
            Deriving the Schrödinger Wave Equation
            Analogue Gravity
    Chapter 23: A Natural Explanation
    Chapter 24: Deriving Bohm's Formalism
            Going Beyond Bohm's Formalism
    Chapter 25: Eras of Symmetry
            Lorentz, Galilean, and Other Symmetries
    Chapter 27: Before the Bang
    Chapter 28: Another Way to Explain Redshift

Acknowledgements

Writing this book has been an epic journey, worthy of its own tale. I would like to thank everyone who played a role in its evolution, from the prison guards that routinely stole its early chapters from my cell during "shake downs", forcing me to rewrite the entire book several times by hand—a process that richly improved my writing skills, to the hundreds of people that requested a pre-print version of the book and sent me valuable feedback. Thank you all.

I thank Benoit Mandelbrot for the insights he gifted the world, and for his encouragement of this work. I thank Sheldon Goldstein, Detlef Dürr, Hans Westman, Garrett Lisi, Christian Wüthrich, Jim Tabery, Franck Laloë, Tim Maudlin, Nazim Bouatta, Frank Brown and George Cassiday, for richly contributing to my intellectual progress, for having patience with my questions, and for daring to struggle against the mainstream sociopolitical current in search of a deeper truth.

I warmly thank those that read the pre-print of this book and offered me feedback, including: Wayne Eskridge, whose insightful comments and questions led to hundreds of back and forth discussions that significantly improved the presentation quality of this work, David Heggli, whose abundant talents and passion for this work have spread it across the globe, Richard Hitchings, Marlene High, Jerry Blanchard, Pierre Rousseau, Ted Allman, Jan Szott, Jay Buzin, David Potschka, Bruce Penney, Mike Ardoline, and Andreas Frickinger, whose thorough examinations of the book's prose and arguments have polished its edges, Paul Brennan, Lynn Nicholson, Chris Tuason, Paul van der Geer, Travis Horlacher, Devyn Hepworth, Craig Joiner, Chris Wilshaw, Jeremy Bromham, Joe Boyce, Norm Peterson, John Lowe, Scott Levy, Roger Backman, Greg Engh, Jonathan Farkasofsky, Fred Goode, John Griggs, Paul van der Geer, Clare Laughran, and Nik Ignjatovic, for their questions and constructive criticisms, and everyone else that took the time to comment.

I thank all of those that have philosophically challenged me, shared the gift of intimacy, and richly added to the beauty of my life, including: Marcus Tofanelli, Matt Emmi, Mike Ritter, Chris Miller, Marie Green, Josh Bross, William Prince, Yoshika Ramanujam, Zia Faculjak, John Dunsmoor, Cindy Gasaway, Bob Gilstrap, Nathan Miller, Marij van Strien, Faye Gosnell, Russ Nickle, Jessica Oldham, Aubrey Spivey, Tomoko Richard, Tanja Traxler, Jamie Lynn Catrett, Dawn Brockett, Lisa Marie Wood, Rebecca Boozan, Caitlin Polansky, Marc Jones, Maria Stanley, Tiffany Fowler, Steven Baird, Cara Elizabeth Yar Khan, Christine Saffell, Dennis Christiansen, Justin Goode, Andrea Wong, Lisa Marie Young, Lindsay Anderson, Ashley Clickner, Patrick Chapin, Wil Biddle, Chris Volk, Ray Archuleta, Terry Milligan, Amji Ramanujam, Dave Nugent, Lucas Mathews, Angie Anderson, Tom Rees, Pashaa Sanwick, Andrew Hyde, Thomas Saffell, Cynthia Stark, Mark Walsh, Eric James, Jasminder Singh, Richard Grote, Lizzy Vincent, Pack Landfair, Mireille Charlotte, Reid Ewing, Jeff Gilbert, Shauna Montgomery, Joey Monteith, Kimberly Dill, Ingo Walter, Andrew Martin Hogsten, and Dan Adams. Thank you all for being the treasures of my journey, for exposing me to new wonders, for your character, for the adventures we've shared, and for the adventures that still lie ahead.

With great adoration, I extend a very special thanks to Elaine and Phil Emmi for their genuine concern for my emotional health and growth, for mentoring me, for creating an atmosphere that allowed me to continue this work, for reintroducing me to the delights of life after prison (like orange juice), and for treating me like a son.

I thank Angela Arvizu for having the courage to face the world with wide eyes, for her boundless sense of adventure, for sharing that sparkling journey across a sea of nighttime glitter, and above all else for her unconditional acceptance and support.

I thank David Cantu for raising the bar of what it means to be a friend. I thank him for his outstanding courage, for his refusal to submit to tyranny, and for the endless months in which his wit and humor kept the fire inside of me alive.

Dave served an extra 52 days in prison for mailing an early draft of this book out to a friend, asking for his editorial advice. Although that punishment was a puritanical act of despotism, Dave never wavered in his stance that he would do it all over again—just for the chance to help this book reach its end goal. Thank you for your sacrifice Dave.

Finally, I thank my dear friend Jeff Chapple for being my Alfred, my man behind the scenes, and for always having my back. Thank you for supporting this project from the very beginning, for challenging each and every argument within it, for the countless hours you spent sending me articles, books, and researching for me. Thank you for creating all the beautiful figures in this book and the videos for my Ted Talk. Most of all, thank you for your friendship. I shall always count you among the best experiences of my life.

Thank you all for making me delight in the idea that the Omniverse will play this song over and over, echoing our journey throughout the infinities.

References

Abbot, E. (1884). *Flatland: A Romance of Many Dimensions*.

Abraham, W. E. (1972). The Nature of Zeno's Argument Against Plurality in DK. 29 BI, *Phronesis* 17: 40-52.

Albert, D. (1996). *Elementary Quantum Metaphysics; Bohmian Mechanics and Quantum Theory: An Appraisal.* Boston Studies in the Philosophy of Science, Volume 184.

Alexander, H.G. (1956). *The Leibniz-Clarke Correspondence.* (ed. And trans.), Manchester University Press.

Anderson, P. W. (1958). *Absence of Diffusion in Certain Random Lattices. Phys Rev* **109** (5): 1492-1505.

Anderson, R. & Brady, R. (2013). *Why quantum computing is hard—and quantum cryptography is not provably secure.* arXiv:1301.7351v1.

Andrews, W. J. H. (2006). A Chronicle of Timekeeping. *Scientific American,* Special Edition—A Matter of Time.

Appell, D. (May 2008). Dark Forces at Work. *Scientific American.*

ar-Razi (1939). *Opera Philosophica.* P. Kraus (ed.), Cairo.

Aristotle (1984). *De Interpretatione*—in Aristotle, *The Complete Works of Aristotle*, Princeton University Press, Esp. Physics Bk. VIII.

Aspect, A., Dalibard, J., & Roger, G. (1982) *Experimental test of Bell's inequalities using time-varying analyzers.* Phys. Rev. Lett 49: 1804-1807.

Atwood, W. B., Michelson, P. F., & Ritz, S. (2007, December). Window on the Extreme Universe. *Scientific American.*

Augustine, St. (1961). *Confessions*, ed. R.S. Pinecoffin, Harmondsworth: Penguin.

Averroes (1962). Long Commentary on the Physics—in Aristotelis *Opera cum Averrois Commentariis*, vol IV, Frankfurt am Main: Minerva, G.m.b.H.

Baez, J., (1998). Spin Foam Models. Class. Quant. Grav. 15(1998)1827, gr-qc/9719052.

—(2000). Spin Foam Models of BF Theory and Quantum Gravity. Lect. N. Physics. 543, Springer N.Y. 2000 gr-qc/9905087.

Barceló, C., Liberati, S., & Visser, M. (2005). Analogue Gravity. Living Rev. Relativity, **8**, 12. [Online Article]: cited [<11-26-2012>], p. 76, http://www.livingreviews.org/lrr-2005-12

—(2001). Analogue gravity from Bose-Einstein condensates. Institute of Physics Publishing, Class. Quantum Grav. **18**, PII: S0264-9381(01)18993-8, p. 1140.

Barrett, J. W., & Crane, L. (1999). A Lorentzian Signature Model for Quantum General Relativity. gr-qc/9904025.

Batterman, R. B. (1993). Defining Chaos. *Philosophy of Science*, 60: 43-66.

Battersby, S. (2006, July 15). Let there be dark. *NewScientist.*

Baxter, G., & Besnard, D. (2004). Cognitive Mismatches in the Cockpit: Will They Ever Be a Thing of the Past? *The Flightdeck of the Future: Human Factors in Datalinks and Freeflight Conference.* UK: University of Nottigham.

Bell, J. S. (1966). On the problem of hidden variables in quantum mechanics. Rec. Mod. Phys. 38, 447-452.

—(1976). The theory of local beables. Epistemological Lett. 9, 11-24.

—(1982). On the impossible pilot wave. Foundations of Physics 12, 989-999.

—(1987). Speakable and unspeakable in quantum mechanics. Cambridge University Press.

Bergson, H. (1911). *Creative Evolution*, A. Mitchell (trans.). New York: Holt, Reinhart and Winston.

Berman, B. (2005, October). What's the Antimatter? *Discover*, p. 22.

Bernard, C. I. (1980), *The Newtonian Revolution*. Cambridge: Cambridge University Press.

Bertone, G., Hooper, D., & Silk, J., (2005). Particle dark matter: Evidence, candidates and constraints. Physics Reports **405** (5–6): 279–390. arXiv:hep-ph/0002126

Bhikku, T. (2006, Summer). Faith in Awakening, *Tricycle*, p. 75.

Bilson-Thompson, S. O., Markopoulou, F., & Smolin, L. (2007). Quantum gravity and the standard model. *Class. Quantum Grav.* **24** (16): 3975–3993, arXiv:hep-th/0603022, Bibcode 2007CQGra..24.3975B, doi:10.1088/0264-9381/24/16/002.

Bishop, R. C. (2002). Deterministic and Indeterministic Descriptions. *Between Change and Choice*, H. Atmanspacher and R. Bishop (eds.), Imprint Academic.

Black, M. (1950). Achilles and the Tortoise. *Analysis*, 11: 91-101.

Blass, T. (1991). Understanding behavior in the Milgram obedience experiment: The role of personality, situations, and their interactions. pp. 398–413. doi: 10.1037/0022-3514.60.3.398.

Bombelli, L., Lee, J., Meyer, D., & Sorkin, R. (1987). Space-Time as a Causal Set. Phys. Rev. Lett. 59, 521.

Bohm, D. (1952). A Suggested Interpretation of the Quantum Theory in Terms of 'Hidden' Variables, I and II. Physical Review 85: 166-193.

—(1953). Proof that probability density approaches $\langle\psi\rangle^2$ in causal interpretation of quantum theory. Physical Review 89, 458-466.

Bohm, D., & Hiley, B. (1993). The Undivided Universe: an Ontological Interpretation of Quantum Theory. Routledge & Kegan Paul, London.

Bond, M. (2007, April 14). They Made Me Do It. *New Scientist*, p.44.

Born, M. (1962). *Einstein's Theory of Relativity*. New York City: Dover Publications.

—(1926). Quantenmechanik der Stossvorgänge. *Zeitschrift für Physik* **38**, 803–827

—(1926). Zur Wellenmechanik der Stossvorgänge. *Göttingen Nachrichten* 146–160.

Bradley, F. H. (1893). *Appearance and Reality*. Swan Sonnenschein. Second edition, with an appendix, Swan Sonnenschein, 1897; ninth impression, corrected, Clarendon Press, 1930.

Brady, R. (2013). The irrotational motion in a compressible inviscid fluid. arXiv:1301.7540 [physics.gen-ph].

Brannigan, M. C. (2000). The Pulse of Wisdom. 2nd ed., Wadsworth.

Brewster, D. (1855). Memoirs of the Life, Writings, and Discoveries of Sir Isaac Newton. Volume II. Ch. 27.

Briggs, J. P., & F. Peat, F. D. (1986). Looking Glass Universe: The Emerging Science of Wholeness.

Brighouse, C. (2014). Geometric Possibility: An Argument from Dimension. Occidental College, 01/2014; 4(1). DOI: 10.1007/s13194-013-0074-1 Available from: www.researchgate.net/publication/263104059

Brightwell, G., & Gregory, R. (1991). The Structure of Random Discrete Spacetime. Phys. Rev. Lett. 66, 260.

Brody, D. C., & Hughston, L. P. (2001). Geometric quantum mechanics. J. Geom. Phys. 38, 19-53.

—(2004). Theory of Quantum Space-Time. arXiv:gr-qc/0406121v1.

Bruno, G. (1584). On the Infinite Universe and Worlds.

Buchanan, M. (2008, March 22). No dice. *New Scientist*, pp. 28-31.

Butterfield. J. (1998). Determinism and Indeterminism. *Routledge Encyclopedia of Philosophy*, Craig. E. (ed). London: Routledge.

—(1999). *The Arguments of Time*. Oxford: Oxford University Press.

Callender, C. (1998). Review, Brit. J. Phil. Sci. 49, 332-337.

Carey, B. (2006, August). Mapping Earth's Fourth Dimension. *Discover*.

Carroll, S. M. (2008). The Cosmic Origins of Time's Arrow. *Scientific American*.

Castelvecchi, D. (2006, August 12). Out of the Void. *New Scientist*, pp. 28–31.

Chaisson, E. (2006, January 7). The Great Unifier. *NewScientist*.

Chalmers, A. F. (1970). Curie's principle. *British Journal for the Philosophy of Science*, **21**, 133-148.

Chalmers, D. J. (1997). *The Conscious Mind—In Search of a Fundamental Theory*.

Chown, M. (2004, December 18). It Came From Another Dimension. *NewScientist*, pp. 30-33.

Chown, M. (2006, June 10). Do the cosmic twist. *NewScientist*, pp. 34-37.

—(2007, September 29). Dimmer outlook for dark matter. *NewScientist*.

Christian, J. (2007). Disproof of Bell's Theorem by Clifford Algebra Valued Local Variables: www.arxiv.org/abs/quawnt-ph/0703179

Clark, S. (2008, March 8). Cosmic enlightenment. *NewScientist*.

Clauser, J. F., Horne, M. A., Shimony, A., & Holt, R. A. (1969). Proposed experiment to test local hidden-variables theories. *Phys. Rev. Lett.* **23**, 880–884.

Clifford, W. K. (1999). The Ethics of Belief. Reprinted from (1877) The Ethics of Belief and other Essays. Amherst, NY: Prometheus Books.

Coleman, S. (1975). Secret symmetry: an introduction to spontaneous symmetry breakdown and gauge field. A. Zichichi (ed.), *Laws of hadronic matter*, New York: Academic Press, pp. 138-215.

Collins, G. P. (2005, June). Making Cold Antimatter. *Scientific American*.

—(2006, July). A Hint of Axions. *Scientific American*.

—(2008, April). Wipeout? *Scientific American*.

Corbin, H. (1969). *Creative Imagination in the Sufism of Ibn 'Arabi*, trans. Ralph Manheim. Princeton, NJ: Princeton University Press.

Crane, L. (2000). Hypergravity and Categorical Feynmanology. gr-qc/0004043.

Crystall, B. (2007, September). Engage the antimatter drive. *NewScientist*.

Curie. P. (1894). Sur la symetrie dans les phenomenes physiques. Symetrie d' un champ electrique et d'un champ magnetique. *Journal de Physique*, 3rf series, vol. 3, 393-417.

Cushing, J. T. (1994). *Quantum Mechanics: Historical Contingency and the Copenhagen Hegemony*. Chicago: University of Chicago Press.

Darwin, C. (1859). *The Origin of Species*.

Daumer, M., Dürr, D., Goldstein, S., & Zanghí, N. (1995). On the Quantum Probability Flux Through Surfaces. Journal of Statistical Physics 88:967-977.

Davies, P. (2002, September). "That Mysterious Flow," *Scientific American—Special Edition: A Matter of Time*.

—(2006, February 11). In Search of a Second Genesis. *New Scientist*, p. 48.

—(2006). How to Build a Time Machine,' Scientific American, Special Edition—*A Matter of Time*.

—(2007, June 30). Laying Down the Laws. *New Scientist*, p. 30–31.

Dawkins, R. (2006). *The God Delusion*. New York: Houghton Mifflin Company.

de Broglie, L. (1924). *Recherches sur la théorie des quanta* (Researches on the quantum theory), Thesis (Paris).

—(1925). *Ann. Phys.* (Paris) **3**, 22.

—(1928). La nouvelle dynamique des quanta. Electrons et Photons: Rapports et Discussions du Cinquieme Conseil de Physique tenu a Bruxelles du 24 au 29 Octobre 1927 sous les Auspices de l'Institut International de Physique Solvay, Gautheir—Villars, Paris, pp. 105-132.

—(1952). La physique quantique restera-t-elle indéterministe? *Revue des sciences et de leurs applications* **5**, 289–311. French Academy of Sciences, 25 April 1953 session, http://www.sofrphilo.fr/telecharger.php?id=74

deGrasse, N. T. (2007). Death By Black Hole—And Other Cosmic Quandaries. W. W. Norton.

DeLanda. M. (2011). Assemblage Theory, Society, and Deleuze.

Deleuze, G. (1980). A Thousand Plateaus.

Diamond, J. (2005). *Guns, Germs and Steel—The Fates of Human Societies*. New York: W. W. Norton & Company.

Douglas, K., & Jones, D. (2007, May 5). How to make better choices. *New Scientist*, p. 35.

Dürr, D., Goldstein, S., & Zanghí, N. (1995). Quantum Physics Without Quantum Philosophy. Physical Review Letters, vol 93, p 090402.

—(1990). Stochastic Processes, Geometry and Physics. (S. Albeverio, G. Casati, U. Cattaneo, D. Merlini, R. Mortesi, eds.), World Scientific, Singapore, pp. 374-391.

—(1992). Quantum equilibrium and the origin of absolute uncertainty. J. Stat. Phys. 67, 843-907;

—(1992). Quantum mechanics, randomness, and deterministic reality,' Phys. Lett. A, 172, 6-12.

—1992a, 'Quantum Chaos, Classical Randomness, and Bohmian Mechanics,' *Journal of Statistical Physics* 68: 259-270.

—(1995). A survey of Bohmian mechanics. Il Nuovo Cimento.

Dürr, D., Goldstein, S., Teufel, S., & Zanghí, N. (2000). Scattering Theory from Microscopic First Principles. Physica A 279: 416-431.

Dzhunushaliev, V., & Zloshchastiev, K. G. (2012). Singularity-free model of electric charge in physical vacuum: Non-zero spatial extent and mass generation. ArXiv:1204.6380.

Eagleman, D. (2007, August). 10 Unsolved Mysteries of the Brain. *Discover*, p. 59.

Earman, J. (1986). *A Primer on Determinism*, Dordrecht: Reidel

—(1995). Recent Work on Time Travel. Savitt, Steven (ed.), *Time's Arrows Today: Recent Physical and Philosophical Work on the Direction of Time*. Cambridge University Press, pp. 268-310.

—(2002). Laws, Symmetry, and Symmetry Breaking; Invariance, Conservation Principles, and Objectivity. PSA, *Proceedings of the Biennial Meeting of the Philosophy of Science Association 2002*.

—(2003). Rough guide to spontaneous symmetry breaking. K. Brading and E. Castellani (eds.), *Symmetries in Physics: Philosophical Reflections*. Cambridge University Press, pp. 334-345.

—(2007). Aspects of Determinism in Modern Physics. http://pitt.edu/~jearman/Earman2007a.pdf

Eddington, A. (1923). The Mathematical Theory of Relativity. Cambridge, England, Cambridge University Press, pp. 37-38.

Edwin A. (1884). *Flatland: A Romance of Many Dimensions*.

Ehrlich, P. (2014, October). Ninetieth Birthday: A Reexamination of Zeno's Paradox of Extension. Philosophy of Science, 81, pp. 654-675. 0031-8248/2014/8104-0005

Einstein, A. (1905). On the Electrodynamics of Moving Bodies. Reprinted and translated in *The Principle of Relativity*, pp. 35-65. New York City: Dover Publications, 1952.

—(1919, December 25). Induction and Deduction in Physics. Berliner Tageblatt, CPAE 7:28.

—(1922). *Sidelights on Relativity—Ether and the Theory of Relativty + Geometry and Experience*. Elegant Ebooks, 2004.

—(1929). Ueber den Gegenwertigen Stand der Feld-Teorie, AEA 4–38.

—(1953). Scientific Papers Presented to Max Born. Oliver Boyd, Edinburgh, pp. 33-40.

—(1954). The World As I See It. In Einstein 1949 & Einstein 1954… Meno, 776–786, trans. Benjamin Jowett in Plato's Meno: Text and Criticism, ed. Alexander Sesonske and Noel Flemins (Belmont Calif: Wadsworth. 1965), pp. 12–13.

—(1973). *Ideas and Opinions*, London: Souvenir Press; first published 1954.

Einstein, A., & de Broglie, L. (1953). Physicien et Penseur. (A. George, ed.), Editions Albin Michel, Paris.

Eliot, T. S. (1921). The Sacred Wood. Tradition and the Individual Talent.

Elwes, R. (2007, July). From e to Eternity. *New Scientist*, p. 38.

Feynman, R. P. (1967). The Character of Physical Law. Cambridge, MA: MIT Press.

—(1988). *QED, The Strange Theory of Light and Matter*. Princeton University Press.

Feynman, R. P., Leighton, R. B., & Sands, M. (1963). The Feynman Lectures on Physics, I. New York: Addison-Wesley.

Filk, T. (2001). Proper Time and Minkowski Structure on Causal Graphs. Class. Quant. Grav. 18, 2785, gr-qc/0102088.

Finkelstein, D. (1986). Hyperspin and Hyperspace. Phys. Rev. Lett. 56, 1532-1533.

Flaubert, G. (1857). Madame Bovary.

Folger, T. (2007, February). The Big Bang Machine. *Discover*, pp. 32–38.

—(2007, June). In No Time. *Discover*, p. 78.

Ford, J. (1989). What is chaos, the we should be mindful of it? *The New Physics*, Davies (ed.), Cambridge: Cambridge University Press, 348-372.

Frank, A. (2008, April). The Day Before Genesis. *Discover*, pp. 54-60.

Furuta, A. (2012). One Thing Is Certain: Heisenberg's Uncertainty Principle Is Not Dead. *Scientific American*.

Gale, R. (1967). *The Philosophy of Time: a Collection of Essays*. Garden City, NY: Doubleday and Company.

Gefter, A. (2005, October 15). The riddle of time. *NewScientist*, p. 30.

—(2007, March 10). Don't mention the F word. *NewScientist*, p. 33.

—(2007, July 14). Lines of attack. *NewScientist*, pp. 30-35.

—(2008, May 3). Which Way Now? *New Scientist*, pp. 28-31.

Gell-Mann, M., & Hartle, J. B. (1990). Complexity, Entropy, and the Physics of Information (W. Zurek, ed.), Addison-Wesley, Reading, pp. 425-485.

Gisin, N. (1991). Propensities in a Non-Deterministic Physics: *Synthese*, 89; 287-297

Godfrey-Smith, W. (1979). Special Relativity and the Present. *Philosophical Studies* **36**, pp. 233-244.

Goldstein, S. (2001). Bohmian Mechanics. Stanford Encyclopedia Of Philosophy, First published Fri Oct 26, 2001; substantive revision Fri May 19, 2006.

Granek, G. (2001). Einstein's Ether: Why did Einstein Come Back to the Ether? *Aperion* **8** (3).

Greene, B. (2003). *The Elegant Universe: Superstrings, Hidden Dimensions, and the Quest for the Ultimate Theory*. New York: Vintage Books.

—(2004). *The Fabric of the Cosmos: Space, Time and the Texture of Reality*. New York: Knopf.

Groleau, R. (2003, July). *Imagining Other Dimensions*. WGBH. http://www.pbs.org/wgbh/nova/elegant/dimensions.html.

Grøn, Ø., & Hervik, S. (2007). *Einstein's general theory of relativity: With modern applications in cosmology*, Springer.

Grünbaum, A. (1967). *Modern Science and Zeno's Paradoxes*, Middletown: Connecticut Wesleyan University Press.

—(1971). The Meaning of Time. *Basic Issues in the Philosophy of Time*, Freeman, E. and W. Sellars (eds.), pp. 195-228. La Salle, IL: Open Court.

Hauser, M. D. (2006). Swappable Minds, The Next Fifty Years. Edited by John Brockman, pp. 53–54. See also Frans de Wall's recent book, *Primates and Philosophers*.

Hawking, S. (1988). *A Brief History of Time*. The updated and expanded tenth anniversary edition. Bantam Books.

—(2005). *God Created the Integers*. Running Press.

Hawley, K. (2001). *How Things Persist*. Oxford University Press.

Heisenberg, W. (1926). Mehrkorperprobleme und Resonanz in der Quantenmechanik. *Zeitschrift fur Physik*, **38**, 411-426.

—(1932). Uber den Bau der Atomkerne. I. *Zeitschrift fur Physik*, **77**, 1-11.

—(1958). Physics and Philosophy. Harper and Row, New York, p. 97, 129.

—(1990). The Physical Principles of the Quantum Theory, Heller, Mark, The Ontology of Physical Objects: Four Dimensional Hunks of Matter. Cambridge University Press.

Hitchcock, C. (1999). Contrastive Explanation and the Demons of Determinism. *British Journal of the Philosophy of Science*. 50; 585-612.

Hobbes, T. (1971). Leviathan. Edited by C. B. Macpherson. Baltimore: Penguin Books.

Hoefer, C. (1996). The Metaphysics of Spacetime Substantivalism. *The Journal of Philosophy*, 93: 5027.

Hoefer, C. (2002). Freedom From the Inside Out. *Time, Reality and Experience*, C. Callender (ed.), Cambridge: Cambridge University Press.

Hofstadter, D. (1979). *Gödel, Escher, Bach: An Eternal Golden Braid*. Basic Books.

—(2007). *I Am A Strange Loop*. Basic Books.

Holland, P. R. (1993). *The Quantum theory of Motion*. Cambridge University Press, Cambridge.

Hughston, L. P. (1979). Some new contour integral formulae, in Complex Manifold Techniques in Theoretical Physics. D. Lerner & P. D. Sommers, eds., Pitman.

Hume, D. (1947). *Dialogues Concerning Human Religion*. ed. Norman Kemp-Smith, Edinburgh: Nelson, Bk II § 1.

Hurd, T. R. (1985). The projective geometry of simple cosmological models. Proc. R. Soc. London A 397, 233-243.

Hutchison, K. (1993). Is Classical Mechanics Really Time-reversible and Deterministic? *British Journal of the Philosophy of Science*, 44: 307-323.

Isham, C. J. (1989, August). An Introduction to General Topology and Quantum Topology. SummerInst. On Physics, Geometry and Topology, Banff, Imperial/TP/88-89/30.

—(2002). Some Reflections on the Structure of Conventional Quantum Theory when applied to Quantum Gravity. quant-ph/0206090.

Isaacson, W. (2007). *Einstein: His Life and Universe*. Simon & Schuster.

Jacobson, T. A., & Parentani, R. (2005, December). An Echo of Black Holes. *Scientific American*.

James, W. (1911). *Some Problems of Philosophy*, New York: Longmans, Green & Co.

Jaspers, K. (1951). *The Way to Wisdom*. Translated by Ralph Manheim. New Haven, Conn.: Yale University Press.

Kaku, M. (1995). *A Scientific Odyssey Through Parallel Universes, Time Warps, and the 10th Dimension*. New York: Anchor Books.

—(2008, April 5). Never Say Never. *New Scientist*. pp. 36-39.

Kane, G. (2005, July). The Mysteries of Mass. *Scientific American* pp. 40-48.

Kant, I. (1963). *The Critique of Pure Reason*, translated by Norman Kemp Smith. Macmillan, esp. pp. 75ff.

Kauffman, S., & Smolin, L. (2000). Combinatorial Dynamics in Quantum Gravity; Towards Quantum Gravity. Proc. Polanica, Ed. J. Kowalski-Glikman, Springer Lect. W. Phys., Springer, N.Y.

Kaufman, M. (2006, August 22). Cosmic puzzle solved?—Dark matter: Scientists claim they have proof it exists, but there are skeptics. *The Washington Post*, reported in *The Denver Post*, p. 1A, 11A.

Keats, J. (1820). Ode on a Grecian Urn.

King, C. (1978, November 30). The gold of El Dorado. *New Scientist*.

Klesius, M. (2004, February). The Mystery of Snowflakes. *National Geographic*.

Koch, R., & Smith, C. (2006, June 26). The Fall of Reason. *New Scientist*, p. 25.

Kochen, S., & Specker, E. P. (1967). The problem of hidden variables in quantum mechanics. *J. Math. Mech.* **17**, 59–87.

Kolak, D., & Martin, R. (1990). The Experience of Philosophy. Belmont, Calif,: Wadsworth, p. 2.

Kosso, P. (2000). The empirical status of symmetries in physics. *British Journal for the Philosophy of Science*, **51**, 81-98.

Kostro, L. (2000). *Einstein and the Ether*. Aperion.

Krauss, L. (2000). quintessence—The Mystery of Missing Mass in the Universe. Basic Books.

Kunzig, R. (2004, September). Testing the Limits of Einstein's Theories. *Discover*.

Laloë, F. (2012). Do We Really Understand Quantum Mechanics? Cambridge University Press.

Laplace, P. (1820). *Essai Philosophique sur les Probabilités* forming the introduction to his *Theorié Analytique des Probabilités*, Paris: V Courcier; repr. F. W. Truscott and F. L. Emory (trans.), A Philosophical Essay on Probabilities, New York: Dover, 1951.

Laurence, W. (1939, March 14). Einstein Sees Key to Universe Near. *New York Times*.

Leavens, C. R. (1996). The 'Tunnelling-Time Problem' for Electrons. Cushing et al.

Leggett, A. J. (1987). *The Problems of Physics*, Oxford University Press.

Leiber, T. (1998). On the Actual Impact of Deterministic Chaos. *Synthese*, 113: 357-379.

Leibniz, G. W., & Clarke, S. (1956) The Leibniz-Clarke Correspondence. Manchester University Press, 3rd paper, §4; Olaf Dryer 'Relational Physics and Quantum Space, arXiv:gr–qc/0404054v1, April 13, 2004.

Levenson, T. (2008, March). Albert the Icon. *Discover*, pp. 44–49.

—(2004, September). Einstein's Gift for Simplicity. *Discover*, p. 45.

Lewis, D. (1986). The Paradoxes of Time Travel. Lewis, David, *Philosophical Papers*, Volume 2. Oxford University Press.

Lightman, A. (2006). A Sense Of The Mysterious: Science and the Human Spirit.

Lisi, A. G. (2007, November 6). An Exceptionally Simple Theory of Everything. arXiv:0711.0770v1 [hep-th].

Liu, C. (2002). The meaning of spontaneous symmetry breaking: From a simple classical model. PSA, *Proceedings of the Biennial Meeting of the Philosophy of Science Association*.

Loewer, B. (2004). Determinism and Chance. *Studies in History and Philosophy of Modern Physics*, 32: 609-620.

Lorentz, A., Einstein, A., Minkowski, H., & Weyle, H. (1908). Space and Time. An address to the 80th Assembly of German Natural Scientists and Physicians, Cologne, Germany, September 21, 1908. Reprinted in *The Principles of Relativity*, New York, Dover, 1952.

Mackenzie, D. (2008, June 14). Don't blame it on the gods. *NewScientist*, pp. 50–51.

Maddox, B. (2008, May). Three words that could overthrow physics: "What is magnetism?"' *Discover*, p. 70.

Mainzer, K. (1996). *Symmetries of nature*, Berlin: Water de Gruyter.

Manuel, F. E. (1968). *A Portrait of Isaac Newton*. Cambridge, Massachusetts: Harvard University Press.

Marcel, G., (1991). Translated in *Existentialism And The Philosophical Tradition* by Raymond, D. B., p. 337.

Markopoulou, F., & Smolin, Lee. (1997). Causal Evolution of Spin Networks. Nucl. Phys. B508 409, gr-qc/9702025.

—(1997). Dual formulation of spin network evolution. gr-qc/9704013.

—(2000). An Algebraic Approach to Coarse Graining. hep-th/0006199.

Markosian, N. (1993). How Fast Does Time Pass? *Philosophy and Phenomenological Research* **53**, pp. 829-844.

Martins, A. A. (2012). Fluidic Electrodynamics: On parallels between electromagnetic and fluidic inertia. arXiv:1202.4611 [physics.flu-dyn].

Matthews, W. (1999). World Religions, 3rd ed. Belmont, California: Wadsworth.

Maudlin, T. (1994). *Quantum Non-Locality and Relativity*, Second Edition, Blackwell Publishing.

—(1995). Why Bohm's Theory Solves the Measurement Problem. Philosophy of Science, Vol. 62, No. 3, pp. 479-483.

—(2007). *The Metaphysics Within Physics*. Oxford University Press.

Maxwell, N. (1985). Are Probabilsim and Special Relativity Incompatible? *Philosophy of Science* **52**, pp. 23-43.

McCall, S. (1994). *A Model of the Universe*. Clarendon Press.

McTaggart, J. M. E. (1993). The Unreality of Time. Le Poidevin, Robin, and McBeath, Murray (eds.), *The Philosophy of Time* (Oxford University Press, pp. 23-34.

—(1908). The Unreality of Time. *Mind*, New Series 68: 457-484.

Mellor, D.H. (1998). *Real Time II*. Routledge.

—(1995). *The Facts of Causation*. London: Rutledge.

Merali, Z. (2006, July 22). Bubble ousts black hole at center of the galaxy. *NewScientist*, p. 11.

—(2007, March 17). The Universe is a String-Net Liquid. *New Scientist*, pp. 8-9.

—(2007, November 17). Is this the theory of everything? *NewScientist*, pp. 8-10.

—(2008, March 8). Has dark fluid saved Earth from oblivion? *NewScientist*, pp. 10-11.

Mermin, N.D. (1968). *Space and Time in Special Relativity*. Prospect Heights, IL: Waveland Press, Inc.

—(1993). Hidden Variables and the Two Theorems of John Bell. *Rev. Mod. Phys.* 65: 803-815.

Michelson, A. A. & Morley, E. W. (1887). On the Relative Motion of the Earth and the Luminiferous Ether. American Journal of Science 34: 333–345; P. A. M. Dirac, Nature 168, 906 (1951); P. A. M. Dirac, Nature 169, 702 (1952).

Miller, P. (2007, July). Swarm Theory. *National Geographic*.

Minkowski, H. (1908). Space and Time. Reprinted and translated in *The Principle of Relativity*, pp 73-91. New York City: Dover Publications, 1952.

Morford, M. (2008, March 5). Survey says we're losing our religion. Let's hope so. San Francisco Chronicle, p. E5.

Muir, H. (2006, August 5). Supernovae make dark matter bloat. *NewScientist*, p. 12.

Muir, H. (2006, June 17). The cosmic controller. *NewScientist*, pp. 46-49.

Mullins, J., (2006, May 13). The stuff of beams. *NewScientist*, pp. 44-47.

Neal, V., Lewis, C. S., & Winter, F. H. (1995). Spaceflight—A Smithsonian Guide. pp. 98–99.

Neil, A. (2002). Relativity and the Global Positioning System. Physics Today 55(5), pp. 41–47.

—(2003). Relativity in the Global Positioning System. Living Reviews in Relativity 6: http://relativity.livingreviews.org/Articels /Irr-2003-1/index.html.

Newton, I. (1934). *Principia, Scholium on Absolute Space and Time* Florian Cajori, trans., Berkeley: University of California Press; reprinted in *The Scientific Background to Modern Philosophy*, Edited by

Michael R. Matthews, Hackett Publishing Company Indianapolis/Cambridge, 1989, pp. 139–146.

Newton-Smith, W.H. (1980). *The Structure of Time*. Routledge & Kegan Paul.

Nietzsche, F. (1882). The Gay Science.

—(1901). Will to Power.

—(1963). Human, All-Too-Human, in *The Portable Nietzsche*, trans. Walter Kaufmann (New York: Penguin, 1968), section 483.

—(2005). *Thus Spoke Zarathustra*. Translated by Clancy Martin. New York: Barnes & Noble Classics.

Norton, J. (1993). General covariance and the foundations of general relativity: eight decades of dispute. *Rep. Prog. Phys.*, **56**, 791-858.

Nounou, A. (2003). A fourth way to the Aharonov-Bohm effect, in K. Brading and E. Castellani (eds.), *Symmetries in Physics: Philosophical Reflections*, Cambridge: Cambridge University Press, pp. 174-200.

Novello, M., Visser, M., & Volovik, G. (2002). Artificial Black Holes, World Scientific, River Edge, USA, p. 391.

Obhi, S. S., & Haggard, P. (2004, July-August). Free Will and Free Won't, Motor activity in the brain precedes our awareness of the intention to move, so how is it that we perceive control? *American Scientist*, Volume 92.

Odegard, D. (1978). Phenomenal Time. Ratio 20, 116-22.

Ohanian, H. C., & Ruffini, R. (1994). Gravitation and Spacetime. W.W. Norton and Company, table 4.1, p. 186.

Ornes, S. (2007). Calculus Was Developed In Medieval India. *Discover*, 100 top discoveries of 2007, p. 52.

Ornstein, D. S. (1974). *Ergodic Theory, Randomness, and Dynamical Systems*. New Haven: Yale University Press.

Overbye, D. (2007 January 2). Free Will: Now You Have It, Now You Don't. *The New York Times*, p. D4.

Ozawa, M. (2003). Universally valid reformulation of the Heisenberg uncertainty principle on noise and disturbance in measurement. *Physical Review A* **67** (4), arXiv:quant-ph/0207121

Pais, A. (1982). Subtle is the Lord. Oxford University Press, New York.

Panek, R. (2008, March). The E Factor. *Discover*.

Penrose, R. (1968). Structure of space-time. In Battelle Rencontres, C. M. DeWitt and J. A Wheeler, eds., New York : W. A. Benjamin.

—(1985). Quantum Concepts in Space and Time. R. Penrose and C.J. Isham, eds., Oxford University Press, Oxford.

—(1989). *The Emperor's New Mind*. Oxford University Press, New York & Oxford.

—(1995). Twistors for cosmological models. In *Further advances in twistor theory*, vol II : Integrable systems, conformal geometry and gravitation, L. J. Mason, L. P. Hughston & P. Z. Kobak, eds., Harlow: Longman.

—(2004). *The Road to Reality: A Complete Guide To The Laws Of The Universe*. Alfred A Knopf.

Peterson, I. (1988). The Mathematical Tourist.

Pincock, S. (2006, July 15). Back to the Future. *Financial Times*.

Plato. (1961). Theatetus. 174 A, translated by F. M. Cornford, in *The Collected Dialogues of Plato: Including the Letters*. New York: Pantheon, p. 879.

Pooley, O. (2003). Handedness, parity violation, and the reality of space. In K. Brading and E. Castellani (eds.), *Symmetries in Physics: Philosophical Reflections*, Cambridge: Cambridge University Press, pp. 250-280.

Poppel, E. (1978). Time Perception. In Richard Held et al., eds., *Handbook of Sensory Physiology*, Vol. VIII: Perception, Berlin: Springer-Verlag.

Powell, C. S. (2006, October). My Three Einsteins. *Discover*.

Price, H. (1996). Time's Arrow and Archimedes' Point: New Directions for the Physics of Time. Oxford University Press.

Prior, A. N. (1970). The Notion of the Present. *Stadium Generale* 23, pp. 245-248.

—(1996). Some Free Thinking About Time. In Copeland, Jack, (ed.) *Logic and Reality: Essays on the Legacy of Arthur Prior*. Clarendon Press, pp. 47-51.

Putnam, H. (1967). Time and Physical Geometry. *Journal of Philosophy* **64**, pp. 240-247.

Randall, L. (2006). *Warped Passages: Unraveling the Mysteries of the Universe's Hidden Dimensions*. Harper Perennial.

Raptis, I., & Zapatrin, R. (2001). Algebraic Description of Space-Time Foam. Class. Quant. Gr. 20, 4187, gr-qc/0102048.

Raymond, D. B. (1991). Existentialism and the Philosophical Tradition.

Requardt, M. (2002). The Translocal Depth-Structure of Space-Time. hep-th/0205168.

—(2003). A Geometric Renormalisation Group in Discrete Quantum Space-Time. arXiv:gr-qc/0110077v3.

Rideaut, D. P., & Sorkin, R. (2000). A Classical Sequential Growth Dynamics for Causal Sets. Phys. Rev. D61, 024002, gr-qc9904062.

—(2000). Evidence for a Continuum Limit in Causal Set dynamics. gr-qc/0003117.

Riordan, M., & Zajc, W. A. (2006, May). The first few Microseconds. *Scientific American*, pp. 34-41.

Roberts, T. (2015). Fluidic Origins of the Magnetic and Electric Fields: A physical interpretation of B and E. Academia.edu. https://www.academia.edu/12637409/Fluidic_Origins_of_the_Magnetic_and_Electric_Fields_A_physical_interpretation_of_B_and_E

Roman, T. (1970). *Hebrew Thought Compared with Greek*. New York: W. W. Norton & Company.

Rosenfeld, L. (1969). Newton's views on Aether and Gravitation. Archive for History of Exact Sciences. 6.1: 29-37. Web. June 4, 2013: Newton, Isaac. Isaac Newton to Robert Boyle, February 28, 1679.

Rozema, L. A., Darabi, A., Mahler, D. H., Hayat, A., Soudagar, Y., & Steinberg, A. M. (2012). Violation of Heisenberg's Measurement—Disturbance Relationship by Weak Measurements. *Physical Review Letters* **109** (10).

Ruelle, D. (1991). Chance and Chaos, London: Penguin.

Russell, B. (1912). On the Notion of Cause. *Proceedings of the Aristotelian Society*, 13: 1-26

—(1915). On the Experience of Time. *Monist* 25, 212-33.

—(1921). *The Analysis of Mind*, London: George Allen and Unwin.

—(1929). *Our Knowledge of the External World*, New York: W. W. Norton & Co. Inc.

—(1957). Why I Am Not a Christian.

Sakharov, A. D. (1968). Vacuum quantum fluctuations in curved space and the theory of gravitation. *Sov. Phys. Dokl.*, **12**. 1040–1041.

Savitt, S. (1995). Time's Arrows Today: Recent Physical and Philosophical Work on the Direction of Time. Cambridge University Press.

Shawhan, P. S. (2004, July-August). Gravitational Waves and the Effort to Detect Them. *American Scientist*, Volume 92, p. 351.

Schilpp, P. A. (1949). Albert Einstein, Philosopher-Scientist. Library of Living Philosophers, Evanston, Ill., pp. 666, 672.

Schneider, K. (2005, November 12). Dream machine. *NewScientist*, pp. 52-5.

Schopenhauer, A. (2001). On Ethics. In Parerga and Paralipomena: Short Philosophical Essays. New York: Oxford University Press, 2:227.

Schrödinger, E. (1935). Die gegenwartige Situation in der Quantenmechanik. Naturwissenschaften 23: 807-812, 823-828, 844-849; English translation by Trimmer, J. D., 1980. The Present Situation in Quantum Mechanics: A Translation of Schrödinger's 'Cat Paradox' Paper. Proceedings of the American Philosophical Society, 124: 323-338. Reprinted in Wheeler & Zurek 1983.

—1935, 23: 807–812, 823–828, 844–849.

Schultz, H. (2005, May). Nobel Efforts. *National Geographic*.

Shanks, N. (1991). Probabilistic physics and the metaphysics of time. *South African Journal of Philosophy*, 10: 37-44.

Shiga, D. (2005, October). Something for Nothing. *New Scientist*: 34-37.

—(2006, April 29). The long arms of the law: did Newton and Einstein get gravity all wrong? *NewScientist*, pp. 52-55.

Shoemaker, S. (1969). Time Without Change. *Journal of Philosophy* **66**, pp. 363-381.

Sinai, Y. G. (1970). Dynamical systems with elastic reflections. *Russ. Math. Surveys* 25: 137-189.

Sklar, L. (1974). *Space, Time and Spacetime*. University of California Press.

Smart, J. J. C. (1949). The River of Time. *Mind* 58, pp. 483-494 (reprinted in Flew, Antony (ed.), *Essays in Conceptual Analysis* (St. Martin's Press, 1966), pp. 213-227).

—(1955). Spatialising Time. *Mind* 64, pp. 239-241.

—(1963). *Philosophy and Scientific Realism*. Routledge & Kegan Paul.

Smith, Q. (1993). *Language and Time*. Oxford University Press.

Smolin, L. (2004, September). Einstein's Lonely Path. *Discover*, p. 40.

—(2006). Never Say Always. *New Scientist*.

Soccio, D. J. (2004). Archetypes Of Wisdom: Introduction to Philosophy, 5th ed., Belmont, Calif.: Wasdwordh.

Soon, C., Brass, M., Heinze, H., & Haynes, J. (2008). Unconscious determinants of free decisions in the human brain. *Nature neuroscience* **11** (5): 543–545. Doi:10.1038/nn.2112.

Sorkin, R. (1995). A Specimen of Theory Construction from Quantum Gravity. Appeared in *The Creation of Ideas in Physics*, ed. J. Leplin, Kluwer, Dordrecht 1995, gr-qc/9511063.

—(1991). A Finite Substitute for Continuous Topology. Int. J. Theor. Phys. 30, 923.

Spinoza, B. (1677). Ethics, part 2, proposition 4B.

Stein, H. (1970). On Einstein-Minkowski Space-Time. *Journal of Philosophy* **67**, pp. 289-294.

Stewart, I. (2006, March). Ride the Celestial Subway. *New Scientist*.

Stewart, I. & Golubitsky, M. (1992). *Fearful symmetry. Is God a geometer?* Oxford: Blackwell.

Stix, G. (2006). Real Time. *Scientific American*, Special Edition,—A Matter of Time.

Stone, A. (2007, June). *Fearful Symmetry—Mathematicians Triumph over 248 Dimensions, Discover*, p. 18; The Taming of the Symmetries, *New Scientist*, March 24, 2007.

—(2007, June). The secret Life of Atoms—Until Recently We Couldn't Even See Them. *Discover*, p. 52.

Strogatz, S. (2002). Fermi's Little Discovery and the Future of Chaos and Complexity Theory—article within *The Next Fifty Years*, p. 124.

Struyve, W., & Valentini, A. (2008, December 27). De Droglie-Bohm Guidance Equations for Arbitrary Hamiltonians. arXiv:0808.0290v3 [quant-ph].

Struyve, W., & Westman, H. (2007). A minimalist pilot-wave model for quantum electrodynamics. Proceedings of the Royal Society A, vol 463, p. 3115.

Suppes. P. (1993). The Transcendental Character of Determinism. *Midwest Studies in Philosophy*, 18: 242-257.

Swarup, A. (2008, March 8). Acid test for alternative to dark matter. *NewScientist*, p. 11.

Swinburne, R. (1966). The Beginning of the Universe. *Proceedings of the Aristotelian Society*, Supplementary Volume **50**, pp. 125-138.

Talbot, M. (1991). *The Holographic Universe*.

Taylor, R. (1992). *Metaphysics*, 4th Edition. Prentice-Hall.

Teller, P. (2000). The gauge argument. *Philosophy of Science*, **67**, S466-S481.

't Hootf, G. (1980, June). Gauge theories and the forces between elementary particles. *Scientific American*, **242**, 90-166.

—(1988). Deterministic and Quantum Mechanical Systems. J. Stat. Phys 53, 323.

—(1990). Quantization of Discrete Deterministic Theories. Nucl. Phys. B342, 471.

—(1999). Quantum Gravity and Dissipative Deterministic Systems. gr-qc/9903084.

Theise, N. (2006, Summer). From the Bottom Up. *Tricycle*, p. 24.

Thorne, K. S. (1995). *Black Holes and Time Warps: Einstein's Outrageous Legacy*. New York: Norton.

Tolstoy, L. (1882). *A Confession*, translated by David Patterson, Chapter IX.

Tooley, M. (1997). *Time, Tense, and Causation*. Oxford: Oxford University Press.

Trefil, J. (2006, July). Where is the universe heading? *Astronomy*, pp. 36-43.

Valentini, A. (1991). Phys. Lett. A 156, 5.

—(2007). Astrophysical and cosmological tests of quantum theory. Journal of Physics A: Mathematical and theoretical, vol. 40, p. 3285.

—(1992). On the Pilot-Wave theory of Classical, Quantum and Subquantum Physics. Ph.D. thesis, International School for Advanced Studies, p. 36.

Valentini, A., & Westman, H. (2005). Dynamical origin of quantum probabilities. Proc. R. Soc. A, 461, 253-272 doi:10.1098/rspa.2004.1394.

van Biema, D. (2006, November 3). God vs. Science. *Time Magazine*.

van Fraasen, B. (1989). *Laws and Symmetry*. Oxford: Claredon Press.

Van Inwagen, P. (1990). Four-Dimensional Objects. *Nous* **24**, pp. 245-255.

Van Kampen, N. G. (1991). Determinism and Predictability. *Synthese*, 89: 273-281.

Visser, M. (1996). *Lorentzian Wormholes: From Einstein to Hawking*. New York: American Institute of Physics Press.

—(2002). Sakharov's induced gravity: A modern perspective. *Mod. Phys. Lett. A*, **17**, 977–992. Related online version (cited on 31 May 2005): http://arXiv.org/abs/gr-qc/0204062.

von Neumann, J. (1932). Mathematische Grundlagen der Quantenmechanik, Berlin: Springer Verlag; English translation by Beyer, R. T., 1955, Mathematical Foundations of Quantum Mechanics, Princeton: Princeton University Press.

Volovik, G. E. (2003). *The Universe in a helium droplet*. Int. Ser. Monogr. Phys. **117**, 1–507.

Wald, R. M. (2000, September 30). The Thermodynamics of Black Holes. arXiv:gr-qc/9912119v2.

Walsh, W. H. (1967). Kant on the Perception of Time. Monist 51, 376-96.

Weatherall, J. O. (2008, May). The Tabletop Universe. *Popular Science*, pp. 72–76.

Weinberg, S. (1989). Phys. Rev. Lett. 62, 485.

Weingard, R. (1972). Relativity and the Reality of Past and Future Events. *British Journal for the Philosophy of Science* **23**, pp. 119-121.

Westfall, R. S. (1980). *Never at Rest: A Biography of Isaac Newton*. Cambridge University Press.

Whitehead, A. N. (1929). *Process and Reality*, New York: The Macmillan Co.

Wigner, E. P. (1967). *Symmetries and reflections*, Bloomington, Indiana: Indiana University Press.

Wigner, E. P. (1976). Interpretation of Quantum Mechanics. In Wheeler and Zurek 1983.

Wilczek, F. (2011, December 29). Beautiful Losers: Kelvin's Vortex Atoms. NOVA: http://www.pbs.org/wgbh/nova/physics/blog/2011/12/beautiful-losers-kelvins-vortex-atoms/

Williams, D. C. (1951). The Myth of Passage. *Journal of Philosophy* **48**, pp 457-472.

Wilson, E. O. (1998). Consilience: The Unity of Knowledge.

Winnie, J. A. (1996). Deterministic Chaos and the Nature of Chance. In *The Cosmos of Science—Essays of Exploration*, Earman, J. and Norton, J. (eds), Pittsburgh: University of Pittsburgh Press, pp. 299-324.

Wright, K. (2005, July). Catch Me IF You Can. *Discover*, pp. 45-47.

Xia, Z. (1992). The existence of noncollision singularities in newtonian systems. *Annals of Mathematics*, 135: 411-468.

Yourgrau, P. (1999). Godel Meets Einstein: Time Travel in the Godel Universe. Open Court.

Zloshchastiev, K. G. (2011). *Spontaneous symmetry breaking and mass generation as built-in phenomena in logarithmic nonlinear quantum theory*, Acta Phys. Polon. B **42**, pp. 261–292, ArXiv:0912.4139.

Zukav, G. (1980). *Dancing Wu Li Masters—An Overview of the New Physics*. Harper Collins.

Zwart, P. J. (1976). *About Time*. North-Holland Publishing Co.

Appendix A — Approaches to Quantum Gravity

The standard loop-variable approach (also known as loop quantum gravity), which is being developed by Lee Smolin, Fotini Markopoulou and others at the Perimeter Institute in Warterloo, Canada, is an example of a theory that is founded on a completely discrete character. But the metric of this framework is still dependent upon a continuous volume of spacetime in which its 'spin networks' are taken to be embedded. These spin networks mathematically represent the intended quantized function, but as of yet they have not been able to generate the fabric of spacetime at large scales. Consequently, this approach doesn't actually provide us with a discrete framework capable of revealing Nature's workings at the tiniest scales. Technically, such an outcome would be an unreasonable expectation of loop quantum gravity. The theory incorporates the mathematics of quantization in an attempt to provide a viable quantum theory of gravitation with a well-defined classical limit in agreement with general relativity. It does not, however, attempt to incorporate the other known fundamental forces into that formalization.

Other models, like the one considered by 't Hooft, Schild, and Snyder, have toyed with the possibility of a periodic lattice spacetime structure—making spacetime crystalline. As an extension of that model Raphael Sorkin suggested a more fluid lattice in his formulation, which results in a causal connection between points. This method, which shares similarities to quantum space theory, enables Lorentz invariance and therefore special relativity to emerge in the set.

In 1965 Ahmavaara suggested that physical reality might actually be a finite field and that the real number system, used fundamentally in mathematics and conventional physics, should be replaced by a discrete number system to accurately map its parameters. Other quantized suggestions appear in quantum set theory, quaternion geometry (pioneered by David Finklestein) and octonionic physics put forth by Corinne Manogue and Tevian Dray.

Stephen Dray of the Institute for Advanced Studies in Princeton, New Jersey, has been working for more than two decades on a structure he calls "emergent quantum theory." This theory depicts pre-quantum fields as being a mixture of both ordinary algebraic properties and noncommutating properties from which a very interesting foundation emerges that may just prove to provide an alternative basis for all of quantum theory.

Renate Loll and her colleagues Jan Ambjorn and Jerzy Jurkiewicz of Utrecht University in the Netherlands have recently developed another formalism known as causal dynamical triangulations (CDT). CDT models spacetime as a composite of tiny higher-dimensional analogues of triangles called 4-simplices that perpetually rearrange themselves. CDT has many strengths. For example, it produces the familiar four dimensions of spacetime on large scales. It also portrays spacetime as a fabric that gains increasingly detailed structure on smaller scales—like a fractal. The limiting factor of CDT may turn out to be that its authors and supporters do not think of these spacetime building blocks as physical entities. Instead, they consider them mathematical and conceptual tools only.

Martin Reuter of the University of Mainz in Germany is developing another formalism called quantum Einstein gravity. The idea behind his formalism is that there is a minimum cut-off scale for the effects of gravity. Technically he is attempting to quantize gravity instead of quantizing spacetime. This approach leads to results that are quite similar to CDT, but it also shares in the conceptual limitations that all of these formalisms are stunted by.

More radical formalisms, known as pre-geometric formalisms go even further and claim that what we perceive as space and time really emerge from pre-geometric interactions. Quantum graphity, developed by Fotini Markopoulou and her colleagues, and internal relativity developed by Olaf Dreyer at MIT are two examples. (Roger Penrose, The Road to Reality. Amanda Gefter, "Which Way Now?," New Scientist, May 3, 2008: 28-31.)

To examine some of the various approaches to quantum gravity and quantum spacetime physics (including Garrett Lisi's *E8 Unification theory*, Alexandre Martins' *Fluidic Electrodynamics*, *Induced gravity*, championed by A. D. Sakharov, D. Berenstein, and the related *Analogue Gravity*, developed primarily by Carlos Barcelo, Stefano Liberati, and Matt Visser, see the references below:

Antonsen, F. (1994). Random Graphs as a Model for Pregeometry. Int. J. Theor. Phys. 33, 1189.

Baez, J. (1998). Spin Foam Models. Class. Quant. Grav. 15, 1827, gr-qc/9719052.

—(2000). Spin Foam Models of BF Theory and Quantum Gravity. Lect. N. Physics. 543, Springer N.Y. gr-qc/9905087.

Barceló, C., Liberati, S. & Visser, M. (2005). Analogue Gravity. Living Rev. Relativity, **8**, 12. [Online Article]: cited [<11-26-2012>], p. 76, http://www.livingreviews.org/lrr-2005-12

Barrett, J. W. & Crane, L. (1999). A Lorentzian Signature Model for Quantum General Relativity. gr-qc/9904025.

Bombelli, L., Lee, J., Meyer, D., & Sorkin, R. (1987). Space-Time as a Causal Set. Phys. Rev. Lett. 59, 521.

Brightwell, G. & Gregory, R. (1991). The Structure of Random Discrete Spacetime. Phys. Rev. Lett. 66, 260.

Brody, D. C. & Hughston, L. P. (2004). Theory of Quantum Space-Time. arXiv:gr-qc/0406121v1.

Crane, L. (2000). Hypergravity and Categorical Feynmanology. gr-qc/0004043.

Filk, T. (2001). Proper Time and Minkowski Structure on Causal Graphs. Class. Quant. Grav. 18, 2785, gr-qc/0102088.

Isham, C. J. (1989). An Introduction to General Topology and Quantum Topology. SummerInst. On Physics, Geometry and Topology, Banff, August 1989, Imperial/TP/88-89/30.

—(2002). Some Reflections on the Structure of Conventional Quantum Theory when applied to Quantum Gravity. quant-ph/0206090.

Kauffman, S. & Smolin, L. (2000). Combinatorial Dynamics in Quantum Gravity. Towards Quantum Gravity. Proc. Polanica, Ed. J. Kowalski-Glikman, Springer Lect. W. Phys., Springer, N.Y.

Lisi, A. G. (2007). An Exceptionally Simple Theory of Everything. arXiv:0711.0770v1 [hep-th].

Markopoulou, F. & Smolin, L. (1997). Causal Evolution of Spin Networks. Nucl. Phys. B508, 409, gr-qc/9702025.

Markopoulou, F. (1997). Dual Evolution of Spin Networks. gr-qc/9704013.

—(2000). An Algebraic Approach to Coarse Graining. hep-th/0006199.

Martins, A. A. (2012). Fluidic Electrodynamics: On parallels between electromagnetic and fluidic inertia. arXiv:1202.4611 [physics.flu-dyn].

Raptis, I. & Zapatrin, R. (2001). Algebraic Description of Space-Time Foam. Class. Quant. Gr. 20, 4187, gr-qc/0102048.

Requardt, M. (2002). The Translocal Depth-Structure of Space-Time. hep-th/0205168.

—(2003). A Geometric Renormalisation Group in Discrete Quantum Space-Time asXiv:gr-qc/0110077v3.

Rideaut, D. P. & Sorkin, R. (2000). A Classical Sequential Growth Dynamics for Causal Sets. Phys. Rev. D61, 024002, gr-qc9904062.

—(2000). Evidence for a Continuum Limit in Causal Set dynamics. gr-qc/0003117.

Sakharov, A.D. (1968). Vacuum quantum fluctuations in curved space and the theory of gravitation. *Sov. Phys. Dokl.*, **12**. 1040–1041.

Sorkin, R. (1991). A Finite Substitute for Continuous Topology. Int. J. Theor. Phys. 30, 923.

—(1995). A Specimen of Theory Construction from Quantum Gravity. Appeared in *The Creation of Ideas in Physics*, ed. J. Leplin, Kluwer, Dordrecht 1995, gr-qc/9511063.

't Hooft, G. (1988). Deterministic and Quantum Mechanical Systems. J. Stat. Phys 53, 323.

—(1990). Quantization of Discrete Deterministic Theories. Nucl. Phys. B342, 471.

—(1999). Quantum Gravity and Dissipative Deterministic Systems. gr-qc/9903084.

Visser, M. (2002). Sakharov's induced gravity: A modern perspective. *Mod. Phys. Lett. A*, **17**, 977–992. Related online version (cited on 31 May 2005): http://arXiv.org/abs/gr-qc/0204062.

About the Author

Thad Roberts is a theoretical physicist, philosopher, and adventurer who passionately explores the possibility that quantum mechanics is not exact, but is, instead, an accurate approximation of a deeper deterministic theory.

Thad has excavated dinosaur fossils, sailed across the Atlantic in a 55' sloop, and lived out of a VW Vanagon for 2 years traveling the world for $10 a day. He has worked as an astrophysicist for NASA, and then a flight lead, training astronauts for their EVAs. His life took an infamous twist when, to impress a girl, he literally stole the moon—a story that is captured in the best-selling book "Sex on the Moon". During his confinement he poured himself into a thorough investigation of Nature's deepest mysteries—an intellectual adventure that gave form to this book.

Thad currently works as a theoretical physicist for a private think tank, and as a motivational speaker for the American Program Bureau, where he shares his story to inspire others to follow their dreams and to let their curiosity propel them to a broader horizon.

www.ingramcontent.com/pod-product-compliance
Lightning Source LLC
Chambersburg PA
CBHW080527170426
43195CB00016B/2490